Statistical Genomics

Linkage, Mapping,
and QTL Analysis

Ben Hui Liu

CRC Press
Boca Raton   New York

Acquiring Editor:	Robert B. Stern
Project Editor:	Debbie Didier
Cover design:	Denise Craig
PrePress:	Kevin Luong

Library of Congress Cataloging-in-Publication Data

Liu, Ben-Hui.
    Statistical genomics : linkage, mapping, and QTL analysis / Ben-Hui Liu.
        p. cm.
    Includes bibliographical references and indexes.
    ISBN 0-8493-3166-8
    1. Genetics--Statistical methods. 2. Gene mapping. I. Title.
QH438.4.S73L55 1997
572.8'633'0727--dc21

97-1863
CIP

No claim to original U.S. Government works
International Standard Book Number 0-8493-3166-8
Library of Congress Card Number 97-1863
Printed in the United States of America 1 2 3 4 5 6 7 8 9 0
Printed on acid-free paper

TO QIANG, MING AND ANNIE

FOREWORD

Statistical Genomics is a new book about an old subject, viewed in a new way. The major focus of the book is genetic mapping, a technique that has been an indispensable part of genetic analysis since Alfred Sturtevant developed the first genetic map in *Drosophila* more than 85 years ago. This book will be a valuable resource to students learning about genomic science and to geneticists and statisticians using genomic mapping for research and development. This book should be useful for genetic analyses across many disciplines, including plant and animal breeding, the genetic analysis of natural populations and human genetic analysis.

There is wide utility for the book both as a text and as a reference. For many students, it will provide their first access to fundamental statistical concepts essential for genomic mapping. For the more experienced scientist, it provides useful explanations of some complex statistical concepts and provides approaches to solving many common problems. In addition, the book provides a considerable amount of new theory and techniques, not yet in the primary literature, related to mating populations and estimation of genetic parameters.

Genetic mapping has become an important component of both fundamental research and practical application in a great many studies of plants, animals and microorganisms. The ability to correlate genetic function and chromosome location has led to the identification of the biological roles of a great many important genes. New technology has created the potential to generate functional genetic maps for entire genomes of virtually any plant or animal, limited only by the degree of genetic diversity and the ability to obtain a modest number of progeny. This technology has resulted in an expanded and integrated view of genomes and led to the expectation that we may be able to understand genomes at both a molecular and a phenotypic level. The acquisition and application of this knowledge is one part of what has come to be called genomic science.

Our current view of genomic science has its origin in the human genome project, an experiment of major importance for our century. Within a period of only 50 years since Watson and Crick proposed that genetic information was contained in a DNA sequence, it is likely that we will have a catalog of most of the expressed human genes and a large fraction of DNA sequence for the entire human genome. This information will lead to the determination of functions for many genes and to the first estimates of the number of genes that are necessary for the known functions of the human organism. The human genome project will profoundly affect our understanding of human biology and disease, as well as leading to new methods of diagnosis and new therapeutic products. Similarly, the extension of the theory and technology to other organisms will have profound effects on agriculture, forestry, bioprocessing, material science and environmental science as well.

Since the application of DNA based molecular markers to genomic mapping, formidable advances have been made in human genetics and in plant and animal breeding. Molecular marker analysis has made possible the integration of many aspects of quantitative and molecular genetics through the genetic dissection of quantitative traits. It is now possible to define what fraction of quantitative variation may be assigned to major qualitative factors and quantitative factors in an individual or in a pedigree. These developments have made marker aided selection a practical tool with widening applications and provided new approaches to breeding based on molecular technology. The methods have been applied to many nontraditional genetic systems because the technology allows development of intensive genetic analysis, even in organisms not previously studied. Detailed genetic analysis, including genetic mapping of forest trees from natural populations, has become possible through these advances in technology. Genomic mapping has allowed greater understanding of many types of complex traits, including traits with incomplete penetrance, heterosis, genetic load, segregation distortion and epistasis, through the correlation of molecular markers and components of a complex phenotype.

Genomic mapping is an essential tool for the discovery of new genes and the definition of their function. Location of a qualitative trait locus may define a new gene. Location of a gene known only by sequence may distinguish between members of a multigene family related directly by descent (orthologs) or through duplication and divergence of function (paralogs). Used in this way, mapping of quantitative effects becomes a tool for molecular genetics.

The primary focus of this book is on new statistical methods needed to use genetic mapping in the context of this new genomic research paradigm. Genomic science represents a new integration of many aspects of genetics related to the determination of the function, location and evolution of all genes in all genomes. This concept brings together molecular genetics, quantitative genetics, cytogenetics, evolutionary biology, statistics and the technology of computers and automated systems for structural and quantitative analysis of nucleic acid sequences. Continued integration of these diverse disciplines requires new statistical methods such as those found in this book.

Ronald Sederoff,
Edwin F. Conger Professor of Forestry and
Director of the Forest Biotechnology Group
North Carolina State University

GENOMICS

Genomics is a new science that studies whole genomes by integrating the traditional disciplines of cytology, Mendelian genetics, quantitative genetics, population genetics and molecular genetics with new technology from informatics and automated robotic systems. The purpose of genomic research is to learn about the structure, function and evolution of all genomes, past and present. As a branch of biological science, genomics has made great progress in the last decade and has helped to fill the gap between the laws of chemistry, physics and biology. Many fundamental biological questions are close to being answered as a result of research in genomics, such as:

- What aspects of genome structure are physically and chemically necessary?
- Do genes have to be located at certain sites in a genome to perform their functions?
- What DNA sequences and structures are needed for genes to perform their specific functions?
- How many functional genes are necessary and sufficient in a biological system, such as an animal or a plant species?
- How many different functional genes are present in the whole biosphere?
- How many genes are essentially the same in terms of their DNA sequences and structures across different species?

What Does Genomics Include?

Classical genomics started in the early 20th century when the discovery of gene linkage (genes on the same chromosome) led to the idea of making a linkage map. This classical discipline was rejuvenated following the spectacular development of molecular marker technology beginning in 1980, especially when the Human Genome Project and the Plant Genome Project were initiated. Classical genomics, based mainly on cytogenetics and genetic mapping, is close to traditional genetics. In a way, classical genomics is a bridge between genetics and the other branches of genomics, as genomics as a whole is a bridge among different branches of genetics.

Genomic informatics was created to manage the massive amount of new information which arose from the advances of genomics. Physical genomics emphasizes the physical composition of genomes, such as nucleotide sequences and DNA clones and includes subjects such as DNA sequencing, DNA library construction, physical map assembly and DNA sequence analysis.

Where Is Genomics Heading?

Genomics has been driven by technology development since its inception. This impetus will continue into the future. Large throughput DNA sequencing equipment, DNA chip technology for quick and automated genotyping, high capacity robotic stations for sample preparation and biochemical assays, sophisticated mathematical algorithms and statistical procedures for information analyses and high power computer software packages for handling data may dominate the development of genomics.

Genomics is driven also by practical applications. It is widely recognized that genomic research has a great potential to benefit the biomedical industry, agriculture and forestry and forensic science. Genome programs (human, plant and animal) will greatly advance the technology for DNA analysis and generate vast amounts of information that will ultimately lead to a deeper understanding of biology. The technology and information from genome research can revolutionize the health-care system by advancing the diagnosis of human diseases and for forensic purposes by enabling more precise identification of human individuals. For agriculture and forestry, the technology and information from genomics can assist plant and animal breeding, disease management, genetic diversity assessment and variety protection. These applications will help to ensure environmentally friendly and sustainable agriculture and forestry.

THIS BOOK

Topics

Statistics plays an essential role in genomics because random sampling and experimental error are greatly involved in genomic research. This book covers the following topics in the area of statistical genomics:

- introduction to genomics (Chapter 1)
- brief biological background for genomics, including basic genetics, mating design and genetic markers (Chapters 2 - 3)
- brief statistical theory for genomics, including probability distribution, hypothesis testing and point and confidence interval estimation (Chapter 4)
- statistical principles and methods for screening genetic markers, linkage analysis, linkage grouping, gene ordering, multilocus models, linkage map construction, linkage map merging and searching for genes using population disequilibrium (Chapters 5 - 11)
- theory and methods for identifying genes controlling complex traits, commonly called Quantitative Trait Loci (QTL) mapping, including single-marker analysis, interval mapping, composite interval mapping, QTL mapping using natural populations and some future considerations (Chapters 12 - 17)
- computer tools for genomic map construction and QTL analysis (Chapter 18)
- resampling techniques and computer simulations in genomics, including jackknife and bootstrap methods, permutation analysis and mapping population simulations (Chapter 19)

Intended Audience

This book can be used as a handbook for biologists interested in statistical issues related to genomics, as a textbook for upper-level undergraduate and graduate students majoring in genetics or in genomic science and as an introduction to genomics for statisticians. For biologists, information is outlined on how statistical analyses of genomic data are implemented and how results are interpreted and presented. Students will find information on the statistical principles and on new developments in the field of statistical genomics. Information that ties statistics and biology together is outlined for statisticians.

The book is intended to reach many scientists in plant, animal and human genome communities. Important topics, such as linkage analysis and QTL mapping, are divided into different parts for experimental populations (controlled crosses) and natural populations (uncontrolled crosses). Experimental populations cover most mapping populations for plants and animals. The natural populations include some of the plant (for example, forest trees) and animal populations and the mapping populations for the humans.

How to Read This Book

As a handbook: This book can be a handbook for genomic data analysis for readers with either a good background in statistics (statisticians) or a good background in biology (biologists). The considerable information on the biological aspects of genomics will help the statisticians in their interpretations of the statistical treatments of genomic data. For the biologists, the detailed discussion on the statistical procedures and derivations will help them to gain a better understanding of the principles behind computer software packages for genomic analysis. I have included 222 tables and 170 figures to explain both the underlying statistics and the biology. This certainly will help the biologists to follow the statistical procedures and the statisticians to understand the kinds of biological data with which they work.

As a textbook: This book can be used as a textbook for students majoring in genetics or genomic science. For undergraduates with basic biology, a one semester introduction to genomics can be covered by:

- Chapters 1, 2, 3, 6, 9, 12, 13 and 18.

For upper level undergraduates or graduates with basic genetics and statistics, a one-semester presentation of statistical genomics can include the following chapters:

- track one for biologists: Chapters 5 to 18 and 4; or
- track two for statisticians: Chapters 2, 3, 5 to 19.

I presume that students in track one already have basic knowledge of genomics, such as basic genetics and DNA structure and that those in track two have a relatively good background in theoretical statistics.

About Exercises

A set of exercises is given at the end of each chapter. Some of the exercises emphasizing statistical derivation usually have defined answers and are relatively easy for readers with a strong background in statistics. Those exercises should be very helpful to biologists in building a firm background in statistical genomics. Some exercises emphasize a comprehensive understanding in biology and these exercises may not have defined answers. I included some of my thoughts on the perspectives of genomics into some of the exercises rather than into the text and I encourage readers to try those exercises in the context of group discussions.

Homepage for This Book

A homepage has been set up for this book by Ms. Ling Li at:

http://www4.ncsu.edu/unity/~benliu

We will post the latest developments regarding this book, errors, answers for the exercises and availability of computer software PGRI in the homepage. Certainly, readers will find interesting links to the world of genomics from this Web site.

Error report: One important function of the homepage is to report errors in this book. Please feel free to report any errors you find in the book to the homepage or contact me directly. Errors and problems will be posted on this homepage (with your permission). Your comments are valuable for the future improvement of this book.

Cyber discussion: I would like to post any comments regarding this book publicly. I am certain that your comments will stimulate discussions on this book and, most importantly, broader interchanges on where genomic research should head.

Computer program: A computer program, PGRI, will be made available at this Web site. I will also collect some other public domain software packages and make them available at this site or at the sites linked to this site.

Role of Statistics in Biology

Statistics is a tool to solve problems which cannot be solved solely through biological observation or qualitative analysis. This is especially true for the statistics used in genomic mapping. A biological conclusion should not be overshadowed by statistics when correctness of the statistical methods is not certain. When biological evidence is significant and no adequate statistics can be applied, it is better not to use statistics.

ACKNOWLEDGMENTS

History

I have many people to thank. I learned biometry from Professor Fan Lian 20 years ago. I wrote my first computer program for testing segregation distortion for my colleague Dr. Armi Ahamad when I worked with Drs. Stan T. Cox and Rollin Sears at Kansas State University. That was my starting point for bringing computation, statistics and genetics together, which is the theme of this book. My exposure to modern genomics started at Oregon State University when I worked with Dr. Steve J. Knapp. My collaboration with Dr. Patrick M. Hayes and Andy Kleinhofs and my involvement in the North American Barley Gene Mapping Project gave me an inside view of large plant genome project. I thank Drs. Hayes and Kleinhofs for providing the opportunity for me to work on the barley project.

Raleigh, North Carolina

This book would never have been written if I have not come to Raleigh. I learned my lessons in molecular biology by working for the Forest Biotechnology Group and on the latest developments in statistical genetics through my association with the Statistical Genetics Program at North Carolina State University. The encouragement and support from Drs. Ronald R. Sederoff and Bruce S. Weir have made this book possible. Ron's vision on the future of genomics gave me the determination to write this book. Thank you, Ron, for many many things you have done for me and especially for writing the foreword for this book. It was Bruce's idea for me to come to Raleigh to work at the interface of molecular genetics and statistics. Thank you, Bruce, for your help and pateince.

Many points in this book are results of my association with the following people at North Carolina State University (unless otherwise indicated): Drs. Henry Amerson, William Atchley, Christopher Basten, Malcolm Campbell (now at Oxford University), Rebecca Doerge (now at Purdue University), Major Goodman, Trudy Mackay, Paul Murphy, David O'Malley, Henry Schaffer, Ronald Sederoff, Charles Stuber, Jeffery Thorne, Bruce Weir, Ross Whetten, Shi-Zhong Xu (now at University of California, Riverside) and Zhaobang Zeng at North Carolina State University. Especially, Dave, Ron and Ross have been my immeasurable resources for almost every topic of this book. It is the influence of Dave, Ron, and Ross that has helped me to learn molecular biology.

Several graduate students of the Forest Biotechnology Group at North Carolina State University, Ling Li, David M. Remington and Shuku Sun, have made constructive comments for this book. Dr. Yiyuan Lu, who was working with me during the book's preparation, has provided assistance on a number of mathematical problems in the text. Some of the chapters have been used for a summer course on genomic mapping using RAPD markers at North Carolina State University. Students in that class have provided comments on the material. I thank all of them for their efforts to improve this book.

During the preparation of this book, I have received full support from the Department of Forestry and the College of Forest Resources at North Carolina State University. I appreciate the organizational support from my College and Department.

Data for Examples

The following colleagues at North Carolina State University (when not noted otherwise) have provided data for examples in this book: Drs. Henry Amerson, Malcolm Campbell (presently at Oxford University), Bailian Li, David O'Malley, Ronald Sederoff, Ross Whetten, Phillip Wilcox (now at the Institute of Forest Research in New Zealand) and Mr. David Remington; Dr. Hongbin Zhong of the Crop Biotechnology Center at Texas A&M University; Dr. Patrick Hayes at Oregon State University; Dr. Andy Kleinhofs at Washington State University; and all the members of the North American Barley Gene Mapping Project.

Reviewers

I have received constructive comments on this book from a number of reviewers, whose advice has greatly improved this book. My colleagues at North Carolina State University have also reviewed this book and their efforts have reduced the errors significantly. Dr. Ronald Sederoff helped me significantly in the writing of this book, especially for the parts dealing with the historical perspective and biological background of genomics. Dr. Bruce Weir helped me in written presentation and in the refining of my knowledge in statistics. Dr. Zhaobang Zeng reviewed chapters on QTL mapping and his constructive comments improved these chapters. During the long process of writing this book, I have continuously received comments and suggestions from Drs. David O'Malley and Ross Whetten. Members of the Statistical Genetics Program at North Carolina State University, led by Dr. Bruce S. Weir, including faculty, visiting scientists, post doctoral scientists and graduate students have read through specific chapters. They are Christopher Basten, James Curran, Trevor Hohls, Yue-Fu Liu, Dahlia Nielsen, Ian Painter, Jennifer Shoemaker, Katy Simonsen, Jeffrey Thorne, Bruce Weir, Dmitri Zaykin and Zhao-Bang Zeng. Ms. Reenah Schaffer has carefully read through the whole manuscript. Her efforts have improved the readability and appearance of this book. I am very grateful for their comments and advice. However, I am responsible for all remaining errors.

Publisher and Copy Editors

Bob Stern and his associates at CRC Press have done a superb job of publishing this volume in a timely fashion. I thank them for their effort, consideration and cooperation.

Grant Support

Through the years, my research work has been supported by the United States Department of Agriculture, the National Institutes of Health, the National Science Foundation, Pioneer Hybrid International and the Industrial Associates Consortium of Forest Biotechnology Group at North Carolina State University.

Family

It is difficult to imagine writing a book without the full support and understanding of one's family. My greatest thanks go to my wife, Qiang Xu, who has given me her wholehearted support, and to my son, Ming and daughter, Annie, who have shown great patience with Dad during the preparation of this book.

Ben Hui Liu
Raleigh, North Carolina
October, 1997

CHAPTER LIST

CONTENTS

CHAPTER 1

INTRODUCTION

Twenty years ago, as an undergraduate student, I was fascinated by many courses in the physical, mathematical and biological sciences, but most of all, by genetics. How does one apply the laws of chemistry, physics and mathematics to biology? How does biological inheritance follow the laws of chemistry, physics and mathematics? These basic questions have been given intense attention since modern biology began about a century ago and will continue to be central questions for some time to come.

Genetics includes five well-structured branches: classical genetics, cytogenetics, population genetics, quantitative genetics and molecular genetics (Figure 1.1). Classical genetics began with the breeding studies of Gregor Mendel. Concepts such as genes, alleles, segregation and dominance were invented to describe the results of Mendel's experiments and of similar ones that were carried out after the turn of the century (1901). However, the Mendelian concepts of genetics did not have a chemical or physical basis. These concepts of Mendelian genetics were physically defined later by cytogenetics and molecular genetics. In many cases, the products of genes that follow Mendelian inheritance are now well defined and have been studied extensively. In the human hemoglobins, the underlying molecular events are well understood and the consequences of these changes in metabolism and development have a plausible molecular basis. In many cases, particularly for complex quantitative traits, we do not yet understand their genetic basis in terms of Mendelian genes, and certainly not at the molecular level. Some individual genes and their effects on complex phenotypic traits are understood. Sometimes, we may be able to identify a particular gene, for example, one controlling grain yield of wheat, or a gene responsible for hypertension in humans. However, much less is known about the basis of the regulation and expression of coordinately and differentially controlled genes in metabolism and development. Modern molecular genetics, a combination of the classical genetics and biochemistry, began in 1953 when J. D. Watson and F. H. C. Crick deduced the double-helical structure of DNA. Their deduction made the conceptual connection between biological inheritance and fundamental chemistry and physics.

1.1 INTRODUCING GENOMICS

1.1.1 GENOMICS AND THIS BOOK

Quantitative and population genetics, cytogenetics and molecular genetics are being integrated rapidly with new advances in genome research, due to worldwide efforts in human, plant and animal genome research. This new frontier of genetics can be called genomics. Genomics is a new science that studies genomes at a whole genome level by integrating the five traditional disciplines of genetics with new technology from informatics and automated systems. The pur-

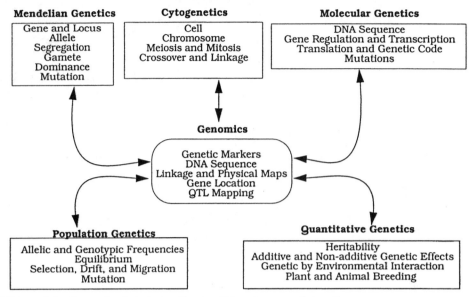

Figure 1.1 Genomics integrates five traditional areas of genetics.

pose of genomic research is to learn about the structure, function and evolution of all genomes, past and present. The ability to fill the gaps between the laws of chemistry, physics and biology has been enhanced by the development of genomics (Figure 1.1). Genomics asks many fundamental biological questions, such as

- How many functional genes are necessary and sufficient in a biological system, such as an animal or plant species?

- How do these genes determine the total phenotype?

- How many different functional genes are present in the whole biosphere?

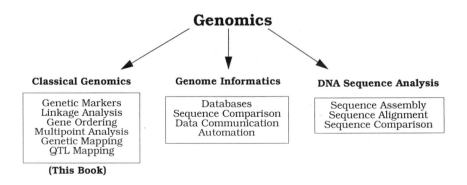

Figure 1.2 Relationships of genomics, informatics and DNA sequence

- What DNA sequences or structures are required for genes to perform their functions?

- What is the essential nature of genome structure?

Genomics can be divided into three major components: classical genomics, genome informatics and physical genomics. These are related to each other in many aspects (Figure 1.2).

The discovery of gene linkage (genes on the same chromosome) (Correns, Bateson, and Punnett 1905; Morgan 1912; Sturtevant 1913) led to the idea of making a linkage map. This classical technique was rejuvenated following the development of molecular marker technology in 1980 (Botstein *et. al.* 1980). Classical cytogenetics and genetic mapping bring together the physical and genetic structure of the chromosome. Today, genomics integrates DNA sequence studies with functional genetics and information sciences.

Genomic information is growing rapidly and computational sciences are needed to manage and analyze the massive amounts of information using improvements in computer technology, database design, networking, graphics and animation. Genomics also focuses on the physical composition of genomes, new methods for nucleotide sequencing and DNA cloning and includes physical map assembly and DNA sequence assembly for genome sequencing. To generate the large amounts of data for genomic studies, automated instruments are needed. To analyze the increasing amounts of data, knowledge of statistics and computers are essential. The statistical and computational knowledge essential to analyze genome data is the focus of this book.

1.1.2 GENOMICS AND MODERN BIOLOGY

In recent years, great progress has been made in both molecular and quantitative genetics. In molecular genetics, many specific genes and proteins have been identified and characterized affecting the growth, metabolism, development and behavior of plants, animals and microorganisms. Genomics will have a great impact on the future development of basic biology (Table 1.1 and Figure 1.3).

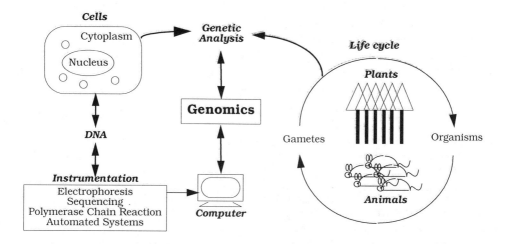

Figure 1.3 From cells to computers through genomics

Table 1.1 Potential impact of genomics on basic biology.

Basic science	Potential impact
Basic biology	1. Better understanding of developmental biology 2. Systematic study of genetic control systems, such as regulation of gene expression 3. An accurate definition of gene structure and function 4. Obtaining a better understanding of biochemical foundation of genetic effects
Population and quantitative genetics	1. Construction of multiple gene models of quantitative traits 2. Obtaining biological definitions of additive and dominant effects and epistatic interactions 3. Better understanding of genetic by environment interactions 4. Development of tools and databases for studying genome evolution

1.1.3 GENOMICS AND ITS PRACTICAL APPLICATIONS

The Potential of Genome Research

It has been widely recognized that genomic research has great potential to benefit the biomedical industry, agriculture, forestry and forensic sciences. The technology and information from genome research should advance the diagnosis of human diseases. Forensic sciences would be improved by enabling more precise identification of human individuals. For agriculture and forestry, the technology and information can assist plant and animal breeding, disease management, genetic diversity assessment and variety protection. These applications should contribute to the development of more sustainable agriculture and forestry.

Population and Quantitative Genetics

New methods of quantitative genetics, which use the genetic dissection of quantitative traits, have identified genetic regions regulating important functions. Quantitative trait analysis has fundamentally changed the conventional view of polygenic inheritance and has led to the identification of loci with quantitative effects on complex traits. Such loci typically have important biological roles, but are usually lacking in molecular identification. Lander and Schork, in their 1994 paper on genetic dissection of complex traits, wrote

> ... one can systematically discover the genes causing inherited diseases without any prior biological clue as to how they function. The method of genetic mapping, by which one compares the inheritance pattern of a trait with the inheritance patterns of chromosomal regions, allows one to find where a gene is without knowing what it is ....

This example shows the importance of genomics to basic biology and and also the limitations of traditional approaches. Traditionally, characterizing one gene could span decades and represent a good career for a scientist. However, many traits require understanding the roles and interactions of many genes acting at the same time. Using new methods, it is now possible for a number of genes to be extensively characterized at the same time. In addition, the genomic approach has revolutionized the ways of finding genes by sequence comparison in very large databases. Genome programs (human, plant, animal and microbial) will greatly advance the technology for DNA analysis and generate a large amount of information for a better understanding of a broad spectrum of biological problems.

Additive, dominance and epistatic interactions are traditional quantitative genetic effects that have been defined in statistical terms. Although some studies have resulted in the biological explanations of genetic effects, in most cases the fundamental biochemistry of the genetic effects is not clear. Genomic research needs to integrate research from quantitative genetics, molecular genetics and biochemistry.

DNA Diagnosis of Human Genetic Disorders

DNA sequence analysis and genomic mapping play an important role in the study of human genetic disorders. The cloning of genes that determine many disease states has resulted from a combination of genetic and physical mapping. Molecular genetic methods are very effective in diagnosis and human genetic disorders may be classified based on how the diagnosis is obtained. In some cases, diagnosis is based on DNA sequence alone. The sequence alterations for the mutations causing the disorder are known and the disease gene may be identified. When there are known linkage relations between genetic markers and affected genes, with or without knowing the genes themselves, genetic linkage can be used to diagnose the disorder. Population genetics based diagnostics uses known gene frequencies and pedigree information to diagnose the disorder. As human genomics advances, multiple-gene models that integrate linkage and sequence information will be developed for diagnosis of many complex human diseases.

Applications in Agriculture and Forestry

Molecular marker analysis and computer simulations provide new technology to assist plant and animal breeding. Lande and Thompson (1990) described a selection index (L-T index) and a method to evaluate the efficiency of progeny selection based on a multiple linear regression model. Molecular markers could help breeders with parental selection and mating design for breeding population synthesis. Markers could also help to design breeding strategies to maximize short-term genetic gain and to preserve genetic diversity for long-term genetic improvement.

Molecular markers are also especially useful for multiple trait breeding. For example, when breeding for two traits, one parent with good performance in one trait and the other parent with good performance in the other trait are chosen to produce offspring with good performance in both traits. The probability of recovering the best genotype based on a conventional approach is small, even when the genes controlling the traits are genetically independent. The probability will be smaller depending on the extent to which the genes are not independent. Parental selection using markers to assist mating design should maximize the probability of recovering the target genotype in the offspring.

Inbred line development for hybrid breeding, such as that used in maize, is a specific case of parental selection with a goal of maximizing the performance of F1 hybrids. Compared to general population improvement, hybrid breeding exploits both additive and non-additive genetic variation. Molecular marker technology creates tools and opportunities to investigate non-additive genetic variation and to use it more efficiently in applied plant breeding.

1.2 STATISTICAL GENOMICS

Statistics plays an essential role in integrating different areas of genetics and making genomic data more biologically meaningful. This book covers the following topics:

- brief biological background for genomics, including basic genetics, mating design and genetic markers. (Chapters 2 - 3)
- brief statistical theory for genomics, including probability distributions, hypothesis testing and point and confidence interval estimation. (Chapter 4)
- statistical principles and methods for screening genetic markers, linkage analysis, linkage grouping, gene ordering, multi-locus models, linkage map construction, linkage map merging and searching for genes using population disequilibrium (Chapters 5 - 11)
- theory and methods for searching for genes controlling complex traits, commonly called Quantitative Trait Loci (QTL) mapping, including single-marker analysis, interval mapping, composite interval mapping, QTL mapping using natural populations and some future considerations (Chapters 12 - 17)
- computer tools for genomic map construction and QTL analysis (Chapter 18)
- resampling techniques and computer simulation in genomics, including jackknife and bootstrap methods and permutation and mapping population simulations (Chapter 19)

Statistics for genomics has some unusual features:

- Some genome data are a mixture of discrete and continuous variables, such as a combination of genotypes of genetic markers (discrete) and values of quantitative traits (continuous).
- Test statistics for some genomic hypotheses have no clear theoretical probability distribution. Empirical distributions of the test statistics for some genomic analyses are needed, such as those needed for QTL analysis and hypothesis tests of locus order.
- Genome databases are becoming unusually large due to the rapid generation of molecular marker and DNA and protein sequence data. To manage the data efficiently, systematic bioinformatics is needed.
- Intensive computation is usually involved in genomic data analysis, for linkage analysis, QTL analysis and for the computationally greedy algorithms in locus ordering and in derivation of empirical distributions of test statistics.
- Linear model approaches, such as the analysis of variance and linear regression, familiar to most biologists, are not sufficient for genomic data analysis. Likelihood approaches are used extensively in genomics.
-

These characteristics are treated thoroughly in this book. Chapter 4 is designed for readers with little background in statistical theory to introduce some essential statistical theory behind statistical genomics. For readers with little background in the area of genetics, some essential biological background needed for genomics is presented in Chapters 2 and 3.

1.3 RELATED BOOKS

As I mentioned earlier in this chapter, much basic knowledge is needed for research in genomics. Books in the following list have helped me to understand genomics and to prepare this book.

General Genetics
Gonick, L. and M. Wheellis. The Cartoon Guide to Genetics. Harper Collins Publishers, New York. 1991. (I recommend this to statisticians for a useful review of molecular genetics).

Griffiths, A.J.F., J.H. Miller, D.T. Suzuki, R.C. Lewontin, and W.M. Gelbart. An Introduction to Genetic Analysis, Fifth Edition, Freeman and Company, New York. 1993.

Swanson, C.P., T. Merz, and W.J. Young. Cytogenetics: The Chromosome in Division, Inheritance, and Evolution. Second Edition. Prentice-Hall, Englewood Cliff, NJ. 1981.

Molecular Biology
Darnell, J., H. Lodish, and D. Baltimore. Molecular Cell Biology. Scientific American Books, New York. 1986.

Lewin, B. Genes VI. Oxford University Press, Oxford. 1997.

Watson, J.D., M. Gilman, J. Witkowski, and M. Zoller. Recombinant DNA. Second Edition. Scientific American Book, New York. 1992.

Population Genetics
Hartl, D.L. A Primer of Population Genetics. Sinauer, Sunderland, MA. 1988.

Li, C.C. Population Genetics. The University of Chicago Press, Chicago. 1955.

Weir, B. Genetic Data Analysis II. Sinauer, Sunderland, MA. 1996.

Quantitative Genetics
Falconer, D.S. and T.F.C. Mackay. Introduction to Quantitative Genetics. Fourth Edition. Longman Scientific & Technical, New York. 1996.

Hallauer, A.R. and J.B. Miranda. Quantitative Genetics in Maize Breeding. Second Edition. Iowa State University Press, Ames, Iowa. 1988.

Kempthorne, O. An Introduction to Genetic Statistics. John Wiley. New York. 1957.

Mather, K. and J.L. Jinks. Biometrical Genetics. Chapman and Hall, London. 1982.

Wricke, G. and W.E. Weber. Quantitative Genetics and Selection in Plant Breeding. Walter de Gruyter, Berlin. 1986.

Genetic Linkage Analysis
Bailey, N.T.J. Introduction to the Mathematical Theory of Genetic Linkage. Oxford. 1961.

Mather, K. The Measurement of Linkage in Heredity. Second Edition. Methuen, London. 1951.

Ott, J. Analysis of Human Genetic Linkage. Revised Edition. The Johns Hopkins University Press, Baltimore. 1991.

Statistical Methods

Agresti, A. Categorical Data Analysis. John Wiley & Sons, New York. 1990.

Gallant, A.R. Nonlinear Statistical Models. John Wiley & Sons, New York. 1987.

Searle, S.R. Linear Models for Unbalanced Data. John Wiley & Sons, New York. 1987.

Sokal, R.R. and F.J. Rohlf. Biometry. Second Edition. Freeman and Company, New York. 1981.

Weir, B. Genetic Data Analysis II. Sinauer, Sunderland, MA. 1996.

Statistical Theory

Edwards, A.W.F. Likelihood. Expanded Edition. The Johns Hopkins University Press, Baltimore. 1992.

Lehmann, E.L. Testing Statistical Hypothesis. Second Edition. Wadsworth & Brooks, Pacific Grove, CA. 1986.

Lehmann, E.L. Theory of Point Estimation. Second Edition. Wadsworth & Brooks, Pacific Grove, CA. 1991.

Mood, A.M., F.A. Graybill, and D.C. Boes. Introduction to the Theory of Statistics, Third Edition. McGraw-Hill, New York. 1974.

Stuart, A. and J.K. Ord. Kendall's Advanced Theory of Statistics, Volume I: Distribution Theory. Edward Arnold, London. 1994.

Mathematics and Algorithms

Abramowitz, M. and I.A. Stegun (editors). Handbook of Mathematical Functions. Dover Publications, New York. 1970.

Baase, S. Computer Algorithms: Introduction to Design and Analysis. Second Edition. Addison-Wesley Publishing Company, Reading, Ma. 1988.

Efron, B. and R.J. Tibshirani. An Introduction to the Bootstrap. Chapman & Hall, New York. 1993.

Shao, J. and D. Tu. The Jackknife and Bootstrap. Springer, New York. 1995.

Computational Biology

Nei, M. Molecular Evolutionary Genetics. Columbia University Press, New York. 1987.

Waterman, M.S. Introduction to Computational Biology. Chapman and Hall, London. 1995.

History of Genome Research

Cook-Deegan, R. The Gene War: Science, Politics, and the Human Genome. W. W. Norton & Company, New York. 1995.

Shapiro, R. The Human Blueprint: The Race to Unlock the Secrets of Our Genetic Code. A Bantam Book/St. Martin's Press, New York. 1992.

CHAPTER 2

BIOLOGY IN GENOMICS

2.1 INTRODUCTION

Mendel discovered the genetic segregation of simple traits using crosses of garden peas expressing different characteristics (1865) (Table 2.1). Only when Mendel's laws were rediscovered (Tschermak, DeVries and Correns 1900) and homologous pairs of chromosomes were observed through cytology (Sutton and Boveri 1902) were the laws validated qualitatively and quantitatively. The discovery of gene linkage (multiple genes on the same chromosome) preceded the idea of a linkage map (Correns, Bateson and Punnett 1905; Morgan 1912; Sturtevant 1913). The concept of a one gene-one enzyme relationship and DNA as the physical material for heredity brought Mendelian genetics, biochemistry and cytogenetics together into a new area of molecular genetics (Beadle and Tatum 1941; Avery, McCarty and McLeod 1944).

The golden age of molecular genetics began with the discovery of DNA structure (Watson and Crick 1953) and the genetic code (Nirenberg and Matthaei 1961). These discoveries were followed by the development of techniques for rapidly obtaining DNA sequences (Sanger 1975), for visualizing specific DNA fragments (Southern 1975) and for DNA amplification (Mullis

Table 2.1 Benchmarks of genetics relevant to genomics.

Discovery	Year (leading researchers)
Cell theory	1666 (Hooke)
Theory of evolution	1858 (Darwin)
Mendelian law of segregation	1865 (Mendel)
Rediscovery of Mendelian law	1900 (Tschermak, DeVies, Correns)
Chromosomal theory of inheritance	1902 (Sutton and Boveri)
Gene linkage	1905 (Correns, Bateson, and Punnett)
Linkage map of *Drosophila*	1912 (Morgan)
First linear linkage map	1913 (Sturtevant)
One gene-one enzyme hypothesis	1941 (Beadle and Tatum)
DNA is physical material for heredity	1944 (Avery, McCarty and McLeod)
DNA structure	1953 (Watson, Crick and Wilkins)
Genetic code	1961 (Nirenberg and Matthaei)
Rapid DNA sequencing method	1975 (Sanger)
Southern blotting	1975 (Southern)
Mapping RFLP in human	1980 (Botstein *et al*)
DNA fingerprinting	1985 (Jeffreys *et al*)
Polymerase chain reaction	1986 (Mullis)

1986). From the mid-1980s, genome research has become a major focus for biological science. High density genetic maps for human and many other important animal and plant species have been obtained.

More mathematically inclined biologists, such as Fisher, Haldane and Wright, laid out the basic theories for population and quantitative genetics early this century. Population and quantitative genetics use mathematical and statistical models to extend Mendelian genetics to multiple-gene models for the analysis of populations and complex traits. After the work of Fisher, Haldane and Wright, and until the late 1970s, development of population and quantitative genetics was largely involved with extension and validation of their pioneering theories, or for practical applications, such as plant and animal breeding. A new frontier for population and quantitative genetics was created with the use of molecular methods of mapping and sequencing that generated the mass of new information for genetic research.

The nature of the gene, its genetic and physical location in the genome, its DNA sequence and physical structure, its effects in terms of biochemistry and quantitative genetics have been major issues. Different branches of genetics have had different ways to approach these issues. A gene is both a segment of a chromosome (cytogenetics) and a stretch of functional DNA (molecular genetics). However, genetic effects are defined more precisely in quantitative genetics than in any other branch of genetics.

In this chapter, genetics basics relevant to genomics will be discussed. These include: 1) meiosis and recombination in cytogenetics, 2) gene frequency and additive and dominant genetic effects in population and quantitative genetics and 3) DNA sequence and gene expression in molecular genetics.

2.2 MENDELIAN GENETICS AND CYTOGENETICS

Although techniques used by genomics are largely generated by molecular, quantitative and population genetics, genetic concepts and terminology are largely derived from Mendelian genetics and cytogenetics. It is essential to review and understand the concepts of Mendelian genetics and cytogenetics.

2.2.1 MENDELIAN GENETICS

Terminology
Mendelian genetics is concerned with classical simple trait segregation the- ory and simple linkage genetics. A gene is defined as a unit of heredity. In a pop- ulation of individuals of a sexually reproducing species, a single gene is passed from generation to generation following simple Mendelian inheritance. Each dip- loid individual has two copies (alleles) of each gene. It is also common to call an allele 'a gene' and a gene 'a gene pair'. For example, gene *A* may have two alleles in a population, *A* and *a*. If an individual has two copies of *A*, then the genotype of the individual is *AA* and it is homozygous. An individual with the Aa genotype is heterozygous and an individual with the *aa* genotype is the other homozygote. The number of alleles for a single gene is not restricted to two. In natural popu- lations, multiple alleles are commonly observed.

The appearance or measurement of a characteristic governed by a gene or a number of genes is defined as the phenotype. If the three possible genotypes (*AA*, *Aa* and *aa*) for a characteristic controlled by a single gene show three distinct phenotypes, then the alleles are described as codominant. If individuals with genotypes AA and Aa show the same phenotype, then A is defined as the dominant allele and a is the recessive allele.

Mendelian genetics was derived mainly from experimental populations obtained by controlled crossing between individuals having distinct phenotypes such as flower color and leaf shape. Controlled crossing is still the most common way to obtain experimental populations for genomic research for many plant and animal species.

"Backcross" and "F2" are commonly used mating schemes for obtaining experimental populations. The mating starts with a cross between two homozygous individuals (AA and aa for a single gene trait). A haploid germ cell produced by a parent, such as a sperm or an egg, is defined as a gamete. In this example, a haploid gamete contains a single copy of each gene. The AA individual produces gametes with a single copy of A and the aa individual produces gametes with a single copy of a. The offspring of the cross is defined as an F1 hybrid and will result from the fusion of gamete A and gamete a to produce a diploid genotype Aa. The F1 crossed with one of the parents (AA or aa) is a backcross. F2 progeny are the result of the F1 crossed with itself (for self-pollinating plant species) or with its sibs (for dioecious species (two sexes)). For multiple-gene models and non-inbred parents, the definitions of backcross and F2 may be different. For example, if we cross parents with genotype $AaBb$ and $Aabb$ for a two-gene model, then the cross between these two parents has a backcross configuration for gene B and an F2 configuration for gene A.

Mendelian Laws

Mendelian laws include the law of segregation and the law of independent assortment. For traits controlled by a single gene and having simple heredity, each somatic cell of an individual carries a pair of alleles. The paired alleles segregate from each other into gametes during meiosis. Among the gametes produced by an individual, one-half carry one allele of the pair and the other one-half carry the other member of the gene pair. Mendelian segregation implies that the genotypic segregation ratio is 1:1 for alleles having codominant inheritance in a backcross population and is 1:2:1 in an F2 population. For alleles having dominant inheritance, the phenotypic segregation ratio is 3:1 in F2 progeny. However, dominant genes could have a phenotypic segregation ratio of 1:1 or have no segregation in backcross progeny, depending on the choice of parent for the backcross.

Table 2.2 Number of genotypes and phenotypes for genes exhibiting codominant and dominant effects.

Number of segregating genes	Number of genotypes or phenotypes for codominant genes	Number of phenotypes for dominant genes	Mixture		
			Number of codominant genes	Number of dominant genes	Number of phenotypes
1	3	2			
2	9	4	1	1	6
3	27	8	2	1	18
4	81	16	2	2	36
n	3^n	2^n	c	d	$3^c 2^d$

The law of independent assortment explains the inheritance of unlinked multiple genes. The law states that each pair of alleles of a gene segregate independently of the segregation of alleles of another gene. Based on the law of independence, two genes A and B, with two alleles for each gene (A and a for gene A and B and b for gene B), can form 9 possible genotypes ($AABB$, $AABb$, $AAbb$, $AaBB$, $AaBb$, $Aabb$, $aaBB$, $aaBb$ and $aabb$). The expected segregation ratio for the 9 genotypes is $(1{:}2{:}1)^2 = 1{:}2{:}1{:}2{:}4{:}2{:}1{:}2{:}1$ in the F2 progeny. If the genes are dominant, then there are 4 possible phenotypes and their expected segregation ratio is 9:3:3:1 for $A\_B\_$, A_bb, $aaB\_$ and $aabb$, where the underline means that the allele can be either dominant or recessive. The ratio means that the frequency of $A\_B\_$ in the F2 population is 9/16. Table 2.2 shows the expected number of genotypes and phenotypes in F2 progeny. The number of genotypes is certain when the number of genes is fixed. The number of possible phenotypes is different for the same number of genes according to the mode of the inheritance.

An application of Table 2.2 is the estimation of the expected frequency of a specific genotype or phenotype in a population. For example, the expected frequency of the genotype $AaBBccDd$ in the F2 progeny of a cross between $AAbbC-Cdd$ and $aaBBccDD$ when the four loci are independent is

$$P\left[AaBBccDd\right] = \frac{1}{2} \times \frac{1}{4} \times \frac{1}{4} \times \frac{1}{2} = \frac{1}{64} \tag{2.1}$$

Equation (2.1) can be used to estimate the expected counts for different genotypes or phenotypes in a sample. A chi-square test can determine how closely the observed segregation ratio fits the expected segregation.

Gene Linkage

Correns, Bateson and Punnett (1905) discovered gene linkage through analyzing two-gene segregation ratios in sweet pea which did not follow Mendelian independent assortment. Morgan (1912) confirmed and extended their results using *Drosophila*. These studies initiated the concepts of linkage and genetic mapping. The first genetic map of six sex linked genes was constructed by Sturevant (1913), one of Morgan's students.

2.2.2 MECHANISMS OF MENDELIAN HEREDITY - CYTOGENETICS

Gene linkage is the foundation of genomic mapping. We will carefully define gene linkage and discuss its physical basis. Knowledge of the principles of genetic and physical mapping is critical for deriving appropriate statistical procedures for genomic analysis.

Cell Division and Chromosomes

There are two regular types of cell division in biological development: mitosis and meiosis. Mitosis occurs during normal biological growth and differentiation. Mitosis is the cell division that produces two genetically identical cells from a single cell. Each of the daughter cells contains the same number of chromosomes as the progenitor cell. Chromosomes are linear arrangements of genes and other DNA; they are visible during cell division. There is a specific number of chromosomes in an organism and a genome is defined as an entire set of chromosomes in the nucleus, plus the genetic material in the cytoplasm. On occasion, the genetic material in the mitochondria or chloroplast may be referred to as a mitochondrial genome or a chloroplast genome.

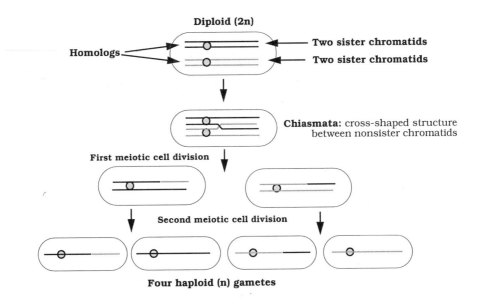

Figure 2.1 Meiosis. One chromosome pair is shown in the diploid cell after DNA replication.

Meiosis

Meiosis is essential for the sexual reproduction of eukaryotes. Meiosis is a specialized type of cell division that produces four daughter cells, each having half of the number of chromosomes of the progenitor cell. These are haploid gametes (Figure 2.1). Meiosis includes one round of chromosome duplication and two rounds of cell division. It begins with cells containing two sets of chromosomes, commonly called diploid (abbreviated as 2n). After DNA replication, each chromosome contains two identical DNA duplex strands, called sister chromatids. A pair of chromosomes are called homologs. The homologs duplicate to form four duplex strands of DNA. The homologs pair during the first phase of meiosis and then two rounds of cell divisions result in four haploid cells, each with one chromatid. Sutton and Boveri (1902) asserted that the Mendelian Laws parallel the cytologically visible events during meiosis. That is, Mendelian genes segregating independently parallel the homologous chromatids segregating independently into gametes. For a single gene located on a chromosome, two copies of the gene are produced by the chromosome duplication and each copy of the gene is contained in a gamete after cell division. The diploid status is restored during fertilization (when a male and a female haploid gamete fuse together) to form the zygote, which can become an embryo.

The chromosome location of a gene or any specific DNA sequence is defined as a locus, an analogy to a point on a line in mathematics. If several genes are located on different chromosomes or distantly located on the same chromosome, then the loci act independently.

Linkage and Recombination

Genetic linkage is defined as the association of genes located on the same chromosome. For these genes, the segregation ratio for the genotypes and phenotypes departs from the Mendelian independent assortment ratios. The paren-

tal (non-recombinant) types are more frequent when the recombination
frequency is low. During meiosis, the homologous chromatids can go through a
process of breakage and reunion. In the first cell division, visible cross-shaped
structures form between nonsister chromatids. They are called chiasmata (the
singular is chiasma). Chiasmata are the visible cytological evidence of homologous crossover between nonsister chromatids. If the reunion results in chromosomal segment exchange between homologs, it is called a crossover. The result
of the recombination is the existence of non-parental chromosomes in two of the
cellular meiotic products. Each crossover event creates two reciprocal recombinant (non-parental) gametes. Crossing over has at least two functions. It
ensures proper chromosomal disjunction and thus preserves the integrity of dip-
loidy at meiosis. It also generates and maintains genetic variation by creating
new combinations of alleles and increases evolutionary flexibility.

Recombination in general occurs randomly on chromosomes and the
recombination between different loci is associated in significant part with the
physical distance between the loci. These inferences are the foundation of
genetic mapping. However, the relationship between recombination and physical
distance varies from organism to organism and even within single organisms.
There is also evidence of genetic control and non-randomness of crossing over,
such as genes controlling the frequency of recombination on specific segments of
a chromosome or on a whole chromosome. Recombination "hot spots" or cases
of site specific recombination have been observed in some organisms. Another
issue of importance is whether recombination occurs preferentially within genes,
rather than distributed within and between genes.

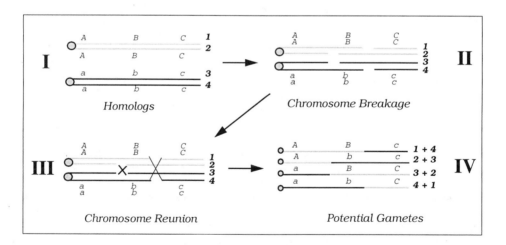

Figure 2.2 The Holliday model (1964) of recombination. I: Homologs pairing (1 and 2
are sister chromatids, 3 and 4 are sister chromatids), II: Random chromosome breakage, III: Two crossovers, one between loci A and B (chromosome exchange between 2
and 3) and the other between loci B and C (chromosome exchange between 1 and 4) as
a result of chromosome reunion between nonsister chromatids, IV: Production of four
gametes. The four gametes are all recombinant with respect to the three loci. However,
combinations of two loci, AB, ab, BC and bc are parental type and Ab, aB, Bc, bC, Ac
and aC are recombinant. There are no parental type gametes for combination A and C
in this illustration.

The Mechanism of Recombination

The genetic basis of linkage mapping is genetic recombination resulting from crossing over between homologs during meiosis. Since genetic recombination was discovered by Bateson, Saunders and Punnett (1905) in *Lathyrus odoratus* (sweet pea) and Morgan (1911a, b) in *Drosophila melanogaster*, research related to the mechanism of crossing over and the applications of genetic recombination has been conducted on bacteria (reviewed by Conley 1992), fungi (reviewed by Hastings 1992) and many higher organisms (Whitehouse 1982; Sall 1990).

Two influential models on the initiation of homologous recombination involve DNA strand exchange beginning at a site where one strand of DNA has been broken (Holliday 1964; Meselson and Radding 1975) (Figure 2.2), although direct evidence that a break is sufficient to initiate recombination has not been presented (McGill *et al.* 1993). A double-strand break model has also been proposed for genetic recombination (Szostak *et al.* 1983). In *Saccharomyces cerevisiae*, there has been genetic and physical evidence for initiation of recombination by double-strand breaks (McGill *et al.* 1993). A site-specific gene-conversion mechanism that transfers a sequence from the locus where it is silent to a mating type control gene (MAT) where it is expressed is initiated by a double-strand cleavage at a MAT. The cleavage is done by an endonuclease encoded by the HO gene (McGill *et al.* 1993).

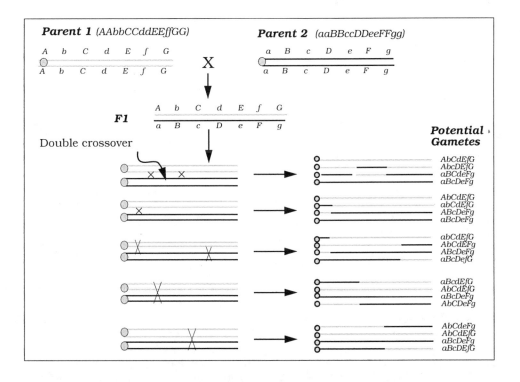

Figure 2.3 An illustration of recombination and 7 genes located on the same chromosome. Two parents are homozygous for the 7 loci (AAbbCCddEEffGG and aaBBccDDeeFFgg). The F1 is heterozygous for all 7 loci. The gametes produced by the F1 will vary depending on where the crossovers have occurred.

Holliday's model of crossing over includes a series of chromosome breakage and reunion events between two homologous chromosomes (nonsister chromatids) during meiosis (Figure 2.2). The crossovers involve the random breakage of the chromatids and the joining of different homologous chromatids. Figure 2.3 illustrates a process similar to that shown in Figure 2.2, with more loci on the chromosomes.

Linkage Phase

Linkage phase is the term used to denote chromatid associations of alleles of linked loci. Two alleles located on the same chromosome are linked in coupling phase. Two alleles located on different chromosomes are linked in repulsion phase. For example, A and B alleles are linked in coupling phase in Figure 2.4.I and in repulsion in Figure 2.4.II.

The linkage phase is usually known or can be assigned for codominant genes in controlled crosses and can be inferred in some cases from experimental data in natural populations (see Chapters 7, 8 and 15 for more on linkage phase). Drawing inferences about the linkage phase in natural populations usually requires parental information.

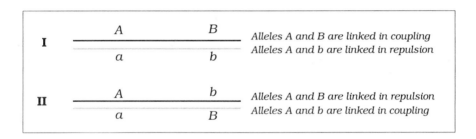

Figure 2.4 Linkage phase for two loci

Factors Affecting Recombination

A better understanding of the mechanisms of crossing over and control of genetic recombination will lead to more precise linkage map construction and greater success in the comparison and integration of linkage maps in different individuals and different species. Crossing over is believed to be influenced by both genetics and the environment. Genetic recombination was found to be affected by environmental agents (Bridges 1915; Levine 1955; Suzuki 1965; Simchen and Stamberg 1969) and Gowen (1919) found that recombination frequency in *Drosophila* was influenced by genetic background. Much subsequent research has confirmed that recombination is under genetic control. For example, great variability of genetic recombination was found in maize and recombination fractions between the Hor 1 and Hor 2 loci on the short arm of chromosome 5 were found to be significantly different among three varieties of barley (Sall 1990 and 1991).

In *Petunia hybrida*, a recombination modulator gene (Rm1) found on chromosome II acts on all seven chromosomes of the species. It often increases recombination, especially for closely linked genes, but its effects vary according to the chromosome segments studied. Many mutations have been found to have effects on recombination in *E. coli* (Hastings 1992; Lanzov *et al.* 1991; Takahashi 1992). In *C. elegans*, a recessive mutation, rec-1, increased recombination fraction at least threefold higher than that found in the wild type. Experiments of two-direction selection on recombination fraction have been carried out in lima bean (Allard 1963) and *Drosophila* (Chinnici 1971; Kidwell 1972a and b; Cederberg 1985; Charlesworth and Charlesworth 1985a and b). Recombination fraction changes significantly as a result of two-direction selection. Genetic control of recombination is likely to be polygenic in nature.

Importance of Manipulation of Genetic Recombination

If genetic recombination is influenced by environmental and genetic factors, it may be amenable to manipulation. When loci are either very closely or very loosely linked, the locus order is difficult to resolve and the recombination fraction estimates may be biased or have large variances (Keats *et al.* 1991; Lathrop *et al.* 1987; Olson and Boehnke 1990; Ott 1991; Smith 1990). By manipulating recombination frequency, the resolution of physical and linkage maps would be increased (see Chapter 17), map-based cloning would be more efficient and more precise and some problems related to quantitative trait loci (QTL) mapping would be partially solved (Dudley 1993) (see Chapter 17). Alternatively, the absence of recombination could be useful to avoid loss of linkage over generations when specific linkages are desirable.

Genetic recombination is a major force for generating genotypes with new gene combinations in breeding populations; it is the foundation of plant and animal breeding (Fehr 1987; Falconer 1982) and many evolution studies (Nei 1987; Felsenstein 1988). It has been a long-term goal of breeders to manipulate genetic recombination (Hanson 1959; Fehr 1987). An intercross or a backcross breeding program would be far more efficient if genes controlling target traits were localized on the genome and the recombination among them controlled.

2.2.3 MEASUREMENT OF GENETIC RECOMBINATION

Statistical treatment of the recombination fraction will be discussed extensively in this book. By way of introduction, I will outline some important terminology and basic concepts for both recombination fraction and mapping functions.

Recombination Fraction

Genetic recombination is measured by the recombination fraction, which is the ratio of recombinant gametes to total gametes. For homologous recombination, we expect that the greater the physical distance between two loci on a chromosome, the greater the chance they will recombine. Many statistical procedures have been used to detect linkage and to estimate recombination fraction at two-point or multi-point levels (Mather 1958; Ott 1991; Chapters 5 and 6). These procedures are the fundamentals of linkage map construction. Recombination fraction is not additive along a chromosome and the departure from additivity increases with distance between loci. Therefore, mapping functions have been developed to correct for this effect. Mapping functions such as Haldane's function, Kosambi's function and some other functions have been developed to make recombination additive and are also suitable for phage and fungi (Haldane 1919; Kosambi 1944; Chapter 10).

For two loci (A and B) on the same chromosome and segregating for two alleles at each locus (A and a for gene A; B and b for gene B), four types of gametes will be produced by meiosis, AB, Ab, aB and ab. If the parental types are AB and ab, then Ab and aB are the recombinant types. If we sample from the population and observe n_r recombinant gametes (Ab and aB) among a total of n samples, then the recombination fraction between the two loci is

$$\hat{r} = n_r/n \qquad (2.2)$$

In practice, what is observed is usually not the gamete frequencies, but rather phenotypic frequencies. Estimation of recombination fraction using the phenotypic data usually involves construction of the likelihood for recombination fraction and estimation of the recombination fraction using a variety of maximum likelihood approaches (Chapters 6, 7 and 8).

Interference

For three linked loci, A, B and C, there are three possible recombination fractions, r_{AB}, r_{BC} and r_{AC}. If the three loci are in ABC order on the chromosome and crossovers occur at random, then there is a relationship between the three recombination fractions based on simple probability theory, which is

$$r_{AC} = r_{AB} + r_{BC} - 2r_{AB}r_{BC} \qquad \frac{r_{12}}{C} \qquad (2.3)$$

where $2r_{AB}r_{BC}$ is the expected double crossover frequency (i.e., crossovers between A and B and between B and C, simultaneously). However, departure from this expectation has been observed. This departure is defined as the result of crossover interference between the two segments (between A and B and between B and C). This phenomenon is quantified by adding a coefficient to Equation (2.3): $r_{AC} = r_{AB} + r_{BC} - 2Cr_{AB}r_{BC}$. C is defined as the coefficient of coincidence and $1 - C$ is defined as interference. If we assume that there is no interference and crossovers occur randomly, then the expected double recombinant frequency will be $2r_{AB}r_{BC}$ and $C = 1$; therefore,

$$\text{Interference} = 1 - C = 0$$

If crossovers in the intervals AB and BC are not independent, the observed double recombinant frequency may not equal the expectation. If we use r_{12} to denote the true double recombinant frequency, the coefficient of coincidence is

$$C = \frac{r_{12}}{2r_{AB}r_{BC}}$$

Some possible values for the coefficient of coincidence and interference and their corresponding interference status are

Coefficient of Coincidence	Interference	Interference Status
$C > 1$	$(1 - C) < 0$	Negative
$C < 1$	$(1 - C) > 0$	Positive
$C = 0$	$(1 - C) = 1$	Complete
$C = 1$	$(1 - C) = 0$	Absent

Negative interference means that the observed double crossovers are more than expected. Positive interference means that the observed double crossovers are less than expected. Absence of interference means that they are equal. Complete interference means that no double crossovers occurred.

High levels of interference have been a limiting factor in the development of complete multiple-locus models. Many models have been developed under the assumption of no interference or no high level of interference (see Chapter 10).

Haldane's Mapping Function

Mapping functions have been commonly used in genomic analysis. When there are more than three loci, the relationship among the possible recombination fractions is complex; the recombination fractions between loci flanking a region are not the simple sum of the recombination fractions for the adjacent loci within the region. In other words, the recombination fractions are not additive. A mapping function is designed to solve this problem. If we assume that the crossovers occur randomly along the length of the chromosome, then the crossover events can be modeled as a Poisson process. If we assume that the average number of crossovers is λ, then the probability of no crossovers occurring in an interval is

$$P[\text{no crossover}] = e^{-\lambda} \tag{2.4}$$

and the probability of a crossover occurring is

$$P[\text{crossover}] = 1 - e^{-\lambda} \tag{2.5}$$

Because each pair of homologs with one crossover event results in one-half recombinant gametes (Figure 2.1), the probability of a recombinant is half the value in Equation (2.5). If we define the expected number of recombinants as a mapping function ($m = 0.5\lambda$), then

$$r = 0.5\,(1 - e^{-2m}) \tag{2.6}$$

This is the form of Haldane's mapping function. The inverse of Equation (2.6) is

$$m = -0.5\log\,(1 - 2r) \tag{2.7}$$

which will convert an estimated recombination fraction (r) to Haldane's map distance (m). When the recombination fraction is small, the quantities of the map distance and recombination fraction are approximately equal (Figure 2.5). More mapping functions will be discussed in Chapter 10.

To show how Equation (2.6) works, we write, for the locus order ABC

$$m_{AC} = m_{AB} + m_{BC}$$

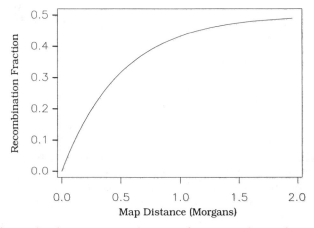

Figure 2.5 Relationship between recombination fraction and map distance according

Because

$$m_{AB} = -0.5\log(1 - 2r_{AB})$$
$$m_{BC} = -0.5\log(1 - 2r_{BC})$$
$$m_{AC} = -0.5\log(1 - 2r_{AC})$$

we can then write

$$-0.5\log(1 - 2r_{AC}) = -0.5\log(1 - 2r_{AB}) - 0.5\log(1 - 2r_{BC})$$

which reduces to

$$r_{AC} = r_{AB} + r_{BC} - 2r_{AB}r_{BC}$$

which is equivalent to Equation (2.3). This verifies that Haldane's mapping function converts recombination fractions to additive map distance for the simple three locus case (Figure 2.6).

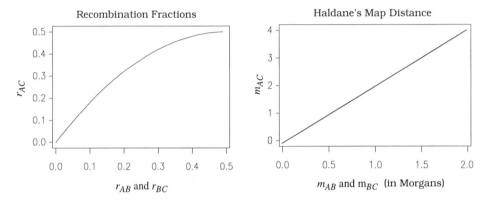

Figure 2.6 Relationship among the three recombination fractions (nonlinear and non-additive) for the three loci, ABC and the relationship after conversion of the recombination fractions to Haldane's function (linear and additive).

Haldane's mapping function works well for situations where crossover interference does not occur. However, in general, mapping functions (see Chapter 10) only work for specific conditions. Mapping functions may not correspond to physical distance for most cases and have little advantage over simple recombination fraction.

Chromosome Rearrangements

Insertions, duplications, deletions, inversions and translocations are major factors in the evolution of chromosome number, structure and genetic content. Detailed analyses of chromosome rearrangements have been carried out in many species, including *Drosophila* (Sturtevant 1931; Bridges and Brehme 1944), maize (Russell and Burnham 1950; Morris 1955), barley (Das 1951; Kasha 1961; Caldecott and Smith 1952) and oats (Koo 1958). These phenomena change the linear structure of chromosomes. This is the cytological basis for linkage map construction and is especially important for comparative mapping. Inversions can change the gene order and greatly alter the map distances for genes located near the breakpoints. Translocations change arrangements of large segments of chromosomes to constitute entire new linkage groups. Insertions, duplications and deletions can also have profound effects on genetic maps. Classical research on these phenomena has been limited to microscopic observation and genetic mapping of morphological markers in a handful of organisms.

For some plant species, such as maize and wheat, cytogenetic variants have been well documented (Carlson 1977). Maize B-A translocations have been successfully used for mapping enzyme encoding genes and defined DNA fragments in maize (Newton and Schwartz 1980; Burr *et al.* 1988; Weber and Helentjaris 1989). Inversions and translocations can be studied at the whole genome level. Recent developments in molecular marker technology have generated a large amount of information on the genome structure of many plant and animal species. For plant species without a rich body of cytogenetic research, molecular marker technology will create an opportunity for significant advances.

Recent developments in biochemistry and molecular biology had a great impact on understanding the effects of transposons and other transposable elements that create insertions and deletions which alter the structure of the genome. These phenomena affect the gene order and may alter the map distance between some genes. DNA fragment based markers are affected greatly by these processes.

2.2.4 APPROACHES USED FOR GENETIC RECOMBINATION STUDIES

Cytology

Both genetic and cytological techniques have been used to study genetic recombination and chiasmata formation. Chiasma counts provide information on overall cellular number and distribution of recombination events. But since chromosomes are relatively condensed during the stages when chiasmata can be detected and since chromosome morphology is generally not finely detailed, specific locations of the crossover events are poorly resolved. Moreover, there are always some instances where it is impossible to distinguish a chiasma from a twist of the chromosome and some organisms have such small chromosomes that detailed chiasma cytology is impossible.

Genetics

A great deal of the information has come from studying a small number of model species and detailed studies have been limited to relatively few chromosome segments. Because of this, the available information is not systematic and has had limited application. The limits on these genetic techniques for studying recombination are numbers of visible or morphological mutations available and the tolerance of the specific organism for the mutant phenotypes.

2.2.5 APPLICATIONS FOR MANIPULATING RECOMBINATION

Application to Fine Genetic Mapping

High density genetic and physical genome maps have been objectives of the human and plant genome initiatives (Watson 1990). One of the keys to achieving these objectives is locus ordering associated with closely and loosely linked loci. Strategies have been proposed to deal with the locus ordering issue by means of statistics and experimentation (Keats *et al.* 1991; Liu and Knapp 1992; Olson and Boehnke 1990; Smith 1990; Reiter *et al.* 1992). Most of the approaches were designed to control recombination fraction and variance of the estimated recombination fraction by screening marker loci for recombinants or by scoring larger numbers of informative genotypes in the population. If we assume a binomial distribution, the standard deviation of the estimated recombination fraction $\hat{r}$ from a sample size of n is estimated as

$$\sqrt{\hat{r}(1-\hat{r})/n}$$

An approximate 95% confidence interval is

$$\hat{r} \pm 1.96\sqrt{\hat{r}(1-\hat{r})/n}$$

The relative confidence interval, which is the ratio of the confidence limits and the recombination fraction, is therefore

$$\frac{1.96\sqrt{\hat{r}(1-\hat{r})/n}}{\hat{r}}$$

which decreases as recombination fraction increases.

Application to Map-Based Cloning

One important application of genome mapping is the cloning and identification of a gene based largely on mapping data. This process is called map-based cloning (Arondel et al. 1992). Resolution of a linkage map with a target locus on a physical map is an important factor for efficient and precise map based cloning. The approaches for precise locus ordering (Keats et al. 1991; Liu and Knapp 1992; Olson and Boehnke 1990; Smith 1990; Reiter et al. 1992; see Chapter 9) may lead to high resolution genetic maps, but may not be sufficient to lead to high resolution linkage maps in physical terms. Reduction of the ratio between physical distance and recombination distance (or genetic distance) may lead to high resolution of both linkage and physical maps. This can be accomplished by selecting genotypes with high recombination for certain segments or chromosomes (see Chapter 17).

Application to QTL Mapping

Genes affecting quantitative traits are often called quantitative trait loci (QTL). Many genes important in selective breeding are QTLs. QTL mapping is one of the most important activities connecting the recent plant and animal genome research to plant and animal improvement. QTL mapping is the key for application to applied breeding and map-based cloning for economically important genes. Generally, when recombination between adjacent loci is high or the linkage map density is low, the power to map QTLs is low. When QTLs are linked to each other, the genetics are complex and statistical models are difficult to fit. These issues will be discussed in greater detail in later chapters.

Application to Plant and Animal Breeding

Manipulating mating design and selection has been the genetic basis of plant and animal breeding. There has been much research conducted on how linkage affects a breeding program (Fehr 1987). In general, more recombination will create more genetic variation. For many breeding programs, intercrossing between two selected parents followed by backcrossing has been the primary mating system. Depending on the stage of backcrossing, high or low recombination is desired. A backcross breeding program would be many times more efficient if one could manipulate genetic recombination. For conventional breeding, the manipulation of specific alleles by genetic recombination could improve the breeding efficiency substantially. But this improvement would be limited as long as the majority of conventional breeding operates on a population and on a whole genome level.

Theory of Genetic Mapping

Numerous theories related to genetic map construction were developed based on the classical understanding of genetic recombination, such as random or semi-random crossover events (Watson *et al.* 1987ab; Ott 1991), Mendelian segregation and classical interference (Bailey 1961; Ott 1991). If recombination is "finely" (at a whole genome level) or "coarsely" (at a specific segment level) genetically controlled (Simchen and Stamberg 1969), then some of the classical theories and assumptions may not be valid and the commonly used procedures for linkage map construction may not be efficient or sufficient.

2.3 POPULATION GENETICS

Population genetics focuses on the frequencies, distribution and origins of genes in populations. It combines Darwin's evolutionary theory with Mendelian genetics and molecular biology to quantify the evolutionary process. At the population level, genetics can be characterized by allelic and genotypic frequencies. Forces changing the allelic and genotypic frequencies are mutation, natural and artificial selection, population admixture (migration) and random genetic drift. Population genetics has contributed significantly to the understanding of evolution.

Many excellent books on population genetics are available (see Chapter 1). In this section, I will briefly cover some concepts of population genetics relevant to genome analysis.

2.3.1 ALLELIC FREQUENCY

The frequency of an allele i in the population is defined as the probability that a haplotype (a particular haploid combination of alleles in a defined region of a chromosome or a genome) carries the allele i. The methodology to estimate allelic frequency will be discussed in Chapters 5 and 7 in greater detail.

For mapping populations commonly used in genomic research which are generated by controlled crosses, allelic frequency is a discrete variable and takes on values from 0 to 1. For example, many crosses in genomic research can be generalized as crosses between two individuals with four different allelic genotypes ab and cd at a locus. Four alleles at the locus are denoted by a, b, c and d in the two parents. If all four alleles are distinct, then each has a frequency of 0.25 in the mapping population composed of progeny from the two parents (Table 2.3). If only two alleles can be identified and the two parents are heterozygous, then the cross is a classical F2 type and allelic frequency is 0.5 for both alleles in the progeny.

Table 2.3 Allelic frequency in progeny of a cross between two parents.

Cross	Comment	Frequency			
		A1	A2	A3	A4
ab X cd	Four alleles	0.25	0.25	0.25	0.25
ab X cc	Three alleles	0.25	0.25	0.5	0
ab X ab	Two alleles (F2)	0.5	0.5	0	0
ab X aa	Backcross	0.75	0.25	0	0
aa X aa	Fixed	1.0	0	0	0

2.3.2 HARDY-WEINBERG EQUILIBRIUM

Genotypic frequencies can be obtained by simple counting in the population. For a gene with n alleles, there are $n(n+1)/2$ possible genotypes. The relationship between gene frequency and genotypic frequency for a single gene at the population level can be used to infer the genetic status of the gene in the population, relative to expected equilibrium.

A population is in equilibrium (often called Hardy-Weinberg equilibrium) if the gene and genotypic frequencies are constant from generation to generation. This implies that the allelic frequency and genotypic frequency have a simple relationship. That is, for a two-allele model where p_A and p_a are allelic frequencies for alleles A and a, respectively and p_{AA}, p_{Aa} and p_{aa} are genotypic frequencies for genotypes AA, Aa and aa, respectively

$$p_{AA} = p_A^2$$
$$p_{Aa} = 2p_A p_a$$
$$p_{aa} = p_a^2$$

or

$$(p_A + p_a)^2 = p_A^2 + 2p_A p_a + p_a^2$$

One generation of random mating will bring a population into Hardy-Weinberg equilibrium at the single locus level. For testing and quantifying Hardy-Weinberg equilibrium and multiple-locus Hardy-Weinberg equilibrium, see Chapter 8.

Figure 2.7 shows the relationship between genotypic and allelic frequencies for one locus with two alleles A and a under Hardy-Weinberg equilibrium. The frequency of heterozygotes is maximum when the two allelic frequencies are equal (0.5).

Another interesting point is the high probability that any rare allele is carried in the heterozygous state, which is

$$\frac{p_{Aa}}{p_{Aa} + p_{AA}} = \frac{2p_A p_a}{2p_A p_a + p_A^2} = \frac{2p_a}{1 + p_a}$$

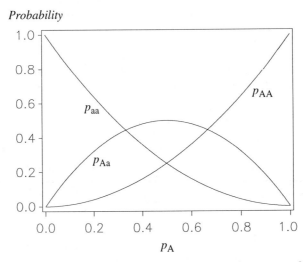

Figure 2.7 Genotypic frequencies p_{AA}, p_{Aa}, and p_{aa} as functions of allelic frequency under Hardy-Weinberg equilibrium.

If the frequency of the allele is 0.01, then 99% of the rare alleles are expected to be carried by heterozygotes. The relative frequencies of heterozygotes and homozygotes carrying a rare allele are even more different. For example, if $p_A = 0.1$, the ratio between the expected heterozygous frequency and homozygous frequency (p_{AA}) is

$$\frac{p_{Aa}}{p_{AA}} = \frac{2p_A(1-p_A)}{p_A^2} = \frac{2(1-p_A)}{p_A} = \frac{2(1-0.1)}{0.1} = 18$$

The heterozygotes carrying the rare allele are 18 times more frequent than the homozygotes carrying the rare allele in the population. If the rare allele has a frequency of 0.01, then the heterozygotes are 198 times more frequent than the homozygotes. When a rare allele is in a population, most likely it is carried by heterozygotes.

For multiple alleles at a single locus, $p_1, p_2, ..., p_i, ...p_n$ are used to denote the frequencies of alleles $A_1, A_2, ..., A_i, ...A_n$. The $n(n+1)/2$ possible genotypes are denoted by $A_{11}, A_{12}, ..., A_{ij}, ...A_{nn}$. Under Hardy-Weinberg equilibrium, the expected genotypic frequencies are

$$(p_1 + p_2 + ... + p_i + ... + p_n)(p_1 + p_2 + ... + p_j + ... + p_n)$$
$$= p_1^2 + 2p_1p_2 + ... + 2p_ip_j + ... + 2p_{n-1}p_n + p_n^2$$

where the combination of the subscripts for the two probabilities in each term correspond to the subscript of the genotype. For example, there are 10 possible genotypes for a four-allele system. Their expected frequencies are listed in Table 2.4. The proportion of heterozygosity in a population, estimated by

$$P_H = 1 - \sum_i p_i^2$$

under Hardy-Weinberg equilibrium, is an important criterion for screening genetic markers in genomic research. The heterozygosity reaches a maximum when alleles are equally frequent in the population. For the four-allele example (Table 2.4), the heterozygosity is 0.75 when the four alleles have equal frequency

Table 2.4 The expected genotypic frequencies for a four-allele system under Hardy-Weinberg equilibrium and the expected proportion of heterozygosity.

Genotype	Expected Frequency	p_i				
		$p_1 = 0.25$ $p_2 = 0.25$ $p_3 = 0.25$ $p_4 = 0.25$	$p_1 = 0.3$ $p_2 = 0.3$ $p_3 = 0.2$ $p_4 = 0.2$	$p_1 = 0.4$ $p_2 = 0.4$ $p_3 = 0.1$ $p_4 = 0.1$	$p_1 = 0.4$ $p_2 = 0.3$ $p_3 = 0.2$ $p_4 = 0.1$	$p_1 = 0.7$ $p_2 = 0.1$ $p_3 = 0.1$ $p_4 = 0.1$
A_1A_1	p_1p_1	0.0625	0.09	0.16	0.16	0.49
A_1A_2	$2p_1p_2$	0.125	0.18	0.32	0.24	0.14
A_1A_3	$2p_1p_3$	0.125	0.12	0.08	0.16	0.14
A_1A_4	$2p_1p_4$	0.125	0.12	0.08	0.08	0.14
A_2A_2	p_2p_2	0.0625	0.09	0.16	0.09	0.01
A_2A_3	$2p_2p_3$	0.125	0.12	0.08	0.12	0.02
A_2A_4	$2p_2p_4$	0.125	0.12	0.08	0.06	0.02
A_3A_3	p_3p_3	0.0625	0.04	0.01	0.04	0.01
A_3A_4	$2p_3p_4$	0.125	0.08	0.02	0.04	0.02
A_4A_4	p_4p_4	0.0625	0.04	0.01	0.01	0.01
	P_H	0.75	0.74	0.66	0.70	0.48

(0.25) and reduces to 0.48 when frequencies of the four alleles are 0.7, 0.1, 0.1 and 0.1. If one allele has a frequency of 0.9, then the upper limit of the heterozygosity is 0.19.

For l loci, there are

$$\frac{1}{2^l}\prod_{i=1}^{l}[n_i(n_i+1)]$$

possible genotypes, where n_i is the number of alleles for locus i. If all l loci are independent and the population is under Hardy-Weinberg equilibrium at a multiple-locus level, then the expected frequency of a multiple-locus genotype is the simple product of the expected frequencies for the single-locus genotypes. However, when the loci are not independent (or are linked), or the population is in disequilibrium, calculation of the expected genotypic frequencies is more difficult. Most analyses in this book are based on expected multiple-locus genotypic frequencies, such as linkage analysis and multiple-locus models (see Chapters 6 and 10).

2.3.3 CHANGES IN GENE FREQUENCY

Migration, mutation and selection are the forces that change gene frequency from generation to generation in large populations. In small populations, random sampling has a greater effect on gene frequency than other forces. We will use a two-allele model to illustrate how these factors affect gene frequency.

If an allelic frequency in a native population is p_{n0}, a proportion m_i relative to the native population migrates from the *ith* population among k populations to the native population every generation and the allelic frequency among the immigrants from the *ith* population is p_i, then the allelic frequency in the mixed population is

$$p_{n1} = \left[1-\sum_{i=1}^{k}m_i\right]p_{n0}+\sum_{i=1}^{k}(m_ip_i)$$

$$= p_{n0}+\sum_{i=1}^{k}[m_i(p_i-p_{n0})]$$

(2.8)

The change in allelic frequency is

$$\delta p = p_{n1}-p_{n0}$$

$$= \sum_{i=1}^{k}[m_i(p_i-p_{n0})]$$

(2.9)

Mutation is the ultimate source of new alleles. If the mutation rate from wild type to mutant is u and the reverse mutation rate is v per generation, then the frequency of the wild type after one generation is

$$p_{n1} = p_{n0}-up_{n0}+v(1-p_{n0})$$

(2.10)

where p_{n0} is the allelic frequency of the wild type in the population of the previous generation, and the allelic frequency change is

$$\delta p = p_{n1}-p_{n0}$$

$$= (v(1-p_{n0})-up_{n0})$$

(2.11)

If we set the allelic frequency change equal to zero, the equilibrium condition is

$$\frac{p_{n0}}{1 - p_{n0}} = \frac{v}{u} \qquad or$$

$$p_{n0} = \frac{v}{u + v}$$

In nature, the forward mutation rate is usually lower than 10^{-5} per locus per generation. The mutation rate from the wild type to mutant is generally much higher than the mutation rate reverting the mutant to the wild type.

Another force changing the gene frequency in a population is selection. The selection can be artificial or natural. For a two-allele system, if fitness (1 - coefficient of selection) for the three genotypes AA, Aa and aa is f_1, f_2 and f_3, then the allelic frequency for allele A after one generation of selection is

$$p_{n1} = \frac{f_1 p_{n0}^2 + f_2 p_{n0}(1 - p_{n0})}{f_1 p_{n0}^2 + 2 f_2 p_{n0}(1 - p_{n0}) + f_3 (1 - p_{n0})^2} \qquad (2.12)$$

where p_{n0} is the allelic frequency for A in the population before selection and the coefficient for selection is the measure of the disadvantage of a given genotype in a population. Depending on the values of f_1, f_2 and f_3, selection has different biological significance and effectiveness in changing the allelic frequency in the population.

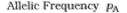

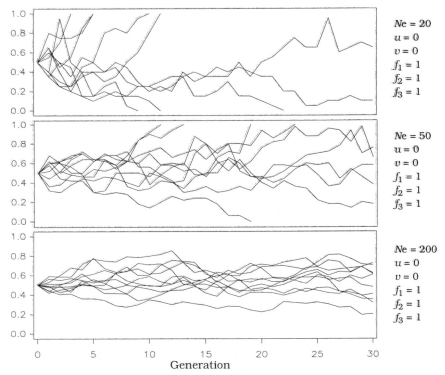

Figure 2.8 Random drift from a 100% heterozygous population for different effective population sizes. The data was simulated by computer with assumptions of no migration, mutation or selection. Ten replications were simulated for each setting. Ne is population size.

Allelic Frequency p_A

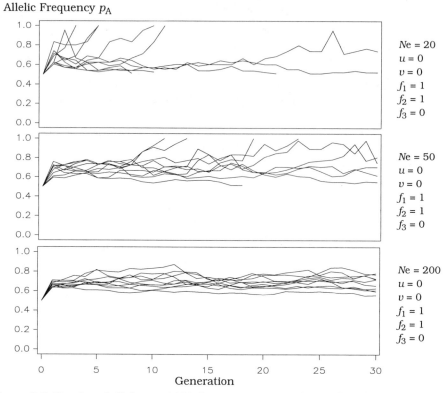

Figure 2.9 Random drift from a 100% heterozygous population for different effective population size. The data was simulated by computer with assumptions of no migration and mutation. Complete selection was applied against the mutant homozygotes. Ten replications were simulated for each setting.

Random sampling plays a more important role in allelic frequency change from generation to generation when the population size is small. In nature, population sizes for the majority of plant and animal species are usually large enough to ignore this random sampling pressure. However, for many experimental populations, such as breeding populations for crop and animal improvement, sizes are relatively small and random drift can be important.

For a single gene having different allelic forms, the passage of a specific allele from generation to generation is a stochastic process. This stochastic process can be described quantitatively by the variance of allelic frequency, which is

$$Var(p) = p_0(1 - p_0)\left[1 - \left(1 - \frac{1}{2N_e}\right)^t\right] \tag{2.13}$$

where p_0 is the allelic frequency for A at the initial generation, t is the number of generations of random mating from the initial generation and N_e is effective population size. The effective population size can be estimated using

$$N_e = t / \sum_i^t \frac{1}{N_i} \tag{2.14}$$

where N_i is the population size at generation i. If the population sizes are equal for all generations, then the effective population size is the population size.

Figure 2.8 shows the effects of random drift on a population starting with 100% heterozygotes for different effective population sizes. The variance of allelic frequency among different replications increases rapidly as the generation proceeds when effective population size is small. A large portion of the replications were fixed (allele A or a lost) within 15 generations of random mating.

Figure 2.10 shows the effects of the combination of random drift and selection against the homozygous mutant (aa). Because the selection favors the wild type A, the possibility of fixation of the wild type is larger than when there is no selection. Fixation of the heterozygote mutant is impossible. Figure 2.9 shows a combination of random drift, mutation and selection. Selection against the homozygous mutant is complete. Mutation decreases the chance of fixation for the wild type and balances the selection effect on fixation. In nature, the combination of all the forces affecting gene frequency may work to maintain the variation and keep the populations from fixation. The factors influencing gene frequency in the population, such as migration, mutation, selection and random drift, will also affect genotypic frequency. There are more factors influencing genotypic frequency, such as the equilibrium status, mating schemes and relationships between the genes of interest. I will discuss changes of genotypic frequencies further in Chapters 7 and 8.

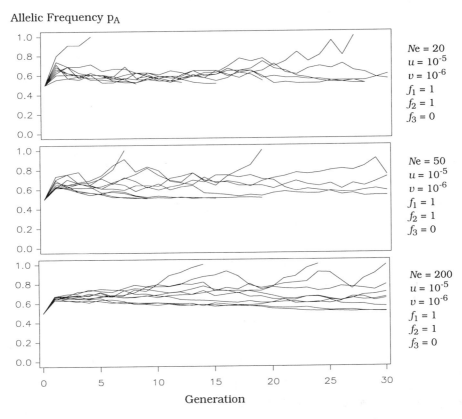

Figure 2.10 Random drift from a 100% heterozygous population for different effective population sizes. The data was simulated by computer with assumptions of no migration. Selection against the homozygous mutant was complete. Mutation rate for A⇒a is 10^{-5} and for a⇒A is 10^{-6}. Ten replications were simulated for each setting.

2.4 QUANTITATIVE GENETICS

Quantitative genetics focuses on the inheritance of quantitative traits. As the number of genes controlling a trait increases and the importance of the effects of the environment on the phenotype of the trait increases, the ability to model the inheritance of the trait using Mendelian genetics diminishes. Quantitative genetics has been one of the foundations of genetic improvement of field crops, forest trees and agricultural animals (Mayo 1987; Fehr 1987).

Excellent books on quantitative genetics are available (see Chapter 1). I will briefly outline some concepts of quantitative genetics relevant to genomic analysis in this section.

2.4.1 SINGLE-GENE MODEL

Notation

Quantitative genetic theory starts with a single-gene model (Figure 2.11). For a single-locus A with two alleles A and a, there are three possible genotypes in the population: AA, Aa and aa. Three values, a, d and $-a$, are assigned arbitrarily to each of the genotypes. The population is assumed to be in Hardy-Weinberg equilibrium and the two alleles have frequencies of p and $(1 - p) = q$. The population can be modeled using the information in Table 2.5.

Figure 2.11 Notation for a single-gene model.

First, we can derive the population mean (μ) in terms of allelic frequencies and the genotypic values in Figure 2.11

$$\mu = p^2 a + 2pqd - q^2 a$$
$$= ((p - q) a + 2pqd)$$

The deviation of genotypic value of AA from the population mean is

$$a - \mu = a - [(p - q) a + 2pqd]$$
$$= 2q (a - pd)$$

Using the same approach, the deviations from the mean for the other genotypes can be obtained (Table 2.5). The deviations can be used as genotypic effects or values.

Average Effect of Gene Substitution

An important concept in quantitative genetics is the average effect of gene substitution (α), which is the average effect on the trait of one allele being replaced by another allele. Several methods can be used to obtain the average effect. For the two-allele system in Table 2.5, a gamete containing allele A will

Table 2.5 Genotypic values for a single locus with two alleles.

	Genotype			Mean
	AA	Aa	aa	
Genotypic Value	a	d	$-a$	
Copy of A Allele (x)	2	1	0	
Frequency	p^2	$2pq$	q^2	
Frequency × Value	p^2a	$2pqd$	$-q^2a$	$(p-q)a + 2pqd$
Genotypic Value as Deviation from Mean	$2q(a-pd)$	$a(q-p) + d(1-2pq)$	$-2p(a+qd)$	
Breeding Value	$2q\alpha$	$(q-p)\alpha$	$-2p\alpha$	
Dominance Deviation	$-2q^2d$	$2pqd$	$-2p^2d$	

result in progeny with the genotype AA and Aa with the frequencies of p and q, respectively. Similarly, a gamete containing allele a will result in progeny with genotype Aa and aa with the frequencies of p and q. Mean values of the genotypes produced for the two types of gametes are

$$A \qquad pa + qd$$

$$a \qquad pd - qa$$

The difference between these two values is the average effect of gene substitution (α). For example, in substituting a with A

$$\alpha = (pa + qd) - (pd - qa)$$

$$= a + (q-p)\,d$$

The average effect can also be obtained using regression of the genotypic value on the number of copies of the target alleles, for example

$$\alpha = \frac{\sum fxy}{\sum fx^2 - (\sum fx)^2}$$

$$= \frac{2p^2q\,(a-pd) + 2pq\,[a\,(q-p) + d\,(1-2pq)]}{4p^2 + 2pq - (2p^2 + 2pq)^2}$$

$$= a + (q-p)\,d$$

Breeding Value

The breeding value of a genotype is defined as the average genotypic value of its progeny. The average effect of gene substitution (α) is the genetic effect of gametes transferred to progeny. Progeny having genotype AA receive two copies of allele A and a genetic effect of 2α. Progeny having genotype Aa receive α and those having aa receive 0. On average

$$2\alpha \times p^2 + \alpha \times 2pq + 0 \times q^2 = 2p\alpha$$

Adjusting by this mean, the breeding values for the three genotypes are estimated

$$AA \qquad 2\alpha - 2p\alpha = 2q\alpha$$

$$Aa \qquad \alpha - 2p\alpha = (1-2p)\,\alpha = (q-p)\,\alpha$$

$$aa \qquad 0 - 2p\alpha = -2p\alpha$$

Breeding value plays an important role in applications of quantitative genetics in plant and animal breeding. The larger the breeding values of selected indi-

Breeding Value

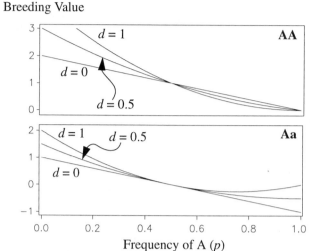

Figure 2.12 Breeding values for genotypes AA and Aa as functions of allelic frequency of A and degree of dominance. In the figures, a = 1.0 and degrees of dominance are d/a = d = 0, 0.5 and 1.0.

viduals, the larger the genetic improvement that can be made. Figure 2.12 shows the relationship between breeding value of a genotype (*e.g.*, AA) and frequency of the allele (*e.g.*, A) in the population. Assume that allele A contributes to the desirable trait value with $a = 1$ and $d = 0$, 0.5 and 1.0 (no dominance, partial dominance and complete dominance). When $d = 0$, the breeding values of AA and Aa have a linear relationship with the frequency of the A allele. In general, breeding values of AA and Aa decrease as the frequency of the A allele increases. This is the rationale behind the idea that a gene with a rare favorable allele has more potential breeding significance than a gene with a favorable allele already at a median or high frequency in the population. However, the rare favorable alleles contribute less than the median and high frequency favorable alleles to the population mean and additive variance (below).

Dominance Deviation

Breeding values for a single locus are additive effects of genotypic values. The dominance deviation is the portion of genotypic value which cannot be explained by breeding values. It can be obtained by subtraction of the breeding value from the genotypic value. For example, the dominance deviation for genotype AA in Table 2.5 is

$$2q\,(a - pd) - 2q\,[a + (q - p)\,d] \;=\; -2q^2 d$$

The dominance deviations for the other genotypes can be obtained in the same way from Table 2.5.

Variance

Variance has been used extensively in quantitative genetics. For a single-locus model, total genetic variance in a population is the variance of the genotypic values, which is

$$\sigma_G^2 = p^2 [2q (a - pd)]^2 + 2pq [a (q - p) + d (1 - 2pq)]^2 + q^2 [-p (a + qd)]^2$$
$$= 2pq\alpha^2 + 4p^2q^2d^2$$

The additive genetic variance is the variance of breeding values

$$\sigma_A^2 = p^2 (2q\alpha)^2 + 2pq [(q - p) \alpha]^2 + q^2 (-p\alpha)^2$$
$$= 2pq\alpha^2$$

The dominance variance is the variance of the dominance deviations

$$\sigma_D^2 = p^2 (-2q^2d)^2 + 2pq (2pqd)^2 + q^2 (-2p^2d)^2$$
$$= 4p^2q^2d^2$$

It is not difficult to see that

$$\sigma_G^2 = 2pq\alpha^2 + 4p^2q^2d^2$$
$$= \sigma_A^2 + \sigma_D^2$$

for a single-locus model in a population under Hardy-Weinberg equilibrium. Figure 2.13 shows the relationships between the additive and dominance vari-

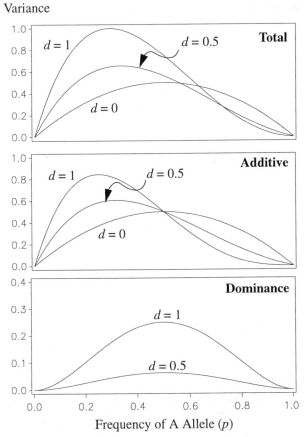

Figure 2.13 Total, additive and dominance genetic variance in the population are plotted against frequency of the A allele. In the figures, a = 1.0 and degrees of dominance are d/a = d = 0, 0.5 and 1.0.

ances and the allelic frequency of the A allele. When $p = q = 0.5$, the additive variance has no relationship to the degree of dominance and dominance variance reaches its maximum. Under complete dominance $d = 1.0$, the additive variance reaches its maximum at $p = 1/3$. In any case, the genetic variance is small when the allelic frequency of A is less than 5%.

2.4.2 TRAIT MODELS

Under quantitative genetic assumptions, a trait may be controlled by a number of genes. However, in the classical quantitative analysis, the number of genes and their genotypic effects are usually unknown. The genetic effects of a number of genes are usually pooled into one genotypic term. A simple model for a continuous trait is

$$y_{ij} = \mu + G_i + \varepsilon_{ij} \tag{2.15}$$

where y_{ij} is the trait value for genotype i in replication j, μ is the population mean, G_i is the genetic effect for genotype i and ε_{ij} is the error term associated with genotype i in replication j. Depending on the mating designs and experimental designs, the resolution for the partitioning of G_i and ε_{ij} changes. The trait models have been the main focus of traditional quantitative genetics. As it will be seen in Chapters 12 to 17, genomic analysis and the concept of QTL are changing the perspectives of modeling quantitative trait inheritance.

If it is assumed that the components in the model are distributed as normal variables

$$y \sim N(\mu, \sigma_p^2)$$
$$G \sim N(0, \sigma_g^2)$$
$$\varepsilon \sim N(0, \sigma_e^2)$$

and the covariance between genetic effects and experimental error is zero, then

$$\sigma_p^2 = \sigma_g^2 + \sigma_e^2$$

If the same genotype is replicated b times in an experiment and phenotypic means are used, then the relationship becomes

$$\sigma_{\bar{p}}^2 = \sigma_g^2 + \frac{1}{b}\sigma_e^2$$

If the same genotypes are tested in several environments, such as locations or years, then the simple model of Equation (2.15) can be extended to

$$y_{ijk} = \mu + G_i + E_j + (GE)_{ij} + \varepsilon_{ijk} \tag{2.16}$$

where E_j and $(GE)_{ij}$ are the environmental effects and genetic by environmental interactions.

An analysis of variance (ANOVA) is commonly used to estimate variance components associated with the model of Equation (2.16). For example, Table 2.6 shows a typical ANOVA table for an experiment with g genotypes evaluated in e environments. The experiment is designed using a completely randomized block design with b replications. It is common to consider Equation (2.16) as a complete random effects model when the objective of the experiment is to estimate population related parameters such as heritability (see next section). When the objective of the experiment is to estimate a genotype specific parameter, such as combining ability, the genotypic effect in Equation (2.16) may be considered a fixed effect.

Table 2.6 A typical analysis of variance (ANOVA) table for multiple environments.

Source	Degrees of Freedom	Expected Mean Square
Environments	$e - 1$	
Blocks	$(b - 1)e$	
Genotypes	$g - 1$	$\sigma_e^2 + b\sigma_{ge}^2 + be\sigma_g^2$
G × E	$(g - 1)(e - 1)$	$\sigma_e^2 + b\sigma_{ge}^2$
Error	$(b - 1)(g - 1)e$	σ_e^2

2.4.3 HERITABILITY

Heritability is defined as the ratio of genotypic to phenotypic variance

$$H = \frac{\sigma_g^2}{\sigma_p^2} = \frac{\sigma_g^2}{\sigma_g^2 + \sigma_e^2} \qquad (2.17)$$

Depending on the relationship among the genotypes, interpretation of the genotypic variance may be different. The genotypic variance may contain complete variances associated with additive, dominance and epistatic interactions. In this case, the heritability of Equation (2.17) is called broad sense heritability. Using some experimental or mating schemes, an additive genetic variance (σ_a^2) may be calculated. Narrow sense heritability is defined as the ratio of the additive portion of genetic variance to the phenotypic variance

$$H = \frac{\sigma_a^2}{\sigma_p^2} = \frac{\sigma_a^2}{\sigma_a^2 + \sigma_d^2 + \sigma_i^2 + \sigma_e^2} \qquad (2.18)$$

where σ_d^2 and σ_i^2 are genotypic variances associated with the dominance effects and epistatic interactions. If phenotypic means are used, then we can obtain the mean-based heritability

$$H = \frac{\sigma_g^2}{\sigma_p^2} = \frac{\sigma_g^2}{\sigma_g^2 + \sigma_e^2/b} \qquad (2.19)$$

where b is the number of replications.

2.4.4 GENETIC CORRELATION

For two related traits, the values can be modeled as

$$\begin{aligned} y_{1j} &= \mu_1 + G_{1i} + \varepsilon_{1j} \\ y_{2j} &= \mu_2 + G_{2i} + \varepsilon_{2j} \end{aligned} \qquad (2.20)$$

where the subscripts '1' and '2' denote trait one and two, respectively, subscript i indicates the gene, j is the indicator for an individual in the population, y is the trait value, μ is the overall mean and ε is the random error. To quantify relationship between the two traits, matrices

$$\Sigma_p = \begin{bmatrix} \sigma_{p1}^2 & \sigma_{p12} \\ \sigma_{p12} & \sigma_{p2}^2 \end{bmatrix} = \Sigma_g + \Sigma_e = \begin{bmatrix} \sigma_{g1}^2 & \sigma_{g12} \\ \sigma_{g12} & \sigma_{g2}^2 \end{bmatrix} + \begin{bmatrix} \sigma_{e1}^2 & \sigma_{e12} \\ \sigma_{e12} & \sigma_{e2}^2 \end{bmatrix} \qquad (2.21)$$

are used, where Σ_p, Σ_g and Σ_e are variance-covariance matrices for phenotypic, genetic and environmental effects, respectively. The relationship between the two traits can be quantified by

$$\rho_p = \frac{\sigma_{p12}}{\sqrt{\sigma_{p1}^2 \sigma_{p2}^2}}$$

$$\rho_g = \frac{\sigma_{g12}}{\sqrt{\sigma_{g1}^2 \sigma_{g2}^2}} \qquad (2.22)$$

$$\rho_e = \frac{\sigma_{e12}}{\sqrt{\sigma_{e1}^2 \sigma_{e2}^2}}$$

where ρ_p, ρ_g and ρ_e are defined as phenotypic, genetic and environmental correlations, respectively.

The genetic correlation between the two traits may be caused by linkage of the genes controlling the traits, or by the same gene actually controlling both traits (pleiotropy). The pleiotropic effects could be explained by a physiological relationship between the traits. For example, plant height and biomass may result from the expression of the same gene products. If traits can be measured at the gene product level, genetic correlation should be shown to be caused by genetic linkage or the co-ordinate control of gene expression or by the expression of a single gene that affects both traits.

2.5 MOLECULAR GENETICS

Molecular genetics is the science of the biological properties of DNA (deoxyribonucleic acid). DNA nucleotide sequence, encoded biological information, function, regulation and structure are the subjects of molecular genetics. Genomic studies with molecular marker technology come from the technology of molecular genetics applied to the traditional problems of classical genetics.

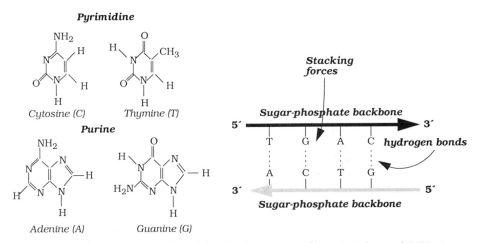

Figure 2.14 The nitrogenous bases for the four types of nucleotides and DNA structure. The nitrogenous bases for each strand are held together by the sugar-phosphate backbone. The backbones are antiparallel. The two strands of DNA are held together by hydrogen bonds between purines and pyrimidines.

2.5.1 DNA

DNA Structure

The subunits of DNA are four basic molecules called nucleotides. The nucleotides have different nitrogenous bases attached to a phosphorylated five carbon sugar and are denoted by those bases: adenine (A), guanine (G), cytosine (C) and thymine (T). The A and G bases are similar in structure and are called purines. The T and C bases are similar and are called pyrimidines (Figure 2.14). The nitrogenous bases for each strand are held together by the sugar-phosphate backbone. The backbones run in opposite directions: one is in the direction of 5´→3´ and the other is antiparallel, 3´→5´. The two strands of DNA are held together by hydrogen bonds between purines and pyrimidines. The hydrogen bonds are always formed in pairs, G-C and A-T, described by the Watson-Crick rules. Each is called a base pair (Figure 2.14). The two strands of DNA twist to form a double helix structure.

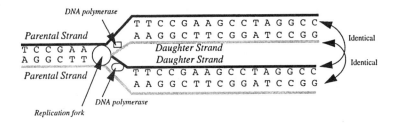

Figure 2.15 DNA replication. The two strands of the parental helix unwind and each of the strands directs the synthesis of a complementary strand.

DNA Sequence

Genetic information is encoded by the sequence of base pairs in the DNA strands. The two strands of DNA are complementary, so the base pair sequence of one strand can be inferred from the other strand. It is the convention to write a DNA sequence for only a single strand and in the direction of 5´→3´.

One of the fundamental functions of DNA is to duplicate itself before cell division (Figure 2.15). The process of DNA replication parallels chromosome duplication. The two strands of the parental helix unwind and each strand directs the synthesis of a complementary strand. The process results in two identical double strands of DNA. A complex of additional enzymes associated with DNA polymerases are involved in the DNA replication process.

2.5.2 DNA-RNA-PROTEIN

Gene Expression

DNA can be considered an information storage bank for biological development. The information can be duplicated and transferred from parental cells to daughter cells in the process of cell division. The consequence of the information transfer is the inheritance of the information and a program for expression from

generation to generation. DNA is also the information resource for protein synthesis. Through control of production in the cells, DNA controls biochemical
pathways, biological development of enzymes and other proteins and responses
to changes in the environment.

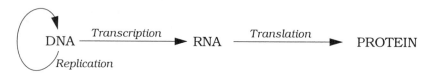

Figure 2.16 Gene expression.

The process from DNA to protein starts with a temporary copy of the information through the transcription of DNA to ribonucleic acid (RNA) (Figure 2.16).
RNA is composed of four bases: A, U(Uracil), C and G. RNA differs from DNA in
that the carbohydrate in the sugar-phosphate backbone is ribose instead of
deoxyribose and one of the four bases is uracil instead of thymine. Uracil forms
base pairs with A (adenosine), much as thymine does. The base pairs for transcription are now A≡U and G≡C. Only a small fraction of the total DNA in cells is
coding DNA, which is transcribed to functionally mature RNA. This process is
initiated by RNA polymerase (Figure 2.17). Only one strand of the double helix is
transcribed and the RNA has a sequence complementary to that strand. The
transcription process is highly regulated. Several kinds of RNA polymerases and
associated protein transcription factors regulate the specificity and rate of transcription. Promoters, which are regions of DNA that include RNA polymerase
binding start sites and transcription start sites, are located 5' to the genes and
control transcription initiation and level of expression. Some of the non-coding
DNA, such as centromeres or telomeres, may have specific chromosomal functions. The non-coding DNA without genetic or chromosomal function is commonly called "junk DNA", but even this DNA may have indirect effects on
biological functions.

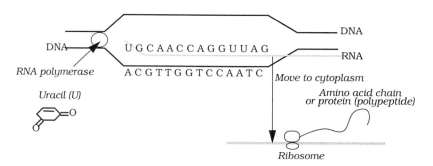

Figure 2.17 DNA transcription and RNA translation.

Table 2.7 The genetic code.

Amino Acid	3 letter symbol	1 letter symbol	Codons (in 5′→3′ order)
alanine	Ala	A	GCU, GCC, GCA, GCG
arginine	Arg	R	CGU, CGC, CGA, CGG, AGA, AGG
aspartic acid	Asp	D	GAU, GAC
asparginine	Asn	N	AAU, AAC
cysteine	Cys	C	UGU, UGC
glutamic acid	Glu	E	GAA, GAG
glutamine	Gln	Q	CAA, CAG
glycine	Gly	G	GGU, GGC, GGA, GGG
histine	His	H	CAU, CAC
isoleucine	Ile	I	AUU, AUC, AUA
leucine	Leu	L	UUA, UUG, CUU, CUC, CUA, CUG
lysine	Lys	K	AAA, AAG
methionine	Met	M	AUG
phenylala-nine	Phe	F	UUU, UUC
proline	Pro	P	CCU, CCC, CCA, CCG
serine	Ser	S	UCU, UCC, UCA, UCG, AGU, AGC
threonine	Thr	T	ACU, ACC, ACA, ACG
tryptophan	Trp	W	UGG
tyrosine	Tyr	Y	UAU, UAC
valine	Val	V	GUU, GUC, GUA, GUG
STOP			UAA, UGA, UAG

RNA Processing

Figure 2.18 shows the process from DNA to polypeptide. The DNA containing the information to be expressed is transcribed into messenger RNA (mRNA).

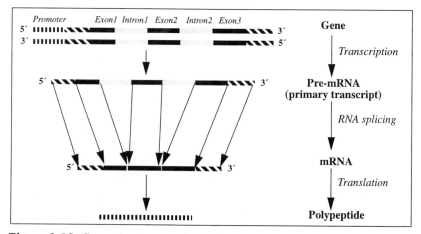

Figure 2.18 Gene structure and gene expression.

Not all of the RNA transcript is translated. An RNA splicing process cleaves the pre-mRNA at specific sites and ligates the exons, producing functional mRNA. The sequences corresponding to introns (non-coding sequence) are discarded (Figure 2.18). The mature functional mRNA moves from the nucleus to the cytoplasm to provide information for protein synthesis. The process of converting the information encoded in the mRNA into protein is called translation. The groups of three nucleotides in mRNA coding for each amino acid are called codons. Table 2.7 shows the codons for the 20 amino acids and the three stop codons for "punctuating" the translation. The process of translation from RNA to protein polypeptide is highly regulated and many components are involved in the process. Most important are the transfer RNAs (tRNAs). tRNAs are small molecules that attach to and activate specific amino acids. There is one or more specific tRNA for each amino acid. These small RNAs bind to the codon by means of complementary base pairing and allow each amino acid to be added at the correct position in a growing peptide chain. The tRNAs actually translate the genetic code.

Reading Frame

Each codon is three nucleotides long, therefore, in any sequence of RNA, there are three possible ways to translate the mRNA sequence to amino acid sequence, depending on the starting point for initiation of the translation. These three ways are called reading frames. Only one initiation site and its corresponding reading frame is used for any protein. Other starting sites and reading frames would lead to the synthesis of completely different proteins.

2.6 EXERCISES

Exercise 2.1

Find a Mendelian trait in any familiar organism. Do you expect that the segregation ratio for the trait follows the Mendelian laws in an F2 population (generated by selfing of a cross between two homozygotes) with distinct phenotypes for the trait? Can you explain why, using knowledge of cytology of the organism, biochemistry of the trait, *etc.?*

Exercise 2.2

The phenotype of an individual in a segregating population is the result of the genotype of the individual and its surrounding environmental conditions and developmental history. The relationship among phenotype, genotype and environment can be quantified using a statistical model, such as

$$P_{ijk} = \mu + G_i + E_j + GE_{ij} + \varepsilon_{ijk}$$

where P_{ijk} is the phenotype for genotype individual i, in environment j and replication k. G, E, GE and ε are genotypic, environmental, genotype by environment interaction and experimental error respectively. Answer the following questions.

(1) Can you think of a trait that can be described using this statistical model? Explain the meaning of each term in the model. Can you propose some data to fit the model?

(2) Can you design an experiment to support or test the model? How might

the model be useful in practical applications, such as plant and animal breeding and human disease diagnosis?

(3) Can you propose an hypothesis for the molecular basis of a trait consistent with the statistical model?

Exercise 2.3

Explain phenotypic differences between a qualitative trait and a quantitative trait. Write down genotypes for a qualitative trait. What are the genotypes for a quantitative trait? Consider biochemical similarities and differences between the two types of traits.

Exercise 2.4

The term "complex traits" has been used in recent years to describe the traits that do not follow simple Mendelian laws. In some aspects, the complex traits include quantitative traits and some qualitative traits. Can you think of a non-quantitative complex trait? Explain its inheritance.

Exercise 2.5

Inheritance of a Mendelian gene at the population level can be studied using population genetic theory. Allelic and genotypic frequencies are essential statistics for describing a population. Factors such as mutation, selection, migration and random sampling (drift) govern the changes of the frequencies from generation to generation. Answer the following questions:

(1) When the population size is large, which of the factors has the largest effect on changing allelic and genotypic frequencies? How about when population size is small?

(2) What is the relationship between the factors that disrupt Hardy-Weinberg equilibrium and the adaptation process of the population to a changing environment?

(3) Can you find a real example to explain the biological adaptation behind the allelic and genotypic frequency changes?

Exercise 2.6

A trait is recorded as the content of a simple compound in a plant. Two enzymes are on the biosynthetic pathway and clones of the DNA encoding the enzymes have been obtained. Let us assume that each cDNA clone, which is a complementary DNA transcript of an mRNA, has two alternative forms in the population. They are denoted by A and a for Enzyme A and by B and b for Enzyme B. An F2 population with 160 individuals was obtained by selfing an F1 from a cross between the AABB and aabb genotypes. Contents of the compound were obtained from an experiment including replications (through vegetative approaches) of the 160 individuals in three different environments. Genotypes in terms of the two forms of the two cDNAs encoding the enzymes were also obtained for the 160 individuals. Surprisingly, the segregation ratio of AABB, AABb, AAbb, AaBB, AaBb, Aabb, aaBB, aaBb and aabb in the 160 individuals follows exactly the Mendelian expectation which is

$$10:20:10:20:40:20:10:20:10$$

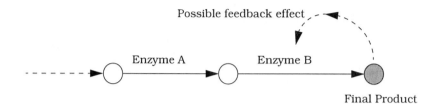

(1) The model of the first analysis of the data only includes the individual (traditionally called genotypic term), environment and individual by environment interaction terms and is

$$P_{ijk} = \mu + I_i + E_j + IE_{ij} + \varepsilon_{ijk}$$

where P_{ijk} is the content of the compound for an individual i in environment j and replication k. I, E, IE and ε are the individual, environmental, individual by environment interaction and experimental error, respectively. Results of the analysis are shown in the following table. The total variation includes variation among the individuals and among the environments. Interpret the result using your knowledge of quantitative genetics in terms of the model.

First analysis. Ex. 2.7 (1)

Term	Total Variation Explained
Individual	60%
Environment	25%
Individual x Environment	15%

(2) The model of the second analysis includes the genotypes of the two cDNAs, the individual within the genotype and the corresponding interaction terms with environment and is

$$P_{g[i(g)]jk} = \mu + G_g + I(G)_{i(g)} + E_j + GE_{gj} + I(G)E_{i(g)j} + \varepsilon_{ijk}$$

where $P_{g[i(g)]jk}$ is the content of the compound for individual i which has a genotype g in environment j and replication k. $G, I(G), E, GE, I(G)E$ and ε are genotypic, individual within the genotype, environment, genotype by environment interaction, individual within genotype by environment interaction and experimental error, respectively. Results of the analysis are shown in the following table.

Table 2.8 Second analysis. Ex. 2.7 (2)

Term	Total Variation Explained
Genotype	30%
Individual within genotype	30%
Environment	25%
Genotype x Environment	10%
Individual (Genotype) x Environment	15%

Can you speculate what are the sources of variation among the individuals within the genotype? Can you interpret the interaction terms?

Table 2.9 Third analysis. Ex. 2.7 (3)

Term	Total Variation Explained
A	15%
B	0%
A by B interaction	15%
Individual within genotype	30%
Environment	25%
A x Environment	10%
B x Environment	0%
Individual (Genotype) x Environment	15%

(3) The model of the third analysis includes the terms for the enzymes A and B and their interactions, which are further partitioned for the genotype in the second analysis and the corresponding interaction terms with environments. The results of the analysis are shown in the above table. Can you interpret the results at the individual enzyme level?

(4) The variation among different cDNA forms is further partitioned into contrasts corresponding to single degrees of freedom. For example, variation among AA, Aa and aa genotypes can be partitioned into two contrasts (-1 0 1) (corresponding to the difference between AA and aa) and (1 -2 1) (corresponding to the difference between the means of AA and aa verses Aa). Similar partitioning can be applied to B. The interaction between A and B can be partitioned accordingly. For example, A (-1 0 1) by B (-1 0 1) interaction is a contrast of (1 0 -1 0 0 0 -1 0 1) and A (-1 0 1) by B (1 -2 1) interaction is a contrast of (-1 2 -1 0 0 0 1 -2 1). The results of the analysis are shown in the following table. Interpret the contrasts for the genotypes.

Table 2.10 4th analysis. Ex. 2.7 (4)

Term	Total Variation Explained
A (-1 0 1)	5%
A (1 -2 1)	10%
B (-1 0 1)	0%
B (1 -2 1)	0%
A x B (1 0 -1 0 0 0 -1 0 1)	0%
A x B (-1 0 1 0 0 0 1 0 -1)	5%
A x B (-1 0 1 2 0 -2 1 0 -1)	0%
A x B (1 -2 1 -2 4 -2 1 -2 1)	10%
A by B interaction	15%
Individual within genotype	30%
Environment	25%
A x Environment	10%
B x Environment	0%
Individual (Genotype) x Environment	15%

(5) Explain the results in the previous tables in relation to classical definitions of additive inheritance, dominance, epistatic interactions and genotype by environmental interactions. Referring to the biosynthetic pathway, do the four analyses make any sense? Do the quantitative definitions of a trait have meaning in terms of biochemistry?

(6) Does the traditional model fit biological reality in this hypothetical experiment?

CHAPTER 3

INTRODUCTION TO GENOMICS

The statistics in this book are intended for the analysis of data from nuclear genomes. The nuclear genome consists of the entire set of chromosomes within the nucleus of a somatic cell and is distinct from the genomes of cytoplasmic organelles, mitochondria and plastids. Genomics studies nuclear genomes at a whole genome level by integrating traditional genetic disciplines with new technology from informatics and automated robotic systems. The purpose of genomic research is to learn about the structure, function and evolution of all genomes. To do this, we study genome variation and similarity among different genotypes within a species and among different species. The major tool in these studies is the genetic marker, which represents variation at a particular site on the genome which is heritable, easy to assay and can be followed over generations. Genetic markers are used extensively to define genes by genome location, to make associations of genome location and function for specific genes and to explore genetic variation during evolution.

3.1 GENOME

The nuclear genome contains most of the genetic information of a cell in a set of chromosomes, each composed of a single DNA molecule. Genomes of different organisms vary in terms of total DNA content (genome size), ploidy level, chromosome number, total recombination distance in the genome and nature and number of functional genes.

Table 3.1 Haploid genome size in terms of chromosome number ($1n$), base pairs (Mb) and total recombination map length (cM) for some extensively studied organisms.

	$1n$	Mbp	cM	Kilobase per cM
Yeast (*Saccharomyces cerevisiae*)	16	14	4,200	3
Nematode (*Caenorhabditis elegans*)	11/12	100	320	300
Fruit Fly (*Drosophila melanogaster*)	4	170	280(female)	600
Mouse (*Mus musculus*)	20	3000	1700	1800
Human (*Homo sapiens*)	23	3000	2800(male), 4800(female)	1100(male) 600(female)
Arabidopsis thaliana	5	100	500	200
Corn (*Zea mays*)	10	2500	1500	1700
Tomato (*Lycopersicon esculentum*)	12	950	1300	750
Barley (*Hordeum vulgare*)	7	5300	1400	4000
Rice (*Oryza sativa*)	12	450	1700	250
Loblolly pine (*Pinus taeda*)	12	20000	2000	10000

3.1.1 GENOME DESCRIPTION

Genome size can be quantified in terms of molecular mass and number of nucleotide pairs (Table 3.1). Traditionally, genomes have been described in terms of ploidy level, number of chromosomes and total recombination units in the genome. The haploid chromosome number is commonly referred to as $1n$. Recombination units are quantified in centi-Morgans (cM), named after the *Drosophila* geneticist Thomas Hunt Morgan. One hundred cMs equal one Morgan.

Chromosome number varies greatly between eukaryotic species. Among well-studied species, haploid chromosome numbers range from 4 for a fruit fly to 23 for a human (Table 3.1). Haploid chromosome number ranges from 2 (*e.g.*, the horse threadworm, *Parascaris equorum* var. *bivalens* and *Haplopappus gracilis*) to several hundreds (*e.g.*, the ancient adder's tongue fern, *Ophioglossum reticulatum*) in eukaryotes.

The physical genome size of the haploid genome ranges from 14 million base pairs (Mbp) for yeast, to approximately 20,000 for loblolly pine (Table 3.1). Some plant species are several times larger. The genome sizes in cM in Table 3.1 are average sizes from published data. The size in terms of recombination units can differ from experiment to experiment for the same organism. One important ratio is the kilobase per recombination unit (cM). It is an indicator for how precisely a target gene can be located in terms of physical distance through linkage mapping approaches. The smaller the ratio, the more precisely a specific gene can be located physically. This knowledge is important for cloning genes based on their genome locations, known as map-based cloning.

3.1.2 GENOME STRUCTURE

At the molecular level, chromosomes are composed of DNA and proteins. The DNA carries the genetic information while the protein components provide

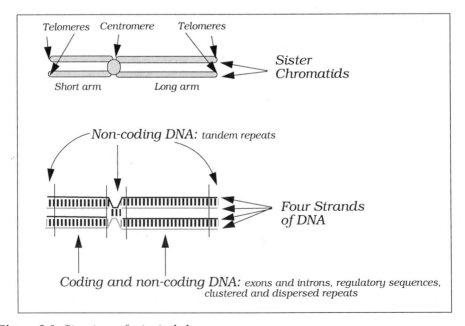

Figure 3.1 Structure of a typical chromosome.

enzymatic and structural functions important to the replication, recombination and segregation of the chromosome. The ends of the chromosomes contain special sequences called telomeres, while a specialized structure called a centromere may be found either at the end of the chromosome or more centrally located (Figure 3.1). The function of the centromere is to ensure the proper disjunction of the chromosomes into daughter cells during meiosis and mitosis. Telomeres ensure the duplication of chromosome ends.

The main part of the chromosome contains a mixture of coding sequences (exons), intragenic (introns) and intergenic non-coding sequences, regulatory sequences, tandem repeats (mainly minisatellites and microsatellites) and other dispersed repeats. Tandem repeats have been used in genomic research; for example, minisatellite markers (commonly referred as VNTR (variable length tandem repeat), see section on molecular markers later in this chapter) have been used for DNA forensics. Microsatellite markers have shown a great potential in genomic mapping (Section 3.4).

3.1.3 GENOME VARIATION AND COLINEARITY

DNA sequence variation among different genotypes within a species is the foundation for genetic analyses. The genetic colinearity among different species is the basis for comparative mapping.

Genome colinearity (also called synteny) is apparent at different levels among different species. Similarity is greater between genomes of closely related species, while distantly related species show greater divergence in genome structure. For example, there is a great similarity among the family *triticeace* at the whole chromosome level. Members of the *triticeace*, such as wheat, barley, rye, oat, *etc.*, share a chromosomal origin within about 60 million years. There is also extensive colinearity of genetic marker order for many segments of the genomes of maize, sorghum and rice (Hulbert *et al.* 1990; Whitkus *et al.* 1992; Ahn *et al.* 1994). Certainly, the well-known homology between human and mouse genomes enables many researchers to use the mouse as a model system for hunting human disease genes.

3.1.4 SOURCES OF GENOME VARIATION

Chromosomal Rearrangement

Chromosomal rearrangement is one important source of genome variation (Figure 3.2). Either due to abnormal pairing or recombination during meiosis, such rearrangements are usually associated with deleterious effects. Rearrangements can be described as insertions, deletions, inversions, translocations or duplications (Figure 3.2). They can be of any size, from a few base pairs to many millions of base pairs. Recombination fraction between genes on either side of a rearrangement may be reduced by a deletion or be increased by a insertion. When an individual having a normal chromosome and an individual having a chromosome with a deletion or a insertion are crossed, some genes or genetic markers may lie in a region that is unable to pair with the homologous chromosome. Recombination will be inhibited in this unpaired region, but deletion and insertion will not change the gene or genetic marker order for flanking chromosome segments.

Inversion and translocation will change not only the recombination fraction but also the linear order of genes or genetic markers on the chromosome. Inversions or translocations can make data analysis complicated. No existing computer software would be able to analyze these types of data. The statistical methodology for detecting the chromosomal rearrangements will be discussed in Chapter 11.

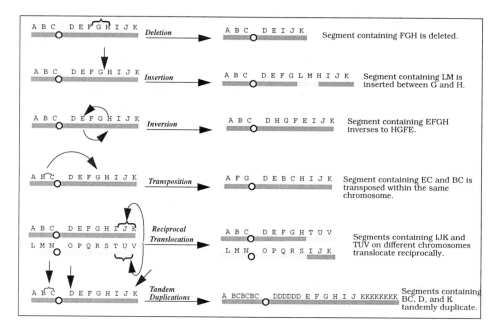

Figure 3.2 Types of chromosome rearrangements.

Tandem duplications of chromosome segments or a short sequence of DNA will increase genome size. Such tandem repeats have been used for genetic markers, such as VNTR and microsatellite markers.

Point Mutation

Polymorphism can also be created by point mutation, which is the substitution of one nucleotide for another. Nucleotide substitution is a common type of mutation and can be classified as a transition or a transversion mutation. A transition results in substitution of a purine nucleotide by a purine nucleotide, or of a pyrimidine by a pyrimidine. Conversion of an A to a G is an example of transition. Transversions result in substitution of a purine nucleotide by a pyrimidine or vice versa. Conversion of an A to a C is an example. Another commonly observed mutation is called a frameshift mutation. This mutation is created by insertion or deletion of one or more (except multiples of three) nucleotides and will cause a shift of the reading frame during the translation of mRNA to protein, usually resulting in a truncated, nonfunctional protein.

3.2 BIOLOGICAL TECHNIQUES IN GENOMICS

In this book, we focus on the statistical side of genomics. However, for readers with little biology background, I will briefly discuss biological techniques of genomics which include genetic mapping, physical mapping, DNA sequencing and genomic informatics.

3.2.1 GENETIC MAPPING

A genetic map of an animal or plant species is an abstract model of the linear arrangement of a group of genes and markers. The gene can be the traditionally defined Mendelian factor or a piece of DNA identified by a known function or by means of a biochemical assay. The marker can be a cytological "marker," a variant based on a change in a known gene or protein or a piece of DNA without known function. Both the gene and the marker should have simple inheritance that can be followed through generations. A gene with known function can be considered a marker if it contains detectable variation.

A genetic map is based on homologous recombination during meiosis, so a genetic map is also a meiotic map. If two or more markers are located close together on a chromosome, their alleles are usually inherited together through meiosis.

Genetic Map Construction

For a data set composed of a small number of genetic markers, a genetic map can be constructed using classical linkage analysis. For example, pairwise recombination fractions among three loci are estimated using the classical linkage analysis from genotypic data of these loci in a doubled-haploid population (Figure 3.3). A linkage map for the three loci can be easily obtained (Figure 3.3).

Genomic analyses rely on the large number of genetic markers. For a large data set with a large number of genetic markers, four steps are needed to construct a genetic map after obtaining marker data. The first step is a pairwise linkage analysis for all possible two-locus combinations. The biological foundation for linkage analysis is the homologous recombination between non-sister chromatids during meiosis. To detect a marker and use the marker to monitor recombination events, the marker must have alternative alleles in the mapping population. Linkage analysis for a two-locus combination is based on comparison of observed and expected frequencies of the possible genotypic classes. The number of possible genotypic classes is a function of the number of alleles at the two loci under consideration and the mating design used to produce the mapping population. Maximum likelihood approaches are commonly used to estimate recombination fraction from the observed genotypic frequencies (see Chapters 6, 7 and 8).

The next step is to group the markers into different linkage groups. Criteria used for linkage grouping are usually recombination fraction, significance level of the recombination fraction and the known genome information (such as number of chromosomes) (see Chapter 9). If many markers are used, a relatively high genome coverage is achieved (see Chapter 10), the data are highly informative, the genetic model for data analysis is adequate and the criteria for the grouping are reasonable, then the number of linkage groups should be close to the haploid number of chromosomes for the organism.

After grouping the markers into different linkage groups, markers in the same linkage group are ordered (the relative position on the genetic map is determined). This is the key step for a high quality genetic map. This is also the most computationally demanding step in genetic map construction. There are controversies on how to quantify the precision of the estimated marker (or gene) order. Please see Chapter 9 for a detailed description of locus ordering.

The last step is to estimate the multipoint recombination fractions among the adjacent loci. See Chapter 10 on statistics for multipoint analysis. The multipoint recombination fraction may differ from the two-point recombination fraction. The multipoint recombination fraction could correspond more closely

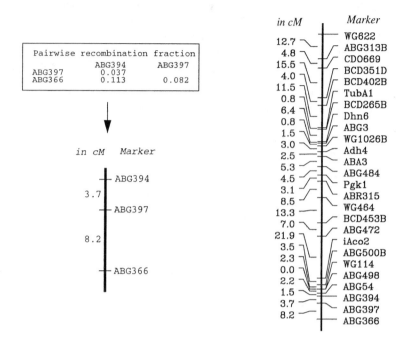

Figure 3.3 Linkage map of barley with three loci (left) and linkage map of chromosome IV (right) with 26 loci generated from data provided by the North American Barley Genome Mapping Project (NABGMP).

to the physical distance between the loci. In general, physical distance and genetic distance do not have a fixed one-to-one relationship, due to the sequence specificity of recombination events. Large stretches of satellite DNA may have actually no recombination events, while recombination "hot spots" have been observed for some organisms.

Figure 3.3 shows a genetic map for one of the seven barley chromosomes. The markers on the genetic map represent DNA fragment, isozyme and morphological markers. Distance between adjacent markers can be written as a percent of recombination between the two loci, or corrected for multiple crossover events and expressed in centiMorgans (cM). See Chapter 10 for discussion on the relation between recombination fraction and mapping functions that correct for multiple recombination events.

Comparative Mapping

As genome information accumulates for many animal and plant species, genetic mapping may be carried out by comparing genome maps among relatives in same species or different species. For example, co-linear genetic maps have been found among maize, rice and sorghum (Bonierbale *et al.* 1988; Tanksley *et al.* 1988; Hulbert *et al.* 1990; Ahn *et al.* 1994) and among mouse, human and other mammals (O'Brien *et al.* 1993). The purposes of comparative mapping are two-fold. One of the purposes is the transfer of mapping information across the species. This is the rationale of the search for human disease genes by using genome information from model organisms, such as the mouse. The other purpose is to achieve a better understanding of the evolution of genome structure.

Highly polymorphic markers are essential for genetic mapping within a species. However, for comparative mapping among different species, highly polymorphic markers are less informative than the more conserved coding gene loci (O'Brien *et al.* 1993). The highly polymorphic markers are usually too variable to identify genome homology among different species, because the variation within the species confounds the variation between species.

Mapping Genes of Interest

One of the most important applications of genetic maps is to locate specific genes of interest, such as those causing human diseases or controlling traits of economic importance in plants and animals. These traits can be controlled by single genes, as are some human diseases, or controlled by a number of genes, as in the case of yield of field crops or growth rate of forest trees. Some human diseases controlled by a single gene are still quite complex, because the analysis of the mechanism underlying the phenotype is confounded by incomplete penetrance. When more than one gene is involved in a quantitatively varying phenotype, the loci are commonly described as Quantitative Trait Loci (QTL) and the mapping procedure is called QTL mapping. We will discuss these issues extensively later in this book.

3.2.2 PHYSICAL MAPPING

One goal of genomic analysis is to obtain DNA sequence and information on how the DNA is transcribed and translated in development. Complete genome DNA sequences have already been obtained for several bacteria and for yeast, since their genomes are small compared to higher organisms, some species with genomes larger by more than two orders of magnitude. However, for most plant and animal species, it is not yet practical to obtain complete genome DNA sequences. Even if complete sequencing were possible, some direct correlations or associations are needed between traits and DNA sequences in order to understand the genes controlling complex traits. Genetic mapping is one way to make the associations to bridge this gap. Genetic maps are still made by traditional Mendelian analysis, although the genetic markers may be generated using recombinant DNA techniques or polymerase chain reaction.

A physical map is another mechanism that can be used to make associations between traits and DNA sequences. A physical map can be the traditional cytogenetic chromosome map, based on chromosome structure or banding patterns observed using modern cytogenetic chromosome mapping, particularly using FISH (fluorescence *in situ* hybridization). The chromosome maps are more frequently available for some of the plant and animal species with large chromosomes and extensive cytogenetic information. Cytogenetic chromosome maps usually have low resolution, compared to genetic maps. A gene can frequently be located to a specific chromosome and often to a segment of one chromosome, but the resolution of at the DNA level is low.

The commonly described physical map is one which contains ordered overlapping cloned DNA fragments. The cloned DNA fragments are usually obtained using restriction enzyme digestion. The restriction map is a common form of a physical map of a chromosomal segment, but the ultimate physical map is the DNA sequence.

DNA Fragmentation

In general, when making the restriction physical maps, the larger the fragment, the better. This is usually achieved by using rare-cutting endonucleases and partial digestion (Figure 3.4). Commonly, a procedure to

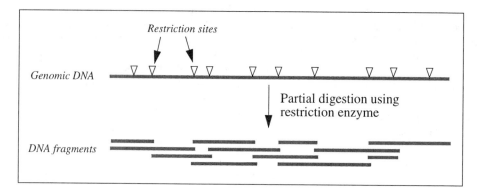

Figure 3.4 Overlapping DNA fragments are obtained using partial digestion with a restriction enzyme for a segment of chromosome DNA.

select fragments with the desired large size is applied after the digestion, using centrifugation or electrophoresis. The commonly used electrophoresis methods are pulsed field gel and contour-clamped homogeneous electrical field electrophoresis.

DNA Vector

To propagate and maintain the DNA fragments, it is necessary to insert the DNA fragments into cloning vectors. A vector is a cloning vehicle for replicating the fragments. Commonly used vectors are phage, cosmids, yeast artificial chromosomes (YAC) and bacterial artificial chromosomes (BAC). The important considerations in choosing a vector system are the size of DNA fragment that can be inserted into the vector and the stability of the product. For phage, cosmids, YAC and BAC, the average sizes of the insert are 5 - 25 kb, 35 - 45 kb, 200 - 2000 kb and ≤300 kb, respectively. Smaller inserts are usually more stable than larger.

Table 3.2 Numbers of clones needed for a single genome and for 95% genome coverage using average sizes of the inserts 700 kb (YAC) and 200 kb (BAC).

	Genome in Mb	YAC (700 kb)		BAC (200 kb)	
		Single Genome	95% Coverage	Single Genome	95% Coverage
Yeast	14	20	59	70	209
Nematode	100	143	427	500	1496
Drosophila	170	243	726	850	2545
Mouse	3000	4286	12838	15000	44935
Human	3000	4286	12838	15000	44935
Arabidopsis thaliana	100	143	427	500	1496
Corn	2500	3572	10698	12500	37445
Tomato	950	1357	4064	4750	14228
Barley	5300	7572	22681	26500	79386
Rice	450	643	1924	2250	6739
Loblolly pine	20000	28571	85591	100000	299572

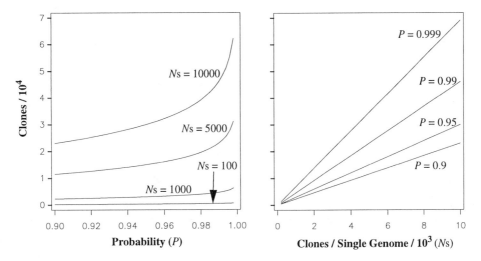

Figure 3.5 Theoretical example of numbers of clones (N$_s$) needed as a function of the probability (P) and number of clones needed to cover a single genome.

The size of the inserts determines how many clones are needed to include essentially every sequence from a genome in the genomic library. The expected number of clones needed can be determined using

$$N = \frac{\log (1 - P)}{\log (1 - 1/N_S)} \tag{3.1}$$

where P is the probability that any given clone is included in the library and N_S is the number of clones needed for covering a single genome if the clones are not overlapping (Clarke and Carbon 1976). The rationale behind Equation (3.1) is that the probability of a clone not in the library with size N is

$$(1 - 1/N_S)^N$$

The probability that a clone is in the library is

$$P = 1 - (1 - 1/N_S)^N$$

A simple manipulation of the probability will result in Equation (3.1).

The numbers of clones needed for a single genome and for a 95% genome coverage can be readily calculated for many species (Table 3.2). The number of clones needed as a function of the probability (P) and number of clones for a single genome equivalent may also be calculated (Ns)(Figure 3.5). The number of clones needed has a nonlinear relation with P and a linear relationship with Ns. As the probability increases from 95%, the number of clones needed increases nonlinearly. As for the relationship between the number of clones needed and genome size, the number of clones needed always increases linearly once the expected probability is determined.

Physical Map Assembly

Once a genome has been disassembled and "parts" (cloned fragments) are in a storage bank (genomic library), the next step of physical mapping is to

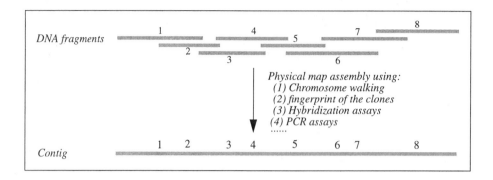

Figure 3.6 Overlapping DNA fragments are assembled into a contig, which resembles linear order of the clones for a region of genome or a whole chromosome.

determine a linear order of the original genome before the digestion (Figure 3.6). There are several methods for assembling physical maps. The goal is to establish which clones overlap by one or more methods. The ultimate way to assemble a physical map is by sequencing all the clones.

Chromosome walking was originally used to isolate a gene of known phenotype but unknown sequence whose genetic location was known (Bender *et al.* 1983). The strategy of chromosome walking is the sequential isolation of adjacent fragments of genomic DNA. A clone is selected and used as a probe to detect the overlapping clones in the genomic library. By repeating this step many times, the organization of a larger segment becomes known. This assembled fragment or sequence is called a contig (Figure 3.6). The problem with this approach is that if it is not a unique sequence, the probe could hybridize with non-contiguous clones. Chromosome walking is usually laborious, time consuming and works best when there are no repeated sequences.

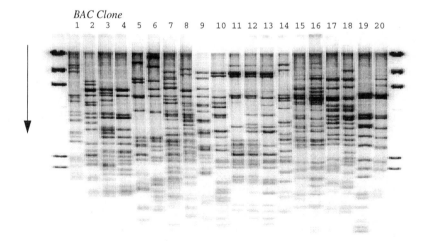

Figure 3.7 DNA fingerprints for 20 BAC clones of rice. (Gel picture provided by Dr Hongbin Zhong of Crop Biotechnology Center at Texas A&M University.)

Another common approach for assembling physical maps is using restriction fragment fingerprints to find overlapping clones. If there is a sufficient number of identical fragments for any pair of the clones, then the clones are likely to overlap. Fingerprints may be obtained for a set of clones and then pairwise comparisons for the cleavage sites can be performed. For example, Figure 3.7 shows DNA fingerprints for 20 BAC clones from a rice DNA library. Each lane contains a different BAC clone and the banding pattern in the lane is determined by the restriction sites within the clone. Overlapping clones may be confirmed by determining the sequences of the ends of clones or by other methods, such as DNA hybridization or presence of a DNA marker.

Physical map assembly is a labor intensive and computationally demanding process. However, automation for physical mapping has been proposed using DNA hybridization (Old and Primrose 1994). For computational aspects of physical mapping, see Waterman (1995).

3.2.3 DNA SEQUENCING

The ultimate goal of genome research is the determination of complete DNA sequences and the biological roles or effects of these sequences. Genetic mapping and physical mapping can be considered to be early steps toward reaching that goal. Development of automated, high throughput DNA sequencing technology has been a focus of biological research, both public and private. New automated technologies, for example, using DNA fixed to a surface in microarrays (DNA chips) or capillary electrophoresis in combination with automated devices, promise to reduce the cost and increase the speed of DNA sequencing. New technology can also be applied to DNA fragment sizing used in genetic and physical mapping.

Even though many improvements have been made, the most widely used method of DNA sequencing is still based on procedures developed by Sanger

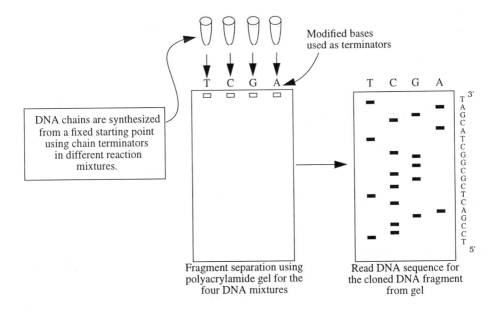

Figure 3.8 Principle of DNA sequencing (Sanger method)

and co-workers in the late 1970's. The basic strategy is to create a population of DNA molecules differing in length by one nucleotide increments. These molecules are then separated by electrophoresis in polyacrylamide gels that resolve the single nucleotide differences (Figure 3.8). As DNA sequencing technology is advanced and more DNA sequencing data is accumulated, our dependence on new technology for managing and analyzing the DNA sequence data has increased rapidly. See Waterman (1995) for more information on DNA sequence analysis.

3.2.4 GENOMIC INFORMATICS

Genomic DNA is an information bank of DNA sequences. The purpose of genetic mapping, physical mapping and DNA sequencing is to explore how the information is organized and how it is used in growth, reproduction and development. Genomics is becoming an information science as the genetic information stored in cells in DNA is transferred from molecules to computers. The amount of data on genetic maps, physical maps and DNA sequences stored in computers is growing exponentially. The well-organized data bank that exists as DNA molecules is not necessarily well organized when entered into computers. How to organize, manage and analyze the information in a computer database is becoming an essential part of genomics. Genome informatics began as a tool for management and analysis of genomic data and now is becoming a science of its own.

Genomic informatics has three functions: information management, analysis and communication. Several genome databases are now available for different organisms, including *E. coli*, yeast, nematode (*C. elegans*), *Drosophila*, mouse, human and some of the higher plants. GenBank is a DNA sequence database maintained by the U.S. National Center for Biotechnology Information (NCBI). EMBL is the European Molecular Biology Laboratory database. DDBJ is the DNA database of Japan. A comprehensive USDA plant genome database is located at the U.S. National Agriculture Library (NAL) and satellite genome databases for several plant species have been established at different institutes (Bigwood *et al.* 1992, Neale 1992). The current genomic data banks have the function of bookkeeping, communication and limited analytical functions. The future development of genomic databases will include more analytical power and greater capacity.

3.2.5 RELATING GENETIC MAPS, PHYSICAL MAPS AND DNA SEQUENCE

Genetic and Physical Maps

Both genetic and physical maps have been obtained for several organisms with small genomes, such as *E. coli*, yeast, *Drosophila*, *C. elegans*, *Arabidopsis thaliana* and rice. High-density genetic maps have been developed for many organisms with relatively large genomes, such as human, barley, maize, tomato and loblolly pine. Physical maps have already been constructed for some of the human chromosomes.

Genetic and physical maps should reflect the same genome structure based on the order of genes or markers and the distances between them. However, the relationship between distance units of genetic recombination and physical length of DNA is often not proportional, due to non-randomness of the crossover events. Moreover, the gene order on a genetic map and a physical map may not agree, due to sampling errors or inappropriate laboratory or data analysis procedures. Attention should be paid, therefore, to the relationship between the genetic and the physical maps. Although distances between genes or sites may vary, relative order should be the same for both genetic and physical maps. Non-corresponding order between the genetic and physical maps usually indicates error in one or both maps.

Resolution of a genetic map is limited by the number of recombination events. If no recombination events can be detected within a region of the genome, then genetic mapping can not resolve the relative positions of genes and markers in that region. To increase the resolution for a genetic map, a larger mapping population is needed. The physical length of a genome has a relatively small impact on the size of a genetic map. For a small genome, a recombination unit may represent a small physical segment of the genome. For a large genome, a recombination unit may represent a relatively large segment of the genome. For example, a recombination unit (one cM) represents on average approximately 200 kb in *Arabidopsis thaliana* and 15 to 20 Mb in loblolly pine.

Physical mapping is based on establishing the order of overlapping DNA fragments, therefore polymorphism and recombination are not essential. The limitation on resolution is not applicable to physical mapping. Instead, physical mapping is limited by the size of the cloned DNA fragments used to obtain the overlaps. If the average size of the DNA fragment is small, then a large number of clones are needed to cover even a small region of the genome. As genome size increases, the number of clones needed to cover the genome increases.

Genetic maps and physical maps can be related to each other by placing the genetic markers on the physical map and the DNA fragments on the genetic map. By doing this, the genetic map and the physical map can cross validate each other. Closely linked genes and markers usually have uncertain order on a genetic map but can be ordered precisely on a physical map. Gaps, common for a low coverage physical map, can be covered using a genetic map. Co-localization of genes on genetic and physical maps is becoming an efficient way to identify and isolate many genes of interest.

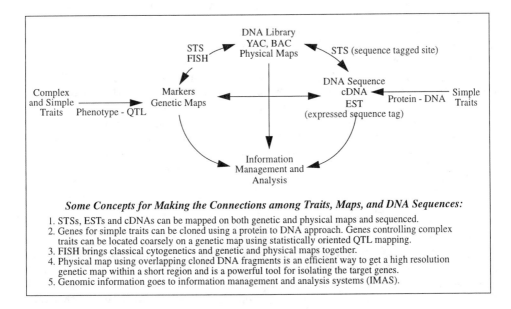

Some Concepts for Making the Connections among Traits, Maps, and DNA Sequences:

1. STSs, ESTs and cDNAs can be mapped on both genetic and physical maps and sequenced.
2. Genes for simple traits can be cloned using a protein to DNA approach. Genes controlling complex traits can be located coarsely on a genetic map using statistically oriented QTL mapping.
3. FISH brings classical cytogenetics and genetic and physical maps together.
4. Physical map using overlapping cloned DNA fragments is an efficient way to get a high resolution genetic map within a short region and is a powerful tool for isolating the target genes.
5. Genomic information goes to information management and analysis systems (IMAS).

Figure 3.9 Connecting traits, maps and DNA sequence.

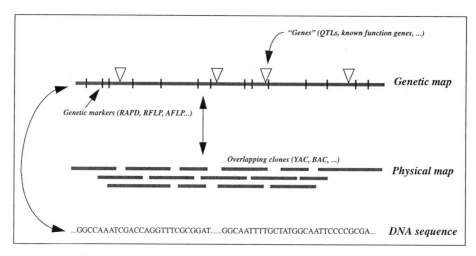

Figure 3.10 Relationship of genetic map, physical map and DNA sequence.

Traits, Maps and Sequence

It is not possible (at least for now) to directly relate most complex traits with the already massive DNA sequence databases, because little is known about the molecular identity of most genes controlling complex traits. However, DNA sequence is the ultimate information resource for the traits. Reverse-genetics, from protein to DNA, is an elegant way to isolate genes with simple functions. Many genes have been identified and cloned using the reverse-genetics approach. The coordination among genetic and physical maps and DNA sequences may provide more efficient ways to isolate genes with simple functions and possibly to isolate genes controlling complex traits (Figure 3.9). Genetic and physical maps are bridges between complex traits and DNA sequences (Figure 3.10).

Association between a segregating marker and a genetic trait suggests that a gene controlling a component of the trait is located near the marker. The resolution of gene mapping may be increased by increasing mapping population size, improving statistical methodology or using linkage disequilibrium accumulated over generations. However, those improvements are usually not enough to increase the precision of mapping in order to identify and isolate a specific gene. For example, a human disease gene may be located within a one centiMorgan (cM) region flanked by two markers (Figure 3.11). In humans, however, 1 cM represents 1 Mb of DNA. This is not precise enough to clone the disease gene. If a physical map for the region can be obtained, then 12 BAC clones will cover the 1 Mb region. The precise location of the disease gene can be located relative to the overlapping BAC clones by sequence and further functional analysis.

It is important for statisticians to be familiar with the biology behind the statistics to help insure that the statistics will be appropriately used. Using biology and statistics together is the best way to solve genomic problems. When complex statistics and intensive computing cannot solve the problem, biological evidence may. There are many difficult statistical and computational problems associated with genetic mapping, physical mapping and DNA sequence analysis. For example, the problems of gene ordering in genetic mapping and ordering DNA fragments produced by simultaneous digestion

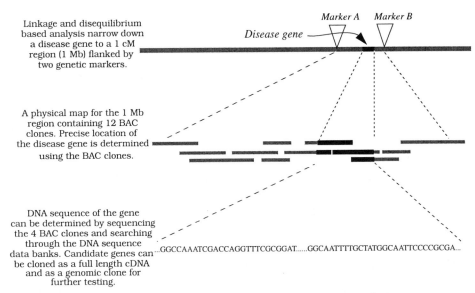

Linkage and disequilibrium based analysis narrow down a disease gene to a 1 cM region (1 Mb) flanked by two genetic markers.

Disease gene Marker A Marker B

A physical map for the 1 Mb region containing 12 BAC clones. Precise location of the disease gene is determined using the BAC clones.

DNA sequence of the gene can be determined by sequencing the 4 BAC clones and searching through the DNA sequence data banks. Candidate genes can be cloned as a full length cDNA and as a genomic clone for further testing.

...GGCCAAATCGACCAGGTTTCGCGGAT......GGCAATTTTGCTATGGCAATTCCCCGCGA...

Figure 3.11 A disease gene may be studied using a combination of genetic mapping, physical mapping, and DNA sequencing.

with two different restriction enzymes are specific examples of a mathematically difficult problem, the NP-complete problem. There is no general solution for the NP-complete problem. However, if we put genetic mapping and physical mapping together, those problems may be solved without a solution to the NP-complete problem.

3.3 MAPPING POPULATIONS

A population used for gene mapping is commonly called a mapping population. Commonly used mapping populations are obtained from controlled crosses or from natural populations.

3.3.1 POPULATIONS FROM CONTROLLED CROSSES

Selection of a population for genomic mapping involves choosing parents and determining a mating scheme. Decisions on selection of parents and mating design, as well as the type of markers, should depend upon the objectives of the experiment (Figure 3.12).

Parents of a mapping population must have sufficient variation for the traits of interest at both the DNA sequence and the phenotypic level. The variation at the DNA level is essential to trace recombination events. The more DNA sequence variation exists, the easier it is to find polymorphic informative markers. When the objective of the experiment is to search for genes controlling a particular trait, genetic variation of the trait between the parents is important. If the parents are greatly different at the phenotypic level for a trait, there is a reasonable chance that genetic variation exists between the parents, although uncontrolled environmental effects could create large phenotypic variation without any genetic basis for the effects. However, lack of

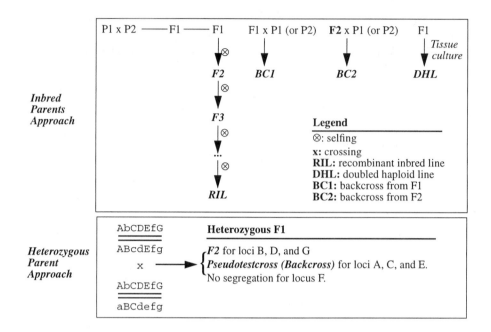

Figure 3.12 Commonly used mating schemes.

phenotypic variation between the parents does not mean that there is no genetic variation. Different sets of genes could result in the same phenotype. In some cases, variation within individuals, based on a high level of heterozygosity, may be exploited in controlled crosses. A highly heterozygous individual may be treated as an F1 in some experimental designs.

Different sets of genetic markers have different levels of resolution for detecting genomic variation. For some species, little genomic variation exists in natural populations. However, technology is available to detect even a single base change in a DNA fragment of interest, for example, by direct DNA sequencing.

3.3.2 NATURAL POPULATIONS

It is sometimes difficult to make a clear distinction between natural and artificial (experimental) populations. Populations obtained using controlled crosses between selected parents can be considered experimental populations. The populations produced by naturally occurring matings (without artificial control) can be considered "natural populations" (Figure 3.13). Population parameters, such as allelic frequencies, genotypic frequencies and disequilibrium at different levels, are commonly used to characterize the genetic architecture of a population. Evolutionary forces, such as mutation, selection, population admixtures (migration), random drift and recombination, may play important roles in the population history (Figure 3.13).

For some species, natural populations may be generated by assortative mating or even complete self mating, instead of random mating. Wheat, for example, is a self-pollinated crop species. However, natural populations of self-pollinated crop species are seldom used for genomic research, because genetic relationships are difficult to identify between individuals within the population. The natural populations to which we refer are usually naturally outbred species.

Samples from "natural populations" can be half-sib families, mixtures of random and self matings or multi-generation pedigrees. They may, to different extents, show characteristics of the "true natural population" (Figure 3.13). Genetic variation among half-sib families is commonly used for classical quantitative genetics in many plant and animal species. For example, populations generated from a single plant pollinated by many unknown or partially known pollen resources or using semen of a bull to inseminate many female animals are typical half-sib families. Genetic variation within a half-sib family is a resource to search for genes controlling traits of interest. In some ways, the matings to produce the half-sib families are controlled instead of completely random, because the inheritance of one parent is determined. However, in genetic terms, the pollen sources and the female animals in the examples given are random sub-sets of the populations.

Another type of sample from a natural population that is common in the plant kingdom is generated by a combination of random outcrossing and self-pollination. For example, seeds on a tree could be the result of outcrossing or selfing. Another common sample design from natural populations has pedigree structure which can be recorded for many generations.

The samples discussed above are obtained by sampling parents. An extra generation of mating is needed to reconstruct the linkage disequilibrium in the samples. The data analysis for these types of samples is linkage based analysis. Another sampling strategy does not need an extra generation of

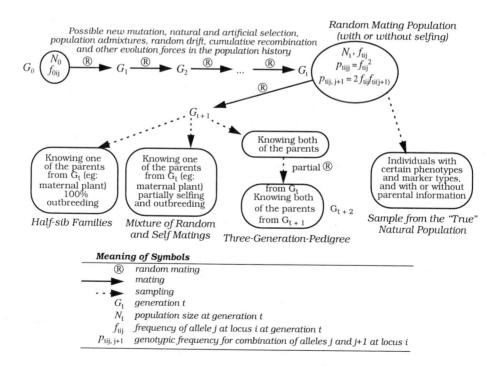

Figure 3.13 "Natural" populations used for genomic research for typical outbred plant and animal species.

mating and is commonly used studies of disease genes in humans. Samples are taken directly from the population of interest according to trait phenotypes and marker types. Analysis of the data is based on unknown linkage disequilibrium in the population. The strategy has been considered a way to increase resolution of gene identification and location because recombination between the gene and the marker has accumulated from many generations of random mating.

The essential feature of a mating scheme is the ability to detect and trace the genomic relationships between genes and markers and between genes and phenotypes. As new technologies for the detection of genetic variation and tracing genetic relationship are developed, different mating schemes may be more appropriate.

3.3.3 MATING SCHEMES AND GENETIC MARKER SYSTEMS

Linkage information obtained from genetic markers can follow different models of inheritance in the common mating systems for plants. F2 populations, where codominant markers segregate in a 1:2:1 ratio and dominant markers in a 3:1 ratio, are often used for inbred plants. The advantage of recombinant inbred lines is that the mapping population may be "immortalized" as a collection of pure lines. Other inbred mapping populations are derived from backcrossing, which yields a 1:1 ratio of markers when a heterozygote is crossed to a homozygous parent or individual.

In outbred plants, three generation pedigrees are often needed to distinguish alleles and to establish linkage phase. The markers can segregate in 1:1, 1:2:1, 3:1 and 1:1:1:1 ratios, because the parental genotypes are heterogeneous. Combining linkage information from different models of inheritance in different kinds of families is straightforward in principle, but difficult in practice.

There are several kinds of genetic marker systems commonly used in plants, but none are ideally suited to all purposes. RFLPs are codominant and multi-allelic and may often be used as gene-specific probes. In practice, however, RFLP probes commonly recognize small gene families and one probe can provide several segregating markers. It may often be difficult to recognize or distinguish specific loci and their allelic forms in different mapping families, or even in the same outbred family. PCR-based markers are frequently dominant and each primer may generate several bands. It can be difficult to determine which segregating bands in different families correspond to the same locus. Microsatellite markers provide hyper-variable codominant markers which have high heterozygosity in many populations. These markers could be useful for "bridging" the other marker systems.

3.4 GENETIC MARKERS

Three types of genetic markers have been used in genomic analysis: morphological markers, protein based markers and DNA based markers. Variation among genotypes within a species is the raw material for genomic analysis. To be a genetic marker, the marker locus has to show experimentally detectable variation among the individuals in the test population. The variation can be considered at different biological levels, from the simple heritable phenotype to detection of variation at the single nucleotide. Once the variation is identified and the genotypes of all individuals in the test population are known, the frequency of recombination events between loci is used to estimate linkage distances between markers.

3.4.1 POLYMORPHISM AND INFORMATIVITY

Polymorphism is defined as detectable heritable variation at a locus. No matter what level of genome variation is used, the polymorphism and informativity of the marker can be defined in the same way. For a single experimental population produced from a controlled cross, the number of alleles and population frequencies of the alleles may be accurately determined. For example, the numbers of alleles can be 1, 2, 3 or 4 (maximum for diploid system) and the allelic frequencies are expected to be either 1.0, 0.5, 0.33, 0.25 or 0.0. For the commonly used classical backcross and F2 types of mapping populations, the numbers of alleles can only be 1 or 2 and the allelic frequencies can only be 0.0, 0.5 or 1.0. The locus polymorphism in these types of populations can be inferred from the marker segregation patterns in the population. If a marker segregates, then it is a heritable polymorphism. If the cross is between two heterozygous individuals in the natural population, then some of the polymorphic marker combinations do not have linkage information (see Chapters 6 and 7 for detailed discussion).

When natural populations are used in genome analysis, both the number of alleles and allelic frequency at any locus are unknown. A marker is considered polymorphic if the most abundant allele in the population has a frequency less than a certain quantity, such as 95%. The quantity can differ depending on the situation. If we choose 95%, this implies that the probability of heterozygosity for the locus in the population is no less than 10%, because

$$H = 1 - \sum_{i=1}^{l} p_i^2 \geq 1 - 0.95^2 - 0.05^2 = 0.1 \tag{3.2}$$

where p_i is the frequency for the ith allele among a total of l alleles. In human genetics, it is also common to use polymorphism information content (PIC) to quantify the information value of the polymorphism (Botstein et $al.$ 1980). PIC is defined as

$$PIC = 1 - \sum_{i=1}^{l} p_i^2 - 2 \sum_{i=2}^{l} \sum_{j=1}^{i-1} (p_i^2 p_j^2) \tag{3.3}$$

See details on screening polymorphic markers in natural population in Chapter 5.

A genetic marker has to be a polymorphic marker. However, the inverse is not true, $i.e.$, a polymorphic marker may not be a genetic marker. A genetic marker may be operationally defined as a heritable polymorphic marker with clear genetic interpretation and repeatability. Genomic analysis using genetic markers should be based on well established genetic models. If the underlying genetics of a marker is not clear, then the analysis may be misleading. It is also important that the marker assay is repeatable at different times in the same or different laboratories. Different types of markers may identify different polymorphisms. The genetic interpretation of a marker strongly depends on the sequence complexity of the genome and the kind of variation the marker identifies.

3.4.2 MORPHOLOGICAL AND CYTOGENETIC MARKERS

Morphological Markers

Early mapping studies concentrated on discrete traits with simple Mendelian inheritance, such as shape, color, size or height. Morphological traits often have one to one correspondence with the genes controlling the traits. In such cases, the morphological characters (the phenotypes) can be used as reliable indicators for specific genes and are useful as genetic markers on chromosomes. A large number of morphological markers have been studied and mapped for human, mouse, *Drosophila*, maize, tomato and many other plant and animal species. Sometimes, morphological variation may have significant economic values.

A subset of the morphological markers may be due to simple point mutations representing changes in single genes. Others may not have a simple genetic basis. However, they often define and identify functional genes. Morphological markers are usually easy to observe, but it is difficult to have a large number of them segregating in a single or few populations. To obtain a reasonable number of polymorphic morphological markers, many mapping populations are needed.

Cytogenetic Markers

Chromosome pairing has been used to study genome evolution within species or among closely related species, because chromosomes that pair are very closely related. Configurations of chromosome pairing have limitations as genetic markers. The resolution of chromosome pairing is low and the methods are usually difficult.

Chromosome banding techniques were originally developed to obtain chromosome-specific staining patterns for chromosome identification. Most of those bands found in mitotic chromosomes correspond to large blocks of repetitive DNA. The chromosome banding patterns can be used for studying chromosome evolution. For example, the genome origin of cultivated polyploid wheat has been studied using the C-band technique (Friebe and Gill 1995). However, the utility of banding patterns as markers within species is limited because the banding variation is low.

In Situ Hybridization (ISH)

ISH combines traditional cytology and DNA hybridization to determine the chromosomal location of a specific DNA sequence using light microscopy. The DNA in metaphase chromosome preparations on microscope slides is denatured gently, without disrupting the chromosome morphology. DNA probes can be labeled with radioactive isotopes and hybridized to the target DNA sequence directly on the chromosome. The method works well for repeated sequence clusters. However, a large set of metaphase chromosomes and appropriate statistical procedures are needed to determine the relatively precise location of the target DNA sequence, particularly for single copy probes.

Advances in hybridization and probe labeling led to a new generation of ISH using fluorescent labeling instead of radioactive labeling, called FISH (fluorescence *in situ* hybridization) (Lichter *et al.* 1990). In FISH, the chromosome preparation is incubated with fluorescently labeled affinity probes which bind to target sequences. Larger probes enhance the hybridization signal. Competitive hybridization with unlabeled repeated DNA sequences is used to solve the problem associated with large probes, which usually contain interspersed repetitive DNA.

For some species, FISH can achieve a resolution of approximately 1 Mb using metaphase chromosomes, within a single chromosome; FISH has been widely used in humans, *Drosophila* and some plant species, such as wheat and barley (Heslop-Harrison and Schwarzacher 1995). FISH is a powerful tool to generate chromosome landmarks to bring the cytogenetic map, genetic map and DNA fragment map together.

3.4.3 PROTEIN MARKERS

Proteins are the products of genes. Different alleles of genes may result in proteins with different amino acid compositions, sizes or modifications. Differences in charge or size can be easily detected using gel electrophoresis and can be used as genetic markers. A commonly used type of protein marker is the isozyme marker, which has been used for several decades. Proteins themselves are usually not visible in gels. Many enzymes can be visualized by using their activity to create a visible product that appears as a band where the enzyme is present. Isozymes are alternative forms of enzymes that often differ in electrophoretic mobility but have the same enzyme activity. The different forms of isozymes can be detected (or visualized) typically using electrophoresis in starch gels or polyacrylamide gels. Isozymes are limited in number and tissue and are developmental-stage dependent. Before DNA markers were discovered, isozymes were extensively used in maize, wheat, barley and many other plant and animal species (Crawford 1989) and are still often used in conjunction with DNA markers.

Another form of protein markers can be obtained by separating proteins according to charge and size by two-dimensional gel electrophoresis. A single two-dimensional gel has the potential to detect the variation among a mixture of 1,000 to 2,000 proteins. These proteins are usually anonymous (*i.e.*, not associated with any known biological function) and the techniques for obtaining data on protein variants from two-dimensional gels and genetic interpretation of those data are rather difficult. However, as more DNA sequence information becomes available, more attention will be paid to structure and function of such anonymous proteins. New technology for detecting protein polymorphism will be developed, for example, using mass spectrometry. Some scientists have predicted that after several plant and animal genomes are completely sequenced, more attention will be paid to the proteins of entire genomes (proteomes), including protein structure, function and protein engineering.

Much of the detectable protein variation identifies allelic sequence variation in the structural gene encoding the protein or at a regulatory sequence. Alternatively, some protein variation is due to post translational modification and is not useful as genetic markers. Isozymes often identify the variation encoded by several related non-allelic genes in a gene family.

3.4.4 DNA MARKERS (RATIONALE)

A DNA marker is typically a small region of DNA showing sequence polymorphism in different individuals within a species. Two basic approaches have been used to detect variation in the small region of DNA. The fragment can be detected by nucleic acid hybridization, which uses another fragment from the same locus which has been isolated and purified from the same or related species. The previously known segment must share considerable DNA sequence homology with the fragment of interest and can be labeled and used as a probe to detect the fragment of interest by complementary base pairing. This is the foundation for RFLP (restriction fragment length polymorphism) markers. In this class, two-dimensional DNA electrophoresis has also been used to identify DNA polymorphism. The second approach is based on the amplification of sequences using PCR (polymerase chain reaction). To amplify a target segment, two primers (flanking the target sequence) designed using known sequence of the segment are needed. Microsatellites, STSs (sequence tagged sites), ESTs (expressed sequence tags), *etc.* have been commonly used as genetic markers based on sequence specific PCR. Shorter, arbitrarily chosen primers have also been used to amplify random polymorphic DNA. Markers of this type include RAPD (random amplified polymorphic DNA) and AFLP (amplified fragment length polymorphism).

Some of the genetic markers can be detected using both the hybridization and the amplification approaches. For example, the hybridization approach using probes flanking the tandem repeats can be used to detect VNTR (variable number of tandem repeats) polymorphism. It is also common to detect VNTR polymorphism by amplification of the repeats using flanking primers. STSs and ESTs can be identified using both the hybridization and the amplification approaches.

Table 3.3 Some restriction enzymes and their restriction sites.

Enzyme	Restriction site
*Eco*RI	5´ GAATTC 3´ 3´ CTTAAG 5´
*Eco*RV	5´ GATATC 3´ 3´ CTATAG 5´
*Hind*III	5´ AAGCTT 3´ 3´ TTCGAA 5´
*Hae*III	5´ GGCC 3´ 3´ CCGG 5´
*Pst*I	5´ CTGCAG 3´ 3´ GACGTC 5´
*Not*I	5´ GCGCGCGCGC 3´ 3´ CGCGCGCGCG 5´

3.4.5 RFLP AND SOUTHERN BLOTTING

The detection of RFLPs involves several steps. First, DNA is digested with a restriction enzyme and the fragments are separated by electrophoresis in an agarose gel. Table 3.3 lists some typical restriction enzymes. Each of the enzymes makes sequence-specific cuts in DNA. Over 150 different enzymes have been found in different microorganisms and are available commercially. If the DNA sequence is random, then the number of sites (Rs) in a genome with a total of N base pairs is

$$R_S = N/4^b \tag{3.4}$$

where b is the number of base pairs of the restriction site and 4^b is average size of the fragment produced by the cuts. For example, for rice, which has a genome size of 4.5×10^9 bp and is cut with a 4 base restriction enzyme, there are $4.5 \times 10^9/4^4 = 17,580,000$ restriction sites. There are $4.5 \times 10^9/4^6 = 10^6$ restriction sites for using a 6-base pair restriction enzyme.

The sizes of the fragments are usually from several hundred base pairs up to 20 kb. During electrophoresis, DNA fragments (with negative charges) move from a negative electrode toward a positive electrode. The smaller fragments are able to move faster through the gel than the large fragments. As a result, the fragments of DNA for each of the samples are spread throughout the lane by size (Figure 3.14). Because there is a very large number of fragments of different sizes in each of the DNA samples, the fragments appear as smears. If the restriction digest is complete and there is an assumption of no random breaks, the number of fragments will be one more than the number of restriction sites for a linear chromosome or equal to the number of restriction sites for a circular chromosome, such as found in bacteria or organelles of eukayotes. In practice, there are many random breaks to DNA during preparation, but those breaks are few compared to the number of sites cut by the restriction enzyme.

The next step is to separate the paired strands of the DNA fragments (denaturation) by putting the gel in alkali. The denatured single-strand DNA fragments are usually transferred to a durable nitrocellulose or a nylon membrane by blotting. The individual DNA fragments are immobilized on the membrane. This blotting procedure was first applied to DNA analysis by E. Southern and is still referred to as a Southern blot (Southern 1975).

The third step is the hybridization of a labelled DNA probe to the blot. A probe is a piece of DNA that may vary in size from several hundred base pairs to 50 kb. The probe can be cloned DNA with known or unknown sequence and can have known or unknown function. The probe also can be obtained by artificial synthesis of oligonucleotides using chemical procedures. The synthetic probes are usually small in size (less than 100 nucleotides). The

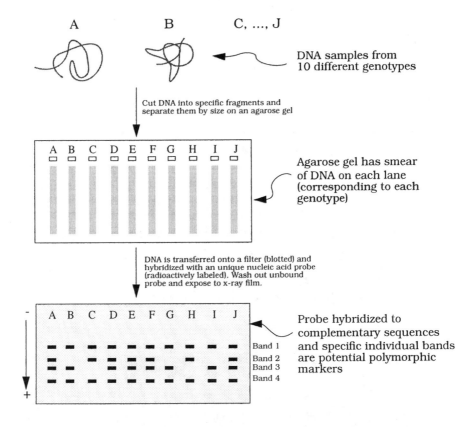

Figure 3.14 Southern blotting to detect restriction fragment length polymorphism (RFLP). Bands 2 and 3 can be interpreted as one polymorphic marker. For the 10 genotypes, A, D, E, F and J are heterozygous at the marker locus, B, G and I are one type of homozygote and C and H are the other type of homozygote.

probe is labeled with some detectable molecule, either radioactive or fluorescent. The labeled probe is added to the hybridization solution already containing the membrane and it will bind to the immobilized DNA fragments containing the complementary sequence on the membrane. The unbound probe can be washed off.

The final step of the Southern blotting is to visualize the hybridized band using x-ray film for a radioactive labeled probe (Figure 3.14) or using special light or image treatment for a fluorescent chemically labeled probe. The procedure for DNA fragment identification also can be used with RNA immobilized to the membrane. The procedure is called Northern blotting when RNA is the target. However, Northern blots are used to study variation in gene expression and are not used to identify polymorphic genetic markers. Gene expression varies greatly with time and phase in development, or as a result of environmental stress, as well as variation between individuals.

Figure 3.14 shows an hypothetical RFLP banding pattern. Scoring the banding pattern is relatively simple. However, in a practical mapping experiment, the banding pattern of the RFLPs is usually more complex. The banding patterns can be experimental. To score the RFLP banding patterns correctly, it is important to understand what kinds of DNA sequence variation the RFLP identifies and the genetics of the experimental materials.

Figure 3.15 shows a hypothetical situation for a small chromosome segment. There are eight restriction sites recognized by the enzyme *EcoRI*. Among the eight sites, A and B are polymorphic. The digestion will result in

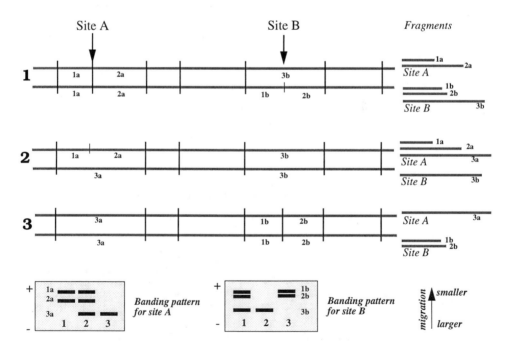

Figure 3.15 Restriction cleavage using enzyme *EcoRI*. Interpretations of the hypothetical banding patterns are: individual 1 is homozygous for the restriction site A and heterozygous for site B, individual 2 is heterozygous for site A and homozygous for site B and individual 3 is homozygous for both site A and site B.

different banding patterns for different allelic forms at these two sites. From Figure 3.15, we may see what kinds of DNA polymorphism RFLPs identify. RFLP analysis identifies variation in DNA sequences at the restriction site. A mutation at the site may result in a long fragment instead of two short fragments due to the loss of a restriction site. Similarly, a mutation in sequence may create a new restriction site. Mutations that lose or gain restriction sites may be identified. RFLP fragment patterns may change due to insertion or deletion of DNA between restriction sites. In these cases, the fragments involved become longer or shorter, accordingly. The sequences between sites recognized by RFLP analysis are only a small fraction of the genome.

Both homozygous genotypes and heterozygous genotypes can be clearly identified with RFLPs. Therefore, RFLPs can typically be scored as codominant markers. In practice, difficulties can arise if the probe used hybridizes to repeated sequences at multiple locations on the genome. In these cases, allelic and non-allelic variations can not be distinguished. These problems commonly occur when the same probe is used to detect RFLPs in progeny of different lineages. Different alleles may be present at the same locus in different pedigrees and different loci may be polymorphic in different pedigrees. If a probe detects more than one locus in different pedigrees, it is a problem to

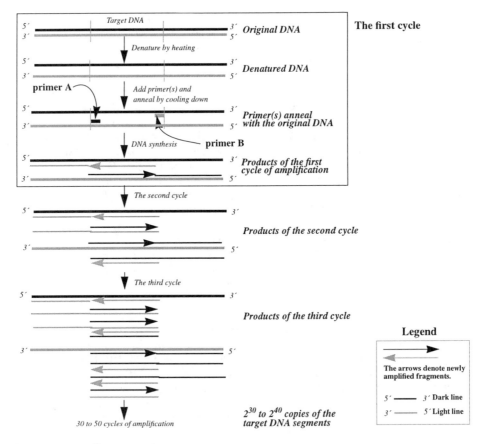

Figure 3.16 Polymerase chain reaction.

determine which locus is truly the same locus across the pedigrees. Therefore, it is important to use probes that detect single polymorphic loci in different pedigrees. However, some loci used for forensics have multiple alleles at a single locus, which increases the statistical power to uniquely identify or exclude individuals. In practice, interpretation of RFLP patterns is difficult when complex banding patterns are present.

3.4.6 PCR

Polymerase chain reaction (PCR) is a DNA synthesis technique that amplifies specific regions of DNA that lie between two sites defined by the complementary sequences of two specific primers. PCR has revolutionized DNA technology through its applications in agriculture, forensic science and diagnosis of human diseases. PCR based procedures to obtain DNA fragment profiles have a higher sensitivity and are less time consuming than the hybridization based procedures. Many commonly used genetic markers, such as microsatellites, AFLPs and RAPD markers, are PCR based.

The principle of PCR is illustrated in Figure 3.16. Two primers (A and B) flank a target DNA sequence to be amplified. The primers are short single stranded DNA molecules that anneal to the denatured DNA strands at sites with complementary sequences. For the first cycle of amplification, two DNA strands extend from each of the primers along the original templates without defined termination sites. In cycle 2, the primers initiate synthesis on the products of the first cycle and two new strands are synthesized. The second strand must terminate at the original primer sites, where the new template ends. Further amplification is therefore restricted to the sequence between sites defined by the primers. As further cycles proceed, the number of DNA fragments increases exponentially. At the end of 30 to 50 cycles, 2^{30} to 2^{40} copies of DNA molecules, having sequences identical to the target segment, are produced.

One type of genetic marker analyzed using PCR is based on loci which have been previously sequenced (Figure 3.16). When the DNA sequence of a target segment is known, primers can be designed to avoid self-complementation within the primer and complementary sequence between the primers. For different plant and animal species, requirements of amplification conditions, such as annealing temperature, primer length, G:C content of the primers and other aspects of the protocol may vary. The polymorphisms identified using PCR based markers where sequence is known represent variation in segment length between the two primers.

3.4.7 MINI- AND MICRO-SATELLITE MARKERS

Minisatellites are tandem repeats of sequences ranging from 9 to 100 bp in the genome. The number of the repeats varies (usually less than 1000). Minisatellites are also referred to as VNTR (Variable Number of Tandem Repeats). VNTR can be detected using either hybridization or PCR approaches. Genomic DNA can be digested using restriction enzymes that recognize restriction sites flanking the tandem repeats. The cutting yields fragments containing cores of the repeats with different numbers of repeats (length variation)(Figure 3.17). Those fragments with different lengths can be detected using a probe designed from DNA sequences flanking the repeats or from the repeat itself. Those fragments can be considered a specific kind of RFLP. For forensic applications such as parentage analysis, multiple probes can be used in one hybridization experiment. If the fragments identified by the probes can be interpreted as independent loci, then the banding pattern for the multiple-probe hybridization is a superimposition of single-locus patterns.

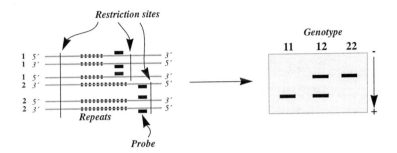

Figure 3.17 VNTR fragment variation is detected by hybridization using a probe designed from the DNA sequences flanking the segments or from the repeat itself. The length differences in the polymorphic bands can be interpreted as variation in the numbers of the tandem repeats.

VNTR can also be detected by amplifying the segments containing different numbers of repeats by using primers flanking the segments for PCR (Figure 3.18) (Weber and May 1987). The polymorphic bands result from the variation in the numbers of the tandem repeats.

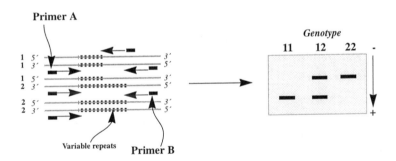

Figure 3.18 VNTR fragment variation is detected by amplifying the segments containing different numbers of repeats by using primers flanking the segments for PCR. The polymorphic bands can be interpreted as the variation in the numbers of the tandem repeats.

VNTR markers used in forensics are usually members of repeated DNA families that are detected using probes hybridizing to the repeats themselves. The probes hybridize with all the members of the repeated DNA family. This results in banding patterns that may represent the repeat variation of that family in the whole genome (multiple loci variation)(Figure 3.19). Those banding patterns have high discrimination power between individuals and are commonly referred to as DNA fingerprinting (Jeffreys *et al.* 1985). For example, in Figure 3.19, genotypes (1) and (3) can be considered as parents and (2) as a child. A single probe containing the core of the tandem repeats hybridizes to

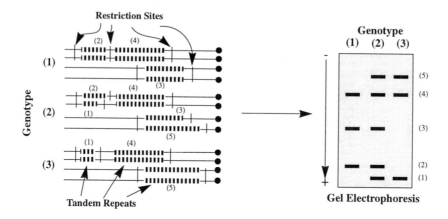

Figure 3.19 An illustration of using the tandem repeats as a probe to detect the repeat array occurring at several loci on the genome. Segments of two chromosome arms are shown, with varying degrees of heterozygosity for tandem repeat loci on each of the homologs

the repeats containing the same core sequence on three loci. Band (1) and (2) can be interpreted as one locus and (3) and (5) as another locus. Band (4) has no polymorphism. So, a single probe detects two polymorphic loci and one monomorphic locus. The gel picture contains DNA fingerprints for the three genotypes.

Microsatellites are direct tandem repeated sequences of DNA with a repeat size ranging from 1 to 6 bp. A microsatellite is also called a simple sequence repeat (SSR). The number of repeats for the microsatellite is usually less than 100.

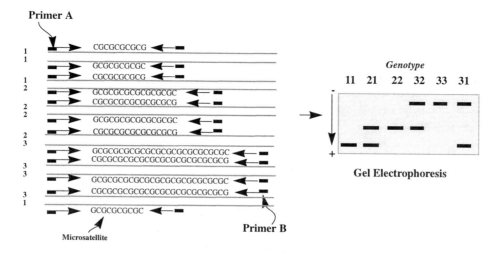

Figure 3.20 Microsatellites are detected by amplifying the segments containing different numbers of repeats using the primers flanking the segments for PCR. Both homologs of a single chromosome arm are shown.

Figure 3.20 shows how microsatellite variation can be detected by PCR using primers flanking the repeats. The microsatellite variation is based on differences in the number of repeats. Variation can be detected by amplifying the tandem arrays and then visualizing them on a gel.

Microsatellites have high levels of variation in many animal and plant species. The common forms of these repeats are simple di-nucleotide repeats, such as $(CA)_n:(GT)_n$, $(GA)_n:(CT)_n$, $(CG)_n:(GC)_n$ and $(AT)_n:(TA)_n$, where n is the number of repeats. Microsatellites with tri- and tetra-nucleotide repeats are also found, but their frequencies are lower than the di-nucleotide repeats (Hearne *et al.* 1992). Different species of animals and plants often have different distributions of microsatellite variation within their genomes.

There are two commonly used ways to identify microsatellite loci suitable for use as genetic markers. For some species, such as human, mouse, *Arabidopsis* and rice, a large amount of DNA sequence data has already been accumulated; microsatellites may be identified by searching through the DNA sequence databases for sequences containing simple repeats. Primers are commonly designed directly from the sequence data. However, for most plant and animal species, a large effort using hybridization and sequencing is needed to identify microsatellites suitable for use as genetic markers. Hybridization using simple repeats as probes to screen genomic clones can be used to identify the microsatellites. The flanking sequences can be obtained by sequencing the cloned fragments containing microsatellites. Once the microsatellite markers are found, the utility of the markers is high. Until now, only a limited number of microsatellite markers have been found in plant species.

3.4.8 STS AND EST

Sequence tagged sites (STSs) were proposed by Olson *et al.* (1989) as chromosome landmarks in the human genome. An STS is a short unique fragment of DNA (~300 bp). Clones containing the same STS must overlap, so STSs are primarily used in physical mapping to join large cloned fragments. If a polymorphism can be detected using the STS as probe, then anchor points between genetic and physical maps can be established (Weissenbach *et al.* 1992; Gyapay *et al.* 1994). Genetic maps using polymorphic STS markers have been constructed for mouse, cow and pig (Dietrich *et al.* 1994; Barendes *et al.* 1994; Archibald 1994). The polymorphic STS markers are also commonly used for genomic analysis in plants (Mazur and Tingey 1995).

Expressed sequence tags (ESTs) are subsets of STSs derived from cDNA clones. ESTs can serve the same purpose as the random STSs, with the advantage that ESTs are derived from expressed genes, *i.e.*, from spliced mRNA which is usually free of introns as well as repetitive DNA.

Many projects of cDNA sequencing in a variety of animal and plant species are underway and a large amount of cDNA sequence data has been accumulated in a relatively short time. ESTs have the advantages of representing real functional genes and are therefore more useful as genetic markers than anonymous nonfunctional sequences. It is likely that polymorphic ESTs will be increasingly available and used more widely. A cDNA sequencing project for loblolly pine is underway to obtain large number of ESTs. Loblolly pine has a huge genome which contains a large amount of repetitive DNA. This makes the STS approach to physical mapping impractical. However, the estimated number of expressed genes may not significantly differ from plants with smaller genome, such as *Arabidopsis* or rice. So, in species having large genomes, cDNA sequencing and ESTs are advantageous for comparative genomic analysis and gain of information on genome structure.

3.4.9 SINGLE-STRAND CONFORMATIONAL POLYMORPHISM (SSCP)

The DNA markers we have described so far detect specific DNA polymorphism in a small fraction of the entire segment being examined, such as changes in specific restriction sites. If the purpose for using the markers is detection of mutations involving a single nucleotide change, then the markers have very low resolution. For this reason, a method that detects changes in nucleotide sequence for an entire fragment of more than 1000 bp has been developed, called single-strand conformational polymorphism (SSCP). SSCP can detect DNA sequence alterations as small as a single nucleotide change (Orita *et al.* 1989). Electrophoretic mobility of single-stranded DNA in non-denaturing polyacrylamide gels depends on both size and sequence characteristics. This method exploits the tendency of single-stranded DNA to form intramolecular base-pairs, resulting in a sequence-dependent conformation with a specific mobility in acrylamide gels. Changes in DNA sequence, even in a single base pair, can cause alterations in the conformation and result in changes in electrophoretic mobility. In practice, SSCPs can be detected using two methods. The DNA can be cut with restriction enzymes, then run on a gel to separate by conformation. A Southern blot of the gel is then done and a specific fragment is used as a probe for hybridization. The other method uses PCR to amplify a specific fragment which is then run on a conformational gel (high resolution acrylamide gel).

3.4.10 RANDOM AMPLIFIED POLYMORPHIC DNA (RAPD) MARKERS

DNA sequence information is needed for many of the markers we have described, such as microsatellites, STSs and polymorphic ESTs. For plant and

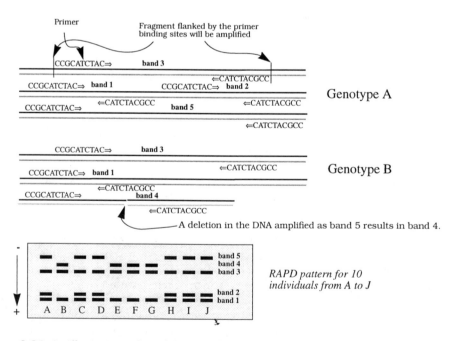

Figure 3.21 An illustration of random amplification of polymorphic DNA (RAPD) using a hypothetical primer 5′-CCGCATCTAC-3′.

animal species that have small amounts of DNA sequence information, screening for polymorphic markers that depend on sequence information can be prohibitive. RAPD markers were designed to solve the problem of lack of pre-existing DNA sequence information. A single, arbitrarily chosen short oligonucleotide can be used as a primer to amplify genome segments flanked by two complementary primer-binding sites in inverted orientation (Williams *et al.* 1990). Short primers with an arbitrary sequence can be complementary to a number of sites within a genome (Figure 3.21). If the sites occur on opposite strands of a segment of DNA in inverted orientation and the distance between the sites is short enough for PCR, then the segment flanked by the sites can be amplified. Amplifications on different segments are independent. If polymorphism exists in the binding sites among different genotypes or the fragment length differs at the same site from genotype to genotype, then a RAPD marker is obtained (Figure 3.21).

In practice, only small amounts of DNA from each genotype are needed as templates. A set of primers is screened for reproducible polymorphisms. Different primers may identify different polymorphisms and have different reproducibility. PCR reactions can be carried out in a 96-well plate using a programmable thermocycler. The amplification products are separated on an agarose gel and visualized using ethidium bromide staining.

Different alleles at the same locus are generally distinguished by the presence or absence of a band of a particular size (Figure 3.21). The band-present phenotype is dominant to the band-absent phenotype. The band is present if the genotypes are homozygous or heterozygous for the locus that is amplified. The band is absent if the genotype is homozygous for a lack of sites to amplify that specific fragment. A testcross is needed to tell the heterozygotes and the homozygotes apart. Dominant markers have low linkage information content in conventional F2 progeny analysis (see Chapter 6). When two dominant markers are linked in repulsion phase, RAPD markers are not recommended for analysis of traditional F2 populations.

If they are used in backcross progeny, recombinant inbred lines or doubled haploid lines, dominant markers have the same amount of linkage information as codominant markers. A pseudo-testcross strategy has been used in genomic analysis in F1 progeny of heterozygous parents using dominant markers (Grattaglia and Sederoff 1994). In this strategy, markers heterozygous in one parent and homozygous for the non-amplified allele in the other parent are selected by screening for 1:1 segregation in a sample of F1 offspring. The result is that all markers used are in backcross configuration. The major advantage of RAPD markers is that it is easy to obtain large numbers of markers and the most informative markers can be selected.

The expected number of amplified products using an arbitrary primer is a function of genome length, number of nucleotides in the primer and the maximum fragment length that can be amplified. If we assume that nucleotide distribution is random, the primer is arbitrarily designed and complete complementation occurs between the primer and the binding sites in the DNA templates, then the expected number of amplified products is

$$b = 2fN/16^n \qquad (3.5)$$

where f is the maximum length of amplified fragment in bp, N is the genome size in bp and n is number of nucleotides in the primer. f can be dependent on the type of enzyme used in PCR. The rationale behind Equation (3.5) is as follows. For a single strand of DNA, the expected number of binding sites within a genome with N bp for a primer with n nucleotides is $N/4^n$. The number is doubled for two strands of DNA. The probability that a binding site

Table 3.4 Expected numbers of PCR products using a single arbitrary primer with different lengths for some well-studied organisms, where $f = 2000$.

	Genome Size (Mb)	Primer Length (bp)			
		8	9	10	11
Yeast	14	13	<1	<1	<1
Nematode	100	93	6	<1	<1
Drosophila	170	158	10	<1	<1
Mouse	3000	2794	175	11	<1
Human	3000	2794	175	11	<1
Arabidopsis thaliana	100	93	6	<1	<1
Corn	2500	2329	146	9	<1
Tomato	950	885	55	4	<1
Barley	5300	4936	309	19	1
Rice	450	419	26	2	<1
Loblolly pine	20000	18626	1164	73	5

occurs at the opposite strand within f bp from the nearest site is $f/4^n$. Table 3.4 lists expected numbers of amplified products using a single arbitrary primer with different numbers of nucleotides for some organisms. Those numbers are expectations under some assumptions of random distribution of nucleotides in the genome and random choice of primer sequence. In practice, the number could be far from the predictions. However, the numbers in the table provide guidance on the size of primer that should be used in searching for reliable RAPD markers. If the number of amplified products is too large, then the interpretation of the gel electrophoresis will be difficult. If the number is small, then a large effort is needed to get adequate numbers of polymorphic markers. Depending on the species and experimental conditions, the number of amplified products may vary. Decamers (10 bp) are widely used.

In practice, optimization for primer choice, PCR conditions and gel reading is needed to obtain RAPD markers with simple genetic interpretation and high repeatability, such as those shown in Figure 3.22 (Williams *et al.* 1990). Since a single primer may generate several polymorphic markers, screening a large

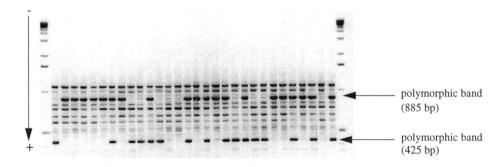

polymorphic band (885 bp)

polymorphic band (425 bp)

Figure 3.22 A RAPD gel using primer OPC4 (5′-CCGCATCTAC-3′) for 30 loblolly pine megagametophytes from an open-pollinated clone 10-5. (Dr. Henry V. Amerson of Forest Biotechnology Group at North Carolina State University.)

number of primers on a small number of genotypes in a mapping population is a useful method to obtain a large group of markers having a high information content.

Figure 3.22 is a gel picture using a 10 base DNA primer, OPC4 (5'-CCGCATCTAC-3'), for amplification of DNA from 30 megagametophytes (haploid tissue) of loblolly pine from an open-pollinated daughter of a selected clone. Two polymorphic markers can be scored on this gel, one at 425 bp and the other at 885 bp.

3.4.11 AMPLIFIED FRAGMENT LENGTH POLYMORPHISM (AFLP)

The AFLP approach uses PCR to amplify a subset of DNA fragments generated by restriction enzyme digestion. Fragment length polymorphism results from standard restriction enzyme digestion as do the RFLP markers. The genomic DNA is digested with two restriction enzymes, usually a rare cutter and a frequent cutter, such as *Eco*RI and *Mse*I. Adapters are ligated to the ends of the genomic DNA at the specific restriction sites. Adapters are short segments of double-stranded DNA with a "sticky end" complementary to that of the restriction site. Separate adapters are needed for each of the different restriction enzymes. This DNA is then used as a template for PCR reactions. Different primers are used to amplify different subsets of the fragments. The primers are specific to the combination of adapter sequence, restriction site and several selective nucleotides (extensions)(Figure 3.23). One of the primers is end-labeled and only amplified fragments that are labeled will be visualized. The use of selective extensions has the function of reducing of the number of fragments amplified. For example, a three-base extension for each primer reduces the number of amplified fragments by a factor of $4^6 = 4096$. By changing the selective extension bases, different subsets of the fragments will be amplified. The PCR products are separated on denaturing polyacrylamide gels.

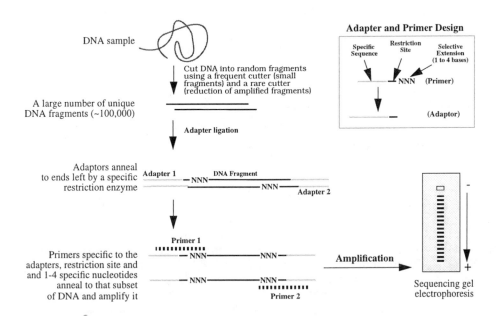

Figure 3.23 An illustration of amplified fragment length polymorphism markers.

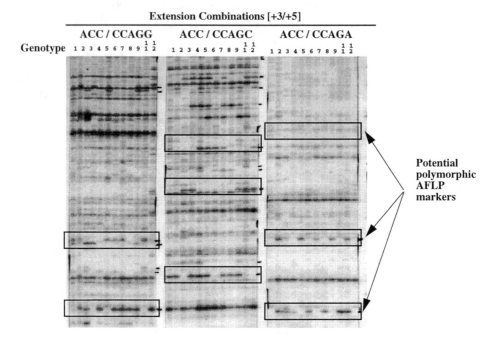

Figure 3.24 A partial amplified fragment length polymorphism (AFLP) gel for haploid megagametophyte DNA from 12 seedlings from the same maternal parent (loblolly pine clone 7-56) using three primer combinations. (Gel image provided by Mr. David L. Remington of the Forest Biotechnology Group at North Carolina State University.)

Figure 3.24 shows a partial AFLP gel image for haploid megagametophyte DNA. The purpose of the experiment is to screen for reliable, polymorphic AFLP markers for loblolly pine. Primers used in the screening experiment are listed in Table 3.5.

Table 3.5 Primer used for gel in Figure 3.24

Primer Sequence	Extension Bases	Restriction Enzyme
5' - GACTGCGTACCAATTC +	-ACC	*Eco*RI
5' - GATGAGTCCTGAGTAA +	- CCAGC	*Mse*I

Caution is needed in scoring the AFLP gel because of the large number of bands. The AFLP bands are usually scored as dominant markers, but occasional length polymorphisms can be used as codominant markers. In addition, differences in band intensity can sometimes be used to distinguish homozygous-present from heterozygous bands, generating codominant markers. To determine the genotype of an AFLP band, a mixture distribution model can be used to fit the band intensity for three possible genotypes (for a di-allelic model). Only a small portion of the bands have band intensities distributed in the middle of the distributions. However, the fraction of bands that can be scored clearly as codominant markers varies from species to species (R. Jansen, personal communication).

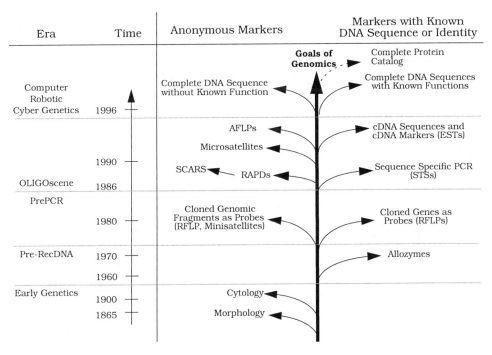

Figure 3.25 "Evolution" tree of genetic markers based on an original drawing by Dr. David O'Malley of Forest Biotechnology Group as North Carolina State University.

3.4.12 COMPARISON AMONG DIFFERENT MARKER SYSTEMS

"Evolution" of Genetic Markers

Figure 3.25 shows an "evolutionary tree" for genetic markers. Markers used by the pioneer geneticists, such as Mendel and Morgan, were morphological. A portion of these morphological markers can now be explained by known DNA sequence based on molecular biology. However, a large portion of these markers remain anonymous in terms of their DNA sequences and identities. Allozyme markers were popular in the 1960s and 1970s, which was an era of pre-recombinant-DNA technology. Allozymes correspond to different alleles of a gene, while isozyme markers can be considered markers with known enzyme activity. As recombinant DNA technology was developed in early 1980s, RFLP marker methods were invented. From the middle 1980s, DNA marker technology entered the OLIGOscene due to the invention and advances in PCR technology. Many types of DNA markers have been invented or developed using PCR and there will surely be more PCR based markers in the future.

The ultimate markers will be complete DNA sequences of genes with known functions. As the speed of DNA sequencing and marker typing increases, information needed to identify all the genes and to define their variation, which is one of the long-term goals of genomics, will be accumulated rapidly. Computers, robotic automation and data communication programs are becoming more and more important as genomics is entering an era of informatics.

Characteristics of Commonly Used Marker Systems

We have discussed commonly used genetic markers in several previous sections. Certainly, those markers have different characteristics in terms of interpretation and implementation. A suitable marker system for any given experiment depends on the objectives of the experiment and pre-existing genome information. Different researchers and different laboratories may have different experiences with their favorite marker systems. In general, high polymorphism, clear genetic interpretation, short time requirement, high repeatability and easy automation are favorite characteristics. However, high polymorphism in one species does not mean that there will be high polymorphism in another species. For species in which a large amount of genome information has been accumulated, many suitable marker systems may be available. For species with little or no genome information, much more effort is often needed to identify suitable marker systems.

Marker Conversion

As DNA cloning technology is developed and genome information is accumulated, it is sometimes possible to convert one type of marker to another without loss of its genetic interpretation and identity. For example, anonymous DNA fragments can be cloned, sequenced and converted to primer specific markers (Paran and Michelmore 1993). Usually, with known DNA sequences, hybridization based markers can be converted to PCR based markers and anonymous markers can be converted to markers with DNA identities. By doing the conversions, disadvantages of a marker system can be overcome. RAPD and AFLP markers based on PCR-amplification of anonymous sequences are cost-effective for genomic analysis at an initial stage. However, they are mostly dominantly inherited. In some cases, these markers can be converted to codominant markers by cloning the polymorphic bands and designing hybridization based markers. Dominant markers can be very useful under many circumstances and are more amenable to automated marker analysis.

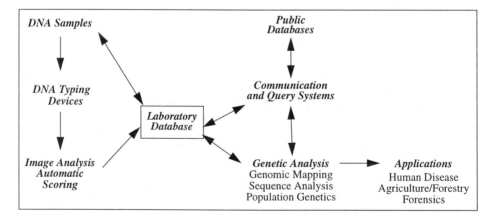

Figure 3.26 Automated genomic research including DNA preparation, DNA typing, image analysis and scoring, laboratory database, communication with public databases, genetic analysis and practical applications in human disease diagnosis, agriculture, forestry and forensics.

3.4.13 AUTOMATION

Different marker systems have been described. Among them, DNA markers play an essential role in genomic analysis. Generating DNA markers involves steps such as DNA preparation, restriction digestion, primer synthesis, PCR, gel electrophoresis, hybridization, visualization and scoring of banding patterns, genetic interpretation, *etc.* Also, a data quality control step is needed as gel electrophoresis patterns are scored.

Automation of every step of marker analysis is essential for reducing the cost and errors in genomic research and for making the applications of genomic information in human disease diagnosis, agriculture, forestry and forensics feasible (Figure 3.26). Among the components shown in Figure 3.26, a laboratory database is essential for making the connection between automated DNA typing devices, data analysis and the public databases. As the amount of information grows, communication among all the components will make it possible to design more efficient genomic experiments. In this subsection, only the three components on the left column of the figure will be discussed.

Robotic-Assisted Assay

In theory, DNA preparation from extraction to purification can be automated. However, commercially available equipment is usually not sufficient for high throughput DNA typing devices. Custom-built instruments for automated DNA preparation exist in some laboratories. For example, an instrument developed at DuPont has the capacity to process several thousand DNA samples a day. But, such custom-built machines are usually not widely available.

After DNA preparation, robotic workstations for liquid DNA handling are the most common forms of semi-automated DNA manipulation, because the they are widely available and relatively easy to operate.

Both the hybridization based and PCR based markers are commonly obtained using gel electrophoresis. Although polyacrylamide gels are more suitable for automation, loading samples on the gels is difficult to automate. The use of capillary electrophoresis makes it possible to automate the complete process from loading the gel to data acquisition.

Automated Scoring Systems

For DNA sequencing, a gel image can be obtained by using the standard sequencing gel approach or from an automated sequencer, such as the images captured by Charge Coupled Device (CCD) cameras. For DNA marker genotyping, images can be a on a piece of film for a Southern blot of hybridized RFLPs, or in a file captured by an automated instrument for some PCR based markers. The currently available image analysis software packages for common laboratories or for special automated DNA typing devices still require intensive human interactions in order to achieve reliable genetic interpretation of the image data. For example, ABI Genotyper software still requires user interaction to assign an allele to a band and can not score closely spaced alleles.

Image analysis includes three steps:

- image enhancement and noise removal

- gel lane correction

- sequence coding or marker type determination

In step one, original images are enhanced and most random noises are removed. The sequence codes and marker genotype are usually difficult to determine from the original image because the gray levels of the original images are unevenly scattered and the contrasts of the images are poor. This is especially true when two bands are closely located. In step two, two kinds of lane corrections are carried out: lane straightening and band correction. In the last step, DNA polymorphic bands are determined and the sequences are coded. To determine each of the DNA bands, the profile of each DNA lane is first obtained along the axis perpendicular to the lane axis and then smoothed by convolving a Gaussian template to remove the random noise. The DNA bands are determined by checking curve peaks of the smoothed profile.

There are four major difficulties with automated scoring systems:

- fuzzy gels

- uneven lighting source

- lane variation problems

- complex banding patterns

For DNA marker genotyping, the banding patterns are simple for most PCR based markers, such as RAPDs, micro-satellites and STSs (sequence tagged sites). Polymorphic bands can be scored as dominant markers for most cases. However, some of the bands can be scored as codominant markers. For those cases, the bands can be scored first as dominant and then tested for allelism by checking the linkage configurations among the markers generated from the adjacent bands in the same gel. If the markers are completely linked in repulsion phase, the markers can be equivalent to a single codominant marker. For markers like AFLPs, codominant markers can be scored based on both presence of bands and density of the bands. Some marker types, such as RFLPs and VNTRs, usually have complex banding patterns, especially for organisms with complex genomes and a large amount of repetitive DNA.

3.5 EXERCISES

Exercise 3.1

We expect that genomes of plants and animals may be very distinct, because different sets of genes are needed for their development. Suppose that a plant genome scientist became interested in animal genome research and found out that large portions of the genomes of the plant and animal species with which she worked shared many common features. Answer the following questions.

(1) 10 out of 100 random cDNAs from the plant and the animal share conserved DNA sequences. Can you explain the significance of the finding?

(2) 10 segments (about 10% of the genome) have similar marker orders using classical genetic mapping. The markers are a mixture of RFLP, RAPD and AFLP markers. Can you explain the significance of the finding?

(3) 10 segments (about 10% of the genome) on the physical maps from the plant have similar sequences to others within the plant genome. The

physical maps are overlapped BAC clones ordered by common restriction sites. Can you explain the significance of the finding?

(4) Do you expect that she would find more similarities between two plant species or between two animal species?

Exercise 3.2

Polymorphism at a DNA sequence level among different genotypes within a species is the foundation of classical genetic mapping and the related applications. What kinds of polymorphisms at the DNA sequence level are ideal for genetic mapping? How can these polymorphisms be revealed and traced in experimental populations?

Exercise 3.3

Contrary to genetic mapping within a species, genome homology (similarity or synteny) among species is the basis for comparative mapping among different species or relatives. Bread wheat (*Triticum aestivum* with A, B and D genomes) and *Triticum tauschii* (with D genome) are relatives. If polymorphic RAPD markers are identified in *Triticum tauschii*, what is the probability that a polymorphic RAPD marker in *Triticum tauschii* will be polymorphic in wheat? Can you speculate on how different and how similar are the genetic maps made using RAPD markers for the D genomes of wheat and *Triticum tauschii*? How about using RFLP markers?

Exercise 3.4

Compare genomes of the following pairs of species. Speculate why they could be similar or different.

(1) corn and rice

(2) humans and mice

(3) mice and rice

CHAPTER 4

STATISTICS IN GENOMICS

4.1 INTRODUCTION

There are two main aspects of statistical interest in genomic map construction: sampling variation and experimental error. Sampling variation is associated with the composition of the mapping population, while experimental error is associated with laboratory procedures. Statistics provides a means to analyze data and to draw genetic inference. Most importantly, it provides ways to evaluate the quality of experimental data and to quantify confidence in genetic inferences. Data are generally discrete in simple linkage analysis, however, the search for genes controlling complex traits (QTL, quantitative trait loci) uses a combination of continuous and discrete variables.

Sometimes, statistics may not be needed. When biological evidence is significant and no adequate statistics can be applied, it is better not to use statistics. Statistics is a tool to solve problems which cannot be solved solely by biology. This is especially true for statistics used in genomic mapping. Biological conclusions should not be overshadowed by statistics when the correctness of the statistical methods is uncertain.

In this chapter, some statistical methods commonly used in genomic mapping will be introduced. The scope of this chapter will include sampling distributions, hypothesis tests and estimation theory. This chapter is written primarily for readers who have little background in statistical theory. Readers who have a good background may find this chapter to be a review of basic theory and methods related to genomic mapping. Most biologists usually have a good background in linear statistics, such as analysis of variance and linear regression. But the statistics used in linear models is not adequate for genomic data analysis, so different kinds of statistics are needed. Biologists who wish to gain some understanding of the theory and concepts of statistical genomics, but who have difficulty reading this chapter, may want to skip this chapter for now and come back to it after reading some of the following chapters.

In the interest of readability and in order to address the needs of a wider audience of biologists, rigorous statistical derivations will not be presented. For rigorous statistics, refer to Lehmann's *Theory of Point Estimation* (1983) and *Test Statistical Hypotheses* (2nd edition, 1983) and to *Kendall's Advanced Theory of Statistics* (Stuart and Ord 1991 and 1994) for general statistical theory and *Categorical Data Analysis* (Agresti 1990) for discrete data analysis. Some detailed methodology will be explained with genomic examples in the next several chapters. The purpose of this chapter is to present readers with a broad perspective on some statistics useful in genomic analysis.

4.2 DISTRIBUTIONS

Statistics is based on probability distributions. Parametric statistics are largely based on theoretical distributions and nonparametric statistics may rely on both theoretical and empirical distributions. In this section, definitions of sample space, random variable, probability distribution and some standard distributions commonly used in genomic data analysis will be introduced.

4.2.1 DISTRIBUTIONS

Example: Data from Mendel

Mendel (1866) crossed inbred pea lines that differed in seed characteristics. One line produced round seeds and the other had wrinkled seeds. When they were crossed, all the resulting seeds (the F1 diploid progeny) were round. The plants that resulted from these seeds were crossed to produce the F2 progeny (seeds on the F1 plants). Both round and wrinkled seeds were observed on all the F1 plants (the seeds themselves being the F2 progeny)(Table 4.1). Mendel hypothesized that the seed characteristic was controlled by a single gene, locus A. Locus A had two alleles: A (dominant, round) and a (recessive, wrinkled). The original cross was between inbred lines AA x aa and resulted in all the F1 progeny being Aa and round. The F1 plants produced two kinds of gametes, A and a. In the F2 generation, there were four possible outcomes for the A locus: AA, Aa, aA and aa. They resulted in two possible seed characteristics, round (AA, Aa and aA) and wrinkled (aa).

Table 4.1 Mendel's experiment on seed character.

Plant	Round Seed		Wrinkled Seed	
	Count	Expectation	Count	Expectation
1	45	42.75	12	14.25
2	27	26.25	8	8.75
3	24	23.25	7	7.75
4	19	21.75	10	7.25
5	32	32.25	11	10.75
6	26	24.00	6	8.00
7	88	84.00	24	28.00
8	22	24.00	10	8.00
9	28	25.50	6	8.50
10	25	24.00	7	8.00
Total	336	327.75	101	109.25

Distributions

When one seed (F2 progeny) is observed on an F1 plant, several outcomes could occur. The seed could have one of four possible genotypes. The four possible genotypes describe the sample space for the genotypes of the F2 progeny. Table 4.1 presents outcomes from these 10 F1 plants. The two seed characteristics, round and wrinkled, define a sample space for the phenotypes of the F2 progeny. Each of the characteristics in the sample space can define an event. An event consists of a set of one or more outcomes from the sample space. An event is said to have occurred if one or more of the outcomes that define the event in a random experiment is observed. If X records the number of dominant A alleles in a randomly chosen genotype then X is a random variable, with sample space given by

$$\Omega = \{0, 1, 2\}$$

The outcomes in the sample space correspond to the events

$$X = \begin{cases} 0 & \text{if} & aa \\ 1 & \text{if} & Aa, aA \\ 2 & \text{if} & AA \end{cases}$$

X is defined as a random variable in the sample space. Any function of X is also a random variable. For example

$$Z = g(X) = \begin{cases} 0 & \text{if} & aa\ (X = 0) \\ 1 & \text{if} & AA, Aa, aA\ (X > 0) \end{cases}$$

is a random variable. Z is a variable for seed character phenotype.

If X is a random variable with a finite countable set of possible outcomes $\{x_1, x_2, \ldots\}$, then the function $p_X(x_i)$, defined by

$$p_X(x_i) = \begin{cases} Pr\{X = x_i\} & \text{if} & x = x_i, i = 1, 2, \ldots \\ 0 & \text{if} & x \neq x_i \end{cases} \qquad (4.1)$$

where

$$\sum_i p_X(x_i) = 1 \qquad (4.2)$$

is called the discrete probability distribution of the random variable X.

Cumulative Distribution

The cumulative distribution function of the random variable X is defined as

$$Pr\{X \leq x_i\} = F(x_i) = \sum_{x \leq x_i} Pr\{X = x_i\} \qquad (4.3)$$

If a random variable X can take any value along an interval on the real number line, then the function, $F(x)$, defined by

$$F(x) = Pr\{X \leq x\} = \int_{-\infty}^{x} f(u)\, du \qquad (4.4)$$

is called a continuous cumulative distribution. If the first derivative

$$F'(x) = \frac{dF(x)}{dx} = f(x)$$

exists, then

$$f(x) = F'(x)$$

is called the continuous probability distribution. For a continuous variable, Equation (4.2) becomes

$$\int_{-\infty}^{\infty} f(x)\,dx = 1 \qquad (4.5)$$

Expectation and Variance

Expectation is an extremely useful concept involving random variables and distribution. For a random variable X, the expectation $E(X)$ is

$$E(X) = \begin{cases} \sum_{x \in \Omega} x_i f(x_i) & \text{discrete} \\ \int_{-\infty}^{\infty} x f(x)\,dx & \text{continuous} \end{cases} \qquad (4.6)$$

and the variance $Var(X)$ is

$$Var(X) = \begin{cases} \sum_{x \in \Omega} [x_i - E(X)]^2 f(x_i) & \text{discrete} \\ \int_{-\infty}^{\infty} [x - E(X)]^2 f(x)\,dx & \text{continuous} \end{cases} \qquad (4.7)$$

For the data in Table 4.1, Z has values

$$Z = \begin{cases} 0 & \text{Round} \\ 1 & \text{Wrinkled} \end{cases}$$

The probability function under the hypothetical genetic model (dominance model) for Z is

$$f(z) = \begin{cases} Pr\{Z = 0\} = 3/4 \\ Pr\{Z = 1\} = 1/4 \end{cases}$$

The expectation of Z for a single observation is $0 \times 3/4 + 1 \times 1/4 = 1/4$, which is the probability that a seed is wrinkled. For each of the plants, the expected frequencies of round and wrinkled seeds, which are the probabilities multiplied by sample size, are given in Table 4.1. The variance of Z is

$$Var(Z) = \sum_i [z_i - E(Z)]^2 f(z_i)$$
$$= (0 - 1/4)^2 \times 3/4 + (1 - 1/4)^2 \times 1/4 = 9/16$$

Joint, Marginal and Conditional Distributions

The concepts of joint, marginal and conditional distributions functions are also important. The joint cumulative distribution, $F(x, y)$, of two random variables X and Y is defined as

$$F(x, y) = Pr\{X \le x, Y \le y\} \qquad (4.8)$$

The marginal cumulative distribution of the random variable X, $F_X(x)$, is defined as

$$F_X(x) = \sum_y Pr\{X \le x, Y \le y\} = F(x) \tag{4.9}$$

which is the distribution of X without regard to Y. The same is true for marginal distribution of variable Y. The joint distribution of X and Y is defined as

$$p(x, y) = Pr\{X = x, Y = y\} \tag{4.10}$$

with

$$\sum_i \sum_j p(x_i, y_j) = 1$$

If X and Y are continuous variables, the joint cumulative distribution is

$$F(x, y) = \int_{-\infty}^{x} \int_{-\infty}^{y} f(u, v)\, du\, dv \tag{4.11}$$

The marginal cumulative distribution for X, $F_1(x)$, is

$$F_1(x) = \int_{-\infty}^{x} \left\{ \int_{-\infty}^{\infty} f(u, v)\, dv \right\} du$$

$$= \int_{-\infty}^{x} f_1(u)\, du = F(x) \tag{4.12}$$

which is the distribution of X without regard to Y. The same is true for variable Y.

The marginal distribution is useful in linkage analysis. As we will see in Chapters 5 and 6, the partitions of hypothesis test statistics are based on marginal distributions. Conditional distributions have been used in linkage analysis for derivation of the EM algorithm, multi-point analysis and QTL mapping and other applications (see Chapters 6, 7, 10, 13 and 14). The conditional distribution of X, given that Y takes the value of y, is

$$p(x|y) = \frac{p(x, y)}{p(y)}$$

$$= Pr\{X = x | Y = y\} \tag{4.13}$$

$$= \frac{Pr\{X = x, Y = y\}}{Pr\{Y = y\}}$$

Similar definitions can be made for the conditional distribution of Y on X. If X and Y are independent, the conditional distributions are the marginal distributions; they are

$$p(x|y) = p(x)$$
$$p(y|x) = p(y) \tag{4.14}$$

and the joint distribution is the product of the two marginal distributions

$$p(x, y) = p(x)p(y) \qquad (4.15)$$

For example, in Mendel's seed characteristic experiment, the probability that a round seed ($Z = 1$) is a homozygote AA ($X = 2$), conditional on round seed ($Z = 1$) is

$$Pr(X = 2|Z = 1) = \frac{Pr(X = 2, Z = 1)}{Pr(Z = 1)}$$

$$= \frac{1/4}{3/4} = \frac{1}{3}$$

4.2.2 STANDARD DISTRIBUTIONS USED IN GENOMIC ANALYSIS

In the previous section, some properties of probability distribution functions have been described. Advanced mathematics are required to derive the distributions and that is beyond the scope of this book. However, knowledge of some of the standard distributions will help in understanding the statistics underlying linkage analysis.

Binomial, Poisson, Normal and chi-square (χ^2) distributions are commonly used in genomic mapping analysis. Table 4.2 lists their probability density functions (pdf), means, variances and moment generating functions (mgf). The pdf is a function from which probability of a variable at a specific point or in a range can be obtained. Distributions of functions of variables, such as the sum, product and ratio of random variables, are useful to develop other statistical concepts such as variance (see Section 4.6).

Table 4.2 Commonly used probability distributions for linkage analysis

	Probability Density Function	Mean	Variance	mgf ($E[e^{tX}]$)
Binomial	$f(x) = \binom{n}{x}p^x(1-p)^{n-x}$	p	$np(1-p)$	$[1-p+pe^t]^n$
Poisson	$f(x) = e^{-\lambda}\lambda^x/x!$	λ	λ	$\exp[\lambda(e^t-1)]$
Normal	$f(x) = \dfrac{1}{\sqrt{2\pi}\sigma}\exp\dfrac{-(x-\mu)^2}{2\sigma^2}$	μ	σ^2	$\exp\left(\mu t + \dfrac{\sigma^2 t^2}{2}\right)$
Chi-square	$f(x) = \dfrac{1}{\Gamma\left(\dfrac{k}{2}\right)}\left(\dfrac{1}{2}\right)^{k/2}x^{k/2-1}e^{-x/2}$	k	$2k$	$\dfrac{1}{(1-2t)^{0.5k}}$ for $t < 0.5$

Moments and Moment Generating Functions

For a random variable X, its rth moment is defined as the expected value of X^r. The first moment of a random variable is its mean. Variance of a variable is the difference between the second moment and the square of the first moment.

$$Var(X) = E(X^2) - [E(X)]^2$$

The moments are usually obtained using a moment generating function (mgf). If X has a density $f(x)$, the mgf is the expected value of e^{tX}. For a continuous variable, it is

$$mgf(X) \; = \; E\,[e^{tX}] \; = \; \int_{-\infty}^{\infty} e^{tx} f(x)\, dx$$

For a discrete variable, it is

$$mgf(X) \; = \; E\,[e^{tX}] \; = \; \sum_{x} e^{tx} f(x)$$

The rth moment of the variable is the rth derivative of the mgf evaluated at $t = 0$.

The Binomial and Multinomial Distributions

Imagine an experiment where the outcome is random. There are only two possible outcomes in each experiment, either "success" or "failure". The probability of a success is a constant, p, for each experiment, and the outcome of one experiment does not affect the outcome of another experiment. Then, if X is a random variable that counts the number of successes in a fixed number n, of experiments, X is said to be binomially distributed with parameters n and p. The probability function for X is given by

$$f(x;n,p) \; = \; Pr\,(X = x \,|\, n, p)$$
$$= \; \binom{n}{x} p^x (1-p)^{\,n-x}$$

The mgf for a binomial variable is

$$mgf(X) \; = \; \sum_{x=0}^{\infty} e^{tX} \binom{n}{x} p^x (1-p)^{\,n-x}$$
$$= \; [\,1 - p + pe^{t}\,]^{\,n}$$

and the first and second moments are

$$E\,(X) \; = \; \left. \frac{d\,[\,1 - p + pe^{t}\,]^{\,n}}{dt} \right|_{t=0}$$
$$= \; \left. npe^{t} (1 - p + pe^{t})^{\,n-1} \right|_{t=0}$$
$$= \; np$$

$$E\,(X^2) \; = \; \left. \frac{d^2\,[\,1 - p + pe^{t}\,]^{\,n}}{dt^2} \right|_{t=0}$$
$$= \; \left. n\,(n-1)\,(pe^{t})^2 (1 - p + pe^{t})^{\,n-2} + npe^{t} (1 - p + pe^{t})^{\,n-1} \right|_{t=0}$$
$$= \; n\,(n-1)\,p^2 + np$$

The variance is

$$Var\,(X) \; = \; E\,(X^2) - [\,E\,(X)\,]^2$$
$$= \; n\,(n-1)\,p^2 + np - (np)^2$$
$$= \; np\,(1-p)$$

The number of recombinant gametes produced by a heterozygous parent for a two-locus model is a binomial variable. If θ is used to denote the probability that a gamete is a recombinant (this will be defined later as recombination fraction, see Chapter 6), then the probability that the gamete is a non-recombinant is $1 - \theta$. The probability of observing r recombinants and $n - r$ non-recombinants for a sample size of n is

$$Pr\{X = r\} = \binom{n}{r}\theta^r (1 - \theta)^{n-r}$$

This is the binomial probability function with parameters θ and n.

If a three-locus (A,B,C) model is considered, there are four possible kinds of gametes referring to possible crossover combinations: non-recombinants, AB recombinants, BC recombinants and double recombinants for loci ABC. The joint probability distribution function for the random variables that count the numbers in each of the four gametic classes is an extension of the binomial distribution called multinomial distribution. The multinomial distribution is also commonly used in linkage analysis. If X_1, X_2, X_3 and X_4 are random variables that count the number of each gamete and there are n gametes altogether, then the multinomial density function for X_1, X_2, X_3 and X_4 is given by

$$Pr\{X_1 = a, X_2 = b, X_3 = c, X_4 = d\} = \frac{n!}{a!b!c!d!}P_1^a P_2^b P_3^c P_4^d$$

where a, b, c and d are numbers in each of the four gametic classes $(a + b + c + d = n)$ and P_1, P_2, P_3 and P_4 are the probabilities of observing a member of each class, respectively $(P_1 + P_2 + P_3 + P_4 = 1)$.

The Poisson Distribution

The binomial distribution approaches the Poisson distribution when $n \to \infty$ and p is small. In this case, np remains constant. If $np = \lambda$, then

$$\binom{n}{x}p^x (1 - p)^{n-x} \to \frac{e^{-\lambda}\lambda^x}{x!}$$

The Poisson distribution has been used in the development of mapping functions. Crossingover has been considered a Poisson event that is randomly distributed in a genome (see Chapter 10). The mgf of a Poisson variable is

$$mgf(X) = \sum_{x=0}^{\infty} e^{tx}\frac{e^{-\lambda}\lambda^x}{x!} = \exp[\lambda(e^t - 1)]$$

The first and second moments for a Poisson variable are

$$E(X) = \frac{d\exp[\lambda(e^t - 1)]}{dt}\bigg|_{t=0} = \lambda e^t\exp[\lambda(e^t - 1)]\big|_{t=0} = \lambda$$

$$E(X^2) = \frac{d^2\exp[\lambda(e^t - 1)]}{dt^2}\bigg|_{t=0} = (\lambda e^t + 1)\lambda e^t\exp[\lambda(e^t - 1)]\big|_{t=0} = \lambda(\lambda + 1)$$

The variance of the Poisson variable is

$$Var(X) = E(X^2) - [E(X)]^2 = \lambda(\lambda + 1) - \lambda^2 = \lambda$$

So, the variance and the mean (the first moment) of a Poisson variable have the same value.

The Normal Distribution

Many statistical techniques are based on the normal distribution. For example, in genomics, the confidence interval of an estimated recombination fraction (see Chapter 6) can be constructed using a normal approximation. If a random variable X is normally distributed, then its probability density function is

$$f(x) = \frac{1}{\sqrt{2\pi}\sigma} \exp\frac{-(x-\mu)^2}{2\sigma^2}$$

and its mgf is

$$\exp\left(\mu t + \frac{\sigma^2 t^2}{2}\right)$$

Its mean and variance are

$$E(X) = \mu$$
$$Var(X) = \sigma^2$$

The Chi-Square Distribution

The chi-square distribution has been widely used for hypothesis tests in genomics. Commonly used test statistics, such as the log likelihood ratio test statistic (see Section 4.4), have chi-square distributions. The chi-square probability density is

$$f(x) = \frac{1}{\Gamma\left(\frac{k}{2}\right)}\left(\frac{1}{2}\right)^{k/2} x^{k/2-1} e^{-x/2}$$

where k is the degrees of freedom for the distribution. From the mgf (Table 4.2), the mean and variance of a chi-square variable can be obtained

$$\left\{ \begin{aligned} E(X) &= k \\ Var(X) &= 2k \end{aligned} \right.$$

So, the mean of a chi-square variable has the same expected value as its degrees of freedom.

4.3 LIKELIHOOD

Likelihood methods have been used extensively in genetic linkage analysis. These methods were developed by R.A. Fisher, using genetic data as examples (Edwards 1992). In this section, definitions of likelihood, support functions and information content will be introduced.

4.3.1 DEFINITIONS

Suppose that X can take on a set of values $x_1, x_2, \ldots$, with

$$L(\theta) \propto Pr\{X = x|\theta\}$$

where θ is a parameter or a vector of parameters that affect the observed x's. For example if $X \sim Normal(\mu, \sigma^2)$, then we could make probability statements about X, assuming we knew $\theta = (\mu, \sigma^2)$. However, often we observe the x's without knowing θ. If we assume that the x's are a random sample from a known distribution, then we can write down a likelihood for θ given x

$$L(\theta) = L(\theta|x_1, x_2, ..., x_n) = \prod_{i=1}^{n} L(x_i|\theta) \qquad (4.16)$$

where n is the sample size. Finding the most likely θ given the data is equivalent to maximizing the likelihood function. The value of θ which maximizes the likelihood function is called the maximum likelihood estimate of θ and is often $\hat{\theta}$.

In this particular example, as it is in the general case, the likelihood could be considered as a function of the probability of the observed data given the parameters. The logarithm of the likelihood is often used. A binomial variable is used as an example to illustrate likelihood. For the example used in Section 4.2, the likelihood function is the probability distribution function

$$L(\underline{\theta}) = L(n, p) = Pr(X = x|n, p)$$
$$= \binom{n}{x} \theta^x (1-\theta)^{n-x}$$

The logarithm is

$$Log\{L(\underline{\theta})\} = Log\binom{n}{x} + xLog\theta + (n-x)Log(1-\theta)$$

Assuming n is constant, the term

$$Log\binom{n}{x}$$

is invariant with respect to the parameter p, therefore the log likelihood function is

$$Log[L(p)] = xLogp + (n-x)Log(1-p)$$

This is also called a support function $[S(\theta)]$ at point p. Maximizing with respect to p gives us

$$\hat{\theta} = \hat{p} = \frac{x}{n}$$

which is the sample proportion.

If the support function is evaluated at the point θ', then the support function for any other point, say θ'', by using the Taylor expansion, will be approximately

$$S(\theta'') = S(\theta') + (\theta'' - \theta')\frac{d[S(\theta')]}{d\theta} + \frac{1}{2}(\theta'' - \theta')^2 \frac{d^2[S(\theta')]}{d\theta^2} + ... \qquad (4.17)$$

This is the basis of the Newton-Raphson iteration to obtain the maximum likelihood estimator.

In practice, both the natural logarithm and the base 10 logarithm have been used to analyze likelihood functions. In this book, natural logarithms are used almost exclusively. It will be clearly indicated when base 10 logarithms are employed. For example, $LogX$ denotes the natural logarithm, while the base 10 logarithm is denoted by $Log_{10}X$.

4.3.2 SCORE

The first derivative of the support function with respect to the parameter is defined as the score. The score for the binomial support function is

$$\frac{d[S(\theta)]}{d\theta} = \frac{x}{\theta} - \frac{n-x}{1-\theta} \qquad (4.18)$$

The score is an important concept for obtaining a maximum likelihood estimator. In practice, the score can be obtained using an equation, such as Equation (4.18) for the binomial support function. When a support function is complex, the score can be obtained numerically by

$$\frac{d[S(\theta)]}{d\theta} \approx \frac{S(\theta + \Delta) - S(\theta)}{\Delta} \qquad (4.19)$$

where Δ is a small value compared to the parameter.

4.3.3 INFORMATION CONTENT

The information content per observation for a single parameter likelihood is

$$\begin{aligned} I(\theta) &= E_\theta\left[\left[\frac{\partial}{\partial\theta}\log L(\theta|x)\right]^2\right] \\ &= -E_\theta\left[\frac{\partial^2}{\partial\theta^2}\log L(\theta|x)\right] \end{aligned} \qquad (4.20)$$

which is -1 times the expectation of the second derivative of the log likelihood function or the support function with respect to the parameter. For a complete explanation of information content, refer to Edward (1992). If the function is evaluated at any arbitrary point, the information is known as the observed information. If it is evaluated at the point of the support function maximum, the information content is defined as the expected information. The expected information content can be used to determine the type of mapping population and sample size needed for designing genomics experiments (see Chapter 6).

4.4 HYPOTHESIS TESTS

Hypotheses are usually the starting points of scientific research. For example, we start with the hypothesis that there is no genetic linkage between the genetic markers and genes when we design a linkage mapping experiment. These kinds of hypotheses may be referred to as biological hypotheses. When data are obtained, the hypotheses must be translated into statistical hypotheses to be tested using statistics. Statistical hypotheses are usually composed of a null hypothesis and an alternative hypothesis. If the null hypothesis is rejected, the alternative hypothesis will be accepted. For linkage experiments, a biological hypothesis can be translated into a group of statistical hypotheses; for example, the null hypothesis for any two-locus linkage experiment could be $\theta = 0.5$ (no linkage) and the alternative could be $\theta = 0.2$ (the two loci are linked with a specified recombination fraction 0.2).

In this section, methods of statistical hypothesis testing, such as chi-square tests, log likelihood ratio tests, the lod score approach and nonparametric approaches, will be explained and the concept of statistical power will be introduced in the context of likelihood support limits and the probability distributions of the test statistics.

4.4.1 METHOD OF HYPOTHESIS TESTING

Goodness-of-fit tests and independence tests are commonly used hypothesis tests in genomic mapping. These tests can be carried out using chi-square statistics, log likelihood ratio test statistics or nonparametric approaches using computer-generated empirical statistical test distributions.

Critical Region

Given a cumulative probability distribution of a test statistic $F(x)$, for example, a chi-square distribution, the critical region for a hypothesis test is defined as the region of rejection in the distribution. The region is the area under the probability distribution where the observed test statistic is unlikely to be observed if the null hypothesis is true. This can be formulated as

$$[1 - F(x)] \leq \alpha \qquad (4.21)$$

for a one-tailed hypothesis test where α is a significance level. This can be formulated as

$$[1 - F(x)] \leq \alpha/2$$
$$or \qquad (4.22)$$
$$1 - \alpha/2 \leq F(x)$$

for a symmetric two-tailed hypothesis test.

The region where the null hypothesis cannot be rejected is called the acceptance region. When the null hypothesis and alternative hypothesis are unidirectional, a one-tailed test is usually used. For example, a null hypothesis is that the parameter is less than a certain value and the alternative hypothesis is that the parameter is greater than the value. In cases like this, if the test statistic falls into the region of Equation (4.21), then the null hypothesis is rejected and the alternative is accepted. When the hypothesis test calls for testing whether two parameters are equal or unequal and bidirectional, a two-tailed test is usually used.

The two-tailed test can be symmetric or non-symmetric. For a significance level α, the critical region for the non-symmetric two-tailed test is

$$[1 - F(x)] \leq a$$
$$or \qquad (4.23)$$
$$1 - b \leq F(x)$$

where

$$0 \leq a \leq \alpha, \ 0 \leq b \leq \alpha \text{ and } a + b = \alpha$$

When $a = 0$ or $b = 0$, the non-symmetric two-tailed test reduces to a one-tailed test.

The values corresponding to the cut-off point between the two critical regions (rejection and acceptance) are usually called critical values. The hypothesis test can be interpreted as the comparison between the critical values and the observed hypothesis test statistic. The rejection regions become

$$x \geq x_\alpha \qquad \text{one} - \text{tailed}$$

$$\left.\begin{array}{c} x \geq x_u \\ or \\ x_l \geq x \end{array}\right\} \qquad \text{two} - \text{tailed}$$

x_α is the critical value for a one-tailed test and x_l and x_u are the lower and upper critical values for a two-tailed test, respectively. For example, when the test statistic is distributed as a Student t variable with 10 degrees of freedom, $\hat{t} \geq 1.812$ can be used to reject a null hypothesis that the population mean is greater than a certain value (one-tailed test) and $-2.228 \geq \hat{t} \geq 2.228$ can be used to reject a null hypothesis that the population mean is not equal to a certain value (two-tailed test). Depending on the values of x_l and x_u, the test can either be symmetric or non-symmetric.

Significance Level
The P-value for a hypothesis test is defined as the probability of observing a sample outcome, assuming that the null hypothesis is true. It can be formulated as

$$P - \text{value} = 1 - F(\hat{x}) \tag{4.24}$$

where $F(\hat{x})$ can be interpreted as the cumulative probability that a statistic is less than the observed test statistic computed for the data under the null hypothesis. If the P-value is equal to or smaller than α, the null hypothesis is rejected at significance level α. The smaller the P-value, the stronger the evidence is to reject the null hypothesis. Chi-square, the normal distribution and F and t distributions are commonly used to obtain the P-values.

The significance level is commonly set at 0.05 to 0.01. However, determining significance levels for many hypothesis tests involved in genomic data analysis is not straightforward, due to the complex probability distributions of the test statistics and the complex nature of the hypothesis tests. The standard hypothesis tests explained in this section are not used directly to determine significance levels for declaring that a group of loci belong to same linkage group, or that a QTL exists (see Chapter 9). However, understanding the standard hypothesis tests is essential for constructing hypothesis tests for complex situations.

Chi-Square Tests
The chi-square probability distribution was given in Table 4.2. Goodness-of-fit tests and independence tests for discrete variables can be formulated as chi-square tests. For a goodness-of-fit test, the test statistic can be calculated using

$$\chi^2 = \sum_{i=1}^{a} \frac{(o_i - e_i)^2}{e_i} \tag{4.25}$$

for one-way classified data, where a is number of classes and o_i and e_i are observed and expected counts for class i, respectively, when the expected values are specified. The expected values are calculated under the null hypothesis. For example, the expected values may be calculated under the assumption of absence of segregation distortion when a hypothesis test is constructed for testing if observed phenotypic frequencies follow certain genetic assumptions. The chi-square statistic is approximately distributed as a chi-square with degrees of freedom $(a-1)$. The rejection region for significance level α is

$$\chi^2 \geq \chi^2_{a-1}(1-\alpha) \qquad (4.26)$$

where $\chi^2_{a-1}(1-\alpha)$ is the critical value for significance level α from a chi-square distribution with degrees of freedom $(a-1)$. The P-value is

$$1 - Pr\{\chi^2_{a-1}(1-\alpha) < \chi^2\} = Pr\{\chi^2_{a-1}(1-\alpha) \geq \chi^2\} \qquad (4.27)$$

When the expected values are not specified and depend on unknown parameters, the degrees of freedom for the distribution of the test statistic equal $a-1-k$ (k is the number of unknown parameters). The expected values can be obtained by replacing the unknown parameters with their estimates, $i.e.$, the maximum likelihood estimates.

A chi-square table is usually included in statistics books. However, the values listed in the tables are usually limited. The P-value can also be obtained from statistical software such as SAS (SAS Institute, 1992).

$$P-\text{value} = 1 - \text{PROBCHI}(\chi^2, v) \qquad (4.28)$$

where PROBCHI is a SAS function to return a cumulative probability from a chi-square distribution and v is the degrees of freedom for the distribution.

When the data can be classified into multiple-way categories, chi-square statistics can be approximately additive and can be partitioned into linear components (Chapters 5 and 6). However, when the classification of the data is not clear, chi-square statistics may not be additive and therefore may not be correctly partitioned.

When two or more variables are involved in an experiment, hypothesis tests can be performed for independence among the variables. A chi-square test statistic may be constructed to test this hypothesis of independence. For two-way classified data, the chi-square test statistic can be computed using

$$\chi^2 = \sum_{i=1}^{a} \sum_{j=1}^{b} \frac{(o_{ij} - e_{ij})^2}{e_{ij}} \qquad (4.29)$$

where a and b are the number of categories for the two variables, respectively and o_{ij} and e_{ij} are the observed and expected counts, respectively. The expected values are obtained using

$$e_{ij} = np_{i\circ}p_{\circ j}$$

where n is total number of observations and $p_{i\circ}$ and $p_{\circ j}$ are marginal probabilities for the two variables, respectively. The chi-square test statistic is distributed as chi-square with degrees of freedom $(a-1)(b-1)$. The test for linkage is one kind of independence test.

Likelihood Ratio Test

The ratio between the likelihood of θ taking a value θ_A which maximizes the likelihood, *e.g.*, a maximum likelihood estimator (see Section 4.5), and the likelihood of θ under the null hypothesis θ_N, *e.g.*, $\theta_N = 0.5$, is

$$\frac{L(\theta_A|x)}{L(\theta_N|x)}$$

Here θ_A is obtained by maximizing the likelihood over the entire parameter space (unconstrained) and $\theta_N = 0.5$ is in a subset of the parameter space (constrained). Twice the natural logarithm of the ratio, denoted by

$$G = 2Log\left[\frac{L(\theta_A|x)}{L(\theta_N|x)}\right] \tag{4.30}$$

is defined as the likelihood ratio test statistic. It is common in statistical texts to define the likelihood ratio test statistic using the null hypothesis in the numerator

$$G = -2Log\left[\frac{L(\theta_N|x)}{L(\theta_A|x)}\right]$$

In this book, the null is used in the denominator to facilitate interpretation. G has an approximate chi-square distribution with degrees of freedom as the difference in the dimensions of the parameter spaces between the two likelihoods, so sometimes it is called the likelihood ratio chi-square. For the goodness-of-fit test, the likelihood ratio test statistic can be computed using

$$G = 2\sum_{i=1}^{n} o_i Log\frac{o_i}{e_i} \tag{4.31}$$

and the statistic for the independence test is

$$G = 2\sum_{i=1}^{a}\sum_{j=1}^{b} o_{ij} Log\frac{o_{ij}}{e_{ij}} \tag{4.32}$$

where the notation is the same as in Equation (4.29). For the conventional goodness-of-fit and independence tests, the chi-square and the log likelihood ratio chi-square provide similar statistical inference in practice. The advantage of using the likelihood ratio approach is due to its power for situations where unknown parameters are involved in the hypothesis tests. For example, to test the hypothesis that an estimated parameter of a binomial variable significantly differs from 0.5, the log likelihood ratio test statistic is

$$G = 2Log\left[\frac{L(\theta_A|x)}{L(\theta_N|x)}\right]$$

$$= 2[S(\hat{\theta}) - S(0.5)]$$

$$= 2[xLog\hat{\theta} + (n-x)Log(1-\hat{\theta}) - nLog0.5]$$

where $\hat{\theta}$ is the maximum likelihood estimate of the binomial parameter. If $\hat{\theta}$ and x are replaced with their expectations or parametric values, we have

$$E(G) = n\{2[\theta Log\theta + (1-\theta)Log(1-\theta) - Log0.5]\}$$

which is defined as the expected likelihood ratio test statistic for a sample of size n and parameter θ. The part in the bracket

$$2\left[\theta Log\theta + (1 - \theta) Log (1 - \theta) - Log 0.5\right]$$

is the expected likelihood ratio test statistic contributed by a single observation.

The Lod Score Approach

For convenience, human geneticists have used the lod score for their hypothesis tests of genetic linkage. The term lod is an abbreviation for log of the odds. The lod score, denoted by Z, is defined as the base 10 logarithm of the likelihood ratio

$$Z = Log_{10}\left[\frac{L(\theta_A|x)}{L(\theta_N|x)}\right] \tag{4.33}$$

For the test of the binomial example, the lod score is

$$Z = Log_{10}\left[\frac{L(\theta_A|x)}{L(\theta_N|x)}\right]$$

$$= xLog_{10}\hat{\theta} + (n - x) Log_{10} (1 - \hat{\theta}) - nLog_{10}0.5$$

and the expected lod score is

$$E(Z) = n\left\{\theta Log_{10}\theta + (1 - \theta) Log_{10} (1 - \theta) - Log_{10}0.5\right\}$$

and

$$\theta Log_{10}\theta + (1 - \theta) Log_{10} (1 - \theta) - Log_{10}0.5$$

is the expected lod score contributed by a single observation.

The interpretation of a lod score, Z, is that the alternative hypothesis is 10^Z times more likely than the null hypothesis. Significant P-values obtained by the likelihood ratio test should be close to 10^{-Z}. There is a one-to-one transformation between lod score and the log likelihood ratio test statistic

$$Z = 0.2172G \tag{4.34}$$

G is distributed as a chi-square, which is more commonly used in biology. Interpretations using lod scores and chi-squares could be different for some cases (Table 4.3). The lod score has been used extensively in genomic analysis.

Table 4.3 A comparison between the statistical significance using the lod score approach and the chi-square approach. The significant P-value for the lod score approach is 10^{-Z}.

Lod Score (Z)	10^{-Z}	$\hat{\chi}^2$	$P(\chi^2 \geq \hat{\chi}^2)$
1	0.1	4.6052	0.031876
2	0.01	9.2103	0.002407
3	0.001	13.8155	0.000202
4	0.0001	18.4207	0.000018
5	0.00001	23.0259	0.000002

Nonparametric Hypothesis Test

For some complex situations, the test statistics may not follow any standard probability distributions. The complexity of some hypothesis tests may be caused by uncertainty about the probability distribution of the data and by nonindependent multiple tests. For example, hypothesis tests of locus orders have complex test statistic distributions. To solve these types of problems, empirical probability distributions of the test statistics are needed. These distributions are usually obtained assuming the null hypothesis using re-sampling techniques such as bootstrap and jackknife methods or permutation methods. A detailed description of resampling and computer simulation is presented in Chapter 19. Let us now assume that an empirical cumulative distribution of the test statistic (x)

$$\widehat{CDF}(x) = P[Y \le x]$$

is obtained under the null hypothesis. If the test statistic calculated from the data is Y', then the rejection region for the nonparametric hypothesis test can be formulated as

$$x_{1-0.5\alpha} \le Y'$$
$$or$$
$$Y' \le x_{0.5\alpha}$$

(4.35)

for a two-tailed symmetric hypothesis test, where α is the significance level for the hypothesis test and $x_{1-0.5\alpha}$ and $x_{0.5\alpha}$ are the $100(1-0.5\alpha)th$ and $100(0.5\alpha)th$ percentiles from the empirical distribution, respectively. Single-tailed or non-symmetric two-tailed hypothesis tests can be constructed in a similar manner.

It is common to believe that many parameters are an inconvenience that can be ignored by using nonparametric statistics. This is not entirely true. Empirical distributions of test statistics are usually obtained under some kind of parametric model. Empirical distributions can be considered parametric distributions with modifications.

4.4.2 THE POWER OF THE TEST

Probability of False Positive and False Negative Errors

Two kinds of error are associated with hypothesis tests (Table 4.4). The probability of a false positive conclusion, usually denoted by α, is the probability that the null hypothesis H_0 will be rejected when it is true. α is the proportion of the test statistic values that fall into the rejection region when the null hypothesis is true and the tests are repeated in infinite replications. For example, a false positive occurs if a linkage between two genes is declared when they are truly unlinked. Figure 4.1 shows three chi-square distributions. Let us look at the distribution under the null hypothesis ($\theta = 0.5$). It is a central chi-square distribution with one degree of freedom. The expected log likelihood ratio test statistic is

$$E(G) = 2n\{\theta Log\theta + (1-\theta)Log(1-\theta) - Log0.5\}$$
$$= 2n\{0.5Log0.5 + 0.5Log0.5 - Log0.5\} = 0$$

(4.36)

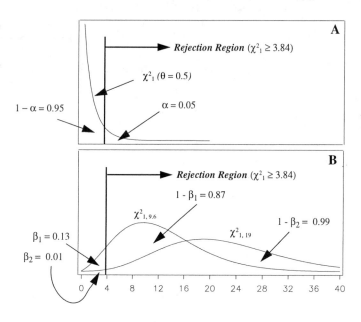

Figure 4.1 Chi-square distributions for the null hypothesis with recombination fraction 0.5 and for the alternative hypothesis with recombination fraction 0.2 with sample size 25 ($\theta = 0.2$, $n = 25$) ($\chi^2_{1, 9.6}$) and 50 ($\theta = 0.2$, $n = 50$) ($\chi^2_{1, 19}$).

If a significance level 0.05 is chosen, then the probability that the test statistic falls into the rejection region is 0.05 when the null hypothesis is true. At the same time, the probability of not rejecting the null hypothesis is 0.95.

Table 4.4 Two types of errors associated with hypothesis tests.

Fact	Hypothesis Test Result	
	Accept Null Hypothesis	Reject Null Hypothesis
Null Hypothesis Is True	$1 - \alpha$	false positive = type I error = α
Null Hypothesis Is False	false negative = type II error = β	power of the test = $1 - \beta$

The probability of a false positive is the same as the significance level when the distribution of the test statistic is used to obtain the P-value. However, when the distribution of the test statistic is not used for obtaining the P-value, the false positive probability and the significance level can differ significantly. Choosing the appropriate distribution is critical for accurate estimates of the probability of a false positive conclusion and for a valid hypothesis test.

The probability of a false negative is defined as the probability, denoted by β, of accepting a null hypothesis when it is false. It is the proportion of the test statistics that fall into the acceptance region when the null hypothesis is false and the tests are repeated in infinite replications. For example, the hypothesis that there is no linkage between two genes is accepted when they are truly linked. Figure 4.1B shows three chi-square distributions for the alternative hypothesis with $\theta = 0.2$ and sample sizes of 25 and 50. The distribution for $\theta = 0.2$ and sample size 25 is a non-central chi-square with one degree of freedom and the non-centrality parameter of

$$
\begin{aligned}
E(G) &= 2n\{\theta Log\theta + (1-\theta)Log(1-\theta) - Log0.5\} \\
&= 50\{0.5Log0.2 + 0.8Log0.8 - Log0.5\} \\
&= 9.6
\end{aligned}
$$

If a significance level 0.05 is chosen, then the rejection region is

$$
\chi^2_{1,9.6} \geq 3.84
$$

As shown in Figure 4.1B, there is a probability of 0.13 that the log likelihood ratio test statistic falls in the acceptance region. 0.13 is the probability of a false negative when $\theta = 0.2$ is true and the sample size is 25. When the sample size increases to 50, the noncentrality parameter becomes 19 and the probability of a false negative is reduced to 0.01.

The Power of the Test

The power of the test is defined as the probability of rejecting a null hypothesis when the alternative is true. It is also commonly called the statistical power. The power is

$$
\text{power} = 1 - \beta \tag{4.37}
$$

The power is defined only when the alternative hypothesis is defined (and is therefore related to it), the experimental conditions are certain and the significance level for the test is chosen. For the binomial example, the powers for rejecting $\theta = 0.5$ at the significance level $\alpha = 0.05$ are 0.87 and 0.99 for the true $\theta = 0.2$ with sample sizes 50 and 100, respectively (Figure 4.1B).

In practice, the statistical power has been used for determining the sample size needed for certain linkage detection power. Here, linkage detection power is the statistical power for linkage detection. To do this usually requires computing what is called the statistical power. For hypothesis tests using chi-square, the power is

$$
\text{power} = Pr[\chi^2_{df,c} \geq \chi^2_\alpha] \tag{4.38}
$$

where χ^2_α is the critical value to reject a null hypothesis at significance level α and $\chi^2_{df,c}$ is a deviate from a non-central chi-square distribution with the degrees of freedom df and non-centrality parameter c.

The critical value can usually be obtained from a chi-square table, for example, the critical value is 3.84 for rejecting a null hypothesis at 0.05 when $df = 1$. For any arbitrary significance level α and degrees of freedom df, the critical value can be obtained using common statistical software. For example, the critical value χ^2_α is

$$
\chi^2_\alpha = CINV(1 - \alpha, df)
$$

using SAS software (SAS Institute, 1992), where $CINV$ is a SAS function to return a quantile from the chi-square distribution.

The non-centrality parameter is the expectation of the log likelihood ratio test statistic under the alternative hypothesis and the specific experimental conditions. For example, 9.6 is the expected log likelihood ratio test statistic when the true parameter for a binomial variable is 0.2, the sample size is 50 and the null hypothesis of the test is H_0:parameter = 0.5. The cumulative probability can be obtained using SAS software by

$$\text{power} = 1 - \text{PROBCHI} \left(\chi^2_\alpha, df, c \right)$$

If computer software is not available, the cumulative probability from the non-central χ^2 distribution can be obtained approximately using

$$Pr\left[\chi^2_{df,\,c} \geq \chi^2_\alpha \right] \approx Pr\left[\chi^2_{df''} \geq \frac{\chi^2_\alpha}{(1+a)} \right] \qquad (4.39)$$

where

$$a = \frac{c}{df + c} \text{ and}$$

$$df'' = \frac{(df + c)}{(1 + b)}$$

Equation (4.39) converts the non-central chi-square cumulative probability approximately into a central chi-square problem. The latter is usually available from a standard chi-square table.

The statistical power can also be obtained using a nonparametric approach such as bootstrapping (see Chapter 19). This involves construction of an empirical distribution for the test statistic under the alternative hypothesis. Then, the statistical power can be the simple count of the statistics which are equal to or greater than the critical value under the null hypothesis.

It is common to obtain statistical powers for a series of experimental conditions before designing an experiment using computer simulation (see Chapter 19 on computer simulations). The statistical power can be an informative guideline for designing an experiment.

4.5 ESTIMATION

Statistical inference includes hypothesis testing and parameter estimation. Hypothesis testing is usually considered to be a qualitative inference. It gives a P-value for the evidence for or against a statistical hypothesis, which is based upon a biological hypothesis. For some experiments, a hypothesis test may be sufficient. However, for many experiments, it is not enough. For example, just knowing that two genes are linked is not sufficient for an experiment with a goal of cloning one or both of those genes. The next logical step is to estimate the parameters and to obtain quantitative inferences. The maximum likelihood (ML) method is the most widely used in genomic data analysis. ML will be the main focus of this section and the method of moments and least squares will be mentioned only briefly.

4.5.1 MAXIMUM LIKELIHOOD POINT ESTIMATION

Let us assume that a random variable, x, has a probability distribution of $f(x;\theta)$ and an unknown parameter θ. The purpose of the maximum likelihood

(ML) point estimation is to estimate the unknown parameter θ based on a random sample, say $X_1, X_2, ..., X_n$, from the population. Interval estimation determines the upper and lower limits for which the probability that the interval contains the unknown parameter can be determined.

The ML estimator is defined as the estimator $\hat{\theta}$ of the parameter θ, which maximizes the likelihood function with respect to the parameter. Commonly used methods to obtain $\hat{\theta}$ are

(1) analytical approach by solving equation $dL(\theta)/d\theta = 0$ or $dS(\theta)/d\theta = 0$ when simple solutions of θ exist
(2) grid search or likelihood profile approach
(3) Newton-Raphson iteration methods
(4) EM (expectation and maximization) algorithm

The basic methodology will be described in this section. The detailed procedures used in genomic data analysis will be given in Chapters 6, 7 and 9, when ML estimation is revisited. In analysis of human genetic linkage, geneticists have used other methods, such as the maximum lod score methods, to obtain the ML estimator. These methods are usually direct analogs of the above methods and therefore will not be discussed in detail here.

4.5.2 ANALYTICAL APPROACH OBTAINING ML ESTIMATOR

For many ML estimation problems, a simple solution for the ML estimator can be obtained by solving

$$\frac{dL(\theta)}{d\theta} = 0 \qquad or$$

$$\frac{dS(\theta)}{d\theta} = 0 \tag{4.40}$$

where $S(\theta)$ has been defined as the support function, which is the natural logarithm of the likelihood function $L(\theta)$. The reasons for taking the logarithm of the likelihood function are

(1) the logarithm of the likelihood reaches its maximum for the same value of θ as the likelihood
(2) the logarithm is usually easier for computation and mathematical manipulation
(3) a close relationship exists between the statistical properties of the maximum likelihood estimator and the logarithm of the likelihood

For the binomial example, the equation is

$$\frac{dS(\theta)}{d\theta} = \frac{x}{\theta} - \frac{n-x}{1-\theta} = 0$$

The first derivative of the support function is also known as the score for the likelihood. The solution of the parameter θ is the ML estimator

$$\hat{\theta} = \frac{x}{n}$$

This approach is effective for many simple and single parameter estimation problems. However, there are many more problems for which simple solutions cannot be obtained.

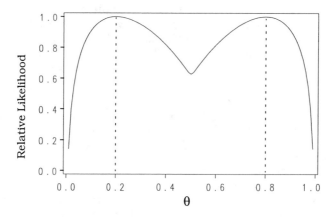

Figure 4.2   Relative likelihood of Equation (4.41) over the parameter space $0 \leq \theta \leq 1$.

4.5.3 GRID SEARCH TO OBTAIN ML ESTIMATOR

When the ML estimator has no simple solution, the estimator can usually be obtained by plotting likelihood or log likelihood against the parameter in the parameter space. For convenience, a relative likelihood, which is likelihood divided by the maximum likelihood (set maximum likelihood = 1), is commonly used. The peak of the relative likelihood can be visually identified or obtained using a searching algorithm. For example, consider the following log likelihood.

$$S(\theta) = Log[\theta^{20}(1-\theta)^{80} + \theta^{80}(1-\theta)^{20}] \qquad (4.41)$$

It is difficult to obtain a simple solution for the parameter θ. However, the likelihood can be numerically evaluated over the ranges of the parameter space $0 \leq \theta \leq 1$. The relative likelihood can be plotted against θ over the range of $0 \leq \theta \leq 1$ (Figure 4.2). The likelihood has two peaks and is symmetrical around $\theta = 0.5$. This is the likelihood profile for the well known mixed linkage phase problem in linkage analysis. If we constrain $0 \leq \theta \leq 0.5$, the maximum likelihood estimate $\hat{\theta}$ is 0.2, which is the recombination fraction between two genes with a possible mixed linkage phase.

This approach to obtain the ML estimator is also sometimes called the profile likelihood method. This method can be implemented using a graphic approach and can also be carried out numerically. Starting with an initial estimate of θ, the direction of the search is determined by evaluating the likelihood for values at both sides of θ. The search goes in the direction of the increase of the likelihood. To save effort, the increment for each step of the search can be large in the first round of the search, such as 0.1. The search stops when the likelihood starts decreasing and the search continues in the upwards direction with reduced increment. Several rounds of the search are repeated using smaller increments, until an appropriate precision is reached.

The numerical approach of a grid search has been effective for problems with single peak likelihood curves. When the likelihood curves have multiple peaks, this approach cannot produce estimates of global maximum likelihood if a large increment is used for the search. If the increment is globally small, the computation needed is intensive.

A grid search can be applied to multiple parameters. However, when the number of parameters is more than two, interpretation of the likelihood profile graph becomes more difficult. For a two-parameter problem, the likelihood profile graph can be constructed as a three-dimensional plot with the relative likelihood as the vertical axis and each of the parameters as one of the horizontal axes.

Example: Mapping a Gene for Resistance to Fusiform Rust Disease

Table 4.10 shows experiment data that was used to show a linkage relationship between a RAPD marker (J7)(see Chapter 3) and a fusiform rust resistance gene (F) in loblolly pine. Fusiform rust resistance in loblolly pine is a penetrance trait with incomplete penetrance. For details on the mating scheme and the experiment designs, see Wilcox *et al.* (1996). The following figure shows a genetic model for the experiment. J7 is a RAPD marker with dominant inheritance. Fusiform rust resistance has dominant inheritance in the host plant. DNA samples for genotyping the RAPD marker were obtained from the haploid megagametophyte tissues, which have genotypes identical with the maternal gametes from which they are derived.

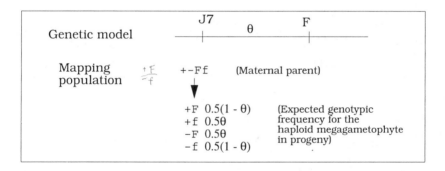

In Table 4.5, the proportion of escapes is ρ and the recombination fraction between the gene and the marker is θ. Escapes are individuals which are genetically susceptible to the disease, but show no disease phenotype under these experimental conditions, due to chance or the conditions of the experiment. $1 - \rho$ is known as penetrance for the disease trait. It is the probability that an individual with the susceptible genotype has the disease phenotype. The purpose of this experiment is to estimate the recombination fraction between the J7 marker and the resistance gene F, using the proportion of escapes in the model. This data set will be used to illustrate the grid search approach, the Newton-Raphson iteration, the EM algorithm for multiple-parameter ML estimation and the moment method.

The support function for the parameters in Table 4.10 is

$$S(\theta, \rho) = 168\log(1 - \theta + \theta\rho) + 3\log(\theta - \theta\rho)$$
$$+ 52\log(\theta + \rho - \theta\rho) + 163\log(1 - \theta - \rho + \theta\rho)$$

By setting the first derivative with respect to θ and ρ equal to zero, we have

$$\frac{-168}{1-\theta+\theta\rho} + \frac{3}{\theta-\theta\rho} + \frac{52}{\theta+\rho-\theta\rho} + \frac{-163}{1-\theta-\rho+\theta\rho} = 0$$

$$\frac{168\theta}{1-\theta+\theta\rho} + \frac{-166}{1-\rho} + \frac{52(1-\theta)}{\theta+\rho-\theta\rho} = 0$$

Simple analytical solutions for θ and ρ cannot be obtained.

Table 4.5 Expected progeny frequency for a marker and a rust resistance gene (F).

G	p(G)	p(P|G)			
		+R	+r	-R	-r
+F	0.5(1 - θ)	1	0	0	0
+f	0.5θ	ρ	1 - ρ	0	0
-F	0.5θ	0	0	1	0
-f	0.5(1 - θ)	0	0	ρ	1 - ρ
	p(P)	0.5(1 - θ + θρ)	0.5(1 - ρ)θ	0.5(θ + ρ - θρ)	0.5(1 - θ - ρ + θρ)
	p(R|P)	θρ/(1 - θ + θρ)	1	θ/(θ + ρ - θρ)	0
	p(E|P)	θρ/(1 - θ + θρ)	0	(1 - θ)ρ/(θ + ρ - θρ)	0
	Count	168	3	52	163

Notes:
G = genotype. *e.g.*, +F is a genotype for the haploid megagametophyte with band for the J7 marker and the resistance allele F. -f is a genotype for megagametophyte without the band and without a resistance gene.
P = phenotype. *e.g.*, +R is a phenotype for megagametophyte with a band for the J7 marker and resistant to the disease. -r is a megagametophyte without the band for J7 marker and susceptible to the disease.
$p(G)$ = the expected genotypic frequency if the marker and the F gene are linked with θ recombination apart.
ρ = probability that an individual with a susceptible genotype is not infected by the disease (escape).
θ = recombination fraction between the marker and the resistance gene F.
$p(P)$ = the expected phenotypic frequency.
$p(R|P)$ = the expected frequency that an individual has a recombination event between the marker and the gene conditional on its phenotype. *e.g.*, We expect that 100% of the individuals with phenotype +r are recombinants.
$p(E|P)$ = the expected frequency that an individual escapes from infection when the individual has a susceptible genotype conditional on its phenotype.
Count = observed count in an experiment with population size 386.

The data were generated by Wilcox *et al.* in the Forest Biotechnology Group at North Carolina State University.

Example: Grid Search

Figure 4.3 shows a three-dimensional plot with the relative log likelihood as the vertical axis and θ and ρ as the two horizontal axes. The relative likelihood is obtained by setting the maximum log likelihood equal to one (all log likelihood values are divided by the maximum log likelihood value). The likelihood reaches a maximum at the point $\theta = 0.02, \rho = 0.22$. Therefore, the maximum likelihood estimates for θ and ρ are

$$\hat{\theta} = 0.02$$

$$\hat{\rho} = 0.22$$

The RAPD marker J7 is linked with the fusiform rust resistance gene F by 0.02 recombination fraction. Among the disease-free individuals, 22% are escapes. The maximum point of the likelihood can be shown more clearly using Figure 4.4. Figure 4.4A shows the plot of the relative likelihood against θ at a fixed point of ρ (0.22) and Figure 4.4B shows a plot of the relative likelihood against ρ at a fixed point of θ (0.02).

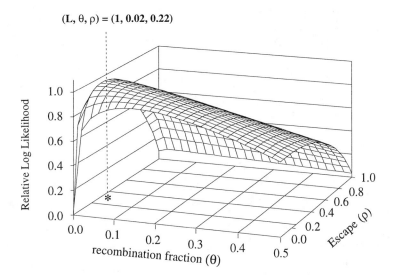

Figure 4.3 A three-dimension plot of relative likelihood against θ and ρ.

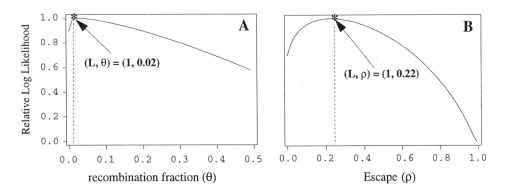

Figure 4.4 (A) Relative likelihood against θ at a fixed point of ρ (0.22) and (B) against ρ at a fixed point of θ (0.02).

4.5.4 NEWTON-RAPHSON ITERATION FOR OBTAINING ML ESTIMATOR

Single Parameter

Newton-Raphson iteration can be implemented to obtain a ML estimator. The equation for obtaining the ML estimator is

$$\text{score}(\theta) = \frac{dS(\theta)}{d\theta} = 0$$

The left side of the equation can be replaced by the linear terms of its Taylor expansion. If θ'' denotes a solution of the equation and θ' is an initial guess of the solution, this leads to the following approximation

$$\frac{d[S(\theta'')]}{d\theta} \approx \frac{d[S(\theta')]}{d\theta} + (\theta'' - \theta')\frac{d^2[S(\theta')]}{d\theta^2} = 0 \qquad (4.42)$$

and

$$\theta'' = \theta' - \frac{\dfrac{d[S(\theta')]}{d\theta}}{\dfrac{d^2[S(\theta')]}{d\theta^2}} \qquad (4.43)$$

The procedure is then iterated by replacing θ' with θ'' and estimating a new θ'' until the iteration converges.

Newton-Raphson iteration has a direct geometrical interpretation. Each iteration fits a parabola for the likelihood function at the point of the parameter. A new parameter is obtained by maximizing the parabola. If the likelihood curve is truly a parabola, then the solution can be obtained by a single iteration.

For most of the common single-parameter problems related to genomic mapping, Newton-Raphson iterations yield maximum likelihood estimates. However, the method may fail when the likelihood curve has multiple peaks or when the information is zero. For problems related to genomic mapping, zero information usually results when the estimate is 0.0 or 1.0 (see Table 6.6). It is suggested that one use different initial values to start the iterations and use bounds for the new estimates.

Multiple Parameters

The Newton-Raphson iteration procedure for multiple parameter estimation is illustrated using the data from Table 4.5. Using matrix notation, scores for the two parameters are

$$S = \begin{bmatrix} \dfrac{\partial S(\theta, \rho)}{\partial \theta} \\ \dfrac{\partial S(\theta, \rho)}{\partial \rho} \end{bmatrix} \qquad (4.44)$$

and elements of this matrix are

$$\frac{\partial S\,(\theta,\rho)}{\partial\theta} = \frac{-168\,(1-\rho)}{1-\theta+\theta\rho} + \frac{3}{\theta} + \frac{52\,(1-\rho)}{\theta+\rho-\theta\rho} - \frac{163}{1-\theta}$$

$$\frac{\partial S\,(\theta,\rho)}{\partial\rho} = \frac{168\theta}{1-\theta+\theta\rho} + \frac{52\,(1-\theta)}{\theta+\rho-\theta\rho} - \frac{166}{1-\rho}$$

The information matrix is

$$I = \begin{bmatrix} -E\{\dfrac{\partial^2}{\partial\theta^2}S\,(\theta,\rho)\} & -E\{\dfrac{\partial^2}{\partial\theta\partial\rho}S\,(\theta,\rho)\} \\[2em] -E\{\dfrac{\partial^2}{\partial\rho\partial\theta}S\,(\theta,\rho)\} & -E\{\dfrac{\partial^2}{\partial\rho^2}S\,(\theta,\rho)\} \end{bmatrix} \qquad (4.45)$$

and its elements are

$$\begin{bmatrix} \dfrac{1}{2}\{\dfrac{(1-\rho)^2}{1-\theta+\theta\rho}+\dfrac{1-\rho}{\theta}+\dfrac{(1-\rho)^2}{\theta+\rho-\theta\rho}+\dfrac{1-\rho}{1-\theta}\} & \dfrac{1}{2}\{\dfrac{-(1-\rho)\theta}{1-\theta+\theta\rho}-\dfrac{(1-\theta)\,(1-\rho)}{\theta+\rho-\theta\rho}\} \\[2em] 0.5\{\dfrac{-(1-\rho)\theta}{1-\theta+\theta\rho}-\dfrac{(1-\theta)\,(1-\rho)}{\theta+\rho-\theta\rho}\} & \dfrac{1}{2}\{\dfrac{\theta^2}{1-\theta+\theta\rho}+\dfrac{(1-\theta)^2}{\theta+\rho-\theta\rho}+\dfrac{1}{1-\rho}\} \end{bmatrix}$$

For the two-parameter problem, Equation (4.43) becomes

$$\begin{bmatrix} \theta'' \\ \rho'' \end{bmatrix} = \begin{bmatrix} \theta' \\ \rho' \end{bmatrix} + \frac{1}{N}I^{-1}S \qquad (4.46)$$

Example: Newton-Raphson Iteration

For the data in Table 4.10, Newton-Raphson iteration was implemented using Equation (4.46). The initial values of the parameters were set to $55/386 = 0.14$, where 55 is the initial guess of the number of recombinants and also the possible number of escapes. The following 4 steps show the computation for the first iteration.

(1) Initialization

$$\begin{bmatrix} \theta' \\ \rho' \end{bmatrix} = \begin{bmatrix} 0.14 \\ 0.14 \end{bmatrix}$$

(2) The first derivatives

$$S = \begin{bmatrix} -160.627 \\ 5.452 \end{bmatrix}$$

(3) The information matrix and its inverse

$$I = \begin{bmatrix} 5.412 & -1.489 \\ -1.489 & 2.013 \end{bmatrix} \text{ and } I^{-1} = \begin{bmatrix} 0.232 & 0.172 \\ 0.172 & 0.624 \end{bmatrix}$$

(4) An update of the estimates

$$\begin{bmatrix} \theta'' \\ \rho'' \end{bmatrix} = \begin{bmatrix} 0.14 \\ 0.14 \end{bmatrix} + \frac{1}{386} \begin{bmatrix} 0.232 & 0.172 \\ 0.172 & 0.624 \end{bmatrix} \begin{bmatrix} -160.627 \\ 5.452 \end{bmatrix} = \begin{bmatrix} 0.0459 \\ 0.0774 \end{bmatrix}$$

The next iteration can be obtained as

$$\begin{bmatrix} \theta' \\ \rho' \end{bmatrix} = \begin{bmatrix} 0.0459 \\ 0.0774 \end{bmatrix}$$

Table 4.6 Newton-Raphson iterations for data in Table 4.10.

Iteration	$\hat{\theta}$	$\hat{\rho}$	$S(\theta, \rho)$
0	0.14	0.14	
1	0.045896	0.077418	-147.039
2	0.12316	0.28776	-147.904
3	0.00001	0.072257	-150.397
...	...	...	-183.617
10	0.022842	0.22526	...
11	0.022425	0.22380	-134.045
12	0.022722	0.22469	-134.044
...	...	...	-134.044
25	0.022618	0.22432	

Table 4.6 shows the complete history of the above iterations. After 25 iterations, the estimates converge to

$$\hat{\theta} = 0.022618$$
$$\hat{\rho} = 0.224320$$

In practice, the iterations after the likelihood reaches a maximum are not needed. For this example, the likelihood reached -134.044 at the 10th iteration. The changes of the estimates after this point are not statistically significant. Using the following likelihood function

$$S(\theta, \rho) = 168\log(1 - \theta + \theta\rho) + 3\log(\theta - \theta\rho)$$
$$+52\log(\theta + \rho - \theta\rho) + 163\log(1 - \theta - \rho + \theta\rho)$$

we have

$$S(\theta, \rho) = \begin{cases} -215.18 & \theta = 0.023, \rho = 0 \\ -265.05 & \theta = 0.5, \rho = 0.22 \\ -134.04 & \theta = 0.023, \rho = 0.22 \end{cases}$$

The null hypotheses of independence between the marker and the gene ($\theta = 0.5$) and no escapes ($\rho = 0$) can be tested using the likelihood ratio test statistics

$$\begin{cases} 2\,[-134.04-(-265.05)\,] \;=\; 262.02 & \text{for} \quad \theta = 0.5 \\ 2\,[-134.04-(-215.18)\,] \;=\; 162.28 & \text{for} \quad \rho = 0 \end{cases}$$

which follow approximately chi-square distribution with one degree of freedom. Both test statistics are highly significant. So, the null hypotheses are rejected.

4.5.5 EXPECTATION-MAXIMIZATION (EM) ALGORITHM

The EM algorithm is another iterative approach for obtaining a maximum likelihood estimator (Dempster *et al.* 1977). The EM algorithm has been used in genomic data analysis and can be considered an iterated counting algorithm (Smith 1957; Ott 1978; Lander and Green 1987 and 1991; Morton and Collins 1990). It has been powerful for obtaining the maximum likelihood estimator when the observations have incomplete data. Most genomic data fits this situation. For example, in linkage analysis using marker genotypes of an F2 progeny, usually only nine categories can be observed for a two-locus and two-allele model. However, there are 16 categories for complete information and 14 categories that provide information about linkage between the two loci. For linkage analysis, 5 categories are hidden (not observed experimentally). However, if the linkage parameter is known, the expected frequencies for the hidden categories can be predicted and the complete data can be restored by means of expectation. The 14 categories for the linkage analysis in an F2 progeny will be sufficient for complete data. This principle is fundamental to the EM algorithm.

Each iteration of the EM algorithm involves two steps: the expectation step (E) and the maximization step (M). The expectation step estimates the statistics for the complete data given the observed incomplete data. The maximization step takes the estimated complete data to obtain the maximum likelihood estimate. By iterating the E and the M steps, the maximum likelihood estimates are reached when the iteration converges.

In practice, the EM can be implemented into the following steps:

(1) Make an initial guess, θ'. For example, 0.25 can be used as an initial value for recombination fraction estimation.
(2) Expectation step: Using θ' as if it were true, estimate the complete data. This usually needs a distribution of the complete data conditional on the observed data. For example, probability that an individual is a recombinant conditional on the observed genotype, $P_i(R|G)$, for recombination fraction estimation.
(3) Maximization step: Compute the maximum likelihood estimate θ''. For example, the maximization step for recombination fraction estimation may involve only a summation of the number of recombinants divided by the total number of informative observations. The maximum likelihood estimate of recombination fraction for a new iteration using EM algorithm is

$$\theta'' = \frac{1}{N}\sum f_i P_i(R|G) \tag{4.47}$$

where $P_i(R|G)$ is the probability of an individual with category i genotype being a recombinant when θ' is the true recombination fraction and f_i is the observed count for category i. This is the reason that the EM algorithm for linkage analysis is called the counting algorithm.

(4) Iterate the E and M steps until the likelihood converges to a maximum ($\theta'' = \theta'$ or $|\theta'' - \theta'| \leq$ tolerance). For a recombination fraction estimation, for example, the tolerance might be 0.00001.

Now, using a two-parameter example, let us illustrate how the EM algorithm can be implemented for genomic data analysis. Readers will find more examples on using the EM algorithm in genomic data analysis in Chapters 6 and 7. It should be easy to understand the single parameter problems once you go through the two-parameter problem.

Example: EM Algorithm

For the data in Table 4.10, it is not difficult to see the conditional probabilities. For example, the probability that an individual with phenotype +R is a recombinant is

$$p_1(R|P) = \frac{0.5\theta\rho}{0.5(1 - \theta + \theta\rho)}$$
$$= \frac{\theta\rho}{1 - \theta + \theta\rho}$$

(4.48)

where $0.5\theta\rho$ is the expected frequency for the hidden category of a recombinant individual with '+' marker genotype for J7 and disease-free phenotype for fusiform resistance (see Wilcox *et al.* 1996) and $0.5(1 - \theta + \theta\rho)$ is the marginal probability of an individual with the '+' marker genotype and disease-free (resistant) phenotype. Equation (4.48) means that if an individual with '+' marker genotype and disease-free phenotype is observed and θ and ρ are the true recombination fraction between J7 and the F gene and the true probability of escapes, respectively, then there is a $[\theta\rho/(1 - \theta + \theta\rho)]$ probability that the individual is a recombinant and a $1 - [\theta\rho/(1 - \theta + \theta\rho)]$ probability that the individual is a non-recombinant. If the observed count for the individual is $f_1 = 168$, then the count for number of recombinants from this observed category is

$$f_1 p_1(R|P) = \frac{168\theta\rho}{1 - \theta + \theta\rho}$$

Similarly, the conditional probabilities for a recombinant on the other categories and for escapes are obtained and given in Table 4.7.

Table 4.7 An EM iteration when $\theta' = 0.14$ and $(\rho' = 0.14)$.

P	Count (f_i)	p(P|G)	$f_i p$(P|G) $\theta' = 0.14$	p(E|P)	$f_i p$(E|P) $\rho' = 0.14$
+R	168	$\theta\rho/(1 - \theta + \theta\rho)$	3.74	$\theta\rho/(1 - \theta + \theta\rho)$	3.74
+r	3	1	3	0	0
-R	52	$\theta/(\theta + \rho - \theta\rho)$	27.96	$(1 - \theta)\rho/(\theta + \rho - \theta\rho)$	24.04
-r	163	0	0	0	0
Sum	386	-	34.70	-	27.79

The expectation step includes computing the number of recombinants (R) and the number of escapes (E) under the assumption that the initial guesses, θ' and ρ', are true. They are

$$R = \sum_{j=1}^{4} f_j p_j (R|P) = \frac{168\theta'\rho'}{1 - \theta' + \theta'\rho'} + 3 + \frac{52\theta'}{\theta' + \rho' - \theta'\rho'}$$

$$E = \sum_{j=1}^{4} f_j p_j (E|P) = \frac{168\theta'\rho'}{1 - \theta' + \theta'\rho'} + \frac{52(1-\theta')\rho'}{\theta' + \rho' - \theta'\rho'}$$

(4.49)

where the notation can be found in Table 4.10. Let us give an initial value 0.14 for both θ' and ρ'. The values for Equation (4.49) are

$$R = \frac{168 \times 0.14 \times 0.14}{1 - 0.14 + 0.14 \times 0.14} + 3 + \frac{52 \times 0.14}{0.14 + 0.14 - 0.14 \times 0.14} = 34.7005$$

$$E = \frac{168 \times 0.14 \times 0.14}{1 - 0.14 + 0.14 \times 0.14} + \frac{52(1 - 0.14) \times 0.14}{0.14 + 0.14 - 0.14 \times 0.14} = 27.7865$$

The maximization step computes new estimates for the parameters, which are

$$\theta'' = \frac{R}{386} = \frac{34.7005}{386} = 0.89898$$

$$\rho'' = \frac{E}{166 + E} = \frac{27.7865}{166 + 27.7865} = 0.14339$$

where 386 is the total number of recombinants and non-recombinants and $166 + E$ is the expected number of susceptible individuals. Table 4.7 shows the computation for the first iteration. Then, set $(\theta', \rho') = (0.89898, 0.14339)$ for the next iteration.

Table 4.8 shows the complete history of the EM algorithm for the data in Table 4.10 with 0.14 as initial value for both parameters. To illustrate, the EM iteration is repeated until the estimates completely converge. In practice, the iteration can be stopped when the likelihood reaches a maximum. That is the 15th iteration for this example. The results are the same as the results of the Newton-Raphson iteration (Table 4.6). However, the EM algorithm has the advantages of being computationally simple and easy to understand.

Table 4.8 EM algorithm for data in Table 4.10

Iteration	Recombination $\hat{\theta}$	Escape $\hat{\rho}$	Log Likelihood $S(\theta, \rho)$
0	0.14	0.14	
1	0.089898	0.14339	-147.039
2	0.069939	0.16163	-140.620
3	0.055773	0.18266	-137.887
...	...	...	-136.156
12	0.023501	0.22335	...
13	0.023203	0.22368	-134.046
14	0.023005	0.22398	-134.045
15	0.022874	0.22404	-134.044
...	...	...	-134.044
28	0.022619	0.22431	

The EM algorithm is commonly used in genomic data analysis because genomic data is always considered as incomplete data when there are a large number of loci in the data set. The EM algorithm is used for two-locus linkage analysis, for gene ordering and for multiple-point map distance estimation.

4.5.6 MOMENT ESTIMATION

If X is a random variable, the rth moment of the variable (m_r) is defined as

$$m_r = E[X^r] \tag{4.50}$$

A function which is representative of all the moments is called the moment generating function (mgf). The mgf for a random variable with density function $f(x)$ is

$$m(t) = E[e^{tX}] = \int_{-\infty}^{\infty} e^{tx} f(x)\, dx \tag{4.51}$$

for continuous variables and

$$m(t) = E[e^{tX}] = \sum_x e^{tx} f(x) \tag{4.52}$$

for discrete variables. Table 4.2 gives the mgf for some commonly used distributions. For example, the mgf for the binomial distribution is

$$m(t) = [1 - p + pe^t]^n \tag{4.53}$$

where p is the parameter for the binomial distribution. The first moment is the mean of the variable. For a binomial variable, the first moment is

$$m_1 = E[X] = \left.\frac{dm(t)}{dt}\right|_{t=0} = npe^t[1-p+pe^t]^{n-1}\big|_{t=0} = np \tag{4.54}$$

and the second moment is

$$m_2 = E[X^2] = \left.\frac{d^2 m(t)}{dt^2}\right|_{t=0}$$

$$= \{n(n-1)pe^t[1-p+pe^t]^{n-2} + npe^t[1-p+pe^t]^{n-1}\}\Big|_{t=0} \tag{4.55}$$

$$= n(n-1)p^2 + np$$

The expectation or mean of a binomial variable is the first moment, np, and the variance of a binomial variable is

$$Var(X) = E[X^2] - [E(X)]^2 = n(n-1)p^2 + np - (np)^2 = np(1-p) \tag{4.56}$$

The moment method for parameter estimation usually replaces the population moment with the sample moments. For example, the binomial parameter, p, can be estimated by

$$\hat{p} = \frac{1}{n} m_1 = \frac{1}{n} \sum_{i=1}^{n} X_i \tag{4.57}$$

Example: Moment Estimate

For data in Table 4.10, the counts in the four categories can be treated as multinomial variables with parameters c_1, c_2, c_3 and c_4. The probability distribution density is

$$f(c_1, c_2, c_3, c_4) = \frac{n!}{f_1! f_2! f_3! f_4!} c_1^{f_1} c_2^{f_2} c_3^{f_3} c_4^{f_4}$$

with constraints of $c_1 + c_2 + c_3 + c_4 = 1$ and $f_1 + f_2 + f_3 + f_4 = n$. The mgf is

$$\{ c_4 + c_1 e^{t_1} + c_2 e^{t_2} + c_3 e^{t_3} \}^n$$

The first moments for the four variables are

$$m_1(t_1) = n c_1$$
$$m_2(t_2) = n c_2$$
$$m_3(t_3) = n c_3$$
$$m_4(t_4) = n c_4$$

and the moment estimates for the four multinomial proportions are

$$\hat{c}_1 = f_1 / n$$
$$\hat{c}_2 = f_2 / n$$
$$\hat{c}_3 = f_3 / n$$
$$\hat{c}_4 = f_4 / n$$

By setting the moment estimates equal to their expectation, $p(P)$ in Table 4.10, we have the following

$$\hat{c}_1 = 0.5 (1 - \theta + \theta\rho)$$
$$\hat{c}_2 = 0.5 (1 - \rho) \theta$$
$$\hat{c}_3 = 0.5 (\theta + \rho - \theta\rho)$$
$$\hat{c}_4 = 0.5 (1 - \theta - \rho + \theta\rho)$$

By adding the constraint $c_1 + c_2 + c_3 + c_4 = 1$, the above equations reduce to

$$1 - \hat{c}_1 - \hat{c}_4 = 0.5 (2\theta + \rho - 2\theta\rho)$$
$$1 - \hat{c}_1 - \hat{c}_3 = 0.5 (1 - \rho)$$

The moment estimates of the two parameters can be obtained by solving the above equations; they are

$$\hat{\theta} = \frac{1.5n - 2f_1 - f_3 - f_4}{2n - 2(f_1 + f_3)} = \frac{1.5 \times 386 - 2 \times 168 - 52 - 163}{2 \times 386 - 2(168 + 52)} = 0.08$$

$$\hat{\rho} = \frac{1}{n}(2f_1 + 2f_3 - n) = \frac{1}{386}(2 \times 168 + 2 \times 52 - 386) = 0.14$$

The moment estimates of the two parameters differ from the maximum likelihood estimates. The moment method has not been widely used in genomic data analysis even though moment estimators usually have analytical solutions and low (or zero) bias. This may be because the large sample statistical properties of the ML estimators are usually superior to the moment estimators and sometimes the simple solutions of the moment estimators are not unique.

4.5.7 LEAST SQUARES ESTIMATION

Least squares is the commonly used method for estimation in linear models. As one of the goals of genomic mapping is to search for genes controlling quantitative (or complex) traits, linear models have been used for modeling the relationship between molecular markers and the putative genes controlling quantitative traits (see Chapters 12 to 17). Least squares estimation is also used to estimate the multipoint map distance for linkage mapping (see Chapter 10).

A simple way to explain least squares estimation is using matrix notation. Suppose we want to find a relationship between a group of markers and phenotype of a trait (see Chapter 13). The problem can be modeled as

$$Y = X\beta + \varepsilon \tag{4.58}$$

where Y is an $N \times 1$ vector containing observed trait values for N individuals in a mapping population, X is an $N \times k$ matrix of the re-coded marker data, β is a $k \times 1$ vector of unknown parameters and ε is an $N \times 1$ vector of residual errors. The constraint for the model is $E(\varepsilon) = 0$. The error sum of squares is

$$\begin{aligned} \varepsilon'\varepsilon &= (Y - X\beta)'(Y - X\beta) \\ &= Y'Y - 2\beta'X'Y + \beta'X'X\beta \end{aligned} \tag{4.59}$$

The least squares estimate of the unknown parameter β is $\hat{\beta}$, which minimizes $\varepsilon'\varepsilon$. Differentiating $\varepsilon'\varepsilon$ with respect to β, we have

$$\frac{\partial \varepsilon'\varepsilon}{\partial \beta} = -2X'Y + 2X'X\beta \tag{4.60}$$

By setting the first derivative equal to zero and replacing β with $\hat{\beta}$, we have the normal equation

$$X'X\hat{\beta} = X'Y \tag{4.61}$$

The solution to the normal equations are given by

$$\hat{\beta} = (X'X)^{-1}X'Y \tag{4.62}$$

Hypothesis tests for the parameters can be performed using the F-statistic, which is the ratio between the residual sum of squares for the reduced model and the full model. For example, the residual sum of squares for Equation (4.58) is

$$SSE_{full} = Y'Y - \hat{\beta}'X'Y \tag{4.63}$$

This sum of squares has $N - k$ degrees of freedom. If we want to test $H_0 : \beta_i = 0$, the residual sum of squares for the reduced model can be obtained by

$$SSE_{reduced} = Y'Y - \hat{\beta}^{R'}X^{R'}Y \tag{4.64}$$

where the dimensions for $\hat{\beta}^R$ and X^R are $(k-1) \times 1$ and $N \times (k-1)$, respectively. So $SSE_{reduced}$ has $N-k+1$ degrees of freedom. The F-statistic, which is

$$F = \frac{\dfrac{SSE_{reduced}}{N-k+1}}{\dfrac{SSE_{full}}{N-k}} \tag{4.65}$$

distributed as $F(N-k+1, N-k)$, tests the hypothesis.

The linear model in Equation (4.58) is a linear regression model. Linear regression can be performed using statistical software such as SAS, SPSS or BMDP. The normal equations can also be solved using simple matrix languages such as S and IML of SAS.

Nonlinear regression is commonly used to find genes controlling quantitative traits. In some situations, the nonlinear regression can be converted to linear regression by re-parameterizing the model. Methodology for nonlinear regression and the nonlinear regression approaches for QTL mapping will be discussed in Chapter 14.

Another group of regression models, which includes multiple dependent variables, has gained attention for QTL mapping problems involving multiple related traits. This topic will be discussed in Chapter 17.

4.6 STATISTICAL PROPERTIES OF AN ESTIMATOR

How can one judge an estimator as "good" or "bad"? How can one evaluate confidence of an estimator as high or low? To answer these questions, some measurement of the statistical properties of the estimator, such as variance, bias, distribution and confidence intervals, are needed.

4.6.1 VARIANCE OF AN ESTIMATOR

If a number of independent estimates can be obtained for a parameter, the variance of the estimator can be estimated using

$$\hat{\sigma}_{\hat{\theta}}^2 = \sum_{i=1}^{k} \hat{\theta}_i^2 - \left[\frac{1}{k} \sum_{i=1}^{k} \hat{\theta}_i \right]^2 = \frac{1}{k-1} \sum_{i=1}^{k} (\hat{\theta}_i - \bar{\theta})^2 \tag{4.66}$$

where $\bar{\theta}$ is the mean of k independent estimates of the parameter and $\hat{\theta}_i$ is the ith estimate. In practice, instead of k independent experiments, one experiment is usually conducted. Equation (4.66) is usually used in conjunction with definition of $\hat{\theta}_i$.

For a large sample, the variance of an ML estimator can be approximated using

$$\hat{\sigma}_{\hat{\theta}}^2 = \frac{1}{nI(\theta)} \tag{4.67}$$

where $I(\theta)$ has been defined as the information content per observation (Equation (4.20)). This is called the Cramer-Rao lower bound for the variance. If an estimator is unbiased and the sample size is sufficiently large, the variance of the estimator can be estimated using Equation (4.67). Mathematical expressions of the conditions are beyond the scope of this book.

In practice, the Cramer-Rao lower bound can be used to judge the merit of an estimator. If an estimator has a variance which is close to the Cramer-Rao lower bound, the estimator is a "good" unbiased estimator.

The variance may also be estimated empirically using one of the re-sampling techniques, such as the bootstrap or jackknife methods (see Chapter 19).

Example: Variance

For data in Table 4.10, the variances for the two estimated parameters are

$$\hat{\sigma}_{\hat{\theta}}^2 = \frac{1}{0.5n\left[\dfrac{(1-\hat{\rho})^2}{1-\hat{\theta}+\hat{\theta}\hat{\rho}} + \dfrac{1-\hat{\rho}}{\hat{\theta}} + \dfrac{(1-\hat{\rho})^2}{\hat{\theta}+\hat{\rho}-\hat{\theta}\hat{\rho}} + \dfrac{1-\hat{\rho}}{1-\hat{\theta}}\right]} = 0.00014$$

$$\hat{\sigma}_{\hat{\rho}}^2 = \frac{1}{0.5n\left[\dfrac{\hat{\theta}^2}{1-\hat{\theta}+\hat{\theta}\hat{\rho}} + \dfrac{(1-\hat{\theta})^2}{\hat{\theta}+\hat{\rho}-\hat{\theta}\hat{\rho}} + \dfrac{1}{1-\hat{\rho}}\right]} = 0.00099$$

4.6.2 VARIANCE OF A LINEAR FUNCTION

It is also common in statistical genomics to find the variance for linear functions of several estimates. For example, we may be interested in the variance of the sum of $\theta_1, \theta_2, ..., \theta_k$, which is $\sum \theta_i$, for $i = 1, 2, ..., k$. The variance can be estimated from the variance for each individual estimate using

$$var\left[\sum_{i=1}^{k} \theta_i\right] = \sum_{i=1}^{k} var[\theta_i] + 2\sum_{i=2}^{k}\sum_{j=1}^{i} cov[\theta_i, \theta_j] \tag{4.68}$$

If the estimates are independent, the covariances are zero and the variance of the sum is the simple sum of the variances. For example, the variance of the sum of the two estimates is

$$var[\theta_1 + \theta_2] = var[\theta_1] + var[\theta_2] + 2cov[\theta_1, \theta_2]$$

If θ_1 and θ_2 are independent, then

$$var[\theta_1 + \theta_2] = var[\theta_1] + var[\theta_2]$$

The variance of the difference between two estimates is

$$var[\theta_1 - \theta_2] = var[\theta_1] + var[\theta_2] - 2cov[\theta_1, \theta_2]$$

If θ_1 and θ_2 are independent, then

$$var[\theta_1 - \theta_2] = var[\theta_1] + var[\theta_2]$$

For a simple linear function of an estimate, say $c\theta$, the variance is

$$var[c\theta] = c^2 var[\theta]$$

where c is a constant.

4.6.3 VARIANCE OF A GENERAL FUNCTION

For a general function of $\theta_1, \theta_2, ..., \theta_k$,

$$F = f(\theta_1, \theta_2, ..., \theta_k)$$

Its variance is approximately

$$var[F] = var[f(\theta_1, \theta_2, ..., \theta_k)]$$

$$\approx \sum_{i=1}^{k} \left[\frac{dF}{d\theta_i}\right]^2 var[\theta_i] + 2\sum_{i=2}^{k}\sum_{j=1}^{i} \left[\frac{dF}{d\theta_i}\right]\left[\frac{dF}{d\theta_j}\right] cov[\theta_i, \theta_j] \qquad (4.69)$$

This has been called the delta method and is based on the Taylor expansion of the first derivatives. A proof of Equation (4.69) is beyond the scope of this book and can be found in statistical theory books. Equation (4.68) is a special case of Equation (4.69). For a function $f(\theta)$, the variance is

$$var[f(\theta)] \approx \left[\frac{df(\theta)}{d\theta}\right]^2 var[\theta] \qquad (4.70)$$

For example

$$var[\theta^2] \approx \left[\frac{d(\theta^2)}{d\theta}\right]^2 var[\theta] = 4\theta^2 var[\theta]$$

If F is a product $\theta_1\theta_2$, its variance is approximately

$$var[\theta_1\theta_2] \approx \left[\frac{dF}{d\theta_1}\right]^2 var[\theta_1] + \left[\frac{dF}{d\theta_2}\right]^2 var[\theta_2] + 2\frac{dF}{d\theta_1}\frac{dF}{d\theta_2} cov[\theta_1, \theta_2]$$

$$= \theta_2^2 var[\theta_1] + \theta_1^2 var[\theta_2] + 2\theta_1\theta_2 cov[\theta_1, \theta_2]$$

If F is a ratio θ_1/θ_2, its variance is approximately

$$var\left[\frac{\theta_1}{\theta_2}\right] \approx \left[\frac{\theta_1}{\theta_2}\right]^2 \{\frac{var[\theta_1]}{\theta_1^2} + \frac{var[\theta_2]}{\theta_2^2} - \frac{2cov[\theta_1, \theta_2]}{\theta_1\theta_2}\}$$

4.6.4 MEAN SQUARE ERROR (MSE) AND BIAS
The MSE of an estimator is defined as

$$MSE = E(\hat{\theta} - \theta)^2 \qquad (4.71)$$

which is the expectation of the square of the difference between the point estimate $\hat{\theta}$ and the true parameter θ. Equation (4.71) can be extended to

$$MSE = E(\hat{\theta} - \theta)^2 = E\{[\hat{\theta} - E(\hat{\theta})] + [E(\hat{\theta}) - \theta]\}^2$$

$$= E[\hat{\theta} - E(\hat{\theta})]^2 - 2[E(\hat{\theta}) - \theta]E[\hat{\theta} - E(\hat{\theta})] + [E(\hat{\theta}) - \theta]^2 \qquad (4.72)$$

$$= E[\hat{\theta} - E(\hat{\theta})]^2 + [E(\hat{\theta}) - \theta]^2$$

$$= \sigma_{\hat{\theta}}^2 + [E(\hat{\theta}) - \theta]^2$$

where

$$E(\hat{\theta}) - \theta$$

is defined as the bias of the estimator. So, the mean square error of an estimator is the sum of variance of the estimator and the square of the bias. The bias can be positive, negative or zero. If $Bias > zero$, the estimator is an over-estimate and if $Bias < zero$ the estimator is an under-estimate. If $Bias = zero$, the estimator is unbiased or the expectation of the estimator is the parameter. If the estimator is unbiased, the mean square error and the variance for the estimator are the same.

The MSE of an estimator can be estimated using analytical and non-parametric approaches. The analytical approach involves the estimation of the variance for the estimator using the moment method or the Cramer-Rao lower bound for the maximum likelihood method. The biases can be estimated using

$$Bias = E(\hat{\theta}) - \theta = \sum_j \hat{\theta}_j f(x) - \theta \qquad (4.73)$$

for discrete variables and using

$$Bias = E(\hat{\theta}) - \theta = \int_{-\infty}^{\infty} \hat{\theta} f(x)\, dx - \theta \qquad (4.74)$$

for continuous variables.

The bias can be estimated using a nonparametric approach, such as bootstrapping. It is

$$Bias_B = \frac{1}{b} \sum_{i=1}^{b} \hat{\theta}_i - \hat{\theta} \qquad (4.75)$$

where b is the number of bootstrap replications, $\hat{\theta}_i$ denotes the bootstrap estimator for the ith replication and $\hat{\theta}$ is the point estimator from the original example. Bootstrap methods will be explained in Chapter 19.

In practice, whether an estimator is biased or not can be determined in several ways. If an estimator has a variance which is close to the Cramer-Rao lower bound, then the estimator is unbiased. If the mean square error for the estimator is the variance for the estimator, it is close to being unbiased.

4.6.5 CONFIDENCE INTERVAL

The bias of an estimator is defined as the difference between the expectation of the estimator and the true parameter. For an individual estimator, $\hat{\theta}$, even when it is the unbiased estimator, a difference between the estimator and the true parameter still exists due to sampling error. To interpret $\hat{\theta}$, a probability statement around $\hat{\theta}$ is needed, such as

$$Pr[T_1 < \hat{\theta} < T_2] = \gamma \qquad (4.76)$$

where (T_1, T_2) is the confidence interval for the estimator $\hat{\theta}$, T_1 and T_2 are defined as the lower and upper limits, respectively and γ is the confidence coefficient. If $\hat{\theta}$ is an unbiased estimator, it is expected that the probability that the true parameter falls into the interval is γ. If

$$Pr[\hat{\theta} \le T_1] = Pr[\hat{\theta} \ge T_2] = 0.5(1 - \gamma) \qquad (4.77)$$

the interval is called the symmetrical two-sided confidence interval. If

$$Pr[\hat{\theta} \le T_1] \ne Pr[\hat{\theta} \ge T_2]$$

$$\text{and} \qquad\qquad\qquad\qquad\qquad\qquad\qquad\qquad (4.78)$$

$$Pr[\hat{\theta} \le T_1] > 0,\, Pr[\hat{\theta} \ge T_2] > 0$$

the interval is called the non-symmetrical two-sided confidence interval. If

$$Pr[\hat{\theta} \leq T_1] = 0 \quad \text{or}$$

$$Pr[\hat{\theta} \geq T_2] = 0 \tag{4.79}$$

the interval is called the one-sided confidence interval. Both two-sided and one-sided intervals have been used in genomic data analysis, for example, a symmetric two-sided confidence interval for the estimated gene effect and an upper limit for an estimated recombination fraction, which is a one-sided confidence interval.

4.6.6 NORMAL APPROXIMATION FOR OBTAINING A CONFIDENCE INTERVAL

Confidence intervals can be determined using parametric and nonparametric approaches. To construct a confidence interval parametrically, a pivotal quantity is usually needed. The pivotal quantity is defined as a variable which is a function of the parameter and the data, but whose distribution does not depend on the parameter. For example

$$(\hat{\theta} - \theta)/\sigma_{\hat{\theta}}$$

can be a pivotal quantity if $\hat{\theta}$ is normally distributed and $\sigma_{\hat{\theta}}$ is its standard deviation (square root of variance), because it will be distributed as a standard normal distribution with a mean of zero and a variance of one.

An important statistical property of the ML estimator is that it is asymptotically normally distributed. The confidence for an ML estimator can be determined using

$$Pr\left[z_{0.5(1-\gamma)} < \frac{\hat{\theta} - \theta}{\sigma_{\hat{\theta}}} < -z_{0.5(1-\gamma)} \right] = \gamma \tag{4.80}$$

$$Pr[\hat{\theta} - z_{0.5(1-\gamma)}\sigma_{\hat{\theta}} < \theta < \hat{\theta} + z_{0.5(1-\gamma)}\sigma_{\hat{\theta}}] = \gamma$$

where $z_{0.5(1-\gamma)}$ is a normal deviate, for example, it is 1.96 for $\gamma = 0.95$. This method for constructing a confidence interval is sometimes called the method of normal approximation, because the pivotal quantity is approximately distributed as a normal variable.

Example: Confidence Interval

For the example in Table 4.10, the 95% confidence intervals are

$$Pr[0.0226 - 1.96 \times \sqrt{0.000136} \leq \theta \leq 0.0226 + 1.96 \times \sqrt{0.000136}] = 0.95$$

$$Pr[0 < \theta < 0.0455] = 0.95$$

$$Pr[0.224 - 1.96 \times \sqrt{0.00099} \leq \rho \leq 0.224 + 1.96 \times \sqrt{0.00099}] = 0.95$$

$$Pr[0.162 < \rho < 0.286] = 0.95$$

using the normal approximation method. The 95% confidence intervals for θ and ρ are

$$0 \leq \theta \leq 0.0455$$

$$0.162 \leq \rho \leq 0.286$$

4.6.7 A NONPARAMETRIC APPROACH TO OBTAIN A CONFIDENCE INTERVAL

Confidence intervals can also be determined using nonparametric approaches. The nonparametric approaches usually involve construction of an empirical distribution for the estimator. The bootstrap approach is commonly used to obtain the empirical distribution. Bootstrapping and other re-sampling techniques will be explained in Chapter 19. For now, assume a cumulative distribution of the estimator is obtained as

$$\widehat{CDF}(x) = P[\hat{\theta}_b \leq x] \tag{4.81}$$

using the bootstrap approach, where b is the number of bootstrap replications and $\hat{\theta}_b$ denotes the bootstrap estimator. The confidence interval with confidence coefficient γ is

$$\{\widehat{CDF}^{-1}[0.5(1-\gamma)], \widehat{CDF}^{-1}[0.5(1+\gamma)]\} \tag{4.82}$$

This is called the percentile confidence interval. The confidence interval can also be determined using a combination of the nonparametric and parametric approaches. The variance of the estimator can be estimated using bootstrapping and then the confidence interval can be obtained using the normal approximation approach.

The bootstrap approach is effective when the distribution of the estimate is unknown or complex. However, it usually involves more computation than the parametric approaches and may fail when sample size of the original experiment is small.

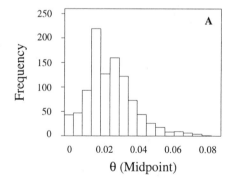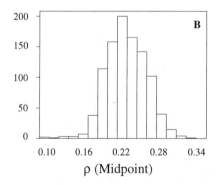

Figure 4.5 (A): The empirical distribution of the ML estimator of the recombination fraction using bootstrapping. (B): The empirical distribution of the ML estimator of the proportion of escapes. The estimates were obtained using the Newton-Raphson iteration, and the number of bootstrap replications is 1000.

Example: Confidence Intervals (Bootstrap Approach)

For an example, 1000 bootstrap samples were drawn from the data in Table 4.10. Figure 4.5 shows the empirical distributions for the bootstrap estimates of recombination fraction between the marker J7 and the rust resistance gene Fr1 and the proportion of disease escapes. Table 4.9 shows the variances, bias and confidence intervals using parametric and bootstrap approaches. The bootstrap variances for the two parameters are close to the parametric variances.

Table 4.9 Statistical properties of ML estimators for the recombination fraction and the frequency of escape from data in Table 4.10.

	Recombination Fraction ($\hat{\theta}$)	Escape ($\hat{\rho}$)
Parametric		
Variance	0.0001357	0.00099
95% Interval	(0, 0.0455)	(0.162, 0.286)
95% Interval (Likelihood approach)	(0.06, 0.056)	(0.17, 0.288)
Bootstrap		
Variance	0.0001666	0.0009025
Bias	0.00008	0.00206
95% Interval (Normal)	(0, 0.048)	(0.1675, 0.2853)
95% Interval (97.5% and 2.5% percentiles)	(0, 0.054)	(0.1815, 0.2826)

The bootstrap biases for the two parameters are small relative to their variances. The ratios between the bias and the standard deviation (square root of variance) for the recombination fraction are

$$0.00008 / \sqrt{0.0001666} = 0.006$$

and

$$0.00206 / \sqrt{0.0009025} = 0.069$$

If the ratio is small, the bias is not important for the estimation relative to the variance. This empirical standard has been used to evaluate the significance of the bias. The confidence intervals obtained parametrically or by using bootstrapping are similar for this example.

4.6.8 A LIKELIHOOD APPROACH FOR OBTAINING A CONFIDENCE INTERVAL

Likelihood approaches are extensively used for obtaining point estimates of parameters and the log likelihood ratio test statistic is commonly used for hypothesis tests in genomic analysis. A confidence interval for a point estimate can also be constructed using a likelihood approach. Consider the following likelihood function

$$L = (1 - \theta)^a \theta^b \qquad (4.83)$$

where θ is the unknown parameter and a and b are observed counts. Four sets of data are observed as

$$
\begin{array}{lll}
A & (a, b) = (8, 2) \\
B & (a, b) = (16, 4) \\
C & (a, b) = (80, 20) \\
D & (a, b) = (400, 100)
\end{array}
\qquad (4.84)
$$

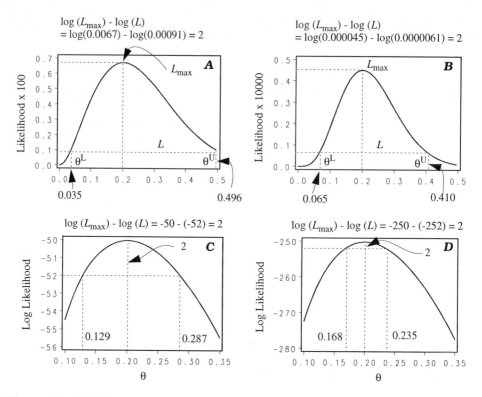

Figure 4.6 Likelihood approach for obtaining confidence intervals. A, B, C and D correspond to the data sets in Equation (4.84) and the likelihood function of Equation (4.85).

Likelihood or log likelihood can be plotted against some possible values of the parameter (Figure 4.6). The maximum likelihood estimate of the parameter is the value corresponding to the peak of the likelihood plots. For the likelihood function of Equation (4.85), the maximum likelihood estimate of the parameter is 0.2 for the four data sets and L_{max} denotes the maximum likelihood value. The confidence interval for the estimate can be obtained using the logarithm of a likelihood and the logarithm of L_{max}. For example, parameter values corresponding to a difference of 2 between the two likelihoods are approximate 95% confidence limits for the parameter (see Chapter 9 in Edwards 1992). In Figure 4.6A

$$L_{max} = 0.0067$$
$$\log L_{max} - \log L = \log(0.0067) - \log(0.00091) = 2 \tag{4.85}$$

and $\theta = (0.035, 0.496)$ and corresponds to $L = 0.00091$. So, 0.035 and 0.496 are the lower and the upper confidence limits for the parameter using the data set A. The likelihood value corresponding to an approximate 95% confidence interval can be determined by manipulating Equation (4.85), resulting in

$$L = 7.389 L_{max} \tag{4.86}$$

It is common to plot the log likelihood against different values of the parameter instead of the absolute likelihood (Figure 4.6C and D). It is easy to obtain the 95% confidence intervals using the four data sets as

$$A \qquad (\theta^L, \theta^U) \; = \; (0.035, 0.496)$$
$$B \qquad (\theta^L, \theta^U) \; = \; (0.065, 0.410)$$
$$C \qquad (\theta^L, \theta^U) \; = \; (0.129, 0.287)$$
$$D \qquad (\theta^L, \theta^U) \; = \; (0.168, 0.235)$$

As sample size increases, the range of the confidence interval decreases and the confidence interval closes to symmetric.

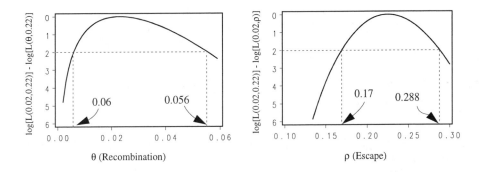

Figure 4.7 Estimation of confidence intervals for the example in Table 4.5 using a likelihood approach. Log likelihood difference of 2 corresponds approximately to a confidence probability of 95%.

Example: Likelihood Approach

For the example in Table 4.10, the log likelihood difference from the maximum likelihood

$$\log L_{max} - \log L \; = \; \log \, [L \, (\theta = 0.02, \rho = 0.22) \,] \, - \log L$$

is plotted against different values of the two parameters (Figure 4.7). The approximate 95% confidence intervals for the two parameters are

$$0.06 \leq \theta \leq 0.056$$
$$0.17 \leq \rho \leq 0.288$$

which differ from the estimates using the normal approximation and the nonparametric approaches (Table 4.9).

4.6.9 LOD SCORE SUPPORT FOR A CONFIDENCE INTERVAL

Instead of a confidence probability or a log likelihood difference, the confidence of an interval is also commonly quantified using lod score support. Lod score support for a confidence interval is the difference in the $\log_{10}$ base of

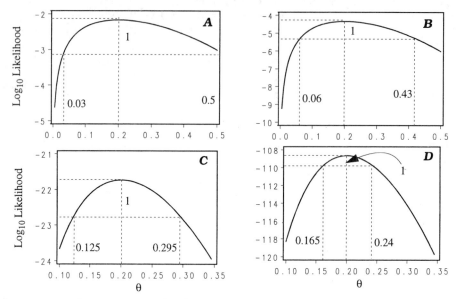

Figure 4.8 Lod score support for the data sets in Equation (4.84).

a likelihood corresponding to the interval from the maximum likelihood. For the likelihood function of Equation (4.83) and the data sets in Equation (4.84), the $\log_{10}$ likelihood is plotted against different values of the parameter (Figure 4.8). By substracting 1 from the maximum $\log_{10}$ likelihood, confidence intervals with a lod score support 1 can be obtained

$$A \qquad (\theta^L, \theta^U) \;=\; (0.03, 0.5)$$
$$B \qquad (\theta^L, \theta^U) \;=\; (0.06, 0.43)$$
$$C \qquad (\theta^L, \theta^U) \;=\; (0.125, 0.295)$$
$$D \qquad (\theta^L, \theta^U) \;=\; (0.165, 0.240)$$

A difference of 1 between two $\log_{10}$ likelihoods is equivalent to a difference of 1 between two lod scores. The ranges of the confidence interval with a lod score support one are wider than the ranges for a confidence probability 95%.

Example: Lod Score Support

For the example in Table 4.10, the $\log_{10}$ likelihood difference from the maximum likelihood

$$Log_{10}L_{max} - Log_{10}L \;=\; Log_{10}\left[L\left(\theta = 0.02, \rho = 0.22\right)\right] - Log_{10}L$$

is plotted against different values of the two parameters (Figure 4.9). The confidence intervals for the two parameters with lod score support one are

$$0.051 \leq \theta \leq 0.058$$
$$0.165 \leq \rho \leq 0.293$$

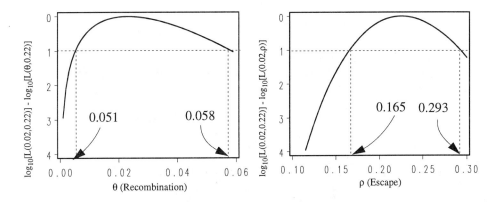

Figure 4.9 Confidence intervals for the example data in Table 4.5 using the concept of lod score support. The indicated values correspond to a lod score support of 1.

4.6.10 WHAT IS A GOOD ESTIMATOR OF A CONFIDENCE INTERVAL

A good confidence interval should have at least two characteristics:

(1) the coverage probability is equal to or does not significantly depart from the stated probability or the confidence coefficient
(2) the interval is biologically meaningful

The coverage probability is defined as the probability that an estimated confidence interval [(T_1, T_2)] covers the parameter. In practice, the coverage probability is usually obtained by computer simulation. If the coverage probability is significantly higher than the stated probability, the interval is wider than the true interval. This may be a result of an over-estimated variance for the estimator or an unappreciated pivotal quantity. If the coverage probability is significantly lower than the stated probability, then the variance may be under-estimated. Both the lower and the higher probabilities are signs of an invalid confidence interval. The biological interpretation of a confidence interval is also an important criterion for the quality of the confidence interval. An interval can be perfect statistically but meaningless biologically, if, for example, the upper or lower limits are out of the range of biologically meaningful information.

4.6.11 WHAT IS A GOOD ESTIMATOR?

Go back to the questions proposed at beginning of Section 4.6. How can one judge an estimator as "good" or "bad"? How can one evaluate confidence of an estimator as high or low?

(1) Consistency: An estimator is called mean square error consistent if

$$\lim_{n \to \infty} E\,(\hat{\theta} - \theta)^2 = 0$$

This means that the expected mean square error for the estimator is zero when the sample size approaches infinity. This also implies that the estimator is a simple consistent estimator, which is

$$\lim_{n \to \infty} Pr\{\theta - \varsigma < \hat{\theta} < \theta + \varsigma\} = 1$$

In practical terms, the estimator is relatively consistent if its variance is sufficiently small. The smaller its variance, the better the estimator will be.

(2) Lack of bias: Absence of bias is a good property for an estimator, that is

$$E(\hat{\theta}) = \theta$$

However, there has been much debate whether an unbiased estimator is always better than a biased estimator. It is commonly agreed that an unbiased estimator may not always be the best estimate. For example, if the variance of an unbiased estimator is greater than that of the biased estimator, the biased estimator could be better. There has also been much discussion on the possibility of obtaining a bias-corrected estimator. However, this topic is beyond the scope of this book. I mention here only that nonparametric approaches such as bootstrap and jackknife methods are good for obtaining a bias-corrected estimator.

(3) Asymptotically Distributed as Normal: If the quantity

$$\frac{\hat{\theta} - \theta}{\sigma_{\hat{\theta}}}$$

is distributed as normal with mean 0 and variance 1 as the sample approaches infinity, the estimator itself $\hat{\theta}$ is said to be asymptotically distributed as normal. This is a good property to have for statistical inferences.

(4) A "Good" Confidence Interval Exists and Can Be Estimated: This is important for making statistical inferences on the estimator. A "good" confidence interval usually can be determined for an ML estimator when the sample size is large.

The above four criteria are based on statistical considerations. The ML estimator for a large sample size has been considered to meet all the four criteria. We can loosely define the ML estimator as the best asymptotically normal estimator (BAN estimator). However, an estimator cannot be judged based only on the statistical properties. Here are two practical considerations:

(5) An estimator should be biologically meaningful. An estimator cannot be a good estimator if the statistical inference about the estimator is biologically meaningless, even though the estimator is perfect in statistical terms.

(6) An estimator should be experimentally and computationally feasible. An estimator is not good if intensive computation and complex analytical reasoning required for obtaining the estimator are not feasible. The estimator is also not good if it is not feasible to acquire the data needed to obtain it.

4.7 SAMPLE SIZE DETERMINATION

To design a genomic mapping experiment, usually mating design (or mapping population type), experimental design (conventional plot design), genetic marker type and sample size are considered. In this section, sample size determination, which has gained a lot attention in statistics, will be discussed. Questions that have been frequently asked are:

(1) What is the statistical power to detect linkage given a progeny size?
(2) What is the precision of the estimated recombination fraction when sample size is N?

The questions can be restated as: What is the sample size needed to have a certain statistical power and precision for the estimator? In this section, some approaches to answer those questions will be explained.

4.7.1 SAMPLE SIZE NEEDED FOR SPECIFIC STATISTICAL POWER

The statistical power for using the likelihood ratio test statistic has been defined as

$$\gamma = 1 - \beta$$
$$= Pr\,[\chi^2_{df,\,c} \geq \chi^2_\alpha] \tag{4.87}$$

where χ^2_α is the critical value to reject a null hypothesis at significance level α and $\chi^2_{df,\,c}$ is a deviate from a non-central chi-square distribution with the degrees of freedom df and non-centrality parameter c. Here, c is usually the expectation of the log likelihood ratio test statistic under the alternative hypothesis and specific experimental conditions.

Given an statistical power (γ), true parameter (θ) and significance level being used (α), the sample size needed can be estimated by manipulating the inequality

$$Pr\,[\chi^2_{df,\,c} \geq \chi^2_\alpha] \geq \gamma \tag{4.88}$$

This implies

$$c \geq (\chi^2_{\gamma,\,df,\,\chi^2_\alpha})^{-1}$$

where c is the expected log likelihood ratio test statistic, which is the expected log likelihood ratio test statistic per observation modified by sample size

$$c = nE\,[G_{unit}] \tag{4.89}$$

and

$$(\chi^2_{\gamma,\,df,\,\chi^2_\alpha})^{-1}$$

is the chi-square value corresponding to probability of γ, df and a non-centrality parameter χ^2_α, which can be obtained using statistical computing packages. For example, it can be obtained using SAS (SAS Institute, 1992)

$$(\chi^2_{\gamma,\,df,\,\chi^2_\alpha})^{-1} = CINV\,(\gamma, df, \chi^2_\alpha)$$

where $CINV$ is a SAS function for obtaining quantiles from the chi-square distribution. The minimum number of observations for statistical power (γ), hypothetical recombination fraction (θ) and significance level (α) is

$$n \geq \frac{(\chi^2_{\gamma, df, \chi^2_\alpha})^{-1}}{E[G_{unit}]} \tag{4.90}$$

These procedures and related examples are contained in later chapters.

4.7.2 SAMPLE SIZE NEEDED FOR A SPECIFIC CONFIDENCE INTERVAL

Using the normal approximation approach, a confidence interval can be calculated. For a 95% confidence coefficient, the confidence interval is

$$\hat{\theta} \pm z_{(1-\gamma)/2}\sigma_{\hat{\theta}} = \hat{\theta} \pm 1.96\sigma_{\hat{\theta}}$$

The range from lower to upper limits

$$d = 3.92\sigma_{\hat{\theta}} \tag{4.91}$$

can be used as a precision measurement for the estimator. Given a true parameter and sample size, the standard deviation for the estimator can be predicted approximately using the expected information content by

$$\sigma_{\hat{\theta}} \approx \sqrt{\frac{1}{nI(\theta)}} \tag{4.92}$$

Given a confidence interval and an true parameter, the minimum sample size can be determined by manipulating the above three equations. These steps are

$$2 \times 1.96\sigma_{\hat{\theta}} \leq d$$

$$3.92\sqrt{\frac{1}{nI(\theta)}} \leq d$$

$$nI(\theta) \geq \left(\frac{3.92}{d}\right)^2$$

$$n \geq \frac{\left(\frac{3.92}{d}\right)^2}{I(\theta)}$$

where d is a target maximum range for a 95% confidence interval and $I(\theta)$ is expected information content per individual. To generalize the interval with confidence coefficient γ, the minimum sample size is

$$n \geq \frac{(2z_{(1-\gamma)/2})^2}{d^2 I(\theta)} \tag{4.93}$$

where $z_{(1-\gamma)/2}$ is the $(1-\gamma)/2$ quantile of a standard normal distribution.

Example: Sample Size Determination

Consider the following likelihood function

$$L = (1-\theta)^a \theta^b$$

where θ is the unknown parameter and a and b are observed counts. For a null hypothesis of $\theta = 0.5$, the expected log likelihood ratio test statistic per observation is

$$E[G_{unit}] = 2[(1-\theta)\log(1-\theta) + \theta\log\theta - \log 0.5]$$

and the average information content per observation is

$$I(\theta) = \frac{1}{\theta(1-\theta)}$$

If the true parameter takes values of 0.05, 0.1, 0.2 and 0.3 respectively, then

θ	G_{unit}	$I(\theta)$
0.05	0.99	21.1
0.10	0.74	11.1
0.20	0.39	6.3
0.30	0.17	4.8

The sample size $(n = a + b)$ needed for a statistical power of 90% using a significance level $\alpha = 0.05$ can be determined using

$$n \geq \frac{(\chi^2_{0.9,\,1,\,3.84})^{-1}}{E[G_{unit}]} = \frac{10.5}{E[G_{unit}]}$$

where 3.84 is the critical value for declaring significance for degrees of freedom 1, $\alpha = 0.05$ and 10.5 is the value of a noncentrality parameter needed for a statistical power of 90%. So

θ	Size
0.05	11
0.10	15
0.20	28
0.30	64

If we want the range of the 95% confidence interval to be equal to or less than the true value of the parameter $(d \leq \theta)$, the sample size needed can be determined using

$$n \geq \frac{(2 \times 1.96)^2}{\theta^2 I(\theta)}$$

where 1.96 is a normal deviate for constructing a 95% confidence interval using the normal approximation approach. So

θ	Size
0.05	292
0.10	139
0.20	62
0.30	36

SUMMARY

In this chapter, some basic concepts of statistics which will be used in genomic analysis have been discussed. The theories include commonly used probability distributions, the concept of likelihood, hypothesis testing, statistical power and point and confidence interval estimations. At the end of this chapter, the basic theory for determining sample sizes was presented. Resampling and computer simulation methods are also commonly used in genomic analysis. However, discussion of these procedures will be delayed until later in of this book (Chapter 19), because it will be easier to explain after introducing more issues in basic genomic analysis.

EXERCISES

Exercise 4.1

Find a maximum likelihood estimate of θ, build a 95% confidence interval for the estimate and test the hypothesis that $\theta = 0.5$ using the following likelihood function and the different values of a, b, c and d.

$$L(\theta) = f_1 \log(3 - 2\theta + \theta^2) + (f_2 + f_3) \log[\theta(2 - \theta)] + 2f_4 \log(1 - \theta)$$

(1) $a = 80, b = 10, c = 10$ and $d = 0$
(2) $a = 90, b = 5, c = 5$ and $d = 0$
(3) $a = 80, b = 9, c = 10$ and $d = 1$

Hints: Simple solutions can be obtained by solving the equation by setting the first derivative of the likelihood function with respect to the parameter equal to zero for the values in (1) and (2). Iterative approaches are needed to obtain a solution for the values in (3). The following table contains the conditional probability needed to use the EM algorithm.

	Probability Contributing to θ
a	$2\theta/(3 - 2\theta + \theta^2)$
b	$1/(2 - \theta)$
c	$1/(2 - \theta)$
d	0

Ex. 4.1 (Hints)

A r B		Genotype	Count
a b		AABb	18
	X →	AAbb	6
		AaBb	23
A r b		Aabb	27
a b		aaBb	5
		aabb	21

Ex. 4.2

Exercise 4.2

A cross between two individuals with genotypes AaBb and Aabb for loci A and B was made (see the above illustration). Assume that the recombination fraction between loci A and B is r.

(1) Derive the expected segregation ratios for single-locus models of A and B and the two-locus model. Using the counts in the illustration, construct a chi-square table to test if A and B follow the expected segregation ratios and if A and B are independent.

(2) Derive a likelihood function for estimating the maximum likelihood estimate of r.

(3) Compute the likelihood ratio test statistic and lod score for testing whether the recombination fraction between A and B is $r = 0.5$ using the likelihood function in (2) and the counts in the illustration. Compare the inferences about linkage using the chi-square approximation in (1), the likelihood ratio test statistic approach and the lod score approach.

(4) Derive the average information content function for the likelihood function and plot the average information content against a range of values of r (e.g., $0 < r < 0.5$). Plot the average information content for the classical backcross model on the same scale. The average information content for the backcross model is $1/[r(1-r)]$. Explain the difference between the average information content in the classical backcross and the mating scheme in this exercise.

(5) Obtain the maximum likelihood estimate of r using the likelihood function derived in (2) and the counts in the illustration. Obtain a 95% confidence interval for the estimate using the normal approximation approach.

(6) Estimate the number of individuals needed to have (a) a significant p-value using the chi-square approximation approach for detecting linkage between A and B that is equal to or less than 0.0001, (b) a significant p-value using the likelihood ratio test statistic approach a detecting linkage between A and B is equal or less than 0.0001, (c) a significant p-value using the lod score approach for detecting linkage between A and B that is equal to or greater than 4 and (d) a 95% confidence limit for the estimated recombination fraction that is equal to or less than 0.03 using the normal approximation approach.

(7) What kind of mating should this scheme be called according to the terms of the classical mating designs (such as backcross and F2)? Can you find an example which has the same mating structure in a real experiment?

Cross 1		Genotype	Count	Cross 2			
A $\,_r$ B		AABb	18	A $\,_r$ B		Genotype	Count
		AAbb	6			AaBb	43
a	b	AaBb	23	a	b	Aabb	5
X	→	Aabb	27	X	→	aaBb	4
A $\,_r$ b		aaBb	5	a $\,_r$ b		aabb	48
a	b	aabb	21	a	b		

Ex. 4.3

Exercise 4.3

Use the information obtained in Exercise 4.2. The above illustration shows two parallel crosses in one experiment. The progeny used in Exercise 4.2 were produced by cross 1 in this illustration. Use the frequencies in this illustration to do the following analyses.

(1) Estimate r independently using the progeny of the two crosses. Estimate r jointly using data from the crosses.

(2) Test the hypothesis that the recombination fraction between A and B is the same for the two crosses.

(3) If the hypothesis test in (2) is significant, can you explain why?

(4) Whether the recombination fraction between A and B is the same or not, it is important for applications of the results of this experiment. You want to reject the null hypothesis at a higher significance level. If the target significant level is 0.0001, how many individuals are needed in the progeny of the two crosses? Assume that the progeny sizes are the same for the two crosses and the true recombination fraction difference is 0.03.

Exercise 4.4

Table 4.10 in this chapter shows experimental data for finding the linkage relationship between a RAPD marker and a fusiform rust resistance gene in loblolly pine. Fusiform rust resistance in loblolly pine can be considered as a trait with incomplete penetrance. Dr. Henry Amerson and co-workers at North Carolina State University and the U.S. Forest Service found ways to eliminate the escapes, or to make the penetrance 100%. However, another problem arose when Dr. Amerson used different fungus genotypes to infect the plants. He hypothesized that some races of the pathogen may not be homozygous for the virulence gene. The following figure shows a hypothetical genetic model for the host and a gene-for-gene resistance model for the experiment.

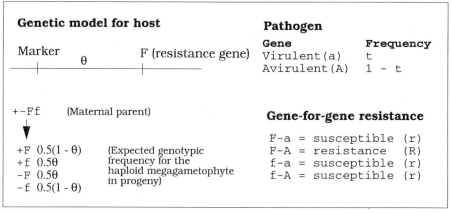

Ex. 4.4

The Table 4.10 shows the expected frequencies and observed counts for the marker and resistant phenotypes in an experiment assuming that the frequency of the virulence gene that overcame the resistance of the host gene F in the pathogen is t. The purpose of this experiment is to estimate the recombination fraction between the marker and the resistance gene F with the the frequency of the virulence gene in the pathogen in the model.

Table 4.10 Expected progeny frequencies for a marker and a fusiform rust resistance gene (F). (Ex. 4.4)

G	p(G)	p(P|G)			
		+R	+r	-R	-r
+F	$0.5(1 - \theta)$	$1 - t$	t	0	0
+f	0.5θ	0	1	0	0
-F	0.5θ	0	0	$1 - t$	t
-f	$0.5(1 - \theta)$	0	0	0	1
	p(P)	$0.5(1 - \theta)(1 - t)$	$0.5(t - t\theta + \theta)$	$0.5(1 - t)q$	$0.5q(1 + t)$
	p(R|P)	0	$\theta/(t - t\theta + \theta)$	1	$\theta v/[\theta(1 + t)]$
	Count	9	12	2	30

(1) Derive a log likelihood function for this data set.

(2) Estimate the recombination fraction between the marker and the gene assuming $t = 0$ and $t = 0.5$, using both EM algorithm and Newton-Raphson iteration approaches.

(3) Test the hypotheses that $t = 0$ and $\theta = 0.5$ using a log likelihood ratio approach.

Exercise 4.5

Maximum likelihood approaches have been used extensively in genomics analysis. List some of the properties of maximum likelihood estimators that are relevant to genomic analysis. Can you find situations in which the maximum likelihood approach cannot be applied? Explain why.

Exercise 4.6

Maximum likelihood estimation and the likelihood ratio test statistics approach for hypothesis testing are based on a well-defined likelihood function for the observed data under null and alternative hypotheses. Explain the consequences if the maximum likelihood estimate is obtained by maximizing a wrong likelihood function. What happens to an hypothesis test under the same situation? Have you ever seen an unappreciated likelihood-based analysis in real data analysis, or can you imagine a situation when the likelihood-based analysis could be incorrectly used?

CHAPTER 5

SINGLE-LOCUS MODELS

In general, a single locus can be characterized by its number of alleles, their frequencies in the population, mode of inheritance and equilibrium status. These characteristics are important, especially when the materials used are natural populations (see Chapters 7 and 14). For controlled crosses, some of the characteristics, such as allelic frequencies and equilibrium status, are specified when the crosses are made.

The purposes of single-locus analyses in genetic mapping are data quality control and single-locus genetic model identification. Given a genetic model, a gene or a genetic marker should follow certain segregation patterns. For example, the heterozygous and homozygous genotypic classes should segregate according to a 1:1 ratio in backcross progeny. AA and Aa genotypic classes segregate according to a 1:1 ratio in backcross progeny of an individual with genotype Aa backcrossed with an individual with genotype AA. In F2 progeny resulting from selfing an individual with genotype Aa, the AA, Aa and aa genotypic classes segregate according to a 1:2:1 ratio. A significant departure (segregation ratio distortion) from the expected segregation ratio may be a sign of the wrong genetic model, low data quality or non-random sampling. For many genomic mapping experiments, screening for genetic markers is part of the experimental strategy. Usually, screening is carried out to determine if the marker genotypes follow expected segregation patterns, such as no segregation or segregation according to a ratio of 1:1, 1:2:1 or 3:1.

For experiments involving more than one population, one purpose of single-locus analysis is to test if genotypic segregation patterns are the same across populations. If the segregation patterns are significantly different, then the populations may have different genetic configurations. If no difference can be detected, then the populations are homogenous in terms of genotypic segregation and can be pooled into one analysis.

Many of the methodologies for genomic analysis have been derived under some assumptions in the single-locus model. For example, commonly used linkage analysis and QTL mapping approaches are derived based on assumption of Hardy-Weinberg equilibrium at a single locus level.

Methods for detecting segregation distortion and determining sample size for marker screening in genomic analysis using controlled crosses will be described in this chapter. Partition of chi-square and likelihood ratio test statistics, which are key concepts for heterogeneity testing among several populations, will be introduced. Statistics for characterizing a single marker in natural populations and the ways to screen markers will be described.

5.1 EXPECTED SEGREGATION RATIOS

5.1.1 SINGLE POPULATION

A test of the expected segregation ratio is a test for goodness of fit. As introduced in Chapter 4, chi-square and log likelihood ratio test statistics can be used. For a simple one way classified data set, the chi-square test statistic is

$$\chi^2 = \sum_{j=1}^{n} \left[\frac{(o_j - e_j)^2}{e_j} \right] \tag{5.1}$$

where n is the number of detectable genotypic classes segregating in the sample and o_j and e_j are the observed and expected counts, respectively. For the majority of genomic mapping data, e_j is the mean of binomial or multinomial distribution.

For a backcross mating, the expected counts for the two genotypic classes in the progeny can be calculated using $0.5n$, where 0.5 is parameter of a binomial distribution. For an F2 mating, the expected frequencies for the two homozygous classes and the heterozygous class are $0.25n$, $0.25n$ and $0.5n$, respectively. For an F2 containing segregants for a dominant gene, the expected counts for the dominant class and the recessive class are $0.75n$ and $0.25n$, respectively. The test statistic is approximately distributed as a chi-square with degrees of freedom $(n-1)$.

The log likelihood ratio test statistic can be calculated using

$$G = 2 \sum_{j=1}^{n} o_j \log\left(\frac{o_j}{e_j}\right) \tag{5.2}$$

The G statistic is also distributed approximately as a chi-square with degrees of freedom $(n-1)$. Use of χ^2 or G usually results in similar statistical inferences in practical data analysis.

Table 5.1 A typical chi-square table for a single-locus segregation analysis, where n is the number of genotypic classes and p is the number of populations.

Source	Degree of Freedom	Chi-square
Total	$np - 1$	χ^2_{Total}
Pooled	$n - 1$	χ^2_{pooled}
Heterogeneity	$n(p-1)$	$\chi^2_{Total} - \chi^2_{pooled}$

5.1.2 MULTIPLE POPULATIONS

For multiple population situations, a total chi-square can be calculated using

$$\chi^2_{Total} = \sum_{i=1}^{p} \sum_{j=1}^{n} \left[\frac{(o_{ij} - e_{ij})^2}{e_{ij}} \right] \tag{5.3}$$

where subscript i indicates population and p is the number of populations. A pooled chi-square can be estimated using

$$\chi^2_{Pool} = \sum_{j=1}^{n} \frac{\left(\sum_{i=1}^{p} o_{ij} - \sum_{i=1}^{p} e_{ij} \right)^2}{\sum_{i=1}^{p} e_{ij}} \tag{5.4}$$

The difference between the total chi-square and the pooled chi-square is a chi-square test statistic for heterogeneity among the populations with degrees of freedom $n(p-1)$. For multiple population data, single-locus analysis can be summarized into a chi-square table (Table 5.1). The populations must have the same number of detectable genotypic classes segregating.

The likelihood ratio test statistic can be partitioned in a similar way. The total log likelihood ratio test statistic is

$$G_{Total} = 2\sum_{i=1}^{p}\sum_{j=1}^{n}\left[o_{ij}\log\left(\frac{o_{ij}}{e_{ij}}\right)\right]$$ (5.5)

and the pooled log likelihood ratio test statistic is

$$G_{pooled} = 2\sum_{j=1}^{n}\left[\left(\sum_{i=1}^{p}o_{ij}\right)\log\left(\frac{\sum_{i=1}^{p}o_{ij}}{\sum_{i=1}^{p}e_{ij}}\right)\right]$$ (5.6)

The heterogeneity log likelihood ratio test statistic is $G_{Total} - G_{pooled}$.

Table 5.2 Mendel's experiment on seed characteristics. The hypothesis test is based on a chi-square test.

Plant	Round Seed		Wrinkled Seed		χ^2	DF	P-value
	Count	Expected	Count	Expected			
1	45	42.75	12	14.25	0.47	1	0.49
2	27	26.25	8	8.75	0.09	1	0.77
3	24	23.25	7	7.75	0.10	1	0.76
4	19	21.75	10	7.25	1.39	1	0.24
5	32	32.25	11	10.75	0.01	1	0.93
6	26	24.00	6	8.00	0.67	1	0.41
7	88	84.00	24	28.00	0.76	1	0.38
8	22	24.00	10	8.00	0.67	1	0.41
9	28	25.50	6	8.50	0.98	1	0.32
10	25	24.00	7	8.00	0.17	1	0.68
Total	336		101		5.30	10	
Pooled	336	327.75	101	109.25	0.83	1	0.36
Heterogeneity					4.47	9	0.88

Example

A seed characteristic experiment by Mendel (1866) is used to illustrate the hypothesis tests and to introduce the concept of heterogeneity among different samples (Table 5.2). He crossed inbred pea lines that differed in seed characteristics. One line produced round seeds and the other had wrinkled seeds. Both round and wrinkled seeds were observed on each of 10 F1 plants. We can use his data to test the hypotheses:

(1) a single gene controls the seed character
(2) the F1 seed is round and heterozygous (Aa)
(3) seeds with genotype aa are wrinkled

(4) the A allele (normal) is dominant to a allele (wrinkled)

Seeds produced by the F1 plants represent the F2 plants. They show phenotypes reflecting their genotypes and are expected to show a ratio of 3 ($A\_$, round) to 1 (aa, wrinkled). Table 5.2 and Table 5.3 show the data and results of hypothesis tests using both the chi-square approximation and likelihood ratio approaches.

Table 5.3 Hypothesis test based on log likelihood ratio test statistic.

Plant	Round Seed		Wrinkled Seed		G	DF	P-value
	Count	Expected	Count	Expected			
1	45	42.75	12	14.25	0.49	1	0.49
2	27	26.25	8	8.75	0.09	1	0.77
3	24	23.25	7	7.75	0.10	1	0.75
4	19	21.75	10	7.25	1.30	1	0.26
5	32	32.25	11	10.75	0.01	1	0.93
6	26	24.00	6	8.00	0.71	1	0.40
7	88	84.00	24	28.00	0.79	1	0.38
8	22	24.00	10	8.00	0.63	1	0.43
9	28	25.50	6	8.50	1.06	1	0.30
10	25	24.00	7	8.00	0.17	1	0.68
Total	336		101		5.34	10	
Pooled	336	327.75	101	109.25	0.85	1	0.36
Heterogeneity					4.50	9	0.88

For each of the 10 plants, a chi-square test statistic can be calculated. The expected frequencies of $A\_$ genotype of seeds on plant#1 is

$$\frac{3}{4}(45 + 12) = 42.75$$

The total chi-square (5.30) is the sum of all 10 chi-squares for the 10 crosses. A pooled chi-square (0.83) can be estimated using the marginal frequencies under assumption of the same segregation ratio for all 10 plants. The heterogeneity chi-square is the difference between the total and pooled chi-squares in this case. No significant departure from the expected frequencies was detected for each of the 10 plants and the pooled frequencies. The heterogeneity chi-square was not significant. Table 5.3 lists results using the log likelihood ratio approach. There are some differences from Table 5.2.

5.2 MARKER SCREENING

5.2.1 SCREENING FOR POLYMORPHISM

Genomic map construction is based on genome variation at locations which can be identified by molecular assay or traditional trait observations. Screening polymorphic genetic markers is the first step of an efficient experiment. This screening is usually done by assaying a large number of possible genetic markers, such as polymerase chain reaction (PCR) and restriction fragment length polymorphism (RFLP) assays, for a small set of progeny randomly sampled from the mapping population. If a marker does not show polymorphism for the set of progeny, then the marker will be a non-informative monomorphic marker and will not be used in data analysis.

To determine progeny size for screening, the convenience of experiment, statistical power and tolerance for probabilities of false negative and false positive events are to be considered. The false negative event in this case is when a polymorphic marker is wrongly determined to be a monomorphic marker. The false positive event in this case is when a monomorphic marker is wrongly determined to be a polymorphic marker. When screening for polymorphic markers, false positives are rare, because a monomorphic marker cannot produce segregating genotypes if the genotypes are determined accurately. However, false negatives can happen with a high frequency if the size of the sample used for screening is small.

The equipment used, such as sizes of PCR reaction plates and gel electrophoresis equipment, is important for determining sample size in terms of convenience of the experiment. For example, screening sample sizes 6, 8, or 12 may be convenient for a 8×12 = 96 well PCR reaction plate.

For markers segregating 1:1, such as for a backcross, recombinant inbred lines and doubled haploid lines, the probability of sampling all n individuals with the same genotype is $2(0.5)^n$. When $n = 5$, the probability of all 5 individuals having the same genotype is

$$2(0.5)^5 = 0.0625$$

For screening in an F2 population using codominant markers, the probability of a false negative for a single marker is

$$2(0.25)^5 + 0.5^5 = 0.0332$$

for $n = 5$. In screening for 200 polymorphic markers, there will be about 7 markers for which polymorphism is not detected. For some experimental

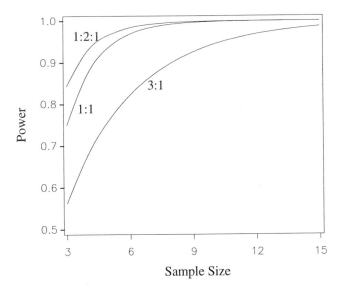

Figure 5.1 Statistical power for screening for polymorphisms for different sample sizes and expected segregation ratios.

populations, missing 7 polymorphic markers may not be significant. However, for some populations, a single missing marker could be significant. For dominant markers with expected phenotypes segregating 3:1, the probability is

$$0.75^n + 0.25^n$$

Figure 5.1 shows the statistical power for detecting polymorphisms with the three common segregation ratios: 1:1, 1:2:1 and 3:1. The statistical power for detecting one polymorphic marker is

$$\text{Power} = 1 - P[\text{ False Negative }]$$

For the same expected statistical power, more samples are needed for screening dominant markers in F2 than codominant markers. For a power of 95% in F2 (meaning that 5 out of 100 polymorphic markers will not be detected), 11 individuals are needed for dominant markers and 5 individuals are sufficient for codominant markers. For markers with an expected segregation ratio 1:1, 9 individuals are needed for the same statistical power.

5.2.2 SCREENING 1:1 OVER 3:1

For some genomic mapping experiments, we need to determine if a marker segregates with a 1:1 or 3:1 ratio. For example, in genomic mapping using populations produced by heterozygous parents, it is desirable to screen markers with a backcross configuration when dominant markers are used. This way, the well-known problem associated with dominant markers in repulsion linkage phase in F2 configuration can be avoided (see Chapter 6). This requires distinguishing between 1:1 and 3:1 segregation ratios, which can be done with a likelihood ratio test. The likelihood ratio test statistic for detecting departure from a 1:1 segregation ratio is

$$G = 2\{o_1\log\left[\frac{o_1}{0.5n}\right] + o_2\log\left[\frac{o_2}{0.5n}\right]\}$$

$$= 2[o_1\log o_1 + o_2\log o_2 - n\log(0.5n)]$$

$$(5.7)$$

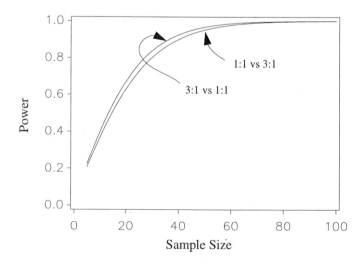

Figure 5.2 Statistical power for rejecting 1:1 and 3:1 segregation ratios when the ratios are 3:1 and 1:1, respectively.

where n is the sample size and o_1 and o_2 are the observed counts for the two genotypic classes. If the true segregation ratio is 3:1 and o_1 is the frequency of the dominant genotype, then the parametric value of this test statistic is approximately

$$G_E = 2\left\{E(o_1)\log\left[\frac{E(o_1)}{0.5n}\right] + o_2\log\left[\frac{E(o_2)}{0.5n}\right]\right\}$$

$$= 2\left\{0.75n\log\left[\frac{0.75n}{0.5n}\right] + 0.25n\log\left[\frac{0.25n}{0.5n}\right]\right\} \tag{5.8}$$

$$= 0.2616n$$

To reject a segregation ratio of 1:1 at 0.05 significance level, a log likelihood ratio test statistic of at least 3.84 (critical value for rejection) is required. The statistical power is

$$Pr[\chi^2_{G_E, 1} \geq 3.84]$$

where $\chi^2_{G_E, 1}$ is a chi-square value in a non-central chi-square-distribution with non-centrality parameter G_E and degrees of freedom 1. For $n = 15$, the power is

$$Pr[\chi^2_{3.924, 1} \geq 3.84] = 0.51$$

for rejecting a 1:1 ratio at the 0.05 level, where 3.924 is the the non-centrality parameter and 3.84 is the critical value for rejecting a 1:1 segregation ratio (Figure 5.2). For a power of 90%, $n \geq 40$ is needed. However, the expected likelihood ratio test statistic is $0.2877n$ for rejecting a 3:1 segregation ratio when the true ratio is 1:1 and $n \geq 35$ is required for a power of 90% (Figure 5.2).

Both false positives and negatives can occur when one screens markers. For example, when 15 individuals are used for screening markers for a 1:1 segregation among a mixture of markers with 1:1 and 3:1 segregations, approximately 50% of the markers with a true segregation ratio 3:1 will not show significant departure from a 1:1 segregation at the significance level of $\alpha = 0.05$ (false positive). Meanwhile, an average one of 10 markers with true 1:1 segregation will be wrongly rejected (false negative).

5.2.3 DISTINGUISHING BETWEEN TWO-CLASS SEGREGATIONS

For distinguishing between two two-class segregations, Bailey (1961) generalized a critical segregation approach to determine the sample size needed. If the two segregation ratios are x:1 and y:1 and observed frequencies for the two classes are a and b, then the lack of fit chi-squares for the two ratios are

$$\begin{cases} \dfrac{(a-xb)^2}{nx} \\[2ex] \dfrac{(a-yb)^2}{ny} \end{cases} \tag{5.9}$$

If the two chi-squares are set to be equal and solved for a and b, we get

$$\begin{cases} a = \dfrac{n(xy)^{0.5}}{1 + (xy)^{0.5}} \\[2ex] b = \dfrac{n}{1 + (xy)^{0.5}} \end{cases} \tag{5.10}$$

The critical ratio of Bailey is

$$\frac{a}{b} = (xy)^{0.5} \tag{5.11}$$

and the number of individuals required for distinguishing the two ratios is

$$n \geq \left[\frac{1 + (xy)^{0.5}}{x^{0.5} + y^{0.5}}\right]^2 \chi^2_{\alpha, 1} \tag{5.12}$$

Bailey gave an example for distinguishing between ratios of 3:1 and 9:7 using $\alpha = 0.05$. The calculation is summarized as follows:

$$\begin{cases} x = 3 \\ y = 9/7 \\ \chi^2_\alpha = 3.841 \\ (xy)^{0.5} = 1.964 \\ n = 94.3 \\ a = 62.5 \\ b = 31.8 \end{cases}$$

The critical ratio is

$$62.5{:}31.8 = 1.964{:}1 \approx 2{:}1$$

The number of individuals needed is 95. Among the 95 individuals, if $b > 32$, then the data will support the 9:7 ratio. Otherwise, the data will support the 3:1 ratio.

5.3 NATURAL POPULATIONS

For a single locus, there may be a number of different detectable forms in nature. These are defined as alleles of the locus. For controlled cross data, the number of alleles and their frequencies are usually not an issue because they are either known before the experiments start or they are easy to determine. The likelihoods for analyzing controlled cross data usually do not involve allelic frequencies. However, allelic frequencies are important components for genomic analysis in natural populations. We will see this in Chapters 8 and 15 for linkage analysis using disequilibrium and QTL mapping using open-pollinated populations. It is important to find out how many common alleles there are in nature in order to design efficient experiments. Certainly, it is important for gene conservation, because the number of alleles and their distribution are indicators of genetic diversity. In this section, the methods for determining the number of alleles and their frequencies, estimating the probability of heterozygosity, the concept of single locus Hardy-Weinberg equilibrium and estimation of a disequilibrium coefficient will be described.

5.3.1 NUMBER OF ALLELES AND THEIR FREQUENCIES

Notation
Use l to denote the number of alleles for a single locus. The alleles and their corresponding allelic frequencies are

$$\text{Alleles} = \begin{bmatrix} A_1 \\ A_2 \\ \cdots \\ A_l \end{bmatrix} \qquad P[\text{Alleles}] = \begin{bmatrix} p_1 \\ p_2 \\ \cdots \\ p_l \end{bmatrix}$$

in matrix notation. For a diploid system, there are $l(l+1)/2$ possible genotypes for the locus with l alleles in nature. These genotypes are denoted by

$$\text{Genotypes} = \begin{bmatrix} A_{11} & A_{12} & \cdots & A_{1l} \\ A_{21} & A_{22} & \cdots & A_{2l} \\ \cdots & \cdots & \cdots & \cdots \\ A_{l1} & A_{l2} & \cdots & A_{ll} \end{bmatrix}$$

The matrix genotype is symmetric about the diagonal $A_{11}, A_{22}, \ldots, A_{ll}$. For example, A_{12} and A_{21} represent the same genotype. Use

$$n[\text{Genotypes}] = \begin{bmatrix} n_{11} & n_{12} & \cdots & n_{1l} \\ n_{21} & n_{22} & \cdots & n_{2l} \\ \cdots & \cdots & \cdots & \cdots \\ n_{l1} & n_{l2} & \cdots & n_{ll} \end{bmatrix}$$

to denote the observed counts for the corresponding genotypes in a sample from a population. The corresponding estimated frequencies are denoted by

$$\hat{P}[\text{Genotypes}] = \begin{bmatrix} \hat{p}_{11} & \hat{p}_{12} & \cdots & \hat{p}_{1l} \\ \hat{p}_{21} & \hat{p}_{22} & \cdots & \hat{p}_{2l} \\ \cdots & \cdots & \cdots & \cdots \\ \hat{p}_{l1} & \hat{p}_{l2} & \cdots & \hat{p}_{ll} \end{bmatrix} = \frac{1}{N} \begin{bmatrix} n_{11} & n_{12} & \cdots & n_{1l} \\ n_{21} & n_{22} & \cdots & n_{2l} \\ \cdots & \cdots & \cdots & \cdots \\ n_{l1} & n_{l2} & \cdots & n_{ll} \end{bmatrix} \qquad (5.13)$$

where

$$N = \sum_{i=1}^{l} \sum_{j=1}^{l} n_{ij}$$

is the total sample size. For N individuals in a sample, there are $2N$ alleles in a diploid system. As with the genotype and the count matrices, the frequency matrix is symmetric about the diagonal.

The observed genotypic count and the estimated genotypic frequency distribution are the starting points for most of the analyses to characterize a single locus in the population. The number of alleles and their frequencies are estimated from the distribution. The relationship between the genotypic distribution and allelic frequencies is commonly used to infer equilibrium status. For a single population, the genotypic frequency is just the frequency of the genotype in the population. The variance of the estimated genotypic frequency is the variance for a multinomial proportion.

Estimating within Population Allelic Frequency

The genotypic frequency in Equation (5.13) is the simple count in a sample from the population. The allelic frequencies can be estimated using

$$\hat{P}\,[Alleles] = \begin{bmatrix} \hat{p}_1 \\ \hat{p}_2 \\ \cdots \\ \hat{p}_l \end{bmatrix} = \textbf{Diag}\left\{ \begin{bmatrix} \hat{p}_{11} & \hat{p}_{12} & \cdots & \hat{p}_{1l} \\ \hat{p}_{21} & \hat{p}_{22} & \cdots & \hat{p}_{2l} \\ \cdots & \cdots & \cdots & \cdots \\ \hat{p}_{l1} & \hat{p}_{l2} & \cdots & \hat{p}_{ll} \end{bmatrix} \begin{bmatrix} 1 & 0.5 & \cdots & 0.5 \\ 0.5 & 1 & \cdots & 0.5 \\ \cdots & \cdots & \cdots & \cdots \\ 0.5 & 0.5 & \cdots & 1 \end{bmatrix} \right\} \qquad (5.14)$$

in matrix notation. In element notation, Equation (5.14) is

$$\hat{p}_i = \hat{p}_{ii} + 0.5 \sum_{j \neq i} \hat{p}_{ij}$$

$$= \frac{1}{N}\left[n_{ii} + 0.5 \sum_{j \neq i} n_{ij} \right]$$

The variance of the estimated allelic frequency is

$$\mathrm{var}\,(\hat{p}_i) = \frac{1}{2N}[p_i + p_{ii} - 2p_i^2]$$

and the estimated covariance between two estimated allelic frequencies is

$$\mathrm{cov}\,(\hat{p}_i, \hat{p}_j) = \frac{1}{4N}[p_{ij} - 4p_i p_j] \qquad (5.15)$$

When the population is random mating and at equilibrium, Equation (5.15) becomes

$$\mathrm{cov}\,(\hat{p}_i, \hat{p}_j) = \frac{1}{4N}[2p_i p_j - 4p_i p_j]$$

$$= -\frac{1}{2N}p_i p_j$$

because under the assumptions $p_{ij} = 2p_i p_j$.

When dominant markers are used, the heterozygote and one of the homozygous genotypes cannot be identified in one generation. A testcross for advanced generations is needed to distinguish the two genotypic classes. The EM algorithm can be used to estimate allelic frequencies under the assumption of Hardy-Weinberg equilibrium. Assume allele A_i is dominant and A_j is recessive. The expected frequencies for the three possible genotypes composed of two alleles are

$$\begin{cases} P\,[A_{ii}] = p_{ii} = p_i^2 \\ P\,[A_{ij}] = p_{ij} = 2p_i p_j \\ P\,[A_{jj}] = p_{jj} = p_j^2 \end{cases}$$

In the dominant case, only $p_{i\circ} = p_{ii} + p_{ij}$ can be observed, where $\circ$ means that the allele can be either the dominant or the recessive allele. The frequencies of alleles A_i and A_j conditional on the observable genotypic frequencies are

$$\begin{cases} P\,[A_i | p_{i\circ}] = p_{i|i\circ} = \dfrac{p_i^2 + p_i p_j}{1 - p_j^2} \\[2ex] P\,[A_j | p_{i\circ}] = p_{j|i\circ} = \dfrac{p_i p_j}{1 - p_j^2} \\[2ex] P\,[A_i | p_{jj}] = p_{i|jj} = 0 \\[1ex] P\,[A_j | p_{jj}] = p_{j|jj} = 1 \end{cases} \qquad (5.16)$$

where $P[A_i|p_{i^\circ}]$ is the probability that an allele is A_i when the individual has a genotype A_{i°. The EM algorithm can be used to estimate the allelic frequencies as

$$\begin{cases} \hat{p}_i = \dfrac{1}{N} p_{i|i^\circ} n_{i^\circ} = \dfrac{n_{i^\circ}[(p_i')^2 + p_i'p_j']}{[1-(p_j')^2]N} \\ \hat{p}_j = \dfrac{1}{N}(p_{j|i^\circ} n_{i^\circ} + n_{jj}) = \dfrac{1}{N}\left[\dfrac{n_{i^\circ}p_i'p_j'}{1-(p_j')^2} + n_{jj}\right] \end{cases} \tag{5.17}$$

where p_i' is an initial value for the allelic frequency for starting the EM algorithm. For each iteration, p_i' is replaced by the new estimate $\hat{p}_i$. The final estimates are obtained when the iteration converges. For two-allele cases, the frequencies can also be obtained by getting the estimate for the recessive allele

$$\hat{p}_j = \sqrt{\frac{n_{jj}}{N}}$$

and the dominant allelic frequency

$$\hat{p}_i = \sqrt{\hat{p}_j^2 + \frac{n_{i^\circ}}{N}} - \hat{p}_j \tag{5.18}$$

For a two-allele example, if $n_{i^\circ} = 5$ and $n_{jj} = 4$ among a sample with 100 individuals, then the estimates of the allelic frequencies are

$$\begin{cases} \hat{p}_j = \sqrt{\dfrac{n_{jj}}{N}} = \sqrt{\dfrac{4}{100}} = 0.2 \\ \hat{p}_i = \sqrt{\hat{p}_j^2 + \dfrac{n_{i^\circ}}{N}} - \hat{p}_j = \sqrt{0.2^2 + \dfrac{5}{100}} - 0.2 = 0.1 \end{cases}$$

Single Allele Detection

The number of alleles at a locus in a population affects the design of experiments. This number is usually estimated by initial screening.

The probability of observing at least one individual with a certain allele in a sample of size N is

$$\gamma_i = 1 - (1-p_i)^{2N} \tag{5.19}$$

under the assumption of Hardy-Weinberg equilibrium, where $(1-p_i)^{2N}$ is the probability of no genotype containing the allele A_i. The sample size needed for detecting an allele with frequency p_i and the power of γ_i can be obtained by manipulating Equation (5.19), which is

$$N \geq \frac{\log(1-\gamma_i)}{\log[(1-p_i)^2]}$$
$$= 0.5\frac{\log(1-\gamma_i)}{\log(1-p_i)} \tag{5.20}$$

If all alleles have the same probability, $1/l$, then Equation (5.19) becomes

$$\gamma = 1 - \left(1 - \frac{1}{l}\right)^{2N} \tag{5.21}$$

The sample size needed for detection of an allele with frequency $1/l$ and a detecting power of γ is

$$N \geq \frac{\log(1-\gamma)}{\log[(1-1/l)^2]}$$

$$= 0.5\frac{\log(1-\gamma)}{\log(1-1/l)}$$

(5.22)

For alleles with a frequency higher than 0.05 (Figure 5.3), a sample size of 20 is usually sufficient to have a power higher than 90%. However, as allelic frequency decreases, the number of individuals needed increases rapidly. For example, a sample size of 100 may not be enough to detect an allele at a frequency of 0.01 for a power 90%.

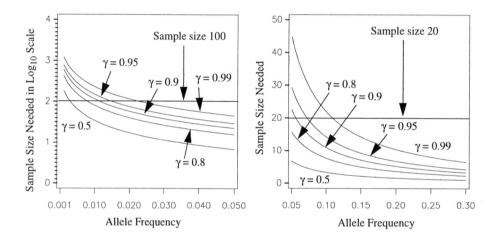

Figure 5.3 Sample size needed for observing at least one individual genotype containing an allele with different frequencies, in log scale for low frequency alleles and in numerical scale for median frequency alleles.

Multiple Allele Detection

The key question is not if just one allele can be detected, but rather how many alleles can be detected. Detection probabilities for each of the alleles are $(\gamma_1, \gamma_2, ..., \gamma_l)$. The average detection probability (γ_m) for detecting at least m alleles among a total of l alleles can be determined by

$$\gamma_m = \frac{1}{c}\sum_{j=1}^{c}\prod_{i=1}^{k_j}\gamma_i \qquad l \geq k_j \geq m$$

(5.23)

where

$$c = \sum_{k_j=m}^{l}\binom{l}{k_j} = \sum_{k_j=m}^{l}\frac{l!}{k_j!(l-k_j)!}$$

is the number of possible combinations.

Table 5.4 Allelic frequency is distributed as a geometric series.

Allele	λ				
	1 / 3	2 / 3	9 / 11	19 / 21	39 / 41
1	0.6667	0.3333	0.1818	0.0952	0.0488
2	0.2222	0.2222	0.1488	0.0862	0.0464
3	0.0741	0.1481	0.1217	0.0780	0.0441
4	0.0247	0.0988	0.0996	0.0705	0.0420
5	0.0082	0.0658	0.0815	0.0638	0.0399
6	0.0027	0.0439	0.0667	0.0577	0.0380
7	0.0009	0.0293	0.0545	0.0522	0.0361
8	0.0003	0.0195	0.0446	0.0473	0.0344
9	0.0001	0.0130	0.0365	0.0428	0.0327
10		0.0087	0.0299	0.0387	0.0311
11		0.0058	0.0244	0.0350	0.0296
12		0.0039	0.0200	0.0317	0.0281
13		0.0026	0.0164	0.0287	0.0268
14		0.0017	0.0134	0.0259	0.0255
15		0.0011	0.0110	0.0235	0.0242
16		0.0008	0.0090	0.0212	0.0230
17		0.0005	0.0073	0.0192	0.0219
18		0.0003	0.0060	0.0174	0.0208
19		0.0002	0.0049	0.0157	0.0198
20		0.0002	0.0040	0.0142	0.0189
21		0.0001	0.0033	0.0129	0.0179
22		0.0001	0.0027	0.0116	0.0171
23			0.0022	0.0105	0.0162
24			0.0018	0.0095	0.0154
25			0.0015	0.0086	0.0147
26			0.0012	0.0078	0.0140
27			0.0010	0.0071	0.0133
28			0.0008	0.0064	0.0126
29			0.0007	0.0058	0.0120
30			0.0005	0.0052	0.0114
31			0.0004	0.0047	0.0109
32			0.0004	0.0043	0.0104
33			0.0003	0.0039	0.0098
34			0.0002	0.0035	0.0094
35			0.0002	0.0032	0.0089
36			0.0002	0.0029	0.0085
37			0.0001	0.0026	0.0081
38			0.0001	0.0023	0.0077
39			0.0001	0.0021	0.0073
40			0.0001	0.0019	0.0069
41			0.0001	0.0017	0.0066
42				0.0016	0.0063
43				0.0014	0.0060
44				0.0013	0.0057
45				0.0012	0.0054
46				0.0011	0.0051
47				0.0010	0.0049
48				0.0009	0.0046
49				0.0008	0.0044
50				0.0007	0.0042

When the number of alleles is large and their frequencies are unequal, this approach may be computationally intensive. In such cases, γ_m can be determined empirically by a Monte Carlo simulation. Use "1" to indicate that an allele is detected and "0" that it is not. For allele i, if a random uniform (0, 1) is less than or equal to γ_i, the allele is detected and set $I_i = 1$. Otherwise, set $I_i = 0$. By doing this for all alleles, the number of alleles detected is

$$\sum I_i$$

If this process is repeated a large number of times, γ_m can be determined by counting the frequency of

$$\sum I_i \geq m$$

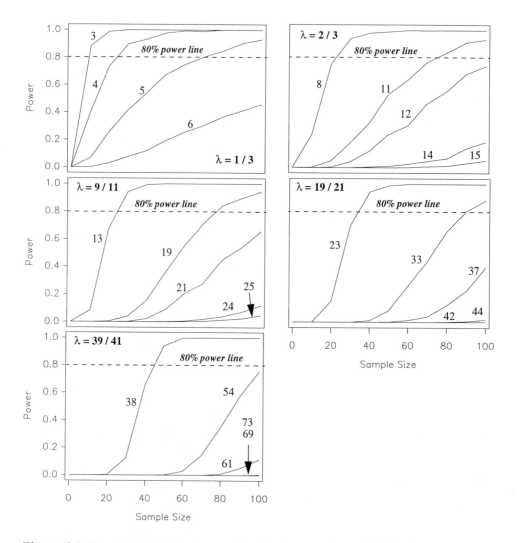

Figure 5.4 Empirical statistical power for detecting numbers of alleles for different screening sample sizes and the distributions of allelic frequencies in Table 5.4. The curves are for approximately 50%, 70%, 80%, 90% and 95% of the total alleles having frequencies higher than 0.001 (exact numbers of alleles are indicated for each of the curves). The power is obtained from 10,000 simulation replications.

which is the probability for detecting at least m alleles

$$\hat{\gamma}_m = Pr\left[\sum I_i \geq m\right] \tag{5.24}$$

A similar approach can be used for determining polymorphic marker detection in the next section and for multiple QTL detection power in Chapter 16.

Multiple allelic frequency can be modeled using a geometric series; it is

$$(1-\lambda)\,[\lambda^0, \lambda^1, \lambda^2, \lambda^3, \ldots, \lambda^{l-1}] \tag{5.25}$$

where λ is the parameter for the geometric series and $0 < \lambda \leq 1$. The sum of the individual frequencies is 1

$$\sum_{i=0}^{l} [(1-\lambda)\,\lambda^i] = 1$$

Table 5.4 lists some examples of the distributions of the allelic frequency as a geometric series. For example, the most frequent allele for $\lambda = 1/3$ is

$$(1-\lambda)\,\lambda^0 = 1 - \frac{1}{3} = \frac{2}{3}$$

Figure 5.4 shows the probability of detecting a number of alleles for different screening sample sizes and the distributions of allelic frequencies in Table 5.4. The data were obtained using the simulation approach described above. For the distributions in Table 5.4, the number of alleles corresponding to the proportion are listed in Table 5.5. For example, 70% of alleles having a frequency higher than 0.001 is $m = 19$ for the distribution corresponding to $\lambda = 9/11$. The detection probability is the proportion of the counts which satisfy

$$\sum I_i \geq 19$$

in the 10,000 simulation replications. For the five allelic frequency distribution patterns, the numbers of loci which have a frequency higher than 0.01 are 6, 15, 26, 46 and 76 for $\lambda = 1/3$, 2/3, 9/11, 19/21 and 39/41, respectively.

Table 5.5 Number of alleles corresponding approximately 50%, 70%, 80%, 90% and 95% of the alleles distributed as shown in Table 5.4.

Percentage	λ				
	1/3	2/3	9/11	19/21	39/41
50%	3	8	13	23	38
70%	4	11	19	33	54
80%	4	12	21	37	61
90%	5	14	24	42	69
95%	6	15	25	44	73
100% (Total)	6	15	26	46	76

From Figure 5.4, we can see that for a sample size 20, half of the alleles for distributions with $\lambda = 1/3$, 2/3 and 9/11 can be detected with reasonable power, while a sample size of 40 is needed for distributions with $\lambda = 19/21$ and 39/41. A sample size of approximately 70 is needed to detect 70% of the alleles for distributions with $\lambda = 1/3$, $2/3$ and $9/11$ and more than 90 individuals are needed for distributions with $\lambda = 19/21$ and $39/41$.

For a fixed detection probability of 80% (see Figure 5.4):
a distribution with $\lambda = 1/3$ detecting:
3 alleles needs a sample size of 10
4 alleles needs a size of 23
5 alleles needs a size of 70
a distribution with $\lambda = 2/3$ detecting:
8 alleles needs a sample size 22
11 alleles needs a size of 75
a distribution with $\lambda = 9/11$ detecting:
13 alleles needs a sample size of 25
19 alleles needs a size of 75
a distribution with $\lambda = 19/21$ detecting:
23 alleles needs a sample size of 32
33 alleles needs a size of 87
a distribution with $\lambda = 39/41$ detecting:
38 alleles needs a sample size of 45.

For all the distributions, it is impossible to detect more than 90% of the alleles with a reasonable detection probability for a sample size under 100.

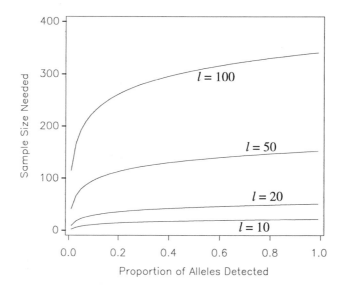

Figure 5.5 Sample size needed for detecting different proportions of alleles when numbers of equal frequent alleles are 10, 20, 50 and 100. The expected statistical power is 90%.

Screening for informative markers is the first step in designing and implementing a successful genomic experiment. How the screening proceeds largely depends on the frequency distribution of the markers. Markers with fewer alleles are usually easier to detect. However, those markers are commonly considered to have low information content. Information content of a marker depends on the design of the experiment and on the approach to analyzing and modeling the data. It is commonly believed that markers with a large number of alleles with equal frequency have the most information. Figure 5.5 shows sample sizes needed for detecting different proportions of alleles. A sample size of 20 is sufficient to detect most of the 10 alleles at a power of 90%. For 20, 50 and 100 alleles, the sample sizes needed to have similar detection probability are 40, 120 and 350, respectively. It is interesting to see from Figure 5.5 that the sample size needed increases slowly after the proportion of detection is 20%. This differs from the conclusion that unevenly distributed alleles follow a geometric series(Figure 5.4).

It is common to assume equal allelic frequency in genomic analysis when markers or genes have a large number of alleles in a population. However, if this assumption is not true, the assumption may lead to a higher number of false positive indications of linkage (Ott 1992). It is essential to have a reasonable sample size to estimate allelic frequencies when natural populations are used for genomic analysis. For a more complete treatment of allelic frequency estimation, see Weir (1996).

5.3.2 HARDY-WEINBERG EQUILIBRIUM FOR A SINGLE LOCUS

Hardy-Weinberg equilibrium has been assumed previously to derive the estimation procedures for allelic frequency and allele detection. It will be used in most derivations for linkage analysis and QTL mapping (see Chapters 6-8 and 11-16). In this section, a disequilibrium coefficient will be defined and methods for estimating disequilibrium will be described.

Di-Allelic System

For a single locus A, with two alleles A and a, there are three possible genotypes, AA, Aa and aa. If frequencies of the three genotypes are p_{AA}, p_{Aa} and p_{aa}, then the allelic frequencies for A and a are

$$p_A = 0.5p_{Aa} + p_{AA}$$

$$p_a = 0.5p_{Aa} + p_{aa}$$

If the mating is random, the following relationship between the allelic frequencies and the genotypic frequencies will be seen

$$p_{AA} = p_A^2$$
$$p_{Aa} = 2p_Ap_a \qquad\qquad (5.26)$$
$$p_{aa} = p_a^2$$

That is

$$(p_A + p_a)^2 = p_A^2 + 2p_Ap_a + p_a^2$$
$$= p_{AA} + p_{Aa} + p_{aa}$$
$$= 1$$

Equation (5.26) is defined as the Hardy-Weinberg law. Populations which meet the Hardy-Weinberg law are commonly referred to as being in Hardy-Weinberg equilibrium. For a single locus on an autosomal chromosome, the population will reach the Hardy-Weinberg equilibrium after one generation of random mating. Disequilibrium is defined as the departure from the Hardy-Weinberg equilibrium. To quantify the disequilibrium, a disequilibrium coefficient, D_A, is used (Weir 1996). D_A is defined by the relationship between the genotypic and allelic frequencies, which is

$$p_{AA} = p_A^2 + D_A$$
$$p_{Aa} = 2p_Ap_a - 2D_A \tag{5.27}$$
$$p_{aa} = p_a^2 + D_A$$

and

$$\begin{aligned} D_A &= p_{AA} - p_A^2 \\ &= p_{aa} - p_a^2 \\ &= 0.5\,(2p_Ap_a - p_{Aa}) \end{aligned} \tag{5.28}$$

The boundary condition for D_A is

$$p_Ap_a \ge D_A \ge -1 \times min\,(p_A^2, p_a^2) \tag{5.29}$$

The log likelihood for the disequilibrium and allelic frequencies is

$$L = f_{AA}\log\,(p_A^2 + D_A) + f_{Aa}\log\,(2p_Ap_a - 2D_A) + f_{aa}\log\,(p_a^2 + D_A) \tag{5.30}$$

where f_{AA}, f_{Aa} and f_{aa} are observed genotypic counts. Because $p_a = 1 - p_A$, there are two unknown parameters in the likelihood equation. By setting the derivatives of the log likelihood function with respect to the two parameters equal to zero

$$\begin{cases} \dfrac{\partial L}{\partial p_A} = \dfrac{2p_Af_{AA}}{p_A^2 + D_A} + \dfrac{(1 - 2p_A)f_{Aa}}{p_A\,(1 - p_A) - D_A} - \dfrac{2\,(1 - p_A)f_{aa}}{(1 - p_A)^2 + D_A} = 0 \\[2ex] \dfrac{\partial L}{\partial D_A} = \dfrac{f_{AA}}{p_A^2 + D_A} - \dfrac{f_{Aa}}{p_A\,(1 - p_A) - D_A} + \dfrac{f_{aa}}{(1 - p_A)^2 + D_A} = 0 \end{cases}$$

we have the maximum likelihood estimates of the two parameters, which are

$$\begin{aligned} \hat{p}_A &= \dfrac{f_{AA} + 0.5f_{Aa}}{N} \\[2ex] \hat{D}_A &= \dfrac{f_{AA}}{N} - \left[\dfrac{f_{AA} + 0.5f_{Aa}}{N}\right]^2 \end{aligned} \tag{5.31}$$

where

$$N = f_{AA} + f_{Aa} + f_{aa}$$

is the total sample size. The hypothesis test for disequilibrium can be implemented using the log likelihood ratio test statistic, which is

$$G = 2\,[L\,(D_A = \hat{D}_A) - L\,(D_A = 0)\,] \tag{5.32}$$

The log likelihood ratio test statistic is approximately distributed as a chi-square variable with one degree of freedom. Estimation of the allelic frequency is independent of the disequilibrium coefficient and will not affect the log likelihood ratio test statistic. However, allelic frequency has effects on the statistical power of detecting disequilibrium.

Multiple-Allelic System

For a diploid system, there are $l(l+1)/2$ possible genotypes for a locus with l codominant alleles. Theoretically, the frequencies of the $l(l+1)/2$ possible genotypes can be determined using frequencies of the l alleles and $l(l-1)/2$ disequilibrium coefficients between all possible combinations of heterozygous genotypes. The frequency (p_{ij}) of genotype A_{ij} is

$$p_{ij} = 2p_ip_j - 2D_{ij} \qquad (5.33)$$

where D_{ij} is the disequilibrium coefficient between alleles A_i and A_j. Each disequilibrium coefficient can be estimated using

$$\hat{D}_{ij} = 0.5\,(2\hat{p}_i\hat{p}_j - \hat{p}_{ij}) \qquad (5.34)$$

and tested using Equation (5.32). The disequilibria for the homozygotes are

$$D_{ii} = \sum_{j \neq i} D_{ij}$$

For a more complete treatment of hypothesis tests using likelihood approaches and the Fisher's exact test, refer to Weir (1996).

5.3.3 HETEROZYGOSITY

One of the most important characteristics of a locus is its heterozygosity. The heterozygosity of a locus is defined as the probability that an individual is heterozygous for the locus in a population. For a genetic marker, a locus with heterozygosity higher than 70% is commonly considered a highly polymorphic marker (Ott 1992). In this subsection, heterozygosity is defined in terms of allelic frequencies, an estimator of heterozygosity will be derived, strategies for screening for highly polymorphic markers will be discussed and approaches for estimating statistical power for detecting heterozygosity will be described.

Definition

For a diploid system, there are $l(l+1)/2$ possible genotypes for a locus with l codominant alleles in a population under random mating. Among them there are l homozygous and $l(l-1)/2$ heterozygous genotypes. The heterozygosity of the locus is defined as

$$H = 1 - \sum_{i=1}^{l} p_{ii} \qquad (5.35)$$

where

$$\sum_{i=1}^{l} p_{ii}$$

is the total frequency for the l homozygotes. If the population is in Hardy-Weinberg equilibrium for the locus, then the heterozygosity can be written in terms of allelic frequency

$$H = 1 - \sum_{i=1}^{l} p_i^2$$

$$= 2 \sum_{i=2}^{l} \sum_{j=1}^{i-1} (p_i p_j) \qquad (5.36)$$

where

$$\sum_{i=1}^{l} p_i^2$$

is the expected frequency for l homozygotes. It is easy to see that the more alleles there are or the more even the allelic frequency distribution is, the larger the heterozygosity. If the l alleles have equal frequencies, then Equation (5.36) reduces to

$$H = 1 - \frac{1}{l} \tag{5.37}$$

The unbiased estimator for the heterozygosity is

$$\hat{H} = \frac{N}{N-1} \left[1 - \sum_{i=1}^{l} \hat{p}_i^2 \right]$$

where N is sample size. The approximate variance of the estimator is

$$Var(\hat{H}) \approx \frac{N}{(N-1)^2} \{ \sum_{i=1}^{l} p_i^3 - \left[\sum_{i=1}^{l} p_i^2 \right]^2 \}$$

Inbreeding will reduce the heterozygosity in the population. If the inbreeding coefficient is F, then the expected heterozygosity in the population reduces to $(1-F)H$.

Gene diversity is a function of the heterozygosity. For m loci, the average gene diversity is

$$D = 1 - \frac{1}{m} \sum_{j=1}^{m} \sum_{i=1}^{l} p_{ji}^2$$

where p_{ji} is the frequency of the ith allele at the jth locus.

It is also common to use polymorphism information content (PIC) to quantify the polymorphism (Botstein et al. 1980). It is believed that the higher the PIC, the higher the linkage information content. PIC is defined as

$$
\begin{aligned}
PIC &= 1 - \sum_{i=1}^{l} p_i^2 - 2 \sum_{i=2}^{l} \sum_{j=1}^{i-1} (p_i^2 p_j^2) \\
&= 2 \sum_{i=2}^{l} \sum_{j=1}^{i-1} (p_i p_j) - 2 \sum_{i=2}^{l} \sum_{j=1}^{i-1} (p_i^2 p_j^2) \\
&= 2 \sum_{i=2}^{l} \sum_{j=1}^{i-1} [p_i p_j (1 - p_i p_j)]
\end{aligned}
\tag{5.38}
$$

If the l alleles have equal frequency, then Equation (5.38) reduces to

$$PIC = 1 - \frac{1}{l} - \frac{1}{l^2} + \frac{1}{l^3}$$

which is smaller than the heterozygosity. When the number of alleles is large, the polymorphism information content approximately equals the heterozygosity.

Table 5.6 shows the heterozygosity and the polymorphism information content values for the allelic frequency distribution in Table 5.4. As the number of alleles increases, the heterozygosity increases.

Table 5.6 Heterozygosity (H) and polymorphism information content (PIC) estimates for the allelic frequency distributions in Table 5.4.

	λ				
	1/3	2/3	9/11	19/21	39/41
H	0.499962	0.800042	0.899991	0.950003	0.975009
PIC	0.474962	0.787740	0.895981	0.948878	0.974712

Table 5.7 H and PIC estimates if all alleles with equal frequency.

Number of alleles	8	18	33	60	100
H	0.875000	0.944444	0.969697	0.983333	0.990000
PIC	0.861328	0.941529	0.968807	0.983060	0.989901

Table 5.7 shows the expected heterozygosity if the allelic frequencies are equal. The heterozygosity is maximized when the allelic frequencies are equal.

As defined by Ott (1992a and 1992b), a locus is considered a polymorphic marker if its heterozygosity $H \geq 0.1$ and a locus is considered highly polymorphic if $H \geq 0.7$. The definitions imply that a marker is considered polymorphic when its most frequent allele has a frequency less than 0.95

$$H \approx 1 - 0.95^2 = 0.1$$

and highly polymorphic when its most frequent allele has a frequency less than 0.55.

Screening Polymorphic Markers

Ott (1992b) has outlined several methods to screen polymorphic markers. Here, a method modified from Ott is introduced. The modified method computes the binomial probability. Given a critical value of a heterozygosity, c, as a cut-off point, the statistical power for screening is the probability that the estimated heterozygosity from the screening sample is higher than c. The statistical power for detecting a locus having heterozygosity higher than c can be obtained using Monte Carlo simulation. For a given sample size, 1,000 repeated samples were simulated and the probability that the estimated heterozygosity is higher than c was computed. For methodologies of computer simulation refer to Chapter 19. The simulation was done to identify the loci with a heterozygosity higher than 0.5, 0.6, 0.7, 0.8, 0.9, 0.95 and 0.99 for the allelic distributions in Table 5.4.

Figure 5.6 shows some partial results of the simulations. The probabilities less than or equal to the true heterozygosity are the statistical power. The curves greater than the true heterozygosity show the false positive. For example, for $\lambda = 1/3$ and the true heterozygosity $H = 0.5$, the curve for c = 0.5 shows the statistical power, and curves for $c \geq 0.6$ show the probability of false positive.

From Figure 5.6, we can see that a proportion of heterozygosity (say 70% of the total heterozygosity) can be detected with reasonable probability for a sample size as small as 15 for all the allelic frequency distributions. However, a large sample size is needed if the goal is to detect 95% of the heterozygosity.

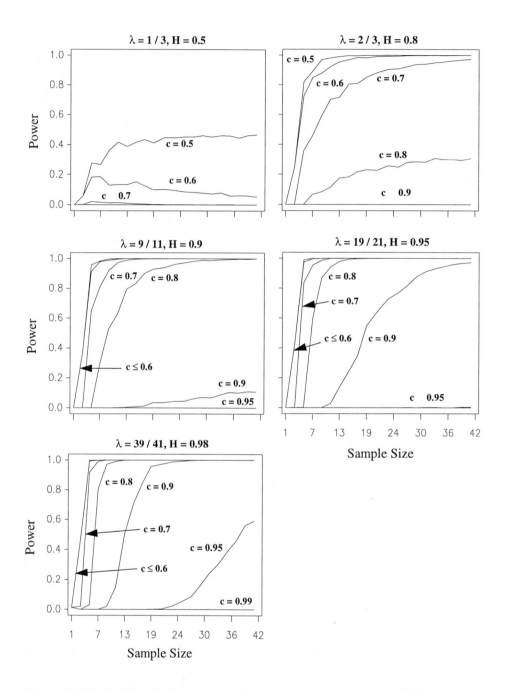

Figure 5.6 Probability for detecting the heterozygosity higher than c $(H \geq c)$ for the allelic distribution in Table 5.4.

EXERCISES

Exercise 5.1

Genotypic frequencies in progeny produced by crossing A_1A_2 and A_3A_4 are shown in the following illustration. Locus A has four distinct alleles A_1, A_2, A_3 and A_4.

	A_1	Genotype	Frequency
Parent 1	A_2	A_1A_3	65
	X	A_1A_4	30
	A_3	A_2A_3	25
Parent 2	A_4	A_2A_4	70

Ex. 5.1

(1) Test the hypothesis that the gametic ratio is 1:1 for A_1 and A_2 produced by the parent 1, and is 1:1 for A_3 and A_4 produced by the parent 2, using both the chi-square and the likelihood ratio approaches.

(2) Test the hypothesis that the genotypic segregation ratio is 1:1:1:1 for A_1A_3, A_1A_4, A_2A_3 and A_2A_4 in the progeny.

(3) Speculate why the genotypic segregation ratio does not follow expectation when the gametes produced by the two parents do follow expectations.

Exercise 5.2

Use the results from Exercise 5.1 for this problem. The same parent 1 with genotype A_1A_2 is crossed with three different parents with genotypes A_3A_4, A_3A_3 and A_4A_4 (see the following illustration).

				Genotype	Frequency
				A_1A_3	65
Cross 1	A_1 / A_2	X	A_3 / A_4	A_1A_4	30
				A_2A_3	25
				A_2A_4	70
Cross 2	A_1 / A_2	X	A_3 / A_3	A_1A_3	75
				A_2A_3	30
Cross 3	A_1 / A_2	X	A_4 / A_4	A_1A_4	40
				A_2A_4	80

Ex. 5.2

(1) Test the segregation ratio heterogeneity for gametes produced by parent 1 (A_1A_2) using both the chi-square and the likelihood ratio tests.

(2) Combine results for all three crosses. Can you draw a biological conclusion about the nature of gametic and genotypic segregation in this experiment?

(3) If we want to extensively test the segregation ratio of gametes produced by the parent 2 (A_3A_4), which crosses should be made?

Exercise 5.3

Screening for polymorphic markers is usually the first step in building a genetic map. An electrophoresis unit is conveniently set up for running 96 samples. For a screening experiment, there are four possibilities in terms of numbers of samples for each primer or probe and numbers of primers or probes: (6, 16), (8, 12), (12, 8) and (16, 6). For example, (6, 16) means that 6 samples for each primer or probe and 16 primers or probes will be screened in each set. The expected probability of a polymorphic marker is p.

(1) If $p = 0.4$ and the markers are codominant and expected to segregate 1:2:1 in the progeny, which screening strategies will give more polymorphic markers: (6, 16), (8, 12), (12, 8) or (16, 6)?

(2) If $p = 0.2$ and the markers are codominant and expected to segregate 1:1 in the progeny, which screening strategies will give more polymorphic markers: (6, 16), (8, 12), (12, 8) or (16, 6)?

(3) If $p = 0.3$ and the markers are codominant and expected to segregate 3:1 in the progeny, which screening strategies will give more polymorphic markers: (6, 16), (8, 12), (12, 8) or (16, 6)?

(4) When the 1:2:1, 3:1 and 1:1 markers are mixed in the progeny with the polymorphic probabilities 0.4, 0.2 and 0.3, respectively, the proportions of the marker types in the progeny are 0.05, 0.65 and 0.3, respectively. If the goal of the screening is to maximize the linkage information content of the identified polymorphic markers, which screening strategies will give more polymorphic markers: (6, 16), (8, 12), (12, 8) or (16, 6)? The linkage information is defined as 2 times the number of 1:2:1 polymorphic markers, plus 1 times the number of 1:1 polymorphic markers.

Exercise 5.4

Determining the expected number of alleles to be sampled in a natural population is useful for designing both genomic mapping experiments using natural populations, and gene conservation experiments. Given that the populations are under random mating, the number of alleles and their distribution frequencies and the size of the samples, determine the expected number of alleles to be sampled. Let us assume that a gene has 4 alleles with frequencies 0.4, 0.3, 0.2 and 0.1 in the population. The population is under Hardy-Weinberg equilibrium for this locus. Answer the following questions:

(1) If we sample 20 individuals in the population, what is the probability that at least 3 of the 4 alleles are sampled?

(2) If the goal is to sample at least 3 alleles with probability of 99%, how many samples are needed?

CHAPTER 6

TWO-LOCUS MODELS: THE CONTROLLED CROSSES

6.1 INTRODUCTION

Data analysis for genomic map construction involves the following five steps:

- single-locus analysis
- two-locus analysis
- linkage grouping
- gene ordering
- multi-point analysis

Single-locus analysis includes estimation of allelic and genotypic frequencies, estimation of the theoretical segregation ratio and detection of segregation ratio distortion. The purpose of the single-locus analysis is to determine if observed data follow the expectation of a hypothetical genetic model at a single locus level. Two-locus analysis includes linkage detection and recombination fraction estimation. Results of the two-locus or "two-point" analysis are the basis for the remainder of the steps of genomic map construction - linkage grouping and locus ordering.

In some of the literature, two-locus linkage analysis is referred to as two-point linkage analysis. The basic methods of two-locus linkage analysis were laid out by Fisher (1935) and Haldane (1919) and have been explained in books such as *The Measurement of Linkage in Heredity* by Mather (1951), a practical treatment, *Introduction to the Mathematical Theory of Genetic Linkage* by Bailey (1961), a more mathematical perspective and more recently, *Analysis of Human Genetic Linkage, Revised Edition* by Ott (1991).

This chapter focuses on linkage detection using goodness of fit statistics, likelihood ratio test statistics and recombination fraction estimation using likelihood approaches. Hypothesis tests among populations, dominant markers in F2 progeny, statistical properties of estimated recombination fraction and violations of assumptions in linkage analysis are also covered in this chapter.

6.2 LINKAGE DETECTION

Genetic linkage is defined as the association of genes located on the same chromosome (see Chapter 2). In statistical terms, linkage is the association or non-independence among alleles at more than one locus. For a two-locus model, linkage is detected by testing the independence between the two loci in segregating populations. For some simple cases, a goodness of fit or a log likelihood ratio test statistic for a two-locus model can be partitioned into components corresponding to the two individual loci and the linkage between

Table 6.1 A backcross (AaBb X aabb) example to illustrate goodness of fit statistic partition for a two-locus model. Linkage phase is coupling. The expected frequencies under the null hypotheses of no segregation ratio distortion for the two loci and no genetic linkage between the two loci are listed in parentheses.

	Genotype	Cross				Pooled
		1	2	3	4	
Frequency	AaBb	310 (300)	36 (30)	360 (300)	74 (60)	780 (690)
	Aabb	287 (300)	23 (30)	230 (300)	50 (60)	590 (690)
	aaBb	288 (300)	23 (30)	230 (300)	44 (60)	585 (690)
	aabb	315 (300)	38 (30)	380 (300)	72 (60)	805 (690)
Marginal A	Aa	597 (600)	59 (60)	590 (600)	124 (120)	1370 (1380)
	aa	603 (600)	61 (60)	610 (600)	116 (120)	1390 (1380)
Marginal B	Bb	598 (600)	59 (60)	590 (600)	118 (120)	1365 (1380)
	bb	602 (600)	61 (60)	610 (600)	122 (120)	1395 (1380)
Sum		1200	120	1200	240	2760

them. However, for some more complex situations, they cannot be partitioned clearly and general likelihood approaches are needed to detect linkage. In this section, goodness of fit statistic and log likelihood ratio test statistic partitioning are illustrated using a simple backcross (Table 6.1). Later, the general likelihood approaches will be introduced.

6.2.1 PARTITION OF TEST STATISTIC

Partition of Goodness of Fit Statistic

For each of the four crosses (Table 6.1), a total goodness of fit statistic is calculated according to the expected segregation ratio 1:1:1:1 under the assumption of no segregation distortion for both loci and no linkage between the loci. For each of the two loci, the goodness of fit statistic is calculated using the marginal counts and assuming the two genotypes segregate 1:1. The difference between the total and the two individual locus goodness of fit statistic is the chi-square test statistic contributed by the association or linkage between the two loci. This can be summarized in the following formulae

$$\chi_{iT}^2 = \sum_j \sum_k \frac{(o_{ijk} - 0.25 o_{i\circ\circ})^2}{0.25 o_{i\circ\circ}} \qquad \chi_{PT}^2 = \sum_j \sum_k \frac{(o_{\circ jk} - 0.25 o_{\circ\circ\circ})^2}{0.25 o_{\circ\circ\circ}}$$

$$\chi_{iA}^2 = \sum_j \frac{(o_{ij\circ} - 0.5 o_{i\circ\circ})^2}{0.5 o_{i\circ\circ}} \qquad \chi_{PA}^2 = \sum_j \frac{(o_{\circ j\circ} - 0.5 o_{\circ\circ\circ})^2}{0.5 o_{\circ\circ\circ}} \qquad (6.1)$$

$$\chi_{iB}^2 = \sum_k \frac{(o_{i\circ k} - 0.5 o_{i\circ\circ})^2}{0.5 o_{i\circ\circ}} \qquad \chi_{PB}^2 = \sum_k \frac{(o_{\circ\circ k} - 0.5 o_{\circ\circ\circ})^2}{0.5 o_{\circ\circ\circ}}$$

and relationships

$$\chi_{iL}^2 = \chi_{iT}^2 - \chi_{iA}^2 - \chi_{iB}^2 \qquad \chi_{PL}^2 = \chi_{PT}^2 - \chi_{PA}^2 - \chi_{PB}^2$$

$$\chi_T^2 = \sum_i \chi_{iT}^2 \qquad \chi_{AT}^2 = \sum_i \chi_{iA}^2$$

$$\chi_{BT}^2 = \sum_i \chi_{iB}^2 \qquad \chi_{LT}^2 = \sum_i \chi_{iL}^2$$

$$\chi_{HT}^2 = \chi_T^2 - \chi_{PT}^2 \qquad \chi_{HB}^2 = \chi_{BT}^2 - \chi_{PB}^2$$

$$\chi_{HA}^2 = \chi_{AT}^2 - \chi_{PA}^2 \qquad \chi_{HL}^2 = \chi_{LT}^2 - \chi_{PL}^2$$

In this set of formulae, o denotes observed counts of genotypes, subscripts i, j, k and $°$ indicate cross, A locus, B locus and summation over these subscripts, respectively. Subscripts for the test statistics usually occur in pairs. T, A, B, L, P and H indicate the goodness of fit statistics for the total, A locus, B locus, linkage, pooled and heterogeneity, respectively. The subscript i sometimes occurs in the goodness of fit statistic to denote a cross-specific test statistic.

Table 6.2 A goodness of fit statistic table for data in Table 6.1.

Cross	Total	Locus A	Locus B	Linkage
1	2.13	0.06(0.86)	0.01(0.91)	2.09(0.15)
2	6.60	0.03(0.86)	0.03(0.86)	6.53(0.01)
3	66.00	0.33(0.56)	0.33(0.56)	65.33(<0.0001)
4	11.60	0.27(0.61)	0.07(0.80)	11.27(0.0008)
Total	86.33	0.66	0.45	85.22
Pooled	61.86	0.15(0.70)	0.33(0.56)	61.38(<0.0001)
Heterogeneity	24.47	0.51(0.92)	0.12(0.99)	23.84(<0.0001)

Values in parentheses are the P-values.

Example: Partition of Goodness of Fit Statistic

Table 6.2 shows the complete goodness of fit statistic analysis for data in Table 6.1. For example, for cross#1, goodness of fit statistics are

$$\chi^2_{1T} = \frac{(310-300)^2}{300} + \frac{(287-300)^2}{300} + \frac{(288-300)^2}{300} + \frac{(315-300)^2}{300} = 2.13$$

$$\chi^2_{1A} = \frac{(597-600)^2}{600} + \frac{(603-600)^2}{600} = 0.03$$

$$\chi^2_{1B} = \frac{(598-600)^2}{600} + \frac{(602-600)^2}{600} = 0.01$$

$$\chi^2_{1L} = 2.13 - 0.03 - 0.01 = 2.09$$

Each follows a chi-square distribution with one degree of freedom under the null hypothesis. Total goodness of fit statistics are sums over the 4 crosses. For example, the total goodness of fit statistic for the complete data set, total goodness of fit statistics for locus A and B and total linkage goodness of fit statistics are

$$\chi^2_T = 2.13 + 6.6 + 66.0 + 11.6 = 86.33$$

$$\chi^2_{AT} = 0.03 + 0.03 + 0.33 + 0.27 = 0.66$$

$$\chi^2_{BT} = 0.01 + 0.03 + 0.33 + 0.07 = 0.45$$

$$\chi^2_{LT} = 2.09 + 6.53 + 65.33 + 11.27 = 85.22$$

Pooled goodness of fit statistics for A, B and linkage are calculated using the marginal frequencies of the 4 genotypic classes over the 4 crosses. This is done under the assumption that there is no heterogeneity among the 4 crosses of segregation ratios for each of the loci and for the linkage relationships between the loci. The pooled goodness of fit statistics are

$$\chi^2_{PT} = \frac{(780-690)^2}{690} + \frac{(590-690)^2}{690} + \frac{(585-690)^2}{690} + \frac{(805-690)^2}{690} = 61.86$$

$$\chi^2_{PA} = \frac{(1370-1380)^2}{1380} + \frac{(1390-1380)^2}{1380} = 0.15$$

$$\chi^2_{PB} = \frac{(1365-1380)^2}{1380} + \frac{(1395-1380)^2}{1380} = 0.33$$

$$\chi^2_{PL} = 61.86 - 0.15 - 0.33 = 61.38$$

The differences between the total goodness of fit statistics and the pooled goodness of fit statistics are due to differences among the segregation ratios and linkage relation over the 4 crosses, which are

$$\chi^2_{HT} = 86.33 - 61.86 = 24.47 \qquad \chi^2_{HA} = 0.66 - 0.15 = 0.51$$

$$\chi^2_{HL} = 85.22 - 61.38 = 23.84 \qquad \chi^2_{HB} = 0.45 - 0.33 = 0.12$$

The total heterogeneity goodness of fit statistic follows a chi-square distribution with (4 - 1)(4 -1) = 9 degrees of freedom under the null hypothesis. The heterogeneity goodness of fit statistics for locus A, locus B and linkage between A and B follow chi-square distribution with 4 - 1 = 3 degrees of freedom under the null hypothesis.

No segregation distortion is found for the two loci for all 4 crosses. Significant linkage is found in three crosses: 2, 3 and 4. In addition, significant heterogeneity among the 4 crosses is found for the linkage relationship between the two loci. The significant goodness of fit statistic for heterogeneity is mainly contributed by the difference between cross#1 and the other crosses. The linkage heterogeneity goodness of fit statistic 0.08 is obtained among the three crosses 2, 3 and 4. In a practical experiment, this result could be evidence to support the hypothesis that cross#1 differs biologically from the other three crosses in terms of the relationship between the two loci A and B. For example, linkage between loci A and B cannot be detected using data of cross# 1. Significant linkage is detected using data from the other crosses.

Partitioning of Log Likelihood Ratio Test Statistic

Log likelihood ratio test statistics (G-statistics) have been commonly used in genomic analyses. The log likelihood ratio test statistics are derived from likelihood functions (Chapter 4). The two approaches (use of chi-square and log likelihood ratio test statistic) yield results that are numerically similar, although they are computed differently. For a set of data that comes from several of the same kind of crosses (e.g., the data in Table 6.1 are classified by genotype in the same way for all the four crosses), the log likelihood ratio test statistic can be partitioned in a similar way to the goodness of fit statistic. Equation (6.1) becomes the following

$$G_{iT} = 2\sum_j \sum_k o_{ijk}\log\frac{o_{ijk}}{0.25o_{i\circ\circ}} \qquad G_{PT} = 2\sum_j \sum_k o_{\circ jk}\log\frac{o_{\circ jk}}{0.25o_{\circ\circ\circ}}$$

$$G_{iA} = 2\sum_j o_{ij\circ}\log\frac{o_{ij\circ}}{0.5o_{i\circ\circ}} \qquad G_{PA} = 2\sum_j o_{\circ j\circ}\log\frac{o_{\circ jk\circ}}{0.5o_{\circ\circ\circ}} \qquad (6.2)$$

$$G_{iB} = 2\sum_k o_{i\circ k}\log\frac{o_{i\circ k}}{0.5o_{i\circ\circ}} \qquad G_{PB} = 2\sum_k o_{\circ\circ k}\log\frac{o_{\circ\circ k}}{0.5o_{\circ\circ\circ}}$$

for the log likelihood ratio test statistics and the relationships become

$$G_{iL} = G_{iT} - G_{iA} - G_{iB} \qquad G_{PL} = G_{PT} - G_{PA} - G_{PB}$$

$$G_T = \sum_i G_{iT} \qquad G_{AT} = \sum_i G_{iA}$$

$$G_{BT} = \sum_i G_{iB} \qquad G_{LT} = \sum_i G_{iL}$$

$$G_{HT} = G_T - G_{PT} \qquad G_{HA} = G_{AT} - G_{PA}$$

$$G_{HB} = G_{BT} - G_{PB} \qquad G_{HL} = G_{LT} - G_{PL}$$

The subscripts here have the same meaning as in Equation (6.1).

Table 6.3 The log likelihood ratio test statistic table for data in Table 6.1.

Cross	Total	Locus A	Locus B	Linkage
1	2.13	0.06(0.86)	0.01(0.91)	2.09(0.15)
2	6.65	0.03(0.86)	0.03(0.86)	6.58(0.01)
3	66.48	0.33(0.56)	0.33(0.56)	65.81(<0.0001)
4	11.77	0.27(0.61)	0.07(0.80)	11.43(0.0007)
Total	87.02	0.66	0.45	85.91
Pooled	62.03	0.15(0.70)	0.33(0.56)	61.55(<0.0001)
Heterogeneity	24.99	0.51(0.92)	0.12(0.99)	24.36(<0.0001)

Values in parentheses are the P-values.

Example: Partition of Log Likelihood Ratio Test Statistic

Table 6.3 lists a log likelihood ratio test statistic table for the same 4 crosses using the log likelihood ratio approach. For cross#1, the G-statistics are

$$G_{1T} = 2\left(310\log\frac{310}{300} + 287\log\frac{287}{300} + 288\log\frac{288}{300} + 315\log\frac{315}{300} \right) = 2.13$$

$$G_{1A} = 2\left(597\log\frac{597}{600} + 603\log\frac{603}{600} \right) = 0.03$$

$$G_{1B} = 2\left(598\log\frac{598}{600} + 602\log\frac{602}{600} \right) = 0.01$$

$$G_{1L} = 2.13 - 0.03 - 0.01 = 2.09$$

Total G-statistics are sums over the 4 crosses. In this example, the total G-statistics for the complete data set, for locus A, locus B and total linkage G-statistic are

$$G_T = 2.13 + 6.65 + 66.48 + 11.77 = 87.02$$

$$G_{AT} = 0.03 + 0.03 + 0.33 + 0.27 = 0.66$$

$$G_{BT} = 0.01 + 0.03 + 0.33 + 0.07 = 0.45$$

$$G_{LT} = 2.09 + 6.58 + 65.81 + 11.43 = 85.91$$

Pooled G-statistics for A and B and linkage are estimated using the marginal frequencies of the 4 genotypic classes over the 4 crosses under the assumption that there is no heterogeneity among the 4 crosses of segregation ratios for each of the loci and the linkage relationship between the loci, which are

$$G_{PT} = 2\left(780\log\frac{780}{690} + 590\log\frac{590}{690} + 585\log\frac{585}{690} + 805\log\frac{805}{690}\right) = 62.03$$

$$G_{PA} = 2\left(1370\log\frac{1370}{1380} + 1390\log\frac{1390}{1380}\right) = 0.15$$

$$G_{PB} = 2\left(1365\log\frac{1365}{1380} + 1395\log\frac{1395}{1380}\right) = 0.33$$

$$G_{PL} = 62.03 - 0.15 - 0.33 = 61.55$$

The log likelihood ratio test statistics for heterogeneity of segregation ratio and linkage among the 4 crosses are

$$G_{HT} = 87.02 - 62.03 = 24.99$$

$$G_{HA} = 0.66 - 0.15 = 0.51$$

$$G_{HB} = 0.45 - 0.33 = 0.12$$

$$G_{HL} = 85.91 - 61.55 = 24.36$$

6.2.2 A GENERALIZED LIKELIHOOD APPROACH

Log Likelihood Approach

The generalized log likelihood ratio approach is not only a necessary procedure for combining data from different kinds of crosses, but also an efficient approach for any kind of data when the number of loci is large. When populations have different mating systems or allelic configurations, the total goodness of fit statistic and log likelihood ratio test statistic cannot be partitioned clearly. The generalized log likelihood ratio approach can be used for the hypothesis tests, but, unlike the two previous approaches (use of chi-square and log likelihood ratio test statistic), an estimate of the parameter is required to construct the tests. Estimation of recombination fraction is presented in Section 6.3, although for now, estimates of the parameter θ will be denoted by $\hat{\theta}$. This approach has been implemented in computer software for genomic analysis, such as PGRI (Liu 1996).

The log likelihood function for multiple populations having different recombination fractions between loci A and B is defined as

$$L = \sum_{i=1}^{c}\sum_{j=1}^{n_{1i}}\sum_{k=1}^{n_{2i}} f_{oijk}\log p_{ijk} \tag{6.3}$$

where subscript i denotes cross, j and k denote genotypic classes for locus A and locus B, respectively, c is the number of crosses, f_{ijk} and p_{ijk} are the observed counts and expected frequencies of the genotypic class and n_{1i} and n_{2j} are the number of genotypic categories for locus A and locus B, respectively,

The log likelihood has different meanings depending on how p_{ijk} is evaluated. A single recombination fraction is assumed for all crosses. If p_{ijk} is evaluated for a particular recombination fraction (θ) between two loci across all crosses, then the log likelihood is written as $L(\theta)$. θ can be estimated by maximizing the overall likelihood function. If p_{ijk} is evaluated assuming different recombination fractions for each cross, then the log likelihood is written as $L(\theta_1, ..., \theta_c)$. $\theta_1, ..., \theta_c$ can be estimated by maximizing the likelihood functions for each cross individually. The likelihood ratio test statistic

$$G = 2 [L (\theta_1, ..., \theta_c) - L (\theta)]$$

will test the hypothesis that the populations share a common recombination fraction between loci A and B. Methods for obtaining the maximum likelihood recombination fractions will be illustrated in the next few sections.

Example: Log Likelihood Approach

The data in Table 6.1, which have a common classification across all populations of the four crosses, are used here to illustrate the log likelihood approach. Later in this chapter, an example of data from two populations, one backcross and one F2, will be used to illustrate how this approach can be used for data with mixed classification.

For the data in Table 6.1, recombination fractions are estimated for each of the four crosses and for the pooled data (Table 6.4) (see Section 6.3 for methodology of recombination fraction estimation). The log likelihoods for computing the log likelihood ratio test statistic for cross#1 are

$$L (\theta = 0.479) = (310 + 315) \log (1 - 0.479) + (287 + 288) \log 0.479 = -830.735$$

$$L (\theta = 0.5) = (310 + 315) \log (1 - 0.5) + (287 + 288) \log 0.5 = -831.777$$

$$G_1 = 2 (- 830.735 + 831.777) = 2.084$$

The total log likelihood ratio test statistic for linkage is the sum over the four crosses

$$G_T = 2.084 + 6.594 + 65.94 + 11.356 = 85.974$$

The log likelihoods for computing the log likelihood ratio test statistic for pooled data are

$$L (\theta = 0.4257) = (780 + 805) \log (1 - 0.4257) + (590 + 585) \log 0.4257$$
$$= -1882.52$$
$$L (\theta = 0.5) = (780 + 805) \log (1 - 0.5) + (590 + 585) \log 0.5 = -1913.086$$
$$G_P = 2 (- 1882.52 + 1913.086) = 61.132$$

The log likelihood ratio test statistic for heterogeneity is

$$G_H = 85.974 - 61.132 = 24.842$$

Table 6.4 The log likelihood ratio test statistic table for data in Table 6.1. Values in parentheses are the P-values.

Cross	$\hat{\theta}$	Log Likelihood		G-statistic for linkage
		$\theta = \hat{\theta}$	$\theta = 0.5$	
1	0.479	-830.735	-831.777	2.084
2	0.383	-79.881	-83.178	6.594
3	0.383	-798.807	-831.777	65.940
4	0.392	-160.677	-166.355	11.356
Total		-1870.100	-1913.087	85.974
Pooled	0.4257	-1882.520	-1913.086	61.132
Heterogeneity				24.842

The Lod Score

Lod score has been used extensively for human genetic linkage analysis (Ott 1992). Lod score is defined as the base-10 log likelihood ratio test statistic, which differs from the natural log likelihood ratio test statistic. The lod score (Z) is

$$Z = Log_{10}\left[\frac{L(\hat{\theta})}{L(0.5)}\right] = Log_{10}[L(\hat{\theta})] - Log_{10}[L(0.5)] \tag{6.4}$$

Example: Lod Score

Lod scores for linkage for the data in Table 6.1 are shown in Table 6.5. The computation of the lod score is straightforward. For example, the lod score for the cross#1 is

$$Z_1 = (310 + 315) Log_{10}\frac{1 - 0.479}{0.5} + (287 + 288) Log_{10}\frac{0.479}{0.5} = 0.4525$$

In practical linkage analysis, lod score 3 has been used by human geneticists as a criterion to define linkage. It means that linkage at $\theta = \hat{\theta}$ is 1,000 times more likely than at $\theta = 0.5$. However, different lod score criteria may be suitable for different organisms or even for different experiments.

Table 6.5   Lod score table for data in Table 6.1.

Cross	$\theta = \hat{\theta}$	Lod Score
1	0.479	0.4525
2	0.383	1.4317
3	0.383	14.3173
4	0.392	2.4657
Total		18.6673
Pooled	0.4257	13.2734
Heterogeneity		5.3939

6.3   RECOMBINATION FRACTION ESTIMATION

The inheritance of two different genes in a cross depends upon their chromosomal location and arrangement in the parent. The parental gene arrangement for the genes was determined by the two gametes that formed the parent. Recombination fraction may be defined as the probability that a parent produces a gamete that carries a recombinant (i.e., non-parental) gene arrangement. For a two-locus model, if genes A and B are linked and the grandparents have genotypes AABB and aabb, then the parent is AaBb. The gametes Ab and aB from AaBb are recombinant and AB and ab are non-recombinant (or parental types). By definition, A and B are linked in coupling. If the grandparents have genotypes AAbb and aaBB, then the AB and ab gametes are recombinant. In the latter case, A and B are linked in repulsion.

The maximum likelihood approach has been used extensively in estimating recombination fractions. In this section, methods for obtaining the maximum likelihood estimate of recombination fraction and the expected average linkage information content will be illustrated using backcross and F2 models.

Table 6.6 Expected genotypic frequencies for a backcross (AaBb X aabb) model. θ is the recombination fraction between A and B. f_{ij} is the observed genotypic count for the ith genotype of locus A and the jth genotype of B.

Genotype	Observed Count (f_{ij})	Expected Frequency (p_{ij})
AaBb	f_{11}	$0.5(1-\theta)$
Aabb	f_{12}	0.5θ
aaBb	f_{21}	0.5θ
aabb	f_{22}	$0.5(1-\theta)$

6.3.1 BACKCROSS MODEL

Table 6.6 shows the expected genotypic frequencies in a population of backcross progeny (AaBb X aabb). The likelihood function is

$$L(\theta) = \sum_{i=1}^{2}\sum_{j=1}^{2} f_{ij}\log p_{ij} = (f_{11}+f_{22})\log(1-\theta) + (f_{12}+f_{21})\log\theta \qquad (6.5)$$

It is also common to use L to denote the log likelihood function. By setting the first derivative of the function, with respect to the recombination fraction, to zero, we have

$$L_{BC}'(\theta) = \frac{f_{12}+f_{21}}{\theta} - \frac{f_{11}+f_{22}}{1-\theta} = 0 \qquad (6.6)$$

The maximum likelihood estimate of recombination fraction can be obtained by solving Equation (6.6)

$$\hat{\theta} = \frac{f_{12}+f_{21}}{N} \qquad (6.7)$$

where N is the total number of individuals in the sample. The average information content for an individual is

$$I_{BC}(\theta) = E_{\theta}\left[\frac{\partial}{\partial\theta}\log L_{\theta}(x)\right]^{2} = \frac{1}{\theta(1-\theta)} \qquad (6.8)$$

The variance of the estimated recombination fraction for sample size of N is

$$Var_{BC}(\hat{\theta}) = \frac{\theta(1-\theta)}{N} \qquad (6.9)$$

6.3.2 F2 MODEL

In many cases, such as the backcross model explained in the previous section, setting the first derivative to zero may yield simple solutions for recombination fractions. Approaches such as grid search, Newton-Raphson method and EM algorithm can be used to estimate recombination fractions for situations where simple solutions cannot be obtained.

Table 6.7 A two-locus model for two loci linked in coupling and recombination fraction θ in F2 progeny.

F1 Gamete Frequency	AB $0.5(1-\theta)$	Ab 0.5θ	aB 0.5θ	ab $0.5(1-\theta)$
AB $0.5(1-\theta)$	AABB $0.25(1-\theta)^2$	AABb $0.25\theta(1-\theta)$	AaBB $0.25\theta(1-\theta)$	AaBb $0.25(1-\theta)^2$
Ab 0.5θ	AABb $0.25\theta(1-\theta)$	AAbb $0.25\theta^2$	AaBb $0.25\theta^2$	Aabb $0.25\theta(1-\theta)$
aB 0.5θ	AaBB $0.25\theta(1-\theta)$	AaBb $0.25\theta^2$	aaBB $0.25\theta^2$	aaBb $0.25\theta(1-\theta)$
ab $0.5(1-\theta)$	AaBb $0.25(1-\theta)^2$	Aabb $0.25\theta(1-\theta)$	aaBb $0.25\theta(1-\theta)$	aabb $0.25(1-\theta)^2$

Table 6.8 Expected genotypic frequency for F2 progeny using codominant markers.

Genotype	Observed Count	Expected Frequency	$P_i(R|G)$
AABB	f_1	$0.25(1-\theta)^2$	0.0
AABb	f_2	$0.50\theta(1-\theta)$	0.5
AAbb	f_3	$0.25\theta^2$	1.0
AaBB	f_4	$0.50\theta(1-\theta)$	0.5
AaBb	f_5	$0.5(1-2\theta+2\theta^2)$	$\theta^2/[(1-\theta)^2+\theta^2]$
Aabb	f_6	$0.50\theta(1-\theta)$	0.5
aaBB	f_7	$0.25\theta^2$	1.0
aaBb	f_8	$0.50\theta(1-\theta)$	0.5
aabb	f_9	$0.25(1-\theta)^2$	0.0

An F2 progeny is produced by crossing with itself the F1 from a cross between two homozygous grandparents (AABB X aabb). The F1 has a genotype of AaBb. Table 6.7 shows the expected frequencies of the four gametes produced by F1 individuals and the expected frequencies of the 16 possible combinations for the four gametes. Each of the 16 combinations corresponds to an F2 genotype, shown in Table 6.8. The last column in Table 6.8 is the probability that the gamete is a recombinant conditional on the genotype for the two loci (this probability will be used for an illustration of EM algorithm later in this section). The log likelihood function for F2 progeny is

$$L(\theta) = \sum_{j=1}^{9} f_j \log p_j$$

which is

$$L(\theta) = 2(f_1+f_9)\log(1-\theta) + (f_2+f_4+f_6+f_8)\log[\theta(1-\theta)]$$
$$+ f_5\log(1-2\theta+\theta^2) + 2(f_3+f_7)\log\theta \tag{6.10}$$

The first derivative of the function with respect to a recombination fraction is

$$L'(\theta) = -\frac{2(f_1+f_9)}{1-\theta} + \frac{(f_2+f_4+f_6+f_8)(1-2\theta)}{\theta(1-\theta)} - \frac{2f_5(1-\theta)}{1-2\theta+2\theta^2} + \frac{2(f_3+f_7)}{\theta} \quad (6.11)$$

The average information content for an individual is

$$I_{F2}(\theta) = E_\theta\left[\left[\frac{\partial}{\partial\theta}\log L_\theta(x)\right]^2\right] = \frac{2(1-3\theta+3\theta^2)}{\theta(1-\theta)(1-2\theta+2\theta^2)} \quad (6.12)$$

The variance of the estimated recombination fraction using F2 population with sample size N is

$$Var_{F2}(\theta) = \frac{\theta(1-\theta)(1-2\theta+2\theta^2)}{2N(1-3\theta+3\theta^2)} \quad (6.13)$$

Setting Equation (6.10) to zero does not provide a simple solution for θ. Alternative approaches are required to solve for θ. Equation (6.10) is for situations where both A and B are codominantly inherited. In the next few sections, the likelihood profile approach, the Newton-Raphson method and the EM algorithm will be illustrated using the example data in Table 6.9.

Table 6.9 Two-locus model with two different mating types.

Population	Genotype	Observed Count
F2	A_B_	140
	A_bb	10
	aaB_	10
	aabb	40
Backcross	AaBb	162
	Aabb	40
	aaBb	40
	aabb	158

Example: Data

Table 6.9 shows an example with two dominant genes segregating in an F2 population and a backcross population. Loci A and B are linked in coupling linkage phase. There are 200 individuals in the F2 progeny and 400 in the backcross progeny. The objectives of the experiment are to test if there is any recombination fraction heterogeneity between loci A and B in the two populations and to estimate the pooled recombination fraction if heterogeneity does not exist.

6.3.3 LIKELIHOOD PROFILE METHOD

The expected genotypic frequencies for an F2 population using dominant markers for coupling linkage phase are shown in Table 6.10. Using the expected frequencies in Table 6.10, the likelihood function can be written as

$$L_{F2D}(\theta) = \sum_{j=1}^{4} f_j\log p_j \quad (6.14)$$

$$= f_1\log(3-2\theta+\theta^2) + (f_2+f_3)\log[\theta(2-\theta)] + 2f_4\log(1-\theta)$$

For the F2 data in Table 6.9, the likelihood is

$$L_{F2D}(\theta) = 140\log(3 - 2\theta + \theta^2) + 40\log[\theta(2 - \theta)] + 80\log(1 - \theta) \quad (6.15)$$

The likelihood function for the backcross data in the table is

$$L_{BC}(\theta) = 320\log(1 - \theta) + 80\log\theta \quad (6.16)$$

The joint likelihood for the whole data set in Table 6.9 is

$$L_{Pool}(\theta) = L_{BC}(\theta) + L_{F2D}(\theta) \quad (6.17)$$

Table 6.10 Expected genotypic frequencies for an F2 population using dominant markers linked in coupling.

Genotype	Observed Count (f_i)	Expected Frequency (p_i)	$P_i(R\|G)$
A_B_	f_1	$0.25(3 - 2\theta + \theta^2)$	$2\theta/(3 - 2\theta + \theta^2)$
A_bb	f_2	$0.25\theta(2 - \theta)$	$1/(2 - \theta)$
aaB_	f_3	$0.25\theta(2 - \theta)$	$1/(2 - \theta)$
aabb	f_4	$0.25(1 - \theta)^2$	0.0

Example: Graphic Approach

The likelihood of Equation (6.17) can be graphed against the recombination fraction (Figure 6.1), resulting in a likelihood profile. The likelihood in Figure 6.1 is adjusted to a 0.0 to 1.0 scale using a constant. The peaks of the curves correspond to maximum likelihood estimates. For the F2

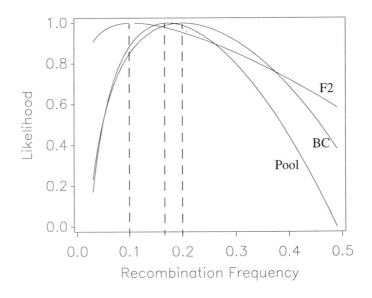

Figure 6.1 Log likelihood profile for the data in Table 6.9.

population, the likelihood peak is at $\hat{\theta} = 0.1$, for the backcross population it is at $\hat{\theta} = 0.2$ and for the pooled likelihood it is at $\hat{\theta} = 0.18$. This approach to estimation of recombination fraction is called the profile likelihood method. In practice, this is done using a grid search. Starting with an initial value of θ, the direction of searching is determined by likelihoods on both sides of θ. The search moves in the direction of increasing likelihood.

The speed with which the maximum likelihood estimates can be obtained depends on the nature of the likelihood function and the searching algorithm. The profile likelihood approach has been effective for problems with single peak likelihood curves. When the likelihood curves have multiple peaks, this approach can produce global maximum likelihood estimates. However, the search increment should be globally small. The amount of computation needed is extensive.

By graphing the likelihood against the parameter, a solution for the likelihood function can be obtained. However, the methods can be computationally intensive when multiple parameters are involved in the models. In addition, the advantages of the profile likelihood method, such as visualization of the solution, disappear for multiple-parameter problems.

6.3.4 NEWTON-RAPHSON ITERATION FOR A SINGLE PARAMETER

Newton-Raphson iteration is a method for solving non-linear equations. It can solve the likelihood equations for obtaining maximum likelihood estimates.

As an example, let us use the likelihood function of Equation (6.15) for dominant markers in coupling linkage phase in F2 progeny (the data in Table 6.9). The score is

$$\frac{dS(\theta)}{d\theta} = \frac{140(2\theta - 2)}{3 - 2\theta + \theta^2} + \frac{20(2 - 2\theta)}{\theta(2 - \theta)} - \frac{80}{1 - \theta} \tag{6.18}$$

Let us use θ' as the first guess of the solution $\hat{\theta}$. By Taylor series expansion, we have

$$\frac{d[S(\theta'')]}{d\theta} = \frac{d[S(\theta')]}{d\theta} + (\theta'' - \theta')\frac{d^2[S(\theta')]}{d\theta^2} = 0 \tag{6.19}$$

where N is number of informative observations and θ'' is the solution for this round of iteration, which is

$$\theta'' = \theta' - \frac{\dfrac{d[S(\theta')]}{d\theta}}{\dfrac{d^2[S(\theta')]}{d\theta^2}} \tag{6.20}$$

if only the first two terms of the Taylor series are considered. The new estimate is obtained by adding to the previous estimate (or initial value for the first iteration) the score divided by the information. The iteration continues until the parameter converges at a certain tolerance or the second term of Equation (6.20) is less than a tolerance (e.g., 0.00001). For dominant markers in coupling linkage phase in F2 progeny, the average information for a single observation is

$$I_{F2D}(\theta) = \frac{2(3 - 4\theta + 2\theta^2)}{\theta(2 - \theta)(3 - 2\theta + \theta^2)} \tag{6.21}$$

Newton-Raphson iteration has a direct geometrical interpretation. Each iteration fits a parabola to the likelihood function at the point of the parameter. A new parameter is obtained by maximizing the parabola. If the likelihood curve is truly a parabola, then the solution can be obtained by a single iteration.

For most of the common single-parameter problems related to genomic mapping, Newton-Raphson iteration yields maximum likelihood estimates. However, the method may fail when the likelihood curve has multiple peaks or when the information is zero. For problems related to genomic mapping, zero information usually happens when the estimate is 0.0, 0.5, or 1.0. It is suggested that different initial values are used as starting points for the iterations and that boundary conditions are used for the new estimates.

Table 6.11 Newton-Raphson iteration for obtaining maximum likelihood estimates of recombination fractions for data in Table 6.9

Iteration	F2	Backcross	Pooled
0	0.25	0.25	0.25
1	0.091667	0.2	0.17159
2	0.10547	0.2	0.17226
3	0.10557		0.17226
4	0.10557		

Example: Newton-Raphson Iteration

For the data in Table 6.9, the Newton-Raphson iteration results are shown in Table 6.11. The initial values for the three estimates are all 0.25. The score for the F2 population is Equation (6.18); for the backcross it is

$$score\,(\theta) \;=\; \frac{80}{\theta} - \frac{320}{1-\theta} \tag{6.22}$$

and for the pool it is the sum of Equation (6.18) and Equation (6.22)

$$score\,(\theta) \;=\; \frac{140\,(2\theta-2)}{3-2\theta+\theta^2} + \frac{20\,(2-2\theta)}{\theta\,(2-\theta)} - \frac{80}{1-\theta} + \frac{80}{\theta} - \frac{320}{1-\theta}$$

Iteration starts with a recombination fraction of 0.25 for the three estimators. The estimator for the backcross population converged with one iteration. The estimator for the F2 population converged at the fourth iteration. The pooled estimator converged at the third iteration. We have the estimated recombination fractions between loci A and B in the two populations and pooled populations as

$$\hat{\theta}_{F2} = 0.10557$$

$$\hat{\theta}_{BC} = 0.2$$

$$\hat{\theta}_{Pooled} = 0.17226$$

6.3.5 EM ALGORITHM

The EM algorithm for recombination fraction estimation involves four steps:

(1) Make an initial guess: θ^{old}.

(2) <u>E</u>xpectation step: Using θ^{old} as if it were the true recombination fraction, compute the expected number of recombinants.

(3) <u>M</u>aximization step: Using the expected value, compute the maximum likelihood estimate θ^{new} for the recombination fraction.

(4) Iterate the E and M steps until the likelihood reaches its maximum, or the estimate converges:

$$\left|\theta^{new} - \theta^{old}\right| \leq tolerance$$

The maximum likelihood estimate of recombination fraction for a new iteration using the EM algorithm is

$$\theta^{new} = \frac{1}{N}\sum f_i P_i (R|G) \tag{6.23}$$

where $P_i(R|G)$ is the probability of an individual with category i genotype being a recombinant when θ^{old} is the true recombination fraction. Tables 6.7, 6.9, 6.11 and 6.12 list the expected genotypic frequencies and the conditional probabilities $P_i(R|G)$ for F2 populations under different genetic marker combinations.

Table 6.12 Expected genotypic frequency for an F2 population using dominant markers in repulsion linkage phase.

Genotype	Observed Count	Expected Frequency (p_i)	$P_i(R|G)$
A_B_	f_1	$0.25(2 + \theta^2)$	$\theta(2 + \theta)/(2 + \theta^2)$
A_bb	f_2	$0.25(1 - \theta^2)$	$\theta/(1 + \theta)$
aaB_	f_3	$0.25(1 - \theta^2)$	$\theta/(1 + \theta)$
aabb	f_4	$0.25\theta^2$	1.0

Table 6.13 Expected genotypic frequency for an F2 population using both codominant and dominant markers.

Genotype	Observed Count	Expected Frequency (p_i)	$P_i(R|G)$
AAB_	f_1	$0.25(1 - \theta^2)$	$\theta/(1 + \theta)$
AAbb	f_2	$0.25\theta^2$	1.0
AaB_	f_3	$0.5(1 - \theta + \theta^2)$	$\theta(1 + \theta)/(1 - \theta + \theta^2)$
Aabb	f_4	$0.5\theta(1 - \theta)$	0.5
aaB_	f_5	$0.25\theta(2 - \theta)$	$1/(2 - \theta)$
aabb	f_6	$0.25(1 - \theta)^2$	0.0

Example: EM Algorithm

Table 6.14 shows the EM steps for estimating recombination fractions for each of the populations and for pooled estimation for data in Table 6.9. The maximum likelihood estimates of recombination fractions using the EM algorithm for the F2, backcross and pooled (over the two populations) are

0.10558, 0.2 and 0.17227, respectively. The estimates differ slightly from the Newton-Raphson solutions (Table 6.11); however, the differences may be due to rounding error. For the data in Table 6.9, the EM algorithm converges more slowly than does the Newton-Raphson iteration. Starting from 0.25, the estimates for the F2 and pooled converge at the 18th and 9th iteration. The advantages of recombination fraction estimation using the EM algorithm over the Newton-Raphson iteration are:

(1) there is no need to know the first and second derivatives of the likelihood functions and

(2) the EM iterations involve simple calculations compared to the Newton-Raphson iterations.

Table 6.14 EM estimation of recombination fractions for data in Table 4.9.

Iteration	F2	Backcross	Pooled
0	0.25	0.25	0.25
1	0.19373	0.2	0.19686
2	0.15771		0.17983
3	0.13577		0.17457
4	0.12284		0.17296
5	0.11537		0.17248
6	0.11111		0.17233
7	0.10869		0.17228
8	0.10733		0.17227
9	0.10656		0.17227
...	...		
18	0.10558		

Example: Heterogeneity Test

The generalized likelihood ratio approach for testing significant differences among estimated recombination fractions is estimated from different populations. This was introduced in Section 6.2.2. For data in Table 6.9, the likelihood ratio test statistic can be calculated for the F2

$$L_{F2}(\hat{\theta}) = 140\log(3 - 2\hat{\theta} + \hat{\theta}^2) + 20\log[\hat{\theta}(2 - \hat{\theta})] + 2 \times 40\log(1 - \hat{\theta})$$
$$= 144.146 - 32.187 - 8.926 = 103.03$$
$$L_{F2}(0.5) = 140\log(2.25) + 20\log(0.75) + 2 \times 40f_4\log(0.5) = 52.33$$
$$G_{F2} = 2[L_{F2}(\hat{\theta}) - L_{F2}(0.5)] = 2(103.03 - 52.33) = 101.4$$

the backcross

$$L_{BC}(\hat{\theta}) = 320\log(1 - \hat{\theta}) + 80\log\hat{\theta} = -200.16$$
$$L_{BC}(0.5) = 320\log(0.5) + 80\log 0.5 = -277.1$$
$$G_{BC} = 2[L_{BC}(\hat{\theta}) - L_{BC}(0.5)] = 2(-200.16 + 277.1) = 154.18$$

and the pooled over the two populations

$$L_{pool}(\hat{\theta}) = 140\log(3 - 2\hat{\theta} + \hat{\theta}^2) + 20\log[\hat{\theta}(2 - \hat{\theta})] + 2 \times 40\log(1 - \hat{\theta})$$
$$+ 320\log(1 - \hat{\theta}) + 80\log\hat{\theta}$$
$$= 138.282 - 23.112 - 15.126 - 60.502 - 140.695 = -101.15$$
$$L_{pool}(0.5) = L_{F2}(0.5) + L_{BC}(0.5) = 52.325 - 277.097 = -224.77$$
$$G_{pool} = 2[L_{pool}(\hat{\theta}) - L_{pool}(0.5)] = 2(-101.15 + 224.77) = 247.24$$

The total log likelihood ratio test statistic is

$$G_{total} = G_{F2} + G_{BC} = 255.6$$

The log likelihood ratio test statistic for heterogeneity is

$$G_{heterogeneity} = G_{total} - G_{pool} = 8.36$$

The test statistics are summarized in Table 6.15. Each statistic is distributed as a chi-square variable with one degree of freedom and all the test statistics are highly significant. A significant heterogeneity means that recombination fractions are different between the estimates using the F2 and backcross populations for the two genes.

Table 6.15 Likelihood ratio test for linkage between loci A and B using the data in Table 6.9.

	F2	Backcross	Pooled	Heterogeneity
G	101.42	154.18	247.237	8.36
$P(\chi^2 \geq G)$	<0.00001	<0.00001	<0.00001	<0.01
$\hat{\theta}$	0.10558	0.20	0.17227	

6.4 STATISTICAL PROPERTIES

Statistical properties of an estimator, such as variance, bias, distribution and confidence intervals, are critical for interpreting estimates of recombination fractions. For the backcross model, the estimated recombination fraction is distributed as a binomial variable. Variance, distribution function and ways to build confidence intervals for a binomial variable can therefore be applied to the estimated recombination fraction. However, statistical properties for estimated recombination fractions using populations other than the backcross are largely based on the optimum properties of a maximum likelihood estimate. For a larger sample, a maximum likelihood estimate is asymptotically normally distributed with a mean given by the parameter and variance given by the inverse of the information content. In this section, the way to obtain statistical properties empirically will be described, using an F2 population as the example. Empirical statistical properties will be compared with theoretical properties.

Table 6.16 Average information content for backcross and F2 models.

Population and Marker	Average Information Content per Observation $I(\theta)$
Backcross model	$\dfrac{1}{\theta(1-\theta)}$
F2 codominant markers	$\dfrac{2(1-3\theta+3\theta^2)}{\theta(1-\theta)(1-2\theta+2\theta^2)}$
F2 dominant markers in coupling phase	$\dfrac{2(3-4\theta+2\theta^2)}{\theta(2-\theta)(3-2\theta+\theta^2)}$
F2 dominant markers in repulsion phase	$\dfrac{2(1+2\theta^2)}{(2+\theta^2)(1-\theta^2)}$
F2 codominant and dominant markers	$2+\dfrac{(1-\theta)^2}{\theta(2-\theta)}+\dfrac{(1-2\theta)^2}{2(1-\theta+\theta^2)}+\dfrac{\theta^2}{1-\theta^2}+\dfrac{(1-2\theta)^2}{2\theta(1-\theta)}$

Table 6.17 Parametric standard deviations for F2 models using the inverse of expected information content.

True θ	Sample Size	CC	CD	DD-c	DD-r
0.01	50	0.010	0.014	0.014	0.141
	100	0.007	0.010	0.010	0.100
	200	0.005	0.007	0.007	0.071
	400	0.004	0.005	0.005	0.050
	1000	0.002	0.003	0.003	0.032
0.05	50	0.022	0.031	0.032	0.141
	100	0.016	0.022	0.022	0.100
	200	0.011	0.016	0.016	0.070
	400	0.008	0.011	0.011	0.050
	1000	0.005	0.007	0.007	0.032
0.1	50	0.032	0.044	0.045	0.140
	100	0.022	0.031	0.032	0.099
	200	0.016	0.022	0.023	0.070
	400	0.011	0.016	0.016	0.049
	1000	0.007	0.010	0.010	0.031
0.2	50	0.046	0.062	0.065	0.135
	100	0.032	0.044	0.046	0.095
	200	0.023	0.031	0.032	0.067
	400	0.016	0.022	0.023	0.048
	1000	0.010	0.014	0.014	0.030
0.3	50	0.057	0.075	0.080	0.127
	100	0.041	0.053	0.057	0.090
	200	0.029	0.038	0.040	0.063
	400	0.020	0.027	0.028	0.045
	1000	0.013	0.017	0.018	0.028
0.4	50	0.067	0.084	0.094	0.117
	100	0.047	0.059	0.066	0.083
	200	0.033	0.042	0.047	0.059
	400	0.024	0.030	0.033	0.041
	1000	0.015	0.019	0.021	0.026

In this table, CC = both codominant markers, CD = one codominant and one dominant marker, DD-c = both dominant markers in coupling linkage phase and DD-r = both dominant markers in repulsion linkage phase.

Table 6.18 Empirical standard deviation for F2 models using parametric bootstrapping. See Table 6.17 for notation.

True θ	Sample Size	CC	CD	DD-c	DD-r
0.01	50	0.010	0.014	0.014	0.026
	100	0.007	0.010	0.010	0.009
	200	0.005	0.007	0.007	0.009
	400	0.004	0.005	0.005	0.009
	1000	0.002	0.003	0.003	0.010
0.05	50	0.023	0.032	0.033	0.054
	100	0.016	0.022	0.023	0.049
	200	0.011	0.016	0.016	0.048
	400	0.008	0.011	0.011	0.044
	1000	0.005	0.007	0.007	0.037
0.1	50	0.032	0.046	0.047	0.095
	100	0.023	0.032	0.032	0.089
	200	0.016	0.022	0.023	0.078
	400	0.011	0.016	0.016	0.064
	1000	0.007	0.010	0.010	0.038
0.2	50	0.046	0.063	0.066	0.156
	100	0.033	0.044	0.046	0.126
	200	0.023	0.031	0.032	0.088
	400	0.016	0.022	0.023	0.054
	1000	0.010	0.014	0.014	0.031
0.3	50	0.059	0.077	0.086	0.177
	100	0.041	0.054	0.058	0.117
	200	0.029	0.038	0.041	0.071
	400	0.020	0.027	0.028	0.046
	1000	0.013	0.017	0.018	0.029
0.4	50	0.067	0.085	0.107	0.163
	100	0.048	0.059	0.069	0.096
	200	0.034	0.042	0.048	0.061
	400	0.023	0.029	0.033	0.042
	1000	0.015	0.019	0.021	0.026

6.4.1 VARIANCE AND BIAS

Parametric Variance

Variances of estimated recombination fractions can be estimated from average information content. The variance of a maximum likelihood estimate from a sample size of N is

$$\sigma^2(\hat{\theta}) = \frac{1}{NI(\theta)} \tag{6.24}$$

where $I(\theta)$ is the average information content per observation. Table 6.16 lists equations for computing the expected information contents for backcross and F2 models using different marker combinations. For the backcross model, the estimated recombination fraction has simple statistical properties. The variance is a simple binomial variance and the estimate is unbiased. The expectation of estimated recombination fraction using the backcross model is

$$E[\hat{\theta}] = \frac{E[f_{12}+f_{21}]}{N}$$

$$= \frac{0.5N\theta + 0.5N\theta}{N} = \theta$$

(6.25)

Table 6.19 Empirical bias for F2 models using parametric bootstrapping. See Table 6.17 for notation.

True θ	Sample Size	CC	CD	DD-c	DD-r
0.01	50	<0.001	<0.001	<0.001	0.008
	100	<0.001	<0.001	<0.001	0.010
	200	<0.001	<0.001	<0.001	0.009
	400	<0.001	<0.001	<0.001	0.009
	1000	<0.001	<0.001	<0.001	0.008
0.05	50	-0.001	-0.001	-0.001	0.039
	100	<0.001	<0.001	<0.001	0.038
	200	<0.001	<0.001	<0.001	0.032
	400	<0.001	<0.001	<0.001	0.027
	1000	<0.001	<0.001	<0.001	0.017
0.1	50	-0.001	<0.001	-0.001	0.064
	100	<0.001	<0.001	-0.001	0.053
	200	<0.001	<0.001	<0.001	0.040
	400	<0.001	<0.001	<0.001	0.022
	1000	<0.001	<0.001	<0.001	0.007
0.2	50	-0.001	-0.001	-0.003	0.076
	100	<0.001	<0.001	-0.001	0.048
	200	<0.001	<0.001	-0.001	0.020
	400	<0.001	<0.001	<0.001	0.008
	1000	<0.001	<0.001	<0.001	0.002
0.3	50	-0.003	-0.003	-0.007	0.061
	100	-0.001	<0.001	-0.002	0.024
	200	<0.001	<0.001	-0.001	0.008
	400	<0.001	<0.001	<0.001	0.004
	1000	<0.001	<0.001	<0.001	0.001
0.4	50	-0.002	-0.002	-0.011	0.037
	100	-0.001	-0.001	-0.004	0.014
	200	<0.001	<0.001	-0.001	0.005
	400	<0.001	<0.001	-0.001	0.002
	1000	<0.001	<0.001	<0.001	0.001

Empirical Variance and Bias

For F2 models, the variance of recombination fraction can be estimated using the inverse of expected information content. However, there is no straightforward way to obtain the bias of the estimates. Resampling techniques such as bootstrapping can be used to obtain empirical bias and variance (see Chapter 19). The empirical estimates were obtained using parametric bootstrapping. The same bootstrapping approach was also used to obtain empirical distributions of the estimated recombination fraction. Cumulative genotypic frequencies for the F2 models can be obtained using the expected frequencies in Tables 6.9, 6.10, 6.12 and 6.13 for different expected recombination fractions. Newton-Raphson iteration was used to obtain the maximum likelihood estimates of recombination fractions. b bootstrap replications were simulated. Mean, variance and bias were empirically estimated using the b bootstrap estimates by

$$Mean = \bar{\theta} = \frac{1}{b}\sum \hat{\theta}_i$$

$$Variance = \frac{1}{b}\sum (\hat{\theta}_i - \bar{\theta})^2 \qquad (6.26)$$

$$Bias = \theta - \bar{\theta}$$

where $\hat{\theta}_i$ is the recombination fraction estimate for the ith bootstrap sample.

Table 6.17 and Table 6.18 show standard deviations of the recombination fraction estimates for several F2 models. The standard deviations for the two approaches are close to each other for the codominant marker model, codominant and dominant marker model and dominant marker coupling model.

For dominant markers in repulsion linkage phase, the two approaches result in similar standard deviation estimates when the true recombination fraction is larger than 0.2 and sample size is greater than 200. It has long been recognized that dominant markers in repulsion linkage phase have low linkage information content in F2 populations (Mather 1951; Allard 1956). This problem is gaining more attention as dominant markers are widely used in genomic research due to advantages of PCR based markers (Tingey *et al.* 1992; Zabeau 1993). Knapp *et al.* (1995) describe a comprehensive study on mapping dominant markers using F2 populations. The variances of estimated recombination fractions can be estimated using the parametric approach for F2 models, except when two dominant markers are linked in repulsion linkage phase. Biases of estimated recombination fractions are small for F2 models, except when two dominant markers linked in repulsion linkage phase (Table 6.19).

6.4.2 DISTRIBUTION AND CONFIDENCE INTERVALS

Distribution

For the backcross model, the estimated recombination fractions are distributed as binomial variables. For F2 models, a parametric bootstrap (see Chapter 19) was used to obtain empirical distributions of the estimated recombination fractions. Figure 6.2 shows the empirical distributions.

The estimated recombination fractions are approximately distributed as binomial variables for the CC, CD and DD-c models. As the true recombination fraction approaches 0.5 and sample size increases, the distributions approach normal. However, the distribution for the DD-r model is irregular.

Confidence Intervals

For $\hat{\theta}$, an estimate of recombination fraction, a confidence interval is defined as (T_1, T_2), for which

$$P_\theta [T_1 < \hat{\theta} < T_2] \equiv \gamma$$

where γ is the confidence coefficient and T_1 and T_2 are lower and upper confidence limits, respectively. Normal approximations, likelihood approaches and nonparametric approaches are commonly used methods for constructing confidence intervals. For a large sample, the maximum likelihood estimate of recombination fraction is approximately normally distributed about the true recombination fraction and the variance of the estimate is the inverse of the

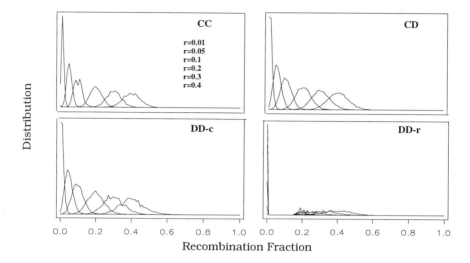

Figure 6.2 Empirical distributions of estimated recombination fractions for F2 models. The distributions from right to left are for true recombination fractions 0.01, 0.05, 0.1, 0.2, 0.3 and 0.4, respectively. In this figure, CC = both codominant markers, CD = one codominant and one dominant marker, DD-c = both dominant markers with coupling linkage phase and DD-r = both dominant markers with repulsion linkage phase. Sample size $N = 100$.

information content. The confidence interval can be obtained using a normal approximation approach. For a large sample

$$\frac{T - \hat{\theta}}{sd_\theta}$$

is distributed approximately standard normal. A confidence interval with confidence coefficient γ can be determined by converting the inequalities

$$-z_{0.5(1-\gamma)} \leq \frac{T - \hat{\theta}}{sd_\theta} \leq z_{0.5(1-\gamma)} \tag{6.27}$$

where sd_θ is the estimated standard error of the maximum likelihood estimate of a recombination fraction and $z_{0.5(1-\gamma)}$ is the normal deviate which satisfies

$$P\left[-z_{0.5(1-\gamma)} \leq z \leq z_{0.5(1-\gamma)}\right] = \gamma$$

$z_{0.5(1-\gamma)}$ can also be defined by

$$\Phi\left[z_{0.5(1-\gamma)}\right] = 1 - 0.5(1-\gamma)$$

$$\Phi\left[-z_{0.5(1-\gamma)}\right] = 0.5(1-\gamma)$$

So

$$\Phi\left[z_{0.5(1-\gamma)}\right] - \Phi\left[-z_{0.5(1-\gamma)}\right] = \gamma$$

The confidence interval is

$$[\hat{\theta} - sd_\theta z_{0.5(1-\gamma)}, \hat{\theta} + sd_\theta z_{0.5(1-\gamma)}] \tag{6.28}$$

A confidence interval can also be determined using nonparametric approaches. A bootstrap approach will be explained here (see Chapter 19 for a full discussion on bootstrap). For linkage data, a bootstrap sample can be obtained by randomly drawing from the data set with replacement. If b independent bootstrap samples are drawn, then b bootstrap estimates of recombination fractions can be obtained. If we use

$$\widehat{CDF}(x) = P[\hat{\theta}_b \leq x]$$

to denote the cumulative distribution of the bootstrap estimates of recombination fraction, then the confidence interval with the confidence coefficient γ is

$$\{\widehat{CDF}^{-1}[0.5(1-\gamma)], \widehat{CDF}^{-1}[0.5(1+\gamma)]\} \tag{6.29}$$

The bootstrap approach is effective when the distribution of the estimate is unknown or complex. This approach is recommended for recombination fraction estimation using complex genetic models.

Example: Confidence Interval (Normal Approximation)

For data in Table 6.9, 95% confidence intervals were determined using both a normal approximation and bootstrapping (Table 6.20). The standard deviation for the F2 population is

$$\hat{sd}_{F2D(\theta)} = \sqrt{\frac{\hat{\theta}(2-\hat{\theta})(3-2\hat{\theta}+\hat{\theta}^2)}{2N_{F2}(3-4\hat{\theta}+2\hat{\theta}^2)}}$$

$$= \sqrt{\frac{0.10558(2-0.10558)(3-2\times0.10558+0.10558^2)}{2\times400(3-4\times0.10558+2\times0.10558^2)}}$$

$$= 0.023$$

for the backcross population it is

$$\hat{sd}_{BC(\theta)} = \sqrt{\frac{\hat{\theta}(1-\hat{\theta})}{N_{BC}}}$$

$$= \sqrt{0.2\times0.8/400} = 0.02$$

and for the pooled data from the F2 and backcross populations it is

$$\hat{sd}_{Pool(\theta)} = \sqrt{\frac{1}{N_{F2}+N_{BC}}\left[\frac{2(3-4\hat{\theta}+2\hat{\theta}^2)}{\hat{\theta}(2-\hat{\theta})(3-2\hat{\theta}+\hat{\theta}^2)}+\frac{1}{\hat{\theta}(1-\hat{\theta})}\right]}$$

$$= 0.011$$

The 95% confidence intervals for the three recombination fraction estimates using the normal approximation are

$$(0.10558 \pm 1.96 \times 0.023) = (0.06, 0.15) \qquad F2$$

$$(0.2 \pm 1.96 \times 0.02) = (0.16, 0.24) \qquad Backcross$$

$$(0.17227 \pm 1.96 \times 0.011) = (0.15, 0.19) \qquad Pooled$$

Table 6.20 Confidence intervals for data in Table 6.9.

	F2	Backcross	Pooled Data
Point Estimate	0.106	0.20	0.172
Standard Deviation	0.023	0.02	0.011
Normal Approximation (Lower-Upper) (95%)	0.06 - 0.15	0.16 - 0.24	0.15 - 0.19
Bootstrap (Lower-Upper) (95%)	0.07 - 0.14	0.16 - 0.24	0.15 - 0.19
Likelihood (95%)	0.065 - 0.159	0.162 - 0.242	0.142 - 0.206
Lod score support 1	0.063 - 0.163	0.159 - 0.245	0.140 - 0.209

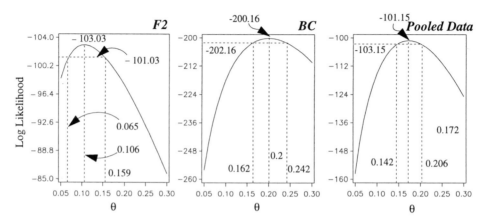

Figure 6.3 Estimation of confidence intervals for the example data in Table 6.9 using likelihood approach. A difference of 2 between the maximum log likelihood and the likelihood at the confidence interval approximately corresponds to a confidence probability 95%.

Example: Confidence Interval (Bootstrap)

Bootstrap samples were drawn from data in Table 6.9 for 1,000 replications. For each replication, bootstrap recombination fraction estimates were obtained for the F2, backcross and pool data. The bootstrap 95% confidence intervals were determined using the 2.5 and 97.5 percentiles of the cumulative distributions of the bootstrap estimates.

Example: Confidence Interval (Likelihood Approach)

An approximate 95% confidence interval can also be obtained by reducing the log likelihood two units from the maximum (Figure 6.3) (see Chapter 4 and see Chapter 9 in Edwards 1992). The lower and upper limits of the confidence interval are the values corresponding to the reduced likelihood. For the example data in Table 6.9, they are

$$(0.065, 0.159) \qquad F2$$
$$(0.162, 0.242) \qquad Backcross$$
$$(0.142, 0.206) \qquad Pooled$$

Example: Confidence Interval (Lod Score Support)

A confidence interval can be also quantified in terms of lod score support. The lod score support for an interval is the difference between the likelihood

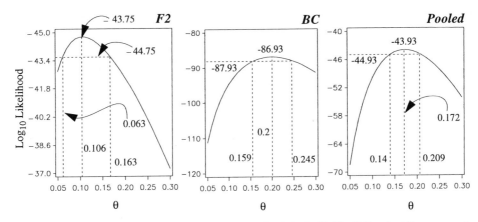

Figure 6.4 Confidence intervals for the example data in Table 6.9 using the concept of lod score support. The arrows correspond to a lod score support of 1.

corresponding to the lower and upper limits of the interval and the maximum likelihood calculated in a $\log_{10}$ scale. For example, the confidence intervals with lod score support 1 for the data in Table 6.9 are (Figure 6.4)

$$(0.063, 0.163) \qquad F2$$
$$(0.159, 0.245) \qquad Backcross$$
$$(0.140, 0.209) \qquad Pooled$$

Quality of a Confidence Interval

The validity of a confidence interval is determined by ranges of the intervals and the probability that the interval includes the parameter or the coverage probability. If the ranges of the intervals are large and have no biological value, then the intervals are not valid in a practical sense. If the coverage probability is significantly lower or higher than the confidence coefficient, then the confidence interval is not valid either.

A computer simulation was done to determine the validity of the normal approximation and the bootstrap approaches for constructing confidence intervals for estimated recombination fraction. Samples were simulated for true recombination fractions of 0.05, 0.1 and 0.3 using the backcross model and four types of F2 models (CC, CD, DD-c and DD-r (Table 6.21) (see Table 6.17 on definitions). Ranges of the confidence intervals were recorded as the differences between the lower and upper limits of the intervals. Coverage probability was calculated by counting the times that the interval contains the true value. Ranges for the bootstrap confidence intervals are consistently narrower than for the normal approximation intervals. This is especially true for the F2-DD-r model. The differences between the two methods could make a significant difference in practical genomic mapping experiments. For example, a confidence interval (0, 0.39) using the normal approximation approach for combination of true recombination fraction 0.1, the F2-DD-r model and sample size 100 may not be a meaningful interval in practice. However, a confidence interval (0, 0.2) using the bootstrap approach could be attractive.

Table 6.22 lists the corresponding coverage probability for the two approaches. All probabilities for the normal approximation intervals are consistently higher than the expected coverage of 95%.

Table 6.21 Average range of 95% confidence interval for F2 models using the normal approximation and the bootstrap approaches. See Table 6.17 for notation.

True θ	Sample Size	CC	CD	DD-c	DD-r
		Normal Approximation:			
.05	100	0.0609	0.0798	0.0805	0.3909
	200	0.0440	0.0607	0.0612	0.2758
.1	100	0.0893	0.1241	0.1273	0.3873
	200	0.0620	0.0861	0.0877	0.2735
.3	100	0.1577	0.2039	0.2170	0.3472
	200	0.1128	0.1468	0.1567	0.2496
		Bootstrap Approach:			
.05	100	0.0412	0.0584	0.0592	0.1480
	200	0.0289	0.0402	0.0406	0.1178
.1	100	0.0569	0.0805	0.0823	0.2028
	200	0.0407	0.0572	0.0583	0.1698
.3	100	0.1047	0.1362	0.1464	0.3357
	200	0.0734	0.0964	0.1031	0.1713

Table 6.22 Coverage probability of estimated confidence intervals for F2 models using the normal approximation and the bootstrap approaches. See Table 6.17 for notation.

True θ	Sample Size	CC	CD	DD-c	DD-r
		Normal Approximation:			
.05	100	1.0	1.0	1.0	0.99
	200	0.98	0.99	1.0	0.99
.1	100	0.97	0.98	0.98	1.0
	200	0.99	1.0	1.0	1.0
.3	100	0.98	0.99	1.0	0.96
	200	0.99	0.99	0.99	0.97
		Bootstrap Approach:			
.05	100	1.0	1.0	1.0	0.92
	200	1.0	1.0	1.0	0.93
.1	100	1.0	1.0	1.0	1.0
	200	1.0	1.0	1.0	1.0
.3	100	1.0	1.0	1.0	1.0
	200	1.0	1.0	1.0	1.0

The bootstrap approach is recommended over the normal approximation approach because of both the range and the coverage probability.

6.5 SAMPLE SIZE

To design a genomic mapping experiment, sample size needs to be determined. Frequently asked questions are: What is the statistical power to detect linkage given a certain progeny size? Are N individuals enough to estimate recombination fraction with a certain precision? An important concept in linkage analysis, the expected likelihood ratio test statistic or expected lod score will be discussed in this section.

6.5.1 EXPECTED LIKELIHOOD RATIO TEST STATISTIC AND POWER

Expected Log Likelihood Ratio Test Statistic

We have discussed methods for detecting linkage in backcross and F2 models. The likelihood ratio test statistic to detect linkage using the backcross model is

$$G = 2 \sum_{i=1}^{2} \sum_{j=1}^{2} f_{ij} \log \frac{p_{ij}}{0.25}$$

(see Table 6.6 for notation). Its expectation is

$$
\begin{aligned}
G_{EBC} &= 2N \left[(1-\theta) \log\left(\frac{1-\theta}{0.5} \right) + N\theta \log\left(\frac{\theta}{0.5} \right) \right] \\
&= 2N \left[\log 2 + (1-\theta) \log (1-\theta) + \theta \log \theta \right] \\
&= N G_{EOBC}
\end{aligned}
\tag{6.30}
$$

Then the expected statistical power for detecting linkage using the backcross model is

$$Pr\left[\chi^2_{G_{EBC},\, df=1} \geq \chi^2_{\alpha} \right]$$

where χ^2_{α} is a critical value for rejecting the hypothesis that the two loci are independent at the α significance level and

$$\chi^2_{G_{EBC},\, df=1}$$

is a variable following a chi-square distribution with a noncentrality parameter G_{EBC} and 1 degree of freedom. The quantity

$$G_{EOBC} = 2 \left[\log 2 + (1-\theta) \log (1-\theta) + \theta \log \theta \right]$$

is the expected likelihood ratio test statistic per observation for the backcross model. For the four F2 models, the expected likelihood ratio test statistic per observation is

$$
\begin{aligned}
G_{EOCC} = 2 \{ & [1 - 2\theta + 2\theta^2] \log [1 - 2\theta + 2\theta^2] \\
& + 2\theta (1-\theta) \log [\theta - \theta^2] + \theta^2 \log \theta \\
& + 0.5 [1 - 2\theta + 2\theta^2] \log [1 - 2\theta + 2\theta^2] \}
\end{aligned}
$$

for both codominant markers and

$$
\begin{aligned}
G_{EOCD} = 2 \{ & 0.2158 + 0.5 (1 - \theta + \theta^2) \log (1 - \theta + \theta^2) \\
& + (1 - \theta^2) \log (1 - \theta^2) + 0.5 \theta^2 \log \theta \\
& + 0.5 (1-\theta) [2 \log (1-\theta) + \theta \log \theta] \\
& + (2\theta + \theta^2) \log (2\theta + \theta^2) \}
\end{aligned}
\tag{6.31}
$$

for one codominant marker and one dominant marker

$$
\begin{aligned}
G_{EODDc} = 2 \{ & 0.25 (3 - 2\theta + \theta^2) \log \left[\frac{0.25}{2.25 (3 - 2\theta + \theta^2)} \right] \\
& + 0.5\theta (2 - \theta) \log \left[\frac{1}{3\theta (2 - \theta)} \right] + 0.5 (1-\theta)^2 \log [2 (1-\theta)] \}
\end{aligned}
$$

for two dominant markers linked in coupling phase and

$$G_{EODDr} = 2\{0.25\,(2 + \theta^2)\log\left[\frac{0.25}{2.25\,(2 + \theta^2)}\right]$$

$$+ 0.5\,(1 - \theta^2)\log\left[\frac{1}{3\,(1 - \theta^2)}\right] + 0.5\theta^2\log\,(2\theta)\ \}$$

for two dominant markers linked in repulsion linkage phase (see Chapter 4 for derivation).

Table 6.23 Minimum sample sizes needed to detect linkage using $\alpha = 0.05$ significance level.   See Table 6.17 for notation.

True θ	Power γ	BC	F2-CC	F2-CD	F2-DDc	F2-DDr
0.05	0.80	16	11	21	21	97
	0.85	18	13	23	24	111
	0.90	21	15	27	28	130
	0.95	26	19	34	35	160
0.1	0.80	21	16	28	30	108
	0.85	24	19	33	34	123
	0.90	29	22	38	40	144
	0.95	35	27	47	49	179
0.2	0.80	41	35	57	63	156
	0.85	47	40	65	73	179
	0.90	55	47	77	85	209
	0.95	67	58	95	105	259
0.3	0.80	95	88	139	166	297
	0.85	109	101	159	190	339
	0.90	128	118	186	223	397
	0.95	158	146	230	275	491

Power and Sample Size

Given statistical power (γ), hypothetical recombination fraction (θ) and significance level being used (α), the sample size needed can be estimated from the inequality

$$Pr\,[\chi^2_{G_E,\,df=1} \geq \chi^2_\alpha] \geq \gamma$$

This implies

$$G_E \geq [\chi^2_{\gamma,\,df,\,\chi^2_\alpha}]^{-1}$$

where G_E is the non-centrality parameter, which is the expected log likelihood ratio test statistic divided by sample size and

$$[\chi^2_{\gamma,\,df,\,\chi^2_\alpha}]^{-1}$$

is the chi-square value corresponding to probability of γ, $df = 1$ and a noncentrality parameter χ^2_α, which can be obtained using statistical computing packages. For example, it can be obtained using SAS (SAS Institute, 1992)

$$[\chi^2_{\gamma,\,df,\,\chi^2_\alpha}]^{-1} = CINV\,(\gamma, df, \chi^2_\alpha)$$

where *CINV* is a SAS function for obtaining a quantile from the chi-square distribution. The minimum number of observations for statistical power (γ), hypothetical recombination fraction (θ) and significance level (α) is

$$N \geq \frac{[\chi^2_{\gamma, df, \chi^2_\alpha}]^{-1}}{G_{EO}} \tag{6.32}$$

Table 6.23 shows the minimum sample size needed to detect linkage using $\alpha = 0.05$ as a significance level for the backcross model and the F2 models with different marker type combinations. As the true recombination fraction increases, the number of individuals required to reach the same statistical power increases. Large numbers of individuals are needed for even moderate statistical power when the F2-DDr model is used. The F2-CC model has the highest statistical power, compared to all other models in the table.

Figure 6.5 shows the statistical power to detect linkage using the backcross model (BC) and F2 models. For the backcross and F2-CC models, a sample size of 50 is enough for a reasonable power for detecting linkage at a recombination fraction 0.2 or less. For F2-CD and F2-DDc models, about 100 individuals are needed to have equal statistical power for detecting similar levels of linkage. For the F2-DDr model, the sample size increases to 200 to have equal statistical power.

When designing a genomic mapping experiment, the type of markers and mating design are important. For F2 models, dominant markers should generally be avoided. Dominant markers in F2 matings usually have a mixture of coupling and repulsion linkage phase. The repulsion linked dominant markers provide the least linkage information and statistical power.

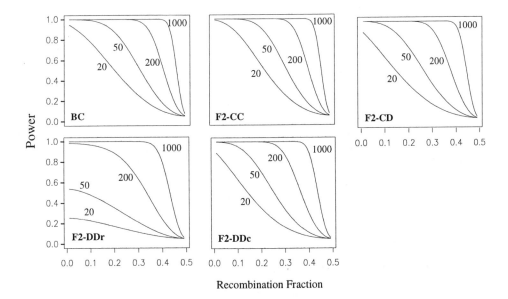

Figure 6.5 Statistical power for detecting linkage with sample sizes of 20, 50, 200 and 1000 using the backcross model (BC) and F2 models with different marker combinations. See Table 6.17 for notation.

Table 6.24 Minimum sample sizes needed to construct a confidence interval less than certain ranges c using a backcross model (BC) and F2 models with different marker combinations. See Table 6.17 for notation.

True θ	c	γ	BC	F2-CC	F2-CD	F2-DDc	F2-DDr
0.05	0.02	0.95	1825	963	1901	1938	38177
		0.99	3162	1668	3294	3358	66150
	0.05	0.95	292	154	304	310	6108
		0.99	506	267	527	537	10584
	0.10	0.95	73	39	76	78	1527
		0.99	126	67	132	134	2646
0.10	0.05	0.95	553	311	604	626	5996
		0.99	959	538	1046	1085	10389
	0.10	0.95	138	78	151	157	1499
		0.99	240	135	262	271	2597
	0.20	0.95	35	19	38	39	375
		0.99	60	34	65	68	649
0.20	0.05	0.95	983	643	1191	1281	5573
		0.99	1704	1114	2065	2220	9656
	0.10	0.95	246	161	298	320	1393
		0.99	426	279	516	555	2414
	0.20	0.95	61	40	74	80	348

6.5.2 MINIMUM CONFIDENCE INTERVAL

Using the normal approximation approach, a confidence interval can be constructed for the recombination fraction from a two-locus model. For a confidence coefficient of 95%, the confidence interval is

$$[\hat{\theta} \pm 1.96 sd_\theta] \tag{6.33}$$

The range from lower to upper limits $3.92 sd_\theta$ can be used as a precision measurement for the estimated recombination fraction. For a given true recombination fraction and sample size, the standard deviation for the estimated recombination fraction can be predicted using the expected information content. So, given a confidence interval and expected recombination fraction, the minimum number of individuals needed can be determined using

$$N \geq \frac{3.92^2}{c^2 I(\theta)} \tag{6.34}$$

where c is a target maximum range for a 95% confidence interval and $I(\theta)$ is expected information content per individual. For interval with confidence coefficient γ, Equation (6.34) becomes

$$N \geq \frac{[2z_{0.5(1-\gamma)}]^2}{c^2 I(\theta)} \tag{6.35}$$

where $z_{0.5(1-\gamma)}$ is the $0.5(1-\gamma)$ quantile of a standard normal distribution.

Table 6.24 shows the minimum sample sizes needed for some combinations of the true recombination fractions and maximum range for 95% and 99% confidence intervals. If the true recombination fraction is around

0.05, at least 963 individuals are needed for the F2-CC model, for example, to ensure the precision of the estimate within a 0.02 range. In other words, the estimate falls between 0.04 to 0.06 with a probability of 95%. If the range increases to 0.1, the minimum number of individuals needed is 126.

To determine an acceptable range for a confidence interval or a target precision for estimated recombination fraction, one should consider the biological basis of the range and experimental reality. For experiments involving map-based cloning of genes, high precision estimation of recombination fraction could be a good starting point. However, the relationship between physical length and genetic recombination varies from organism to organism and varies even for different genome positions of the same organism. For experiments involving QTL mapping (see Chapters 12 - 17), high precision of estimation of recombination fractions may not be necessary because the resolution of QTL locations is most likely much lower than marker locations. The minimum numbers of individuals listed in Table 6.24 are only reference points. Readers should be cautious in using these numbers for specific experiments.

6.6 DOMINANT MARKERS IN F2 PROGENY

Dominant markers are gaining more and more attention in genomic research due to the recent developments in automated generation of PCR based markers, such as random amplified polymorphic DNA (RAPD) and amplified fragment length polymorphism (AFLP) markers (Tingey *et al.* 1992; Rafalski and Tingey 1993; Zabeau 1993). Strategies such as pseudo-testcross have been used to avoid dominant markers in repulsion phase in progeny of heterozygous F1, because dominant markers have the same amount of information as codominant markers in backcross models. This strategy has been successfully used to build a linkage map for *Eucalyptus* (Grattapaglia and Sederoff 1994). A big effort has been made to score AFLP markers as codominant markers. In this section, the disadvantages of dominant markers used in F2 progeny will be reviewed and a strategy using trans dominant linked markers (TDLM) will be discussed.

6.6.1 DISADVANTAGE OF DOMINANT MARKERS IN F2 PROGENY

Low Linkage Information Content
Relative linkage information content for different marker configurations is shown in Figure 6.6 (Allard 1956). However, dominant markers provide less information on linkage than do codominant markers when F2 progeny are used (Allard 1956; Knapp *et al.* 1995). This is especially true when dominant alleles are in repulsion linkage.

In practical mapping experiments, linkage phase for dominant markers is random, that is, approximately half of the markers are linked in coupling and the other half are also linked in coupling. However, the linkage phase between the two groups is repulsion. Let us assume that the two marker groups can be clearly identified. Linkage maps may be constructed independently for the two groups. The markers within each of the linkage maps are linked in coupling. However, if we want to make a connection between the two maps, repulsion linkage phase will be a barrier. Dominant markers will also have low information content for detecting genes controlling quantitative traits (see Chapters 13 and 14).

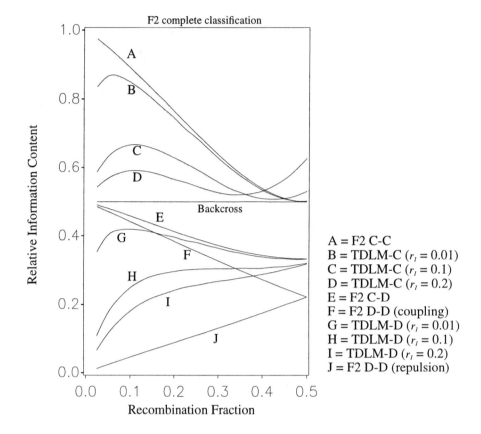

Figure 6.6 Relative information content per observation for linkage between: A: two codominant markers (C-C); B,C and D: a TDLM and a codominant marker (TDLM-C), E: a codominant and a dominant markers (C-D); F: two dominant markers in coupling phase (D-D coupling); G, H and I: a TDLM and a dominant marker (TDLM-D); and J: two dominant markers in repulsion phase (D-D repulsion).

Table 6.25 Expected genotypic frequency for an F2 population using dominant markers in repulsion linkage phase.

Genotype	Observed Count	Expected Frequency
A_B_	f_1	$0.25 (2 + \theta^2)$
A_bb	f_2	$0.25 (1 - \theta^2)$
aaB_	f_3	$0.25 (1 - \theta^2)$
aabb	f_4	$0.25\theta^2$

Bias Estimator for Recombination Fraction

In Section 6.4.1, biases for recombination fraction estimation using different marker configurations were discussed. Among the marker configurations, the estimator using dominant markers in repulsion phase in F2 is seriously biased (Table 6.19). Knapp *et al.* (1995) explored possible reasons for the bias. One reason is the low frequency of the double recessive class (aabb) in the progeny (Table 6.25). When the observed frequency of the double recessive class is zero, the maximum likelihood estimator of the recombination fraction is zero, regardless of the frequencies of the other classes.

The log likelihood function for a pair of dominant markers in repulsion phase in F2 progeny is

$$L = f_1\log(2 + \theta^2) + (f_2 + f_3)\log(1 - \theta^2) + 2f_4\log\theta$$

The maximum likelihood estimate of the recombination fraction is the solution of

$$\frac{dL}{d\theta} = \frac{2\theta f_1}{2 + \theta^2} - \frac{2\theta(f_2 + f_3)}{1 - \theta^2} + \frac{2f_4}{\theta} = 0$$

which can be simplified as

$$-[2(f_1 + f_2 + f_3) + f_4]\theta^4 + [2f_1 - 4(f_2 + f_3) - f_4]\theta^2 + 2f_4 = 0$$

The solution is

$$\hat{\theta} = \left\{\frac{-[2f_1 - 4(f_2 + f_3) - f_4] \pm \sqrt{[2f_1 - 4(f_2 + f_3) - f_4]^2 + 8[2(f_1 + f_2 + f_3) + f_4]f_4}}{-2[2(f_1 + f_2 + f_3) + f_4]}\right\}^{\frac{1}{2}}$$

If $f_4 = 0$, the solution reduces to

$$\hat{\theta} = \sqrt{\frac{-[f_1 - 2(f_2 + f_3)] \pm [f_1 - 2(f_2 + f_3)]}{-2(f_1 + f_2 + f_3)}}$$

The solution can take two values

$$\begin{cases} \hat{\theta}^{(1)} = 0 & for \quad f_1 \le \dfrac{2N}{3} \\[2ex] \hat{\theta}^{(2)} = \sqrt{\dfrac{f_1 - 2(f_2 + f_3)}{f_1 + f_2 + f_3}} & for \quad f_1 > \dfrac{2N}{3} \end{cases}$$

where

$$N = f_1 + f_2 + f_3 + f_4$$

is the progeny size. To make the second solution meaningful, f_1 has to be equal to or greater than $2(f_2 + f_3)$. This implies that the estimate of the recombination fraction is greater than $\sqrt{2/3}$, which is outside the parameter space. So, the maximum likelihood estimate of recombination fraction is zero for the observed frequency of the double recessive class regardless of the frequencies for the other classes.

As given by Knapp *et al.* (1995), the probability of no double recessive class being observed in a sample of N in F2 progeny is

$$P[f_4 = 0] = \binom{N}{0}(0.25\theta^2)^0(1 - 0.25\theta^2)^N = (1 - 0.25\theta^2)^N$$

Figure 6.7 shows that the probability of no double recessive class being observed decreases as sample size and recombination fraction increase.

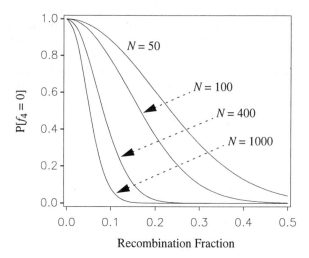

Figure 6.7 Probability of no double recessive class observed in F2 progeny when two dominant markers are linked in repulsion linkage phase for different sample sizes and recombination fractions between the two markers.

Table 6.26 Joint genotypic frequency for a pair of TDLMs and a codominant marker C and approximate frequency of recombinants conditional on the genotype.

Genotype	p_j	$P_i(R\|G)$
ABCC	$0.5r_2(1-r_1-r_2)$	0.5
ABCc	$0.5[1-r_1-2r_2+r_2^2+(r_1+r_2)^2]$	$r_2^2/[r_2^2+(1-r_2)^2]$
ABcc	$0.25[2r_1+2r_2-r_2^2-(r_1+r_2)^2]$	0.5
AbCC	$0.25(1-2r_2-r_1+r_2^2)$	0.0
AbCc	$0.5r_2(1-r_2)$	0.5
Abcc	$0.25r_2^2$	1.0
aBCC	$0.25r_2(2r_1+r_2)$	1.0
aBCc	$0.5(1-r_1-r_2)(r_1+r_2)$	0.5
aBcc	$0.25(1-r_1-r_2)^2$	0.0
abCC	$0.25r_1^2$	NA

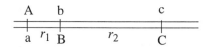

Figure 6.8 Linkage relation among a pair of TDLMs and a codominant marker C.

Table 6.27 Joint genotypic frequency for a pair of TDLMs and a codominant marker C and approximate frequency of recombinants conditional on the genotype.

Genotype	p_j	$P_i(R\|G)$
ABD_	$0.5\,(1 - r_1 - r_2 + r_1^2 + r_2^2 + r_1 r_2)$	$r_2(1 + r_2)\,/\,(1 - r_2 + r_2^2)$
ABdd	$0.25\,[2r_1 + 2r_2 - r_2^2 - (r_1 + r_2)^2]$	0.5
AbD_	$0.25\,(1 - r_1^2 - r_2^2)$	$r_2/\,(1 + r_2)$
Abdd	$0.25\,r_2^2$	1.0
aBD_	$0.25\,[2r_1 + 2r_2 - r_1^2 - (r_1 + r_2)^2]$	$1/\,(2 - r_2)$
aBdd	$0.25\,(1 - r_1 - r_2)^2$	0.0
abDD	$0.25\,r_1^2$	NA

6.6.2 USE OF TRANS DOMINANT LINKED MARKERS (TDLM)

TDLM

In the previous section, the disadvantages of using dominant markers in F2 progeny were described. However, if two dominant markers are closely linked in repulsion phase, they can be approximately re-coded as a codominant marker. These two markers are defined as a pair of TDLMs (Plomion *et al.* 1996). Table 6.26 shows the joint genotypic frequency for a pair of TDLMs and a codominant marker C. The linkage orientation among these three loci is shown in Figure 6.8. Markers A and B are linked in repulsion phase with r_1 recombination. A and B denote the dominant alleles. Table 6.27 shows the joint frequencies for a pair of TDLMs with a dominant marker D.

If we assume the linkage is tight between the two dominant markers, then the expected frequencies are 0.5, 0.25, 0.25 and 0.0 for the four genotypic classes (AB, Ab, aB and ab). Therefore, the pair of the TDLMs can be re-coded as a single codominant marker. Table 6.28 shows the corresponding coding scheme.

Table 6.28 TDLM (A and B) genotypes and their corresponding re-coded codominant marker genotypes and their marginal frequencies.

TDLM	Expected Frequency	Re-coded as C Locus	Approximate Frequency When $r_1 \rightarrow 0$
AB	$0.25\,(2 + r_1^2)$	Cc	0.5
Ab	$0.25\,(1 - r_1^2)$	CC	0.25
aB	$0.25\,(1 - r_1^2)$	cc	0.25
ab	$0.25\,r_1^2$		

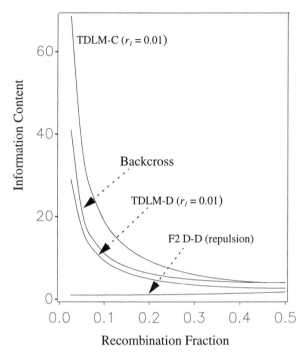

Figure 6.9 Information content per observation for linkage between two dominant markers in repulsion linkage phase in F2 (F2 D-D), a TDLM pair and a codominant marker (TDLM-C), a TDLM pair and a dominant marker (TDLM-D) and a pair of markers in backcross.

Linkage Information Content for TDLM

For linkage analysis using the TDLM, the information content per observation can be computed by

$$I(r_1, r_2) = \sum \frac{1}{p_i}\left(\frac{\partial p_i}{\partial r_2}\right)^2 \tag{6.36}$$

where p_i is the probability for the TDLM-marker genotype class i. Figure 6.6 shows the information content for the TDLM-codominant marker and the TDLM-dominant marker relative to some standard marker configurations. When the recombination fraction between the pair of TDLMs is small, the information content is high. The decreases of the relative information contents when the TDLM and the marker are linked with less than 0.1 recombination may be misleading, in that it concludes not informative for those situations (Figure 6.6). However, this conclusion is not true if we look at the absolute information content. Figure 6.9 shows information contents per observation for linkage. The information contents shown are the absolute values. The information content is always large when linkage is tight, except when the F2 dominant markers are linked in repulsion.

Estimate of Recombination Fraction between a TDLM and a Marker

The log likelihood function for estimating recombination fraction between a TDLM and a marker is

$$L = \sum f_i \log p_i \tag{6.37}$$

where f_i is the observed frequency for the TDLM-marker genotype class i. The analytical solution for the recombination between the TDLM and a marker is difficult to obtain. An iterative approach is needed to obtain the estimate of recombination fraction between the TDLM and a marker.

An EM algorithm can be used to estimate the recombination fraction between a TDLM and a marker. The probability of an individual being a recombinant conditional on the TDLM-marker genotype is given for a codominant marker in Table 6.26 and for a dominant marker in Table 6.27. The EM algorithm can be implemented as

$$r^{new} = \frac{1}{N}\sum f_i P_i (R|G) \Big|_{r^{old}} \tag{6.38}$$

where r^{new} is a new estimate

$$P_i (R|G) \Big|_{r^{old}}$$

is the probability that the individual is a recombinant conditional on the genotype and the value is obtained using the "old" estimate. For the first iteration, r^{old} is an initial guess of the recombination fraction. For each additional iteration, the r^{old} is replaced by the new estimate. The iteration stops when

$$|r^{new} - r^{old}| \leq tolerance$$

The tolerance is a sufficiently small number relative to the estimated recombination fraction, say 0.000001.

6.7 VIOLATION OF ASSUMPTIONS

In previous sections, linkage analysis has assumed no errors in marker genotypes, no segregation distortion at single loci and known linkage phases. However, errors cannot be completely avoided; segregation distortion has been common in mapping data due to sampling or biological selection and sometimes linkage phases are not known. In this section, linkage analysis using data with marker determination error, segregation distortion and lethal genes will be explained.

6.7.1 SEGREGATION RATIO DISTORTION

Segregation ratios of observed genotypic frequencies may depart from the expected frequencies. Systematic marker or genotype classification error could be one of the reasons that segregation ratio distortion occurs. Random classification error, a kind of sampling error, may not affect the segregation ratio and linkage analysis. Classification error will be explained in this section as a type of systematic error. In terms of data analysis, segregation ratio distortions may be caused by systematic classification errors or by known biological reasons, such as selection, and both will be treated in the same way.

Additive Distortion

First, let us consider a situation where distortion can be described as an additive process. For a backcross model, the expected genotypic frequencies can be found in Table 6.6. If we assume that the distortion parameters for the two loci are a and b, respectively, and they are independent of each other, the expected genotypic frequencies become the frequencies listed in Table 6.29 using an additive model. The parameter a means that frequency of genotype Aa has a change of a. a can be a positive or negative number, depending on whether the mis-classifications or selection favor or disfavor genotype Aa. The distortion parameters and recombination fraction can be estimated by setting the observed frequency equal to the expectation. That is

$$0.5\,(1-\theta) + a + b = f_{11}/N$$
$$0.5\theta + a - b = f_{12}/N$$
$$0.5\theta - a + b = f_{21}/N \qquad (6.39)$$
$$0.5\,(1-\theta) - a - b = f_{22}/N$$

The solutions for the parameters are

$$\hat{\theta} = (f_{12} + f_{21})\,/\,N$$
$$\hat{a} = 0.25\,[\,(f_{11} + f_{12}) - (f_{21} + f_{22})\,]\,/\,N \qquad (6.40)$$
$$\hat{b} = 0.25\,[\,(f_{11} + f_{21}) - (f_{12} + f_{22})\,]\,/\,N$$

Table 6.29 Expected frequencies for a backcross (AaBb X aabb) model. θ is the recombination fraction under additive distortion with parameters a and b.

Genotype	Observed Count (f_{ij})	Expected Frequency (Distortion) (p_{ij})
AaBb	f_{11}	$0.5\,(1-\theta) + a + b$
Aabb	f_{12}	$0.5\theta + a - b$
aaBb	f_{21}	$0.5\theta - a + b$
aabb	f_{22}	$0.5\,(1-\theta) - a - b$

This approach of obtaining the maximum likelihood estimates was first introduced by Bailey (1961). When the number of unknown parameters equals the number of degrees of freedom, the maximum likelihood estimates of the parameters can be obtained by solving the equations of the expectations equal the observations. This is commonly referred to as Bailey's method (Weir 1996).

The solution for the recombination fraction is the same as the solution for the model without distortion. The estimate is an unbiased estimate for the recombination fraction

$$E(\hat{\theta}) = E\left[\frac{f_{12} + f_{21}}{N}\right]$$
$$= (0.5\theta + a - b) + (0.5\theta - a + b)$$
$$= \theta$$

Using a backcross model, the additive distortion has no effect on estimation of recombination fraction. The distortion parameters are nuisance parameters for the model. The same is true for the F2 models (Table 6.30).

The statistical powers for linkage detection and the expected likelihood ratio test statistic for the backcross model are also the same for both with or without an additive segregation distortion. However, they differ for the F2 models. Figure 6.10 shows the differences of statistical power between the models with and without additive segregation distortions for the F2-CC model.

Table 6.30 Expected genotypic frequency for an F2 population using codominant markers under genotypic segregation distortion.

Genotype	Observed Count	Expected Frequency
AABB	f_1	$0.25(1-\theta)^2 + a + b$
AABb	f_2	$0.5\theta(1-\theta) + a$
AAbb	f_3	$0.25\theta^2 + a - b$
AaBB	f_4	$0.5\theta(1-\theta) + b$
AaBb	f_5	$0.5(1 - 2\theta + 2\theta^2)$
Aabb	f_6	$0.5\theta(1-\theta) - b$
aaBB	f_7	$0.25\theta^2 - a + b$
aaBb	f_8	$0.5\theta(1-\theta) - a$
aabb	f_9	$0.25(1-\theta)^2 - a - b$

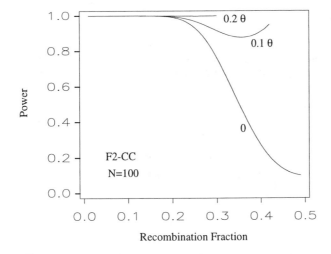

Figure 6.10 Statistical power for detecting linkage with sample size 100 using the F2 model with both codominant markers. The distortion parameters for both loci A and B are 0 (no distortion), 0.1 θ (10% of the recombination fraction) and 0.2 θ using the additive distortion model.

Segregation ratio distortion in an additive fashion increases the false positive rate for linkage analysis in F2. In other words, segregation distortion increases the expected likelihood ratio test statistic for detecting linkage.

However, as mentioned above, segregation ratio distortion has no effect on the estimation of recombination fraction. In practical genomic mapping experiments, care should be taken in interpreting the linkage hypothesis test when segregation ratio distortion exists. If the hypothesis test is statistically significant and the estimated linkage is loose, then most likely the statistically significant linkage is a false positive. For most linkage analysis models, if the data have no segregation ratio distortion, the estimated recombination fraction and the likelihood ratio test statistic have a one-to-one relationship.

Table 6.31 Expected genotypic frequencies for a backcross (AaBb X aabb) model. θ is the recombination fraction between A and B under gametic and genotypic distortion.

Genotype	Observed Count	Expected Frequency under Distortion
AaBb	f_{11}	$0.5\,[\theta\,(a+b)+(1-\theta)\,(1+ab)\,]$
Aabb	f_{12}	$0.5\,[\theta+a\,(1-\theta)\,]\,(1-b)$
aaBb	f_{21}	$0.5\,[\theta+b\,(1-\theta)\,]\,(1-a)$
aabb	f_{22}	$0.5\,(1-\theta)\,(1-a)\,(1-b)$

Penetrance Distortions

Let us consider situations where segregation ratio distortions cannot be described as additive models. Instead, the distortion can be explained by penetrance. Table 6.31 shows the expected genotypic frequencies under a penetrance distortion model for a backcross population. In the model, $(100 \times a)$ % of genotype aa is mis-classified as genotype Aa and $(100 \times b)$ % of genotype bb is mis-classified as Bb. In other words, relative fitnesses for Aa, aa, Bb and bb are $1+a$, $1-a$, $1+b$ and $1-b$, respectively. By setting the observed observations equal to their expectations, estimates for recombination fraction and the distortion parameters are

$$\hat{\theta} = 1 - \frac{2f_{22}}{N\,(1-a)\,(1-b)}$$

$$= \frac{2f_{12}f_{21}+f_{22}\,(N-2f_{11})}{2\,(f_{12}+f_{22})\,(f_{21}+f_{22})}$$

$$\hat{a} = \frac{(f_{11}+f_{12}) - (f_{21}+f_{22})}{N} \tag{6.41}$$

$$\hat{b} = \frac{(f_{11}+f_{21}) - (f_{12}+f_{22})}{N}$$

The recombination fraction estimate is unbiased with a large-sample variance (Bailey 1961)

$$V(\hat{\theta}) = \frac{N\,[f_{22}N\,(f_{22}^2+f_{12}f_{21}) - f_{22}\,(f_{12}+f_{22})\,(f_{21}+f_{22})\,]}{4\,(f_{12}+f_{22})^3\,(f_{21}+f_{22})^3} \tag{6.42}$$

If the recombination fraction is estimated using a non-distortion model, then the estimate will be biased by

$$Bias = \theta - E(\hat{\theta})$$
$$= \theta(a + b - ab) - 0.5(a + b - 2ab)$$

(6.43)

As far as statistical power is concerned for the penetrance distortion model, Figure 6.11 shows the statistical power for detecting linkage with a sample size of 100 using a backcross population. Segregation ratio distortion in a penetrance fashion decreases the statistical power for linkage detection for the range of true recombination fraction having significant meaning in genomic mapping ($\theta \le 0.4$), which is contradicted by the additive distortion models.

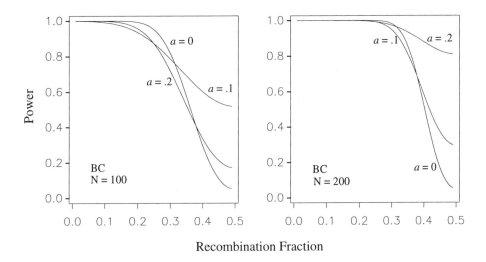

Figure 6.11 Statistical power for detecting linkage with sample sizes 100 and 200 using a backcross model (BC). The distortion parameters for both loci A and B are 0 (no distortion), 0.1 (10% of aa mis-classified as Aa) and 0.2 (20% of aa mis-classified as Aa) using the penetrance distortion models (Table 6.30).

Impact of Segregation Ratio Distortion in Practical Linkage Analysis

Commonly used computer software for linkage analysis and genomic map construction, such as Mapmaker (Lander *et al.*, 1987), Gmendel (Liu and Knapp, 1992), JoinMap (Stam 1993), Linkage (Ott 1985) and PGRI (Liu 1994), cannot analyze data using distortion models. When segregation ratio distortion exists and the distortion cannot be modeled in an additive fashion, one should be very careful in interpreting the output from the software packages.

Figure 6.12 shows biases of estimated recombination fractions using non-distortion models when penetrance distortion exists for a backcross model. As the true recombination fraction goes up, the tolerance for segregation distortion increases rapidly. However, tight linkage is far more important than loose linkage for linkage grouping and gene ordering. Tolerance for segregation ratio distortion is usually low for practical genomic mapping experiments.

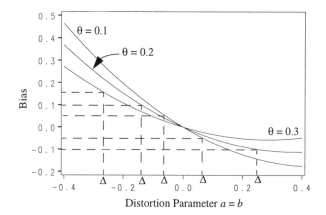

Figure 6.12 Bias of estimated recombination fraction using non-distortion model when distortion exists for a backcross model. Given a tolerance on bias there is a range of the distortion parameter corresponding to the tolerance. For example, for $\theta = 0.1$ and tolerance 0.05 for each direction, the range is $-0.07 \le a \le 0.07$, for $\theta = 0.2$ and for tolerance 0.1, the range is $-0.13 \le a \le 0.24$ and for $\theta = 0.3$ and tolerance 0.15, the range is $-0.26 \le a$.

6.7.2 LINKAGE ANALYSIS INVOLVING LETHAL GENES

Single Gene Defect

Lethality could be the result of a single gene defect. In this case, the data becomes a special case of segregation ratio distortion. For example, a recessive lethal gene a will cause homozygous aa to die before the individuals can be genotyped. The expected genotypic frequencies for the remaining individuals are listed in Table 6.32. The log likelihood function for the F2 recessive lethal model is

$$
\begin{aligned}
L_{F2RL}(\theta) &= \sum_{j=1}^{6} f_j \log p_j \\
&= 2f_1 \log(1-\theta) + (f_2 + f_4 + f_6) \log[\theta(1-\theta)] \\
&\quad + (f_5 \log(1 - 2\theta + \theta^2) + 2f_3 \log\theta)
\end{aligned}
\tag{6.44}
$$

which is similar to the likelihood function for the non-lethal model. The first derivative of the function with respect to recombination fraction is

$$
L'_{F2RL}(\theta) = -\frac{2f_1}{1-\theta} + \frac{(f_2 + f_4 + f_6)(1-2\theta)}{\theta(1-\theta)} + \frac{2f_5(1-\theta)}{1 - 2\theta + 2\theta^2} + \frac{2f_3}{\theta}
\tag{6.45}
$$

The average information content for an individual is

$$
\begin{aligned}
I_{F2RL}(\theta) &= E_\theta \left[\frac{\partial}{\partial\theta} \log L_\theta(x)\right]^2 \\
&= \frac{20}{3} + \frac{2(1-2\theta)^2}{\theta(1-\theta)} - \frac{8(1-\theta)(1-2\theta)}{3(1-2\theta+2\theta^2)}
\end{aligned}
\tag{6.46}
$$

Table 6.32 Expected genotypic frequency for a F2 population when aa is lethal.

Genotype	Observed Count (f_i)	Expected Frequency (p_i)
AABB	f_1	$(1/3) (1 - \theta)^2$
AABb	f_2	$(2/3) \theta (1 - \theta)$
AAbb	f_3	$(1/3) \theta^2$
AaBB	f_4	$(2/3) \theta (1 - \theta)$
AaBb	f_5	$(2/3) (1 - 2\theta + 2\theta^2)$
Aabb	f_6	$(2/3) \theta (1 - \theta)$

The large-sample variance of the estimated recombination fraction using F2 single locus recessive lethal model for sample size of N is

$$Var_{F2}(\theta) = \frac{1}{N I_{F2RL}(\theta)} \tag{6.47}$$

Two-Locus Recessive Lethal

When the lethality is caused by two or more genes, the nature of the data will be a combination of segregation ratio distortion and gene interaction. Here, only two-locus lethal models will be discussed. Under normal circumstances, the recombination fraction cannot be estimated for a two-locus recessive lethal model using backcross progeny, because the two lethal genes, if they are linked, have to be linked in repulsion phase and progeny of the cross Aabb X aaBb does not have linkage information. To illustrate, a special situation is assumed where the double recessive lethal is a result of the environment interacting with the two genes. So, a coupling linkage phase is possible for the two-locus lethal model in backcross progeny. In practical linkage experiments dealing with lethal genes, a large number of possible genetic models exist.

Table 6.33 Expected genotypic frequencies for a backcross (AaBb X aabb) model when the double recessive homozygous are lethal.

Genotype	Observed Count	Expected Frequency
AaBb	f_{11}	$(1 - \theta) / (1 + \theta)$
Aabb	f_{12}	$\theta / (1 + \theta)$
aaBb	f_{21}	$\theta / (1 + \theta)$

Table 6.33 shows the expected genotypic frequencies for the hypothetical backcross model. The log likelihood function for recombination fraction estimation is

$$\begin{aligned} L_{B C\overline{DRL}}(\theta) &= f_{11}\log\left[\frac{1 - \theta}{1 + \theta}\right] + (f_{12} + f_{21}) \log\left[\frac{\theta}{1 + \theta}\right] \\ &= f_{11}\log(1 - \theta) + (f_{12} + f_{21}) \log\theta - N\log(1 + \theta) \end{aligned} \tag{6.48}$$

By setting the score equal to zero, the maximum likelihood solution for recombination fraction is

$$\hat{\theta} = \frac{f_{12} + f_{21}}{f_{11} + N} \tag{6.49}$$

which is an unbiased estimate. The large-sample variance for the recombination fraction estimate is

$$Var_{BCDRL}(\theta) = \frac{\theta(1 - \theta^2)(1 + \theta)}{2N} \tag{6.50}$$

Table 6.34 Expected genotypic frequency for an F2 population using codominant markers when the double recessive homozygous are lethal.

Genotype	Observed Count (f_i)	Expected Frequency (p_i)	$P_i(R\|G)$
AABB	f_1	$\dfrac{0.25(1-\theta)^2}{1-0.25(1-\theta)^2}$	0.0
AABb	f_2	$\dfrac{0.5\theta(1-\theta)}{1-0.25(1-\theta)^2}$	0.5
AAbb	f_3	$\dfrac{0.25\theta^2}{1-0.25(1-\theta)^2}$	1.0
AaBB	f_4	$\dfrac{0.5\theta(1-\theta)}{1-0.25(1-\theta)^2}$	0.5
AaBb	f_5	$\dfrac{0.5(1-2\theta+2\theta^2)}{1-0.25(1-\theta)^2}$	$\dfrac{\theta^2}{(1-\theta)^2+\theta^2}$
Aabb	f_6	$\dfrac{0.5\theta(1-\theta)}{1-0.25(1-\theta)^2}$	0.5
aaBB	f_7	$\dfrac{0.25\theta^2}{1-0.25(1-\theta)^2}$	1.0
aaBb	f_8	$\dfrac{0.5\theta(1-\theta)}{1-0.25(1-\theta)^2}$	0.5

For F2 double recessive lethal models, the estimation of recombination fraction using the EM algorithm is straightforward because the probabilities of a recombinant conditional on the genotype for the lethal model are the same as the conventional F2 (Table 6.34). However, the average information content per observation for the lethal model has a complex form, which is

$$I_{F2DRL}(\theta) = \frac{1}{1-0.25(1-\theta)^2}\{\frac{\theta}{2} + \frac{0.5+\theta+\theta^2}{(1-\theta)^2} + \frac{2(1-\theta)(1-2\theta)}{1-2\theta+2\theta^2}$$

$$-\frac{0.5(1-\theta)^2}{1-0.25(1-\theta)^2} - 2.5\} \tag{6.51}$$

The large-sample variance for the estimate is

$$Var_{F2DRL}(\theta) = \frac{1}{NI_{F2DRL}(\theta)} \tag{6.52}$$

As with the linkage analysis with segregation ratio distortion, the commonly used computer software packages cannot directly handle linkage analysis involving lethal genes. However, the estimation of recombination fraction may be obtained approximately by using software packages for the F2 model. For the backcross model, the estimated recombination fraction should be approximately twice the output value from the software using the conventional backcross model for tightly linked loci. The factor becomes 1.333 for loosely linked loci. Special caution is needed for analyzing data involving lethal genes using conventional genetic models.

EXERCISES

Exercise 6.1

Crossover between nonsister chromatids of two homologs within a segment between two loci (genes or markers) is the basis of linkage detection and recombination fraction estimation. The crossover can be observed through polymorphic loci segregating in progeny of a certain mating. The gametes resulting from the crossover are defined as the recombinants. The non-recombinants are usually defined as parental types. If there is no crossover between the loci, then the two loci are linked completely. When two loci are linked completely, there are no recombinants in the progeny. If numbers of recombinants and non-recombinants are the same or close to the same, then the loci are not linked. Statistical procedures are used to implement the hypothesis test for detecting linkage and to estimate recombination fraction. Statistics are applied only when random sampling takes place in the process of collecting the data. Furthermore, many parametric statistical procedures depend on the theoretical distributions of the data. Please describe where sampling takes place in counting the number of recombinants in an experiment for obtaining recombination fraction between two loci. What kinds of theoretical distributions are associated with estimating recombination fraction?

Exercise 6.2

In progeny of a cross between two individuals AaBb and aabb (traditionally called double-backcross mating), counts of the four possible genotypes are given in the following illustration.

	A	r	B		Genotype	Count
Parent 1					AaBb	18
	a		b	X	Aabb	6
	a	r	b		aaBb	3
Parent 2					aabb	27
	a		b			

Ex. 6.2

(1) Detect linkage between loci A and B and estimate the recombination fraction.

(2) Estimate the variance of the estimated recombination fraction and build a 95% confidence interval for the recombination fraction using approaches of normal approximation and likelihood.

(3) Estimate confidence interval with lod score support 1.5.

(4) Sample randomly with replacement from the data for 10 replications. For each replication, 54 individuals are sampled and a recombination fraction can be estimated. Estimate a variance from the 10 estimates and compare this variance with the variance in (2). Explain why they are the same or different.

(5) If the range of the 95% confidence interval less or equal to 0.05 is targeted, what is the minimum progeny size required?

(6) If the true recombination between loci A and B differs between the two parents, can the recombination fraction be estimated using the backcross? What does the estimate mean? Can you find this type of situation in real-life experiments?

Exercise 6.3

In the progeny of a cross between two individuals AaBb and AaBb (traditionally called F2 mating), counts of the nine possible genotypes are given in the following illustration.

		Genotype	Count
Parent 1	A B	AABB	35
	‾‾‾‾‾	AABb	14
	a b	AAbb	5
	X	AaBB	10
	A B	AaBb	20
Parent 2	‾‾‾‾‾	Aabb	11
	a b	aaBB	3
		aaBb	12
		aabb	40

Ex. 6.3

(1) Detect linkage between loci A and B and estimate the recombination fraction. Please use all possible approaches described in this chapter.

(2) Estimate the variance of the estimated recombination fraction and build a 95% confidence interval for the recombination fraction.

(3) If the range of the 95% confidence interval less than or equal to 0.05 is targeted, what is the minimum progeny size required?

(4) If the true recombination between loci A and B differs between the two parents, can the recombination fraction be estimated using the F2? Can you design a strategy to solve the problem? (Hint: The strategy can be a biological one or a statistical one.)

Exercise 6.4

Two backcrosses were made to estimate recombination fraction between loci A and B (see the following illustration). The recombination fractions between loci A and B could be the same or different for the parents in the two crosses and are denoted by r_1 and r_2, respectively.

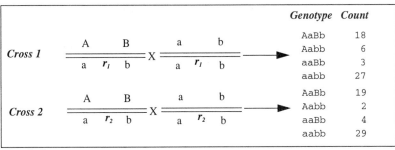

Ex. 6.4

(1) Perform a complete single-locus and two-locus goodness of fit statistic and likelihood ratio analyses. The analyses should include the partition of the total test statistic into pooled and heterogeneity.

(2) Estimate the recombination fraction between loci A and B using each of the crosses independently and using the pooled data.

(3) Test the heterogeneity between the estimates of the recombination fraction between the two crosses. Compare the conclusion with the conclusion in (1). Explain why the results are the same or different.

(4) If the true difference between r_1 and r_2 is 0.05, what is the expected minimum progeny size needed for both crosses, assuming that progeny sizes are the same for both crosses to detect the difference at a 0.0001 significance level?

Exercise 6.5

Two F2 crosses were made to estimate the recombination fraction between loci A and B (see the following illustration). The recombination fractions between loci A and B could be the same or different for the parents in the two crosses and are denoted by r_1 and r_2, respectively.

Cross 1	Genotype	Count	Cross 2	Genotype	Count
A B	AABB	35	A B	AABB	38
———	AABb	14	———	AABb	15
a b	AAbb	5	a b	AAbb	1
X	AaBB	10	X	AaBB	11
A B	AaBb	20	A B	AaBb	22
———	Aabb	11	———	Aabb	8
a b	aaBB	3	a b	aaBB	0
	aaBb	12		aaBb	10
	aabb	40		aabb	45

Ex. 6.5

(1) Perform a complete single-locus and two-locus goodness of fit statistic and likelihood ratio analysis. The analysis should include the partition of the total test statistic into pooled and heterogeneity.

(2) Estimate the recombination fraction between loci A and B using each of the crosses independently and using the pooled data.

(3) Test the heterogeneity between the estimates of the recombination fraction between the two crosses. Compare the conclusion with the conclusion in (1). Explain why the results are the same or different.

(4) If the true difference between r_1 and r_2 is 0.05, what is the expected minimum progeny size needed for both crosses, assuming that progeny sizes are the same for both crosses to detect the difference at a 0.0001 significant level?

Exercise 6.6

One F2 cross and one backcross were made to estimate recombination fraction between loci A and B (following illustration). The recombination fractions between loci A and B could be the same or different for the parents in the two crosses and are denoted by r_1 and r_2, respectively.

Cross 1		Genotype	Count	Cross 2			
A B		AABB	35	A B		Genotype	Count
$\overline{}$		AABb	14	$\overline{}$			
a r_1 b		AAbb	5	a r_2 b		AaBb	38
X		AaBB	10	X		Aabb	2
A B		AaBb	20	a b		aaBb	1
$\overline{}$		Aabb	11	$\overline{}$		aabb	34
a r_1 b		aaBB	3	a r_2 b			
		aaBb	12				
		aabb	40				

Ex. 6.6

(1) Estimate recombination fraction between loci A and B using each of the crosses independently and using the pooled data.

(2) Test the heterogeneity between the estimates of the recombination fraction between the two crosses.

(3) If the true difference between r_1 and r_2 is 0.05, what is the expected minimum progeny size needed for both crosses to detect the difference at a 0.0001 significant level? Here we assume that the progeny size for the F2 cross is double the size of the backcross.

(4) If the true difference between r_1 and r_2 is 0.05, what is the expected minimum progeny size needed for both crosses to detect the difference at a 0.0001 significant level? Here we do not assume the relative progeny size for the F2 cross and the backcross. (Hint: Please take the information content for the two types of matings into consideration.)

Exercise 6.7

Four crosses were made to detect linkage between loci A and B (see the illustration). The crosses are between four pairs of double heterozygotes for the loci A and B. In the progeny of cross 1, the two homozygous genotypes can be distinguished from the heterozygous for both loci A and B. So, nine classes of genotypes were observed from the progeny of cross 1. For cross 2, one of the homozygotes cannot be distinguished from the heterozygotes for both loci A and B. Four classes of genotypes were observed in the progeny of cross 2. For crosses 3 and 4, one of the homozygotes cannot be distinguished from the heterozygotes for either locus A or B. Six classes of genotypes were observed in the progeny of the crosses 3 and 4.

Cross 1	Genotype	Count		Cross 3	Genotype	Count
A B	AABB	35		A B	A_BB	40
————	AABb	14		————	A_Bb	10
a b	AAbb	5		a b	A_bb	6
X ——→	AaBB	10		X ——→	aaBB	8
A B	AaBb	20		A B	aaBb	15
————	Aabb	11		————	aabb	32
a b	aaBB	3		a b		
	aaBb	12				
	aabb	40				

Cross 2				Cross 4		
A B				A B		
————				————	AAB_	50
a b	A_B_	70		a b	AAbb	10
X ——→	A_bb	20		X ——→	AaB_	32
A B	aaB_	18		A B	Aabb	9
————	aabb	30		————	aaB_	16
a b				a b	aabb	45

Note: Where A_ is mixed with AA and Aa and B_ is mixed with BB and Bb.

Ex. 6.7

(1) Develop likelihood functions for analyzing the data.

(2) Do complete single-locus analysis for the data using the goodness of fit statistic and the likelihood ratio approaches. (Hints: The data can be analyzed for each of the crosses independently and also can be analyzed according to some pairs. For example, segregation among AA, Aa, and aa can be tests using both crosses 1 and 4 and segregation among B_ and bb. The heterogeneity of segregation ratios among the crosses can be tested.)

(3) Detect linkage between loci A and B using data from each of the crosses, using both the goodness of fit statistic and the likelihood ratio approaches.

(4) Obtain the maximum likelihood estimate of recombination fraction independently for each of the crosses and jointly, using all four crosses.

(5) Test the heterogeneity of recombination fraction between loci A and B among the four crosses.

(6) Test the hypothesis that the recombination fraction estimate from crosses 1 and 3 is same as the estimate from crosses 2 and 4.

(7) Estimate the linkage information contents of the crosses 2, 3 and 4 relative to cross 1, assuming that the pooled estimate of the recombination fraction is the unbiased estimate (Hint: Set the information content of cross 1 to one.).

(8) Estimate the variance of the maximum likelihood estimate of the recombination fraction for each of the crosses independently and for the four crosses jointly.

Exercise 6.8

Dominant markers linked in repulsion in F2 progeny provide a relatively small amount of linkage information. This is especially true when the two markers are closely linked and the linkage information is taken relative to codominant markers in F2 progeny.

(1) Review the problems and strategies to solve the problems associated with dominant markers linked in repulsion in F2 progeny.

(2) If the double recessive class is not observed in the F2 progeny for two linked markers in repulsion, estimate the 99% upper bounds for sample sizes of 20, 50, 100, and 1,000.

(3) How can a dominant marker be scored as a codominant marker?

Exercise 6.9

Sampling errors cannot be avoided and should be expected. Statistical procedures are designed to derive inference from data with sampling errors. However, a large sampling error and non-sampling errors may cause problems in resolving an inference from data. Please answer the following questions related to sampling and non-sampling errors in experiments to determine recombination fraction between two loci.

(1) Segregation ratio distortion is common in linkage analysis. In this exercise, the observed segregation ratio is significantly different from the expectation and is referred to as segregation ratio distortion. This may differ from the classical definition of segregation distortion, which refers only to distortion during meiosis. Explain the consequences of segregation ratio distortion on detection of linkage and estimation of recombination fraction.

(2) Sampling is one of the reasons for segregation ratio distortion. Given that the expected segregation ratio is 1:1, what is the probability that the observed segregation ratio is significantly different from the expectation at the 0.05 level when the progeny size is 20? How about when the progeny size is 200?

(3) Besides sampling, what are the sources of segregation ratio distortion? Can segregation ratio distortion be prevented? If the answer is "yes", what are the consequences on detection of linkage and estimation of recombination fraction?

(4) Marker classification error is a common experimental error in linkage experiments. This is usually a result of a complex banding pattern or an unclear band. What action should you take when you have this kind of situation? a) do nothing, b) treat the band as a missing value, or c) discard the marker entirely. Justify your action.

Exercise 6.10

In Exercise 6.3, F2 progeny were obtained for detecting linkage between loci A and B and determining recombination fraction between the two loci. In this exercise, two hypothetical lethal models are assumed. Please implement the hypothesis test on linkage and estimation of the recombination fraction under the hypothetical models.

Single-locus recessive lethal model: Locus B is a recessive lethal gene (see the following illustration). Individuals with homozygous genotype bb cannot be observed in the progeny.

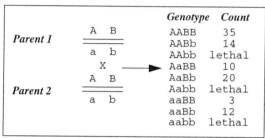

	Genotype	Count
	AABB	35
	AABb	14
	AAbb	lethal
	AaBB	10
	AaBb	20
	Aabb	lethal
	aaBB	3
	aaBb	12
	aabb	lethal

Ex. 6.10 (1)

Two-locus recessive lethal model: In this model, the individuals with double recessive genotype aabb cannot be observed in the progeny (see the following illustration).

	Genotype	Count
	AABB	35
	AABb	14
	AAbb	5
	AaBB	10
	AaBb	20
	Aabb	11
	aaBB	3
	aaBb	12
	aabb	lethal

Ex. 6.10 (2)

(1) Estimate the variance of the estimated recombination between A and B for the two models.

(2) Compare the results of this exercise with the results obtained in Exercise 6.3. Explain why they are the same or different. (Hint: Take significant p-values, recombination fraction estimates, and variances of the estimates into consideration).

Exercise 6.11

One important assumption for estimating recombination fraction between two loci is that the crossover event occurs randomly in the interval between the two loci. However, some experimental evidence indicates that the crossover event may occur non-randomly.

(1) Coarse Control Model: Data was obtained to estimate recombination fraction between loci A and C (see the following illustration). Assume you found that the crossover between loci B and C occurs non-randomly and is controlled by another locus R. When the parent of the progeny had genotype rr, there were no recombinations between B and C

observed in the progeny. When the parent had genotypes RR or Rr, the recombinants between B and C were observed to be normal. Please answer the following:

a) How can the recombination fraction between loci A and C be estimated?

b) How does one interpret the estimated recombination fraction?

c) Can you speculate on the biological basis of the so-called crossover coarse control model?

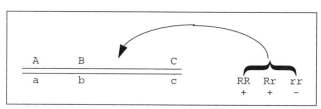

Ex. 6.11 (1)

(2) Fine Control Model: In this model, locus R has a quantitative effect on crossover for the whole genome (see the following illustration). The effect is evenly distributed for every segment. If the parent of the progeny has a genotype RR, crossover occurs more frequently on the whole genome. If the parent has a genotype of rr, crossover occurs less frequently on the whole genome. If the parent has a genotype Rr, the chance of that crossover occurs is in between that of RR and rr. Please answer the following:

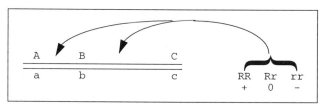

Ex. 6.11 (2)

a) How can the recombination fraction between loci A and C be estimated?

b) How can the estimated recombination fraction be interpreted?

c) Can you speculate on the biological basis of the so-called crossover fine control model?

(3) Design an experiment to determine the effect of locus R on the recombination fraction between loci A and C.

CHAPTER 7

TWO-LOCUS MODELS: NATURAL POPULATIONS

In Chapter 6, linkage analysis using controlled crosses was introduced. However, for many economically important plant and animal species, it is impossible, difficult or even unethical to obtain controlled crosses. For example, doing controlled crosses would involve a very long time frame for many forestry species. This is especially true for the most important animal species, human. For these species, natural populations are usually widely available.

The main focus of this chapter is how linkage analysis can be done using naturally available populations. In controlled crosses, linkage phases are usually known when the crosses are made. In natural populations, the linkage phase is usually unknown unless extensive pedigree information is available. In Section 7.1, the linkage phase problem is dealt with. In Section 7.2, populations produced by both self and random mating will be discussed. This kind of population is widely available in the plant kingdom and is called an open-pollinated population.

7.1 THE LINKAGE PHASE PROBLEM

In many practical linkage mapping experiments, linkage phases are unknown. For some cases, the linkage phase is either coupling or repulsion and in other cases, the linkage phase is a mixture of both. The two linkage phases, coupling and repulsion, were defined in Chapter 3. In this section, methodologies for determining linkage phase will be described when the phase is either one or the other. Analyses using a mixture phase model will be introduced for situations when the linkage phase is biologically mixed.

7.1.1 LINKAGE PHASE CONFIGURATIONS FOR TWO-LOCUS MODELS

For backcross progeny, linkage is detected by the linkage relationship of segregating gametes from a single parent. In other words, for a two-locus model, one of the parents has to be heterozygous at both loci to give linkage information. For two loci A (with two alleles A and a) and B (with two alleles B and b), one of the parents has to have genotype AaBb to give linkage information. If one of the parents is heterozygous for one of the loci and the other parent is heterozygous for the other locus (Aabb X aaBb), the progeny provide no linkage information for the two loci, even if each locus is in a backcross configuration. For the backcross progeny, either coupling (AB/ab) or repulsion (Ab/aB) could exist for a single heterozygous parent. The linkage phase cannot be a mixture of both phases. For multiple cross data, the linkage phases could be different from cross to cross. However, once the linkage phase is determined for each cross, the linkage analysis for the multiple crosses can be performed using the generalized likelihood approaches explained in Chapter 6.

For F2 progeny, linkage phases have three possible configurations for two-locus models (Table 7.1). Here, the F2 has been shown as crosses between any two heterozygous parents. If the two parents were themselves produced by a cross between two inbred lines, then the progeny is a conventional F2.

Table 7.1 Linkage phase configurations for F2 two-locus model.

Linkage Phase Configuration	Cross (Parent1 x Parent2)
Coupling-Coupling (CC)	$\dfrac{AB}{ab} \times \dfrac{AB}{ab}$
Coupling-Repulsion (CR)	$\dfrac{AB}{ab} \times \dfrac{Ab}{aB}$
Repulsion-Repulsion (RR)	$\dfrac{Ab}{aB} \times \dfrac{Ab}{aB}$

7.1.2 LINKAGE PHASE DETERMINATION

For many cases, linkage phase is biologically certain and cannot be a mixture, for example, a single population of a backcross or a conventional F2. For some cases of backcross and F2, linkage phases are known by biology, such as when genotypes of progeny are determined simultaneously with those of the parents and grandparents.

In some cases, linkage phases cannot be determined biologically and statistical approaches are needed. For such cases, new likelihood functions are not needed for the repulsion phase because they are direct transformations of the coupling likelihood function, with θ replaced by $(1 - \theta)$

$$L_{Coupling} = L(\theta)$$
$$L_{Repulsion} = L(1 - \theta)$$

For cases of the backcross model and the F2 models with at least one of the two markers codominant, the estimates of recombination fraction are direct transformations

$$\hat{\theta}_{Repulsion} = 1 - \hat{\theta}_{Coupling}$$

For example, if an estimate of recombination fraction is $\hat{\theta}$ using a coupling phase model, but the true phase is repulsion, then the estimate using the repulsion phase model should be $(1 - \hat{\theta})$.

For the cases in which linkage phase is certain but unknown, an *ad hoc* approach can be used to determine it. For a backcross model and F2 models with at least one marker codominant, a single analysis using coupling linkage phase can be used to determine possible linkage phase. Because likelihood surfaces are symmetric about $\theta = 0.5$, the log likelihood ratio test statistics are the same for both coupling and repulsion linkage phases and the estimates of recombination fractions can be directly transformed for the two phases. For those models, if

$$Pr[\chi^2_{df=1} \geq G] \leq \alpha$$

and   (7.1)

$$\hat{\theta} < 0.5$$

then the linkage is determined to be in coupling phase, where α is a significance level, G is the log likelihood ratio test statistic for linkage and $\hat{\theta}$ is the estimated recombination fraction using the coupling model. If

$$Pr[\chi^2_{df=1} \geq G] \leq \alpha$$

$$\hat{\theta} > 0.5$$

(7.2)

then the linkage is determined to be in repulsion phase. However, if the log likelihood ratio test statistic is not significant

$$Pr[\chi^2_{df=1} \geq G] > \alpha$$

the linkage phase cannot be determined and the linkage is not significant.

For the F2 model with both markers dominant for a single data set, two analyses using both linkage phases are needed to determine the linkage phase. This is because the likelihood surfaces are not symmetric for the two phases; the log likelihood ratio test statistics are different and the estimates of recombination fractions cannot be directly transformed for two phases. For these dominant marker cases, if

$$\left.\begin{array}{c} Pr[\chi^2_{df=1} \geq G_C] \leq \alpha \\ \text{or} \\ Pr[\chi^2_{df=1} \geq G_R] \leq \alpha \end{array}\right\} \text{ and } G_C > G_R$$

then the linkage is determined to be in coupling phase, where G_C and G_R are the log likelihood ratio test statistics obtained using coupling and repulsion models, respectively. If

$$\left.\begin{array}{c} Pr[\chi^2_{df=1} \geq G_C] \leq \alpha \\ \text{or} \\ Pr[\chi^2_{df=1} \geq G_R] \leq \alpha \end{array}\right\} \text{ and } G_C < G_R$$

then the linkage is determined to be in repulsion phase. Otherwise, if the log likelihood ratio test statistic is not significant for either models

$$Pr[\chi^2_{df=1} \geq G_C] > \alpha$$

or

$$Pr[\chi^2_{df=1} \geq G_R] > \alpha$$

the linkage phase cannot be determined and the linkage is not significant.

A computer simulation study was done to validate the *ad hoc* approach to determine linkage phase. Simulations were done for factorial combinations of

$$\theta = 0.01, 0.05, 0.1, 0.2, 0.3, 0.4$$

$$N = 20, 50, 100, 200, 400$$

$$\alpha = 0.05, 0.01$$

and models of backcross (BC), F2-CC (both markers codominant), F2-CD (one marker codominant and the other dominant) and F2-DD (both markers dominant). For each of the combinations, 10,000 replications were simulated (Table 7.2). For $\theta \leq 0.3$ and $\alpha = 0.05$, there is almost no chance of making an incorrect linkage phase determination for BC, F2-CC and F2-CD models, even with a sample size of 20. However, for the F2-DD model, a sample size of at

Table 7.2 Numbers of times the linkage phases were correctly and incorrectly (in parentheses) determined among the 10,000 simulated replications.

α	True θ	N	BC	F2-CC	F2-CD	F2-DD
0.05	0.01	20	10000(0)	10000(0)	10000(0)	6673(449)
		50	10000(0)	10000(0)	10000(0)	9609(53)
		100	10000(0)	10000(0)	10000(0)	9991(8)
	0.05	20	9990(0)	10000(0)	9924(0)	6404(605)
		50	10000(0)	10000(0)	9994(0)	9524(104)
		100	10000(0)	10000(0)	9999(0)	9982(10)
	0.1	20	9899(0)	9990(0)	9421(1)	5883(736)
		50	10000(0)	10000(0)	9998(0)	9225(154)
		100	10000(0)	10000(0)	10000(0)	9945(31)
	0.2	20	8024(0)	9700(2)	6723(1)	4351(697)
		50	9981(0)	9998(0)	9636(0)	7900(271)
		100	10000(0)	10000(0)	9996(0)	9601(90)
		200	10000(0)	10000(0)	10000(0)	9971(19)
		400	10000(0)	10000(0)	10000(0)	9999(1)
	0.3	20	4167(0)	7770(25)	3385(5)	2409(534)
		50	8621(0)	9675(1)	6710(0)	4739(267)
		100	9883(0)	9992(0)	9240(0)	7490(158)
		200	10000(0)	10000(0)	9982(0)	9552(56)
		400	10000(0)	10000(0)	10000(0)	9987(9)
	0.4	20	1270(17)	4213(234)	1160(51)	1054(318)
		50	3404(5)	6228(57)	2194(15)	1658(151)
		100	5341(0)	8230(7)	3714(1)	2556(88)
		200	8244(0)	9603(1)	6413(0)	4692(36)
		400	9827(0)	9990(0)	9145(0)	7573(3)
0.01	0.2	100	10000(0)	10000(0)	9971(0)	8899(87)
		200	10000(0)	10000(0)	10000(0)	9939(19)
		400	10000(0)	10000(0)	10000(0)	9999(1)
	0.3	20	2368(0)	6500(9)	1481(0)	629(415)
		50	5692(0)	9307(0)	4267(0)	2651(171)
		100	9484(0)	9977(0)	7917(0)	5509(123)
		200	9997(0)	10000(0)	9878(0)	8722(56)
		400	10000(0)	10000(0)	10000(0)	9932(9)
	0.4	20	516(3)	2730(82)	325(4)	145(219)
		50	972(0)	4755(21)	787(1)	592(59)
		100	2994(0)	7029(3)	1769(1)	1124(51)
		200	5817(0)	9156(0)	4068(0)	2494(33)
		400	9326(0)	9968(0)	7639(0)	5349(3)

In this table, BC = backcross, F2-CC = both markers codominant, F2-CD = one marker codominant and one marker dominant, F2-DD = both markers dominant.

least 100 is needed to have a small probability of making an uncorrected linkage phase determination. When $\alpha = 0.01$ is used, the sample size needed to have a small chance of making a mistake decreases.

For practical linkage mapping with a reasonable sample size, linkage phase determination should not be a problem. The algorithm for determining linkage phase has been built into linkage analysis computer software packages such as PGRI (see Chapter 18). Once linkage phase is determined, linkage analysis can be performed using approaches described in previous sections. For these cases, the linkage analysis for the data reduces to that for phase known, as described in Chapter 6. For marker combinations where the linkage phase cannot be determined, it is usually true that the markers are loosely linked and their linkage information contribution for the whole experiment is small. For these cases, phase-unknown linkage analysis is used and will be described in the next subsection.

7.1.3 PHASE-UNKNOWN LINKAGE ANALYSIS

For a backcross model, the phase unknown likelihood can be written as

$$L_{BC\overline{PUK}}(\hat{\theta}) = \theta^{f_{12}+f_{21}}(1-\theta)^{f_{11}+f_{22}} + \theta^{f_{11}+f_{22}}(1-\theta)^{f_{12}+f_{21}} \tag{7.3}$$

where θ is the recombination fraction and the superscripts are the observed counts for the two-locus genotypes (see Chapter 6). A grid search scheme can be used to find the recombination fraction that maximizes the likelihood. The likelihood curve has two peaks in the range of parameter space between zero and one. For example, in three experiments, A, B and C with 100 individuals in each with $f_{12}+f_{21}$

$$f_{12}+f_{21} = 10 \qquad A$$
$$f_{12}+f_{21} = 20 \qquad B$$
$$f_{12}+f_{21} = 30 \qquad C$$

Three maximum likelihood recombination fraction estimates can be obtained using the three likelihood profiles (Figure 7.1).

The recombination fraction also can be obtained using a Bayes procedure (Nordheim *et al.* 1983 and 1984). The Bayes estimate is

$$\hat{\theta} = \frac{\int_{0}^{0.5} \theta L_{BC\overline{PUK}}\hat{\theta}d\theta}{\int_{0}^{0.5} L_{BC\overline{PUK}}\hat{\theta}d\theta} \tag{7.4}$$

$$= \frac{k+1}{N+2}I_{0.5}(k+2, N-k+1) + \frac{N-k+1}{N+2}I_{0.5}(N-k+2, k+1)$$

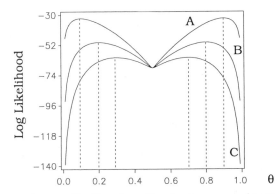

Figure 7.1 Phase-unknown likelihood profiles for true recombination fractions 0.1, 0.2 and 0.3 with 100 backcross progeny.

where $I_{0.5}(k+2, N-k+1)$ and $I_{0.5}(N-k+2, k+1)$ are the incomplete beta function, as defined by Equation 26.5.1 in Abramowitz and Stegun (1972) and k is the smallest of $f_{11}+f_{22}$ and $f_{12}+f_{21}$.

For the F2 models, the phase-unknown likelihood function can be complex. The EM algorithm can be implemented for obtaining the recombination fraction (see Chapter 6). Table 7.3 shows the probability that a genotype is a recombinant conditional on the marker genotype for F2 progeny using codominant markers with unknown linkage phase.

Table 7.3 Probability that a genotype is a recombinant conditional on the marker genotype for F2 progeny using codominant markers with linkage phase unknown.

Genotype	Observed Count (f_i)	$P_i(R|G)$
AABB	f_1	$\theta^2/[(1-\theta)^2+\theta^2]$
AABb	f_2	0.5
AAbb	f_3	$\theta^2/[(1-\theta)^2+\theta^2]$
AaBB	f_4	0.5
AaBb	f_5	0.5
Aabb	f_6	0.5
aaBB	f_7	$\theta^2/[(1-\theta)^2+\theta^2]$
aaBb	f_8	0.5
aabb	f_9	$\theta^2/[(1-\theta)^2+\theta^2]$

Table 7.4 A two-locus model for a generalized F2 (AB/ab X Ab/aB) progeny with coupling linkage phase in one parent, repulsion phase in the other parent and recombination fraction θ.

Parent 2 Gamete	Parent 1 Gamete			
	AB	**Ab**	**aB**	**ab**
	$0.5(1-\theta)$	0.5θ	0.5θ	$0.5(1-\theta)$
AB 0.5θ	AABB $0.25\theta(1-\theta)$	AABb $0.25\theta^2$	AaBB $0.25\theta^2$	AaBb $0.25\theta(1-\theta)$
Ab $0.5(1-\theta)$	AABb $0.25(1-\theta)^2$	AAbb $0.25\theta(1-\theta)$	AaBb $0.25\theta(1-\theta)$	Aabb $0.25(1-\theta)^2$
aB $0.5(1-\theta)$	AaBB $0.25(1-\theta)^2$	AaBb $0.25\theta(1-\theta)$	aaBB $0.25\theta(1-\theta)$	aaBb $0.25(1-\theta)^2$
ab 0.5θ	AaBb $0.25\theta(1-\theta)$	Aabb $0.25\theta^2$	aaBb $0.25\theta^2$	aabb $0.25\theta(1-\theta)$

7.1.4 LINKAGE ANALYSIS WITH A MIXTURE OF LINKAGE PHASES

Linkage estimation with phase-unknown is different from a mixture of linkage phases. Phase-unknown means that linkage phase is either coupling or repulsion, but is unknown. The mixture linkage phase problem being explained here also differs from multiple-population mixture where linkage phases for different populations could be different. For each individual population, the linkage phase is certain and linkage analysis is the same as described in Chapter 6.

Mixture linkage phase problems occur when linkage phases at the two loci are different in the two parents of a cross. For example, in the maternal parent, they are in coupling phase and in the paternal parent, in repulsion phase. In Table 7.4, progeny from the cross of two heterozygous parents, one in coupling and the other in repulsion linkage phase, are shown and the expected genotypic frequencies are summarized in Table 7.5. Table 7.5 also lists the conditional probability needed for implementing the EM algorithm.

Table 7.5 Expected genotypic frequency for F2 progeny with mixture linkage phase using codominant markers.

Genotype	Count	Expected Frequency (p_i)	$P_i(R\|G)$
AABB	f_1	$0.25\theta(1-\theta)$	0.5
AABb	f_2	$0.25(1-2\theta+\theta^2)$	$\theta^2/[(1-\theta)^2+\theta^2]$
AAbb	f_3	$0.25\theta(1-\theta)$	0.5
AaBB	f_4	$0.25(1-2\theta+\theta^2)$	$\theta^2/[(1-\theta)^2+\theta^2]$
AaBb	f_5	$\theta(1-\theta)$	0.5
Aabb	f_6	$0.25(1-2\theta+\theta^2)$	$\theta^2/[(1-\theta)^2+\theta^2]$
aaBB	f_7	$0.25\theta(1-\theta)$	0.5
aaBb	f_8	$0.25(1-2\theta+\theta^2)$	$\theta^2/[(1-\theta)^2+\theta^2]$
aabb	f_9	$0.25\theta(1-\theta)$	0.5

The log likelihood function for a two-locus model using progeny of a cross between two heterozygous parents with different linkage phases is

$$L_{MP}(\theta) = (f_1+f_3+f_5+f_7+f_9)\log\theta + (N+f_2+f_4+f_6+f_8)\log(1-\theta) \qquad (7.5)$$

The first derivative with respect to the recombination fraction (the score) is

$$L_{MP}'(\theta) = \frac{f_1+f_3+f_5+f_7+f_9}{\theta} - \frac{N+f_2+f_4+f_6+f_8}{1-\theta} \qquad (7.6)$$

By setting this to zero, the analytical solution of the maximum likelihood estimate of the recombination fraction is

$$\hat{\theta} = \frac{f_1+f_3+f_5+f_7+f_9}{2N} \qquad (7.7)$$

which is an unbiased estimate. The large-sample variance is

$$Var_{MP}(\theta) = \frac{\theta(1-\theta)^2}{N(1-\theta+\theta^2)} \tag{7.8}$$

7.2 MIXTURES OF SELFS AND RANDOM MATING

For many organisms, controlled crosses are not available, but open-pollinated populations (OP) are widely available and represent a mixture of self and random mating (MSR). In this section, methodology for mapping using populations of MSR will be discussed. Commonly used mapping populations, such as backcross and F2, can be regarded as special cases of populations with MSR with different configurations of outcrossing rates and allelic frequencies in the pollen pool.

For some plant species, maternal gametic tissue passes to the progeny. For those species, a haploid approach, which is the same as the backcross approach in terms of data analysis, can be used for linkage analysis. For example, in conifer species, haploid megagametophyte tissue in the seeds (see Glossary) is genetically identical to the maternal gamete and has been used to construct genetic maps (Conkle 1981; Grattapaglia *et al.* 1991; Neale and Sederoff 1991; Plomion *et al.* 1995). This approach cannot be applied to the available open-pollinated families in the field unless megagametophyte tissue was obtained from the seedlings. In this section, linkage analysis is based on data from diploid tissue of open-pollinated progeny.

Table 7.6 Expected and observed genotypic frequency for codominant markers A and B in an open-pollinated population where θ is the recombination fraction between A and B.

Genotype	Count	Expected Frequency	
		Outcrossed progeny	Selfed progeny
AABB	f_1	$0.5uv(1-\theta)$	$0.25(1-\theta)^2$
AABb	f_2	$0.5u[(1-v)(1-\theta)+v\theta]$	$0.5\theta(1-\theta)$
AAbb	f_3	$0.5u(1-v)\theta$	$0.25\theta^2$
AaBB	f_4	$0.5v[(1-u)(1-\theta)+ur]$	$0.5\theta(1-\theta)$
AaBb	f_5	$0.5[1-\theta-(1-2\theta)(u+v-2uv)]$	$0.5(1-2\theta+\theta^2)$
Aabb	f_6	$0.5(1-v)(u-2u\theta+\theta)$	$0.5\theta(1-\theta)$
aaBB	f_7	$0.5(1-u)v\theta$	$0.25\theta^2$
aaBb	f_8	$0.5(1-u)(v-2v\theta+\theta)$	$0.5\theta(1-\theta)$
aabb	f_9	$0.5(1-u)(1-v)(1-\theta)$	$0.25(1-\theta)^2$

7.2.1 MODEL

Suppose loci A and B are linked with a recombination of q and are in coupling linkage phase. Locus A has alleles A and a. Locus B has alleles B and b. All alleles are codominant. Let u and v be frequencies of allele A and B in the pollen pool. For the two-allele model, frequencies of the alternative alleles a and b are $(1 - u)$ and $(1 - v)$, respectively. Linkage equilibrium in the pollen pool and coupling linkage phase of the mother tree are assumed. The expected genotypic frequencies among outcrossed and selfed progeny are listed in Table 7.6. The log-likelihood equation for estimating the recombination fraction between codominant markers A and B is

$$L_{MSRC}(\theta) = \sum_{i=1}^{9} f_i \log\left[tp_{oi} + (1-t)p_{si}\right] \tag{7.9}$$

where t is the outcrossing rate and p_{oi} and p_{si} are the expected frequencies of the ith genotype for open-pollinated and selfed progeny, respectively. For dominant A and B alleles, the log-likelihood function can be obtained by pooling together corresponding genotypes in Table 7.6

$$L_{MSRD}(\theta) = \sum_{i=1}^{4} f_i \log\left[tp_{oi} + (1-t)p_{si}\right] \tag{7.10}$$

where p_{oi} and p_{si} are shown in Table 7.7. To estimate the recombination fraction, the maternal parent has to be heterozygous because linkage equilibrium is assumed in the pollen pool.

Table 7.7 Expected and observed genotypic frequency for dominant markers A and B.

Genotype	Count	Expected Frequency	
		Outcrossed progeny	Selfed progeny
A_B_	f_1	$0.5\left[(1+uv)(1-\theta) + (u+v)\theta\right]$	$0.25(3 - 2\theta + \theta^2)$
A_bb	f_2	$0.5(1-v)(u - u\theta + \theta)$	$0.25(2\theta - \theta^2)$
aaB_	f_3	$0.5(1-u)(v - v\theta + \theta)$	$0.25(2\theta - \theta^2)$
aabb	f_4	$0.5(1-u)(1-v)(1-\theta)$	$0.25(1-\theta)^2$

Table 7.8 Expected genotypic frequency in open-pollinated population of a heterozygous maternal parent for a marker A with a frequency of u for allele A in the pollen pool.

Genotype	Count	Expected Frequency		
		Outcrossed	Selfed	Total
Codominant Marker				
AA	n_{AA}	$0.5u$	0.25	$0.25\left[1 - t(1 - 2u)\right]$
Aa	n_{Aa}	0.5	0.5	0.5
aa	n_{aa}	$0.5(1-u)$	0.25	$0.25\left[1 + t(1 - 2u)\right]$
Dominant Marker				
A_	$n_{A\_}$	$0.5(1+u)$	0.75	$0.25\left[3 - t(1 - 2u)\right]$
aa	n_{aa}	$0.5(1-u)$	0.25	$0.25\left[1 + t(1 - 2u)\right]$

7.2.2 ALLELIC FREQUENCY IN POLLEN POOL AND OUTCROSSING RATE

Allelic Frequency in Pollen Pool

For most experimental populations used for genomic mapping, such as classical backcross, F2 and recombinant inbred lines, allelic frequencies are expected to be 0, 0.5 or 1. Allelic frequency is not an issue in terms of linkage analysis for these mating types. However, it is an issue when open-pollinated populations are used for linkage map construction.

Assuming a diploid with two alleles segregating in a population and codominant markers, the log likelihood function for estimating allelic frequency for allele A in the population is

$$L(u) = n_{AA}\log[1 - t(1 - 2u)] + n_{aa}\log[1 + t(1 - 2u)] \tag{7.11}$$

where n_{AA} is the count of AA genotypes, n_{aa} is the count of aa genotypes and t is the outcrossing rate. The count of Aa genotype, n_{Aa}, is not in the likelihood function because the expected frequency of Aa in the population does not depend on either the allelic frequency in the pollen pool or the outcrossing rate (Table 7.8). The maximum likelihood estimate of the allelic frequency in the pollen pool is

$$\hat{u} = \frac{1}{2}\left[1 - \frac{n_{aa} - n_{AA}}{t(n_{AA} + n_{aa})}\right] \tag{7.12}$$

with a large sample variance

$$Var(\hat{u}) = \frac{1/t^2 - (1 - 2\hat{u})^2}{2(n_{AA} + n_{aa})}$$

The outcrossing rate and the allelic frequency cannot be estimated simultaneously using the same likelihood function. Here, the outcrossing rate is assumed to be a known constant. Also, the allelic frequency cannot be estimated if the progeny is generated by complete selfing without outcrossing $(t = 0)$.

For dominant markers, allelic frequencies can be estimated using the log likelihood function

$$L = n_{A\_}\log[3 - t(1 - 2u)] + n_{aa}\log[1 + t(1 - 2u)] \tag{7.13}$$

where $n_{A\_}$ is count of phenotype $A\_$. The maximum likelihood estimate of the allelic frequency in the pollen pool is

$$\hat{u} = \frac{1}{2}\left[1 - \frac{3n_{aa} - n_{A\_}}{t(n_{A\_} + n_{aa})}\right] \tag{7.14}$$

with a large sample variance

$$Var(\hat{u}) = \frac{3/t^2 + 2(1 - \hat{u})/t - (1 - 2\hat{u})^2}{4(n_{A\_} + n_{aa})}$$

Outcrossing Rate

Outcrossing rate, allelic frequencies in the pollen pool and recombination fraction cannot be estimated simultaneously. The methods explained in this section will require a pre-estimated outcrossing rate, which can be obtained by using methods developed by Ritland and Jain (1981). As the number of loci included in the estimation procedure increases, the bias of the estimated outcrossing rate decreases. For linkage mapping experiments, the number of loci is usually large, so the bias of the estimated outcrossing rate is small. The outcrossing rate is assumed equal for all loci.

The reason outcrossing rate and recombination fraction cannot be estimated simultaneously is that to estimate recombination fraction, a heterozygous maternal parent is required. However, the heterozygous maternal parent provides the least information for estimating the outcrossing rate. This logic can be illustrated using a single-locus model. For a single-locus outcrossing model, given outcrossing rate t and A allelic frequency in the pollen pool u, the expected genotypic frequencies in the progeny for the three types of maternal parents are listed in Table 7.9. The maximum likelihood estimates of outcrossing rates using the three maternal parents and the average information content per observation are given in Table 7.10.

Table 7.9 The expected genotypic frequencies in the progeny for the three genotypes of maternal parents given outcrossing rate t and A allelic frequency in the pollen pool u.

Maternal Genotype	Progeny		
	AA	Aa	aa
AA	$1 - t(1 - u)$	$t(1 - u)$	0
Aa	$0.25[1 - t(1 - 2u)]$	0.5	$0.25[1 + t(1 - 2u)]$
aa	0	tu	$1 - tu$

Table 7.10 Maximum likelihood estimates of outcrossing rates using the three maternal parents and the average information content per observation.

Maternal Genotype	$\hat{t}$	Average Information
AA	$n_{AA} / [n_{AA} + n_{Aa}(1 - u)]$	$(1 - u) / \{t[1 - t(1 - u)]\}$
Aa	$(n_{AA} - n_{aa}) / [(n_{AA} + n_{aa})(1 - 2u)]$	$(1 - 2u)^2 / [1 - t^2(1 - 2u)^2]$
aa	$n_{aa} / (n_{aa} + n_{Aa}u)$	$u / [t(1 - tu)]$

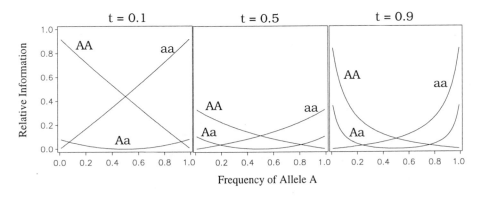

Figure 7.2 Relative information contents contributed by a homozygous maternal parent (AA or aa) and a heterozygous maternal parent (Aa) for outcrossing rate estimation using a single-locus model for true outcrossing rates 0.1, 0.5 and 0.9 and different A allelic frequencies in the pollen pool.

Figure 7.2 plots the relative information contents contributed by a homozygous maternal parent (AA) and a heterozygous maternal parent (Aa). The curve for maternal parent aa is symmetric with AA. Compared to the homozygous parent, the heterozygous parent contributes minimal information for the estimation of outcrossing rate.

In practical genomic mapping experiments using populations of MSR, determination of the outcrossing rate is rarely a problem. For example, for most of the economically important tree species, outcrossing rates have been extensively studied. For some species, outcrossing rates are considered to be 100% because of self incompatibility.

7.2.3 ESTIMATION OF RECOMBINATION FRACTION

There are several ways to estimate parameters from the likelihood functions in Equations (7.9) and (7.10) for given outcrossing rates. Two methods will be described in this section. One method will take the outcrossing rate and allelic frequencies in the pollen pool (estimated using the approach described in Section 7.2.2) as constants and estimate the recombination fraction. The other method estimates recombination fraction and allelic frequencies simultaneously. Both methods can use either an EM algorithm or Newton-Raphson iteration.

Table 7.11 Expected gametic frequencies for pollen pool and from the maternal plant conditional on (u, v, θ, t) , where p_i is genotypic frequency for the progeny (Table 7.6).

Genotype	Gamete	Pollen Pool	Maternal Plant
AABB	AB	1.0	1.0
AABb	AB	$0.5uv\theta/p_2$	$0.5u(1-v)(1-\theta)/p_2$
	Ab	$0.5u(1-v)(1-\theta)/p_2$	$0.5uv\theta/p_2$
AAbb	Ab	1.0	1.0
AaBB	AB	$0.5uv\theta/p_4$	$0.5(1-u)v(1-\theta)/p_4$
	aB	$0.5(1-u)v(1-\theta)/p_4$	$0.5uv\theta/p_4$
AaBb	AB	$0.5uv(1-\theta)/p_5$	$0.5(1-u)(1-v)(1-\theta)/p_5$
	Ab	$0.5u(1-v)\theta/p_5$	$0.5(1-u)v\theta/p_5$
	aB	$0.5(1-u)v\theta/p_5$	$0.5u(1-v)\theta/p_5$
	ab	$0.5(1-u)(1-v)(1-\theta)/p_5$	$0.5uv(1-\theta)/p_5$
Aabb	Ab	$0.5u(1-v)(1-\theta)/p_6$	$0.5(1-u)(1-v)\theta/p_6$
	ab	$0.5(1-u)(1-v)\theta/p_6$	$0.5u(1-v)(1-\theta)/p_6$
aaBB	aB	1.0	1.0
aaBb	aB	$0.5(1-u)v(1-\theta)/p_8$	$0.5(1-u)(1-v)\theta/p_8$
	ab	$0.5(1-u)(1-v)\theta/p_8$	$0.5(1-u)v(1-\theta)/p_8$
aabb	ab	1.0	1.0

Method I

For an EM algorithm, the conditional probability of a recombinant gamete on marker genotypes is given in Table 7.11. Given an initial estimate of recombination fraction θ^0, a new estimate is

$$\hat{\theta}^1 = \frac{1}{N} \sum_{i=1}^{9} f_i \{ [p_{PiAb} + p_{PiaB}] t + [p_{MiAb} + p_{MiaB}] (1-t) \} \tag{7.15}$$

where f_i is the observed count for genotype i shown in Table 7.6 and $p_{PiAb}, p_{MiAb}, p_{PiaB}$ and p_{MiaB} are the probabilities of recombinant gametes Ab, Ab, aB and aB for outcrossed, selfed, outcrossed and selfed individuals conditional on the marker genotype i. In Table 7.11, p_i is the expected frequency of genotype i pooled for both the outcrossed and selfed progeny in Table 7.6. All the conditional probabilities of Equation (7.15) are evaluated based on the specific t, u, v and the initial estimate of recombination fraction. By letting the new estimate be an initial value, the iterations are repeated until the estimate converges.

For Newton-Raphson iteration, the score is

$$L'_{MSR}(\theta) = \sum_{i=1}^{9} f_i \frac{d \{ \log [(1-t) p_{Si} + t p_{Oi}] \}}{d\theta}$$

The notation can be found in Table 7.6. The average information per observation is

$$I_{O\overline{MAR}}(\theta) = -\sum_{i=1}^{9} E(f_i) \frac{d^2 \{ \log [(1-t) p_{Si} + t p_{Oi}] \}}{d\theta^2}$$

and the large-sample variance is

$$Var_{MAR}(\theta) = \frac{1}{N I_{O\overline{MAR}}(\theta)}$$

Given an initial estimate of recombination fraction θ^0, the new estimate using Newton-Raphson iteration is

$$\hat{\theta}^1 = \theta^0 + \frac{L'_{MSR}(\theta^0)}{N I_{O\overline{MAR}}(\theta^0)}$$

The iteration is repeated until the estimate converges. Analytical equations for the score and the information can be very complex. In practical linkage analysis, these equations can be evaluated numerically. Readers should refer to *Numerical Recipes* (Press *et al.* 1986) for more information on numerical methods.

Method II Using EM Algorithm

Recombination fraction and allelic frequencies for the two markers in the pollen pool can be estimated using an EM algorithm (Dempster *et al.* 1977). By letting t be constant, u, v, θ are unknown parameters in the likelihood functions of Equation (7.9) and Equation (7.10). The EM approach to solve the equations includes 4 steps:

(1) Initial step: Let the initial estimates of Method I be

$$(u^0, v^0, \theta^0)$$

(2) Expectation step: Estimate expected gametic frequency for the pollen pool and maternal plant from initial estimates and the observed frequency

$$tf_i p_{Pi\overline{AB}}$$

for gamete $\overline{AB}$ for outcrossed progeny and

$$(1-t)f_i p_{Mi\overline{AB}}$$

for maternal plant (Table 7.11).

(3) Maximization step: The maximum likelihood estimate of $\hat{u}^1$ is

$$\hat{u}^1 = \frac{1}{N}\sum_{i=1}^{9} f_i \{ [p_{PiAB} + p_{PiAb}]t + [p_{MiAB} + p_{MiAb}](1-t) \}$$

where

$$\hat{v}^1 = \frac{1}{N}\sum_{i=1}^{9} f_i \{ [p_{PiAB} + p_{PiaB}]t + [p_{MiAB} + p_{MiaB}](1-t) \}$$

and

$$\hat{\theta}^1 = \frac{1}{N}\sum_{i=1}^{9} f_i \{ [p_{PiAb} + p_{PiaB}]t + [p_{MiAb} + p_{MiaB}](1-t) \} \qquad (7.16)$$

(4) Iteration: Repeat step (2) and (3) until the estimates converge. Equation (7.15) and Equation (7.16) are the same. However, the conditional probabilities are evaluated under different values of the initial recombination fraction and allelic frequencies in the pollen pool.

Method II Using Newton-Raphson Iteration

For Newton-Raphson iterations, a matrix notation is needed for the iteration procedures. The score vector for the unknown parameters u, v, θ is

$$S = \begin{bmatrix} \dfrac{L(\theta, u, v)}{du} \\[2mm] \dfrac{L(\theta, u, v)}{dv} \\[2mm] \dfrac{L(\theta, u, v)}{d\theta} \end{bmatrix}$$

and the information matrix is

$$I = \begin{bmatrix} \dfrac{\partial^2}{\partial u^2}L(\theta, u, v) & \dfrac{\partial^2}{\partial v \partial u}L(\theta, u, v) & \dfrac{\partial^2}{\partial \theta \partial u}L(\theta, u, v) \\[3mm] \dfrac{\partial^2}{\partial u \partial v}L(\theta, u, v) & \dfrac{\partial^2}{\partial v^2}L(\theta, u, v) & \dfrac{\partial^2}{\partial \theta \partial v}L(\theta, u, v) \\[3mm] \dfrac{\partial^2}{\partial u \partial \theta}L(\theta, u, v) & \dfrac{\partial^2}{\partial v \partial \theta}L(\theta, u, v) & \dfrac{\partial^2}{\partial \theta^2}L(\theta, u, v) \end{bmatrix}$$

The new estimate of the parameter vector is

$$\begin{bmatrix} \hat{u}^1 \\ \hat{v}^1 \\ \hat{\theta}^1 \end{bmatrix} = \begin{bmatrix} u^0 \\ v^0 \\ \theta^0 \end{bmatrix} + \frac{1}{N} I^{-1} S$$

The score vector S and the information matrix I are evaluated using $(\hat{u}^0, \hat{v}^0, \hat{\theta}^0)$. The iteration is repeated until all three parameters converge.

7.2.4 EFFICIENCY AND VARIANCES

Information Content for Codominant Markers
For codominant markers, the average linkage information content for each observation of the selfed F2 population ($t = 0$) is

$$-\sum_{i=1}^{9} E(f_i) \frac{d^2 [\log (p_{Si})]}{d\theta^2}$$

and of the complete outcrossing population ($t = 1$) is

$$-\sum_{i=1}^{9} E(f_i) \frac{d^2 [\log (p_{Oi})]}{d\theta^2}$$

When the population is obtained by both outcrossing and selfing, the average

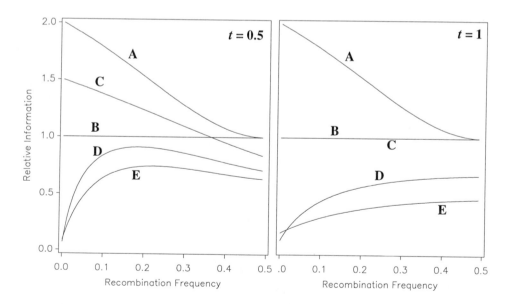

Figure 7.3 Relative linkage information for the codominant markers using a population of mixed self and random matings (MSR) under different allelic frequencies in pollen pool compared to the classical F2 and backcross. The information contents are shown for half outcrossing and complete outcrossing. Curve A = F2, B = backcross, C = mixes self and random (MSR) with u = v = 0.0, D = MSR with u = v = 0.1 and E = MSR u = v = 0.2, respectively.

linkage information per observation is

$$-\sum_{i=1}^{9} E(f_i) \frac{d^2 \{\log [\ (1-t)\, p_{Si} + t p_{Oi}]\ \}}{d\theta^2} \tag{7.17}$$

The large-sample variances of the estimators are the inverse of information functions. Explicit formulae for variances of the estimators are not given here due to the complexity of the second derivative of the likelihood function. Instead, solutions are given numerically for different experimental conditions.

Figure 7.3 shows linkage information provided by populations of MSR for different combinations of outcrossing rate and allelic frequency in the pollen pool, relative to conventional F2 and backcross populations. When $t = 0.5$, $u < 0.2$ and $v < 0.2$, the MSR model provides almost as much linkage information as a backcross for true recombination fractions above 0.2. When $t = 1.0$ (complete outcrossing), MSR is informationally equivalent to a backcross for $u = v = 0.0$. When the recombination fraction is below 0.2, MSR has a small amount of linkage information relative to an F2. When $u = v = 0.0$, MSR with $t = 0.5$ is a mixture of conventional backcross and F2 in terms of information provided. When allelic frequencies are high, ($u > 0.2$ and $v > 0.2$), MSR has a small amount of linkage information relative to an F2. The linkage information content is sensitive to allelic frequency in the pollen pool when the outcrossing rate is high and decreases rapidly as the allelic frequencies get close to 0.5. Figure 7.4 shows the absolute information content for the settings in Figure 7.3. The relative information content may be misleading for cases of tight linkage. The relative information content for MSR is low for tight linkage and high for loose linkage. Absolute information contents for tight linkages are always greater than for loose linkages (Figure 7.4).

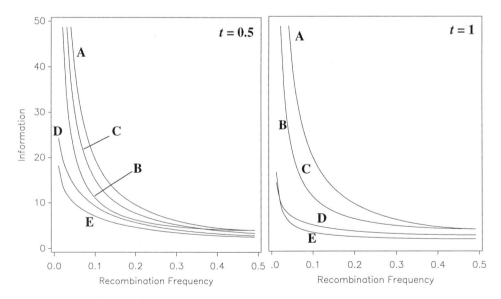

Figure 7.4 Linkage information for codominant markers. Curve A = F2, B = backcross, C = mixes self and random (MSR) with u = v = 0.0, D = MSR with u = v = 0.1 and E = MSR u = v = 0.2, respectively.

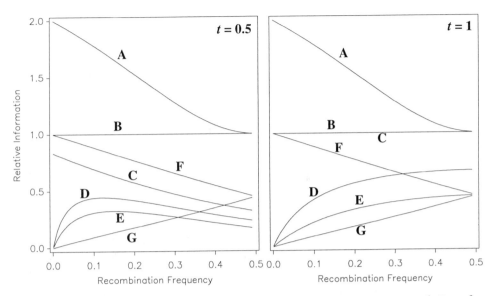

Figure 7.5 Relative linkage information for dominant markers using a population of mixed self and random matings (MSR) under different allelic frequencies in a pollen pool compared to the classical F2 and backcross. The information contents are shown for half outcrossing (t = 0.5) and complete outcrossing (t = 1.0). Curve A = F2 (codominant markers), B = backcross, C = mixes self and random (MSR) with u = v = 0.0, D = MSR with u = v = 0.1, E = MSR u = v = 0.2, F = F2 (dominant markers coupling linkage phase) and G = F2 (dominant markers repulsion linkage phase), respectively.

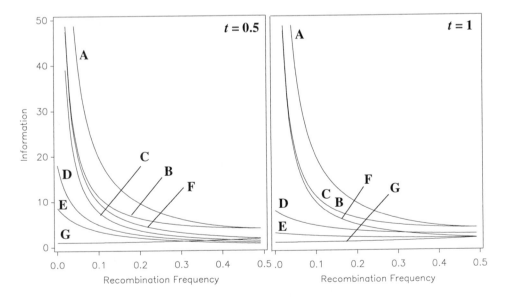

Figure 7.6 Linkage information for dominant markers using a population of mixed self and random matings (MSR) under different allelic frequencies in pollen pool comparing the classical F2 and backcross. See Figure 7.5 for notation.

Information Content for Dominant Markers

For dominant markers, when $u = v = 0.0$, MSR provides more linkage information than does an F2 (Figure 7.5). In these cases, when $t = 1.0$, the MSR is equivalent to the backcross in terms of linkage information. When $0 < t < 1.0$, MSR is a mixture of backcross and F2 in terms of information provided. For dominant markers, the backcross provides more linkage information than F2. The linkage information content is very sensitive to dominant allelic frequency in the pollen pool and decreases quickly as the dominant allelic frequency increases. Similarly to codominant markers, when recombination fraction is low, MSR provides less linkage information relative to F2 than when it is high. When recombination is low, MSR provides more absolute linkage information than when it is high (Figure 7.6).

Empirical Variance and Bias

Computer simulations were done for some of the combinations of t, u, v, θ and number of progeny (n). Both codominant and dominant markers were considered in this study. Given t, u, v and θ, expected genotypic frequencies

Table 7.12 Empirical bias and mean square error for estimated recombination fractions using codominant markers from simulation of 1,000 replications for each of the combinations.

θ	t	u and v	Bias	Sample Sizes		
				50	200	2,000
0.05	0.0	0.0	0.0	0.0005	0.0001	0.0
	0.5	0.0	-0.006	0.0014	0.0008	0.0001
		0.1	0.001	0.0027	0.0013	0.0002
		0.2	0.002	0.0037	0.0014	0.0001
		0.5	0.001	0.004	0.0007	0.0001
	1.0	0.0	-0.004	0.0014	0.0007	0.0001
		0.1	0.0	0.0038	0.0013	0.0002
		0.2	0.001	0.0056	0.0015	0.0001
		0.5	0.0	0.0053	0.001	0.0001
0.20	0.0	0.0	0.0	0.0022	0.0006	0.0
	0.5	0.0	-0.005	0.005	0.0016	0.0001
		0.1	0.0	0.0057	0.0019	0.0003
		0.2	0.0	0.0061	0.0019	0.0002
		0.5	0.001	0.0067	0.0012	0.0001
	1.0	0.0	-0.003	0.0052	0.0012	0.0001
		0.1	0.001	0.0071	0.0018	0.0002
		0.2	0.001	0.0096	0.0023	0.0003
		0.5	0.0	0.0142	0.0033	0.0003
0.40	0.0	0.0	-0.001	0.0041	0.0012	0.0001
	0.5	0.0	-0.004	0.0076	0.0019	0.0002
		0.1	-0.001	0.0086	0.0021	0.0002
		0.2	0.0	0.0085	0.0021	0.0002
		0.5	0.001	0.0087	0.0024	0.0002
	1.0	0.0	0.0	0.0055	0.0012	0.0001
		0.1	0.0	0.0079	0.0018	0.0002
		0.2	0.0	0.0113	0.0026	0.0002
		0.5	-0.002	0.0202	0.0044	0.0005

can be determined. Random samples were drawn from the expected frequency distributions. Allelic frequencies and recombination fractions were estimated from the samples using the methods described in Section 7.2.3. For each of the combinations, 1,000 replications were simulated. Bias and mean-square error (MSE) of the estimated u, v and θ were calculated for each of the combinations. Bias of the estimated recombination fraction on the parametric values used in the simulations showed that the bias is independent of population size. Population size contributes to variance of the estimates and is part of the mean square error of the estimates (Table 7.12 and Table 7.13).

Table 7.13 Empirical bias and mean square error for estimated recombination fraction using dominant markers from simulation of 1,000 replications for each of the combinations.

				Sample sizes		
θ	t	u and v	Bias	50	200	2,000
0.05	0.0	0.0	0.000	0.001	0.000	0.000
	0.5	0.0	-0.004	0.002	0.001	0.000
		0.1	0.017	0.003	0.001	0.000
		0.2	0.031	0.005	0.002	0.001
		0.5	0.035	0.009	0.003	0.001
		0.9	-0.049	0.003	0.002	0.002
		1.0	-0.049	0.002	0.002	0.002
	1.0	0.0	-0.050	0.002	0.002	0.003
		0.1	-0.042	0.003	0.002	0.002
		0.2	-0.015	0.006	0.002	0.000
		0.5	0.073	0.030	0.013	0.006
		0.9	-0.013	0.014	0.003	0.000
0.20	0.0	0.0	0.001	0.004	0.001	0.000
	0.5	0.0	-0.003	0.005	0.001	0.000
		0.1	0.014	0.006	0.002	0.000
		0.2	0.023	0.007	0.002	0.001
		0.5	0.022	0.010	0.003	0.001
		0.9	-0.129	0.016	0.013	0.017
		1.0	-0.188	0.028	0.027	0.035
	1.0	0.0	-0.068	0.013	0.006	0.005
		0.1	-0.037	0.011	0.004	0.002
		0.2	-0.011	0.011	0.003	0.000
		0.5	0.039	0.020	0.009	0.002
		0.9	-0.159	0.024	0.015	0.025
0.40	0.0	0.0	-0.001	0.006	0.002	0.000
	0.5	0.0	-0.002	0.008	0.002	0.000
		0.1	0.007	0.008	0.002	0.000
		0.2	0.014	0.007	0.003	0.000
		0.5	0.005	0.010	0.003	0.000
		0.9	-0.153	0.030	0.025	0.024
		1.0	-0.237	0.052	0.052	0.057
	1.0	0.0	-0.025	0.007	0.002	0.001
		0.1	-0.014	0.009	0.003	0.000
		0.2	-0.004	0.011	0.003	0.000
		0.5	-0.004	0.017	0.006	0.001
		0.9	-0.354	0.100	0.097	0.125

For codominant markers, the results of simulation are symmetric from $u = v = 0.5$. Bias and mean square errors are low for all simulated situations. As allelic frequencies around 0.5, the bias and mean square error increase. As sample size increases, the mean square error decreases. Allelic frequencies in the pollen pool can also be estimated with low bias and mean square error.

For dominant markers, the bias and mean square error for the estimated recombination fraction are generally much larger than for codominant markers. Bias for recombination fraction, relative to the magnitude of true recombination fraction, is large when the true recombination is low. The bias and mean square errors are acceptable for low true recombination ($\theta < 0.1$) only when the dominant allelic frequency is less than 0.1. In general, when $\theta < 0.2$, it is required that the dominant allelic frequency in the pollen pool be less than the true recombination fraction to get an acceptable recombination fraction estimation. When the true recombination fraction is high and dominant allelic frequency in the pollen is less than 0.5, recombination fraction can be estimated with low bias and mean square errors.

When the dominant allelic frequency in the pollen pool is greater than 0.5, the estimate of recombination fraction has a high negative bias and therefore the estimated recombination fraction is much less than the true recombination fraction. Caution is needed for estimating the recombination fraction if the dominant allelic frequency in the pollen pool is greater than 0.2. Allelic frequency in the pollen pool cannot be estimated efficiently using single or two locus models when the true dominant allelic frequency is less than 0.1 or greater than 0.5 and when outcrossing rate is low.

7.2.5 MAPPING USING CROSS BETWEEN TWO HETEROZYGOTES

Controlled crosses between two non-inbred lines, also known as full-sib mating, have been used in linkage mapping for outcrossed plant species. A full-sib population is generated by crossing two heterozygous plants. For this mating scheme, according to the conventional definition of mating, there are 81 possible mating configurations for a two-locus codominant model (Table 7.14). Among the 81 mating configurations for codominant markers, 17 provide information about linkage and the rest do not contain linkage information; 7 of 9 possibilities provide linkage information for dominant markers (see also Ritter *et al.* 1990). The principle of two-point linkage analysis for the composite mating is straightforward. For different loci combinations, there are corresponding likelihood functions. For codominant markers, the linkage phase can be easily determined when linkage is relatively tight (say, less than 30% recombination fraction) for even a small sampling size. It is not critical to determine linkage phase when linkage is loose, because the loci will be located far away on the genome. When loci are loosely linked, they have little or no influence on locus order and multipoint map distance estimation (see Chapters 9 and 10).

For a mixture of codominant and dominant markers or for the dominant marker alone, the repulsion linkage phase could create many problems for linkage map construction. This is because of the low linkage information content for repulsion linkage between a codominant marker and a dominant marker or between two dominant markers. There are several ways to solve the problem. One possible way is to determine the linkage phase before linkage map construction. Two linkage maps can be constructed with markers in coupling phase. Then, the two linkage maps can be pooled together based on the common codominant markers. A pseudo-testcross strategy has been used for forest tree linkage map construction (Grattapaglia *et al.* 1992; Grattapaglia and Sederoff 1994). By screening 1:1 segregation in progeny, data achieved

Table 7.14 Conventional mating type for single-locus and two-locus models in a population produced by a cross between two heterozygous parents.

Single-locus model:

Codominant markers: $\frac{A}{A}x\frac{A}{a}$, $\frac{A}{a}x\frac{A}{A}$, $\frac{A}{a}x\frac{A}{a}$, $\frac{A}{a}x\frac{a}{a}$, and $\frac{a}{a}x\frac{A}{a}$

Conventional mating type: BC,   BC,   F2,   BC,   and BC

Dominant markers: $\frac{A}{a}x\frac{A}{a}$ $\frac{A}{a}x\frac{a}{a}$ $\frac{a}{a}x\frac{A}{a}$

Conventional mating type: F2   BC   BC

Two-locus model (codominant markers):

	$\frac{A}{A}x\frac{A}{a}$	$\frac{A}{a}x\frac{A}{A}$	$\frac{A}{a}x\frac{A}{a}$	$\frac{A}{a}x\frac{a}{a}$	$\frac{a}{a}x\frac{A}{a}$
$\frac{B}{B}x\frac{B}{b}$	BC	no	BCF2	no	BC
$\frac{B}{b}x\frac{B}{B}$	no	BC	F2BC	BC	no
$\frac{B}{b}x\frac{B}{b}$	BCF2	F2BC	F2	F2BC	BCF2
$\frac{B}{b}x\frac{b}{b}$	no	BC	F2BC	BC	no
$\frac{b}{b}x\frac{B}{b}$	BC	no	BCF2	no	BC

Two-locus model (dominant markers)

	$\frac{A}{a}x\frac{A}{a}$	$\frac{A}{a}x\frac{a}{a}$	$\frac{a}{a}x\frac{A}{a}$	
$\frac{B}{b}x\frac{B}{b}$	F2	F2BC	BCF2	
$\frac{B}{b}x\frac{b}{b}$	F2BC	BC	no	Pseudo-testcross
$\frac{b}{b}x\frac{B}{b}$	BCF2	no	BC	Scheme

belong to the bottom right 4 cells in Table 7.14. Two linkage maps will be built for each of the parents. This way, the two cells without linkage information can be eliminated from data analysis.

EXERCISES

Exercise 7.1

Linkage phase is defined arbitrarily in the two-locus model when one of the two loci is codominant in terms of recombination fraction estimation. However, linkage phase has its biological and statistical meanings when multiple-locus models are in consideration or the loci are dominant. In later chapters, it will be shown that linkage phase is essential for using genetic markers to predict the performance of an individual. Answer the following questions.

(1) List possible linkage phase configurations for a two-locus model.

(2) Explain advantages and disadvantages for the approach using analysis of the phase-unknown data after the phase is determined by an *ad hoc* procedure. (Hint: Take biology and statistics into consideration).

(3) When loci A and B are linked in the coupling phase with true recombination fraction r, compute the probability that $(b + c) \geq 10$ when

   a) $n = a + b + c + d = 20$ and $r = 0.4$
   b) $n = 20$ and $r = 0.2$
   c) $n = 100$ and $r = 0.4$
   d) $n = 100$ and $r = 0.2$

Refer to the following illustration for mating design and notations.

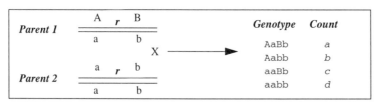

Ex. 7.1 (3)

Exercise 7.2

In the progeny of a cross between two individuals AaBb and AaBb, counts of the nine possible genotypes are given in the following illustration. The linkage phase is the same for both parents and is unknown.

	Genotype	Count
	AABB	35
	AABb	14
	AAbb	5
AaBb X AaBb ⟶	AaBB	10
	AaBb	20
	Aabb	11
	aaBB	3
	aaBb	12
	aabb	40

Ex. 7.2

(1) Detect linkage between loci A and B.

(2) Estimate recombination fraction between A and B based on both the coupling and repulsion models.

(3) Infer the linkage phase between loci A and B and describe the risk of an incorrect inference on the linkage phase.

Cross 1	Genotype	Count	Cross 2	Genotype	Count
A B	AABB	35	A b	AABB	1
‗‗‗	AABb	14	‗‗‗	AABb	15
a b	AAbb	5	a B	AAbb	38
X	AaBB	10	X	AaBB	11
A B	AaBb	20	A b	AaBb	22
‗‗‗	Aabb	11	‗‗‗	Aabb	8
a b	aaBB	3	a B	aaBB	45
	aaBb	12		aaBb	10
	aabb	40		aabb	0

Ex. 7.3

Exercise 7.3

Two F2 crosses were made to estimate recombination fraction between loci A and B (see the above illustration). The recombination fractions between loci A and B could be the same or different for the parents in the two crosses and are denoted by r_1 and r_2, respectively. However, the linkage phase of loci A and B differs between the two crosses.

(1) Perform complete single-locus and two-locus chi-square and likelihood ratio analyses. The analyses should include the partition of the total test statistic into pooled and heterogeneity.

(2) Estimate recombination fraction between loci A and B using each of the crosses independently and using the pooled data.

(3) Test the heterogeneity between the estimates of the recombination fraction between the two crosses.

Exercise 7.4

Two F2 crosses were made for estimating recombination fraction between loci A and B (see the following illustration). The recombination fractions between loci A and B could be the same or different for the parents in the two crosses and are denoted by r_1 and r_2, respectively. However, the linkage phase of loci A and B differs between the two crosses. In the first cross, the linkage phase for the loci is coupling for both parents. In the second cross, the linkage phase differs between the two parents.

Cross 1	Genotype	Count	Cross 2	Genotype	Count
A B	AABB	35	A b	AABB	4
‗‗‗	AABb	14	‗‗‗	AABb	25
a b	AAbb	5	a B	AAbb	2
X	AaBB	10	X	AaBB	33
A B	AaBb	20	A B	AaBb	11
‗‗‗	Aabb	11	‗‗‗	Aabb	30
a b	aaBB	3	a b	aaBB	4
	aaBb	12		aaBb	36
	aabb	40		aabb	5

Ex. 7.4

(1) Estimate recombination fraction between loci A and B using each of the crosses independently and using the pooled data.

(2) Test the heterogeneity between the estimates of the recombination fraction between the two crosses.

(3) Compare the linkage information contents in the progeny of the two crosses.

Exercise 7.5

When progeny is produced by a maternal parent and pollen from a number of pollen resources, the progeny is produced from a mixture of self and random mating. This is common in plant species. If the pollen comes from the maternal parent, then the mating is a self. If the pollen comes from a plant other than the maternal parent, the mating is an outcrossing. Because the outcrossing pollen is contributed randomly from the surrounding plants, the mating is considered random. To estimate recombination fraction in a two-locus model using the progeny of a mixture of self and random mating, the maternal parent has to be a double heterozygote for the two loci. The likelihood equation for estimating recombination fraction using these progeny is a function of the recombination fraction between the two loci, outcrossing rate and allelic frequencies for the two loci in the pollen pool.

(1) Explain the rationale behind estimation of recombination fraction using progeny of mixture of self and random mating.

(2) Why can't outcrossing rate and recombination fraction be efficiently estimated from the same progeny?

(3) Given that the outcrossing rate is 100%, estimate allelic frequency for loci A and B and recombination fraction between loci A and B using the genotypic counts shown in the following illustration. Assume that the pollen pool is under complete linkage equilibrium and loci A and B are linked in coupling phase in the maternal parent.

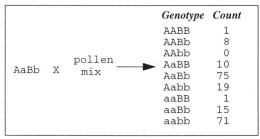

Ex. 7.5 (3)

Exercise 7.6

When progeny is produced by a cross between two heterozygous parents, the mating type, in terms of the classical definitions, is a mixture of backcross and F2. We have described the possible mating configurations for a two-locus model in this chapter.

(1) Prove that the progeny of a double backcross with mating configuration Aabb X aaBb have no linkage information.

(2) Derive a recombination fraction estimation procedure using progeny of a backcross and F2 mixture (e.g., AaBb X Aabb).

Exercise 7.7

A pseudo-testcross strategy has been used for forest tree linkage map construction. By screening markers for 1:1 segregation in progeny, backcross is the only mating configuration for the markers in the progeny. Consider a cross between two parents in the following illustration.

$$\frac{\text{A B c d E f G H I j}}{\text{a b c D e f g h I J}} \quad X \quad \frac{\text{a B C D e f g h I j}}{\text{a b c D e f g H I j}}$$

(1) If markers A to J are all codominant, show a half triangle matrix for the pairs among the 10 loci. Please write down the mating system and linkage phase (if applicable) in each of the 45 cells.

(2) If markers A to J are all codominant, show the markers that will be picked by the pseudo-testcross screening and the half triangle matrix, draw the parental maps for the markers (similar to the illustration) and draw the expected linkage maps that will be made by the pseudo-testcross markers. Explain the advantages of screening for the pseudo-testcross markers.

(3) If markers A to J are all dominant and upper case letters denote dominant alleles, show the markers that will be picked by the pseudo-testcross screening and the half triangle matrix, draw the parental maps for the markers (similar to the illustration) and draw the expected linkage maps that will be made by the pseudo-testcross markers. Explain the advantages of screening for the pseudo-testcross markers when dominant markers are used.

(4) We can summarize the markers picked by the pseudo-testcross screening in the following illustration in terms of a two-locus model.

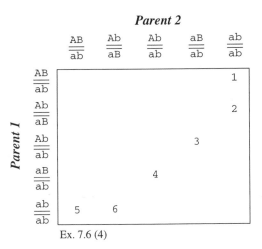

Ex. 7.6 (4)

Types 1 and 2 are informative for a linkage map for parent 1 and types 5 and 6 are informative for a linkage map for parent 2. Types 3 and 4 are not informative.

There are two problems in practical mapping experiments. One is that the two linkage maps for the two parents cannot be connected using the data. This problem will be discussed in the chapter concerning linkage map merging. The other problem is caused by the fact that some of the computer software for linkage analysis can recognize only one linkage phase in one analysis. In this case, types 1 and 2 marker combinations have different linkage phases and they are in the same analysis. To solve this problem, Drs. David O'Malley and Ronald Sederoff of the Forest Biotechnology Group at North Carolina State University use an approach to "trick" the computer. In this approach, the data is recoded and duplicated. Locus A is duplicated in the data set by recoding marker genotype Aa to aa and aa to Aa. So, two loci A's are in the data set. They are the original locus A and the recoded locus A called Ar. For example, loci A and B are duplicated in the data as following.

```
A    11112222121212222211112221
Ar   22221111212121111122221112
B    22211111212112222221112221
Br   11122222121221111112221112
```

Note: 1 denotes "Aa" or "Bb" and 2 denote 'aa' or "bb".

Can you explain the validity of this approach? How many linkage maps will this approach produce? Is this approach necessary if the computer software can recognize linkage phases?

CHAPTER 8

TWO-LOCUS MODELS: USING LINKAGE DISEQUILIBRIUM

In the human species, linkage analysis has limitations compared to most experimental organisms, due to the small number of individuals in a single family. To overcome this limitation, methods have been developed based upon the association of genes at the population level. This association may be due to linkage disequilibrium, which is defined as the departure of loci from equilibrium.

Conventional linkage analysis is based on the cosegregation of loci. For each cosegregation pattern, a likelihood function can be constructed and used to estimate the magnitude of the linkage. These cosegregation patterns are usually applied within families. Association of genes is usually based on the correlation between loci in a population.

Implementation of association analysis usually needs a case-control study. When searching for disease-causing genes, the case-control study is composed of the marker allelic frequencies in a population with normal individuals. If an allele of a genetic marker occurs at a significantly higher frequency among the affected individuals than among the normal individuals, then the marker may be linked with the disease-causing gene. Linkage between a marker and a disease-causing gene is not the only source of the association, but it is an important factor for reducing the speed of decay of the association in natural populations. Association analysis is comprised of a simple chi-square test for departure from independence using a 2 by 2 contingency table. However, the interpretation of the results is usually complicated by selection, drift, migration, new mutation or nonrandom mating.

In this chapter, the concept of disequilibrium, the relationship between disequilibrium and linkage, the transmission/disequilibrium test (TDT) and other disequilibrium-based tests will be introduced.

8.1 LINKAGE DISEQUILIBRIUM

The purpose of controlled crosses for mapping populations is to maximize linkage disequilibrium within the progeny of the crosses. Linkage equilibrium is commonly assumed in natural populations under a large number generations of random mating. However, disequilibrium is commonly found in natural populations, especially for closely linked genes, within families or populations related by decent.

8.1.1 TWO-LOCUS DISEQUILIBRIUM MODEL

Genomic analyses are usually based on multiple-locus models. Methods for two-locus disequilibrium analysis are needed to define disequilibrium in a multiple-locus context. There are several different measures for two-locus disequilibrium (Hedrick 1987; Lewontin 1988). Considering two loci, A and B, each with two alleles (A and a; B and b), the allelic frequencies are

$$p_A, 1 - p_A \qquad for \qquad A, a$$
$$p_B, 1 - p_B \qquad for \qquad B, b$$

A diallelic model is assumed throughout the discussion. For some analyses, the multiple allele problem can be an easy extension from the diallelic model by pooling some of the allelic classes together. If the two loci are independent and each of the loci is in Hardy-Weinberg equilibrium, then the expected gametic frequencies are

$$
\begin{aligned}
p_{AB} &= p_A p_B \\
p_{Ab} &= p_A (1 - p_B) \\
p_{aB} &= (1 - p_A) p_B \\
p_{ab} &= (1 - p_A) (1 - p_B)
\end{aligned}
\tag{8.1}
$$

Departure from equilibrium is commonly measured by a two-locus coefficient of gametic disequilibrium

$$D_{AB} = p_{AB} p_{ab} - p_{Ab} p_{aB}$$

Equation (8.1) becomes

$$
\begin{aligned}
p_{AB} &= p_A p_B + D_{AB} \\
p_{Ab} &= p_A (1 - p_B) - D_{AB} \\
p_{aB} &= (1 - p_A) p_B - D_{AB} \\
p_{ab} &= (1 - p_A) (1 - p_B) + D_{AB}
\end{aligned}
$$

The gametic disequilibrium coefficient can be positive or negative. The general range of D_{AB} depends on the allelic frequencies of the two loci

$$min [p_A (1 - p_B), (1 - p_A) p_B] > D_{AB} > -1 \times max [p_A p_B, (1 - p_A) (1 - p_B)] \tag{8.2}$$

8.1.2 DETECTION AND ESTIMATION

Detection

Commonly used methods of detection and estimation of disequilibrium in a natural population are based on the probability of the data as a function of allelic frequencies and the disequilibrium coefficient. The log likelihood function for a two-locus model is

$$
\begin{aligned}
L = {}& f_{AB} \log (p_A p_B + D_{AB}) + f_{Ab} \log [p_A (1 - p_B) - D_{AB}] \\
& + f_{aB} \log [(1 - p_A) p_B - D_{AB}] + f_{ab} \log [(1 - p_A) (1 - p_B) + D_{AB}]
\end{aligned}
\tag{8.3}
$$

where f_{AB}, f_{Ab}, f_{aB} and f_{ab} are the observed two-locus gametic counts. There are three unknown parameters, p_A, p_B and D_{AB}, in the likelihood equation. The three derivatives for the likelihood with respect to the parameters are

$$\frac{\partial L}{\partial p_A} = \frac{p_B f_{AB}}{p_{AB}} + \frac{(1-p_B) f_{Ab}}{p_{Ab}} - \frac{p_B f_{aB}}{p_{aB}} - \frac{(1-p_B) f_{ab}}{p_{ab}}$$

$$\frac{\partial L}{\partial p_B} = \frac{p_A f_{AB}}{p_{AB}} + \frac{(1-p_A) f_{Ab}}{p_{Ab}} - \frac{p_A f_{aB}}{p_{aB}} - \frac{(1-p_A) f_{ab}}{p_{ab}} \qquad (8.4)$$

$$\frac{\partial L}{\partial D_{AB}} = \frac{f_{AB}}{p_{AB}} - \frac{f_{Ab}}{p_{Ab}} - \frac{f_{aB}}{p_{aB}} + \frac{f_{ab}}{p_{ab}}$$

The maximum likelihood estimates are

$$\hat{p}_A = \frac{f_{AB} + f_{Ab}}{N} = \hat{p}_{AB} + \hat{p}_{Ab}$$

$$\hat{p}_B = \frac{f_{AB} + f_{aB}}{N} = \hat{p}_{AB} + \hat{p}_{aB}$$

$$\hat{D}_{AB} = \frac{f_{AB}}{N} - \left[\frac{f_{AB} + f_{Ab}}{N}\right]\left[\frac{f_{AB} + f_{aB}}{N}\right] \qquad (8.5)$$

$$= \hat{p}_{AB} - \hat{p}_A \hat{p}_B$$

where $N = f_{AB} + f_{Ab} + f_{aB} + f_{ab}$ is the total sample size and $\hat{p}_{AB}, \hat{p}_{Ab}, \hat{p}_{aB}$ and $\hat{p}_{ab}$ are the maximum likelihood estimates for the corresponding gametes, which are simple counts from the population divided by the total number of gametes. Hypothesis testing for disequilibrium for the two-locus model can also be implemented using the log likelihood ratio test statistic, which is

$$G = 2\left[L\left(D_{AB} = \hat{D}_{AB}\right) - L\left(D_{AB} = 0\right)\right] \qquad (8.6)$$

The log likelihood ratio test statistic is asymptotically distributed as a chi-square variable with one degree of freedom. As in the single locus model, estimation of the allelic frequencies is independent from the disequilibrium coefficient. However, allelic frequencies have effects on the statistical power for detecting disequilibrium.

Detection Power

The probability value for declaring significance for a hypothesis test on disequilibrium is

$$P\left[G \geq \chi^2_{(1)}\right]$$

As discussed in Chapter 4, the detection power of a test is

$$Power_D = Pr\left[\chi^2_{df, c} \geq \chi^2_\alpha\right] \qquad (8.7)$$

where χ^2_α is the critical value to reject a null hypothesis at a significance level α and $\chi^2_{df, c}$ is a deviate from a non-central chi-square distribution with the degrees of freedom df and non-centrality parameter c. Here, c is the expected log likelihood ratio test statistic

$$c = 2\left[L\left(D_{AB}\right) - L\left(0\right)\right]$$

It is calculated by replacing the estimate ($\hat{D}_{AB}$ and the allelic frequencies) with the parametric values of D_{AB} and the allelic frequencies in Equation (8.6).

Given the sample size N, allelic frequencies p_A and p_B and the disequilibrium coefficient D_{AB}, the non-centrality parameter is

$$c = N\left\{ (p_A p_B + D_{AB}) \log\left[1 + \frac{D_{AB}}{p_A p_B} \right] \right.$$

$$+ \left[p_A (1 - p_B) - D_{AB} \right] \log\left[1 - \frac{D_{AB}}{p_A (1 - p_B)} \right]$$

$$+ \left[(1 - p_A) p_B - D_{AB} \right] \log\left[1 - \frac{D_{AB}}{(1 - p_A) p_B} \right]$$

$$+ \left. \left[(1 - p_A) (1 - p_B) + D_{AB} \right] \log\left[1 + \frac{D_{AB}}{(1 - p_A) (1 - p_B)} \right] \right\} \tag{8.8}$$

Figure 8.1 shows the statistical power under different settings of magnitude of the true disequilibrium, sample sizes and allelic frequencies. The power curves are symmetric for negative and positive disequilibrium coefficients. As the sample size increases, the power increases. When disequilibrium coefficients are the same, the statistical power for low allelic frequencies (or high frequencies) is higher than for median frequencies.

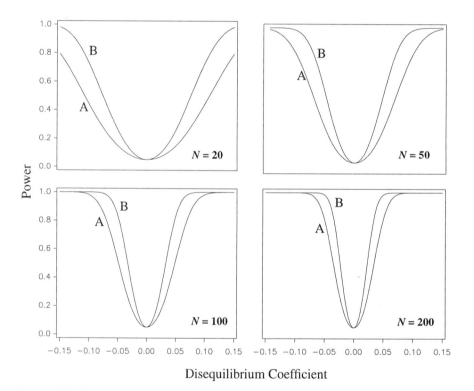

Figure 8.1 The statistical power for detecting disequilibrium under different levels of true disequilibrium in the population, allelic frequencies and sample size. Curve A is for $p_A = p_B = 0.5$ and B is for $p_A = p_B = 0.2$. The significance level is 95%.

8.1.3 DISEQUILIBRIUM AND LINKAGE

Disequilibrium can be the result of many phenomena. For example, a new mutation of an existing gene or recent population admixture can produce disequilibrium. However, genetic linkage is one of the most important sources for reducing the speed of disequilibrium decay from generation to generation. This is shown using examples in Table 8.1 to Table 8.3. There is a frequency difference between individuals with genotype AB/ab and individuals with genotype Ab/aB in the tables, because it will clarify the disequilibrium analysis. AB/ab means that A and B are linked in coupling phase and Ab/aB in repulsion linkage phase.

When the two loci are linked with a recombination fraction of r, the disequilibrium coefficients decrease by a factor of $(1-r)$ each generation of random mating.

$$D_{AB}^{t+1} = (1-r) D_{AB}^t \qquad (8.9)$$

where the superscript denotes the generation. Equation (8.9) is for populations under complete random mating. Weir et al. (1972) derived the relationship of disequilibria between generations when the population is partial self mating as

$$D_{AB}^{t+1} = 0.5 \{ 0.5 (1 + \lambda + S) + \sqrt{(1 + \lambda + S)^2 - 2S\lambda} \} D_{AB}^t \qquad (8.10)$$

where $\lambda = 1 - 2r$ and S is the proportion of selfing. When $S = 0$, Equation (8.10) reduces to Equation (8.9). In this section, we assume that the populations are mating at random. If the population starts with a disequilibrium coefficient of D_{AB}^0, the coefficient will be

$$D_{AB}^t = (1-r)^t D_{AB}^0 \qquad (8.11)$$

after t generations of random mating. If a population begins with individuals with genotype AB/ab, the disequilibrium coefficient can be estimated as

Table 8.1 Change of disequilibrium coefficient between generations when two loci are independent for large population under random mating.

		Expected Frequency	Generation 0	1	2	12
Genotype	AABB	$f_1 = p_{AB}^2$	0.5	0.3905	0.3525	0.3164
	AABb	$f_2 = 2p_{AB}p_{Ab}$	0	0.1562	0.1855	0.2109
	AAbb	$f_3 = p_{Ab}^2$	0	0.0156	0.0244	0.0352
	AaBB	$f_4 = 2p_{AB}p_{aB}$	0	0.1562	0.1855	0.2109
	AB/ab	$f_5 = 2p_{AB}p_{ab}$	0.5	0.1562	0.1113	0.0703
	Ab/aB	$f_6 = 2p_{Ab}p_{aB}$	0	0.0312	0.0488	0.0703
	Aabb	$f_7 = 2p_{Ab}p_{ab}$	0	0.0312	0.0293	0.0234
	aaBB	$f_8 = p_{aB}^2$	0	0.0156	0.0244	0.0352
	aaBb	$f_9 = 2p_{aB}p_{ab}$	0	0.0312	0.0293	0.0234
	aabb	$f_{10} = p_{ab}^2$	0	0.0156	0.0088	0.0039
Gamete	AB	$\hat{p}_{AB}$	0.6250	0.5938	0.5781	0.5625
	Ab	$\hat{p}_{Ab}$	0.1250	0.1562	0.1719	0.1875
	aB	$\hat{p}_{aB}$	0.1250	0.1562	0.1719	0.1875
	ab	$\hat{p}_{ab}$	0.1250	0.0938	0.0781	0.0625
D_{AB}			0.1875	0.0625	0.0312	0.0

Table 8.2 Disequilibrium coefficient when two loci are linked and recombination fraction is 0.2 for large population under random mating.

		Count or Frequency	Generation 0	1	2	34
Genotype	AABB	f_1	0.5	0.4900	0.4232	0.3164
	AABb	f_2	0	0.0700	0.1294	0.2109
	AAbb	f_3	0	0.0025	0.0099	0.0352
	AaBB	f_4	0	0.0700	0.1294	0.2109
	AB/ab	f_5	0.5	0.2800	0.1958	0.0703
	Ab/bB	f_6	0.0	0.0050	0.0198	0.0703
	Aabb	f_7	0	0.0200	0.0299	0.0234
	aaBB	f_8	0	0.0025	0.0099	0.0352
	aaBb	f_9	0	0.0200	0.0299	0.0234
	aabb	f_{10}	0	0.0400	0.0227	0.0039
Gamete	AB	$\hat{p}_{AB}$	0.6725	0.6505	0.6329	0.5625
	Ab	$\hat{p}_{Ab}$	0.0775	0.0995	0.1171	0.1875
	aB	$\hat{p}_{aB}$	0.0775	0.0995	0.1171	0.1875
	ab	$\hat{p}_{ab}$	0.1725	0.1505	0.1329	0.0625
	D_{AB}			0.1875	0.1375	0.1100	0.0

Table 8.3 Disequilibrium coefficient when two loci are linked and recombination fraction is 0.01 for large population under random mating.

		Count or Frequency	Generation 0	1	2	100	∞
Genotype	AABB	f_1	0.5	0.5588	0.5560	0.3980	0.3164
	AABb	f_2	0	0.0037	0.0065	0.1503	0.2109
	AAbb	f_3	0	0.0	0.0	0.0142	0.0352
	AaBB	f_4	0	0.0037	0.0065	0.1503	0.2109
	AB/ab	f_5	0.5	0.3700	0.3663	0.1652	0.0703
	Ab/bB	f_6	0	0.0	0.0	0.0284	0.0703
	Aabb	f_7	0	0.0012	0.0021	0.0312	0.0234
	aaBB	f_8	0	0.0	0.0	0.0142	0.0352
	aaBb	f_9	0	0.0012	0.0021	0.0312	0.0234
	aabb	f_{10}	0	0.0613	0.0603	0.0171	0.0039
Gamete	AB	$\hat{p}_{AB}$	0.7457	0.7438	0.7420	0.6302	0.5625
	Ab	$\hat{p}_{Ab}$	0.0043	0.0062	0.0080	0.1198	0.1875
	aB	$\hat{p}_{aB}$	0.0043	0.0062	0.0080	0.1198	0.1875
	ab	$\hat{p}_{ab}$	0.2457	0.2438	0.2420	0.1302	0.0526
	D_{AB}			0.1875	0.1850	0.1832	0.0691	0.0

follows: the gametic frequencies in the first generation are

$$\hat{p}_{AB}{}^* = f_1 + 0.5f_2 + 0.5f_4 + 0.5f_5$$

$$\hat{p}_{Ab}{}^* = f_3 + 0.5f_2 + 0.5f_7 + 0.5f_6$$

$$\hat{p}_{aB}{}^* = f_8 + 0.5f_4 + 0.5f_9 + 0.5f_6$$ (8.12)

$$\hat{p}_{ab}{}^* = f_{10} + 0.5f_7 + 0.5f_9 + 0.5f_5$$

The disequilibrium coefficient for a population in which 50% of the individuals have genotype AABB and 50% of the individuals having genotype AB/ab ($f_1 = 0.5$ and $f_5 = 0.5$) is

$$D_{AB}^0 = p_{AB}{}^* - p_A{}^* p_B{}^* = 0.75 - 0.75 \times 0.75 = 0.1875$$

It is important to pay attention to how the frequencies of gametes differ from those in the population of the next generation. If loci A and B are linked with recombination fraction r, the gametic frequencies produced by the population are

$$p_{AB} = f_1 + 0.5f_2 + 0.5f_4 + 0.5(1-r)f_5 + 0.5rf_6 = p_A p_B + (1-r)D_{AB}$$

$$p_{Ab} = f_3 + 0.5f_2 + 0.5f_7 + 0.5rf_5 + 0.5(1-r)f_6 = p_A(1-p_B) - (1-r)D_{AB}$$

$$p_{aB} = f_8 + 0.5f_4 + 0.5f_9 + 0.5rf_5 + 0.5(1-r)f_6 = (1-p_A)p_B - (1-r)D_{AB}$$

$$p_{ab} = f_{10} + 0.5f_7 + 0.5f_9 + 0.5(1-r)f_5 + 0.5rf_6 = (1-p_A)(1-p_B) + (1-r)D_{AB}$$

which will be same as the gametic frequencies for the next generation. The relationships are

$$p_A = p_{AB} + p_{Ab}$$

$$p_B = p_{AB} + p_{aB}$$

$$D_{AB}{}^* = p_{AB}p_{ab} - p_{Ab}p_{aB}$$

$$= [p_A p_B + (1-r)D_{AB}][(1-p_A)(1-p_B) + (1-r)D_{AB}]$$

$$\quad - [p_A(1-p_B) - (1-r)D_{AB}][(1-p_A)p_B - (1-r)D_{AB}]$$

$$= (1-r)D_{AB} = (1-r)(p_{AB} - p_A p_B)$$

where the disequilibrium coefficient for the next generation is $D_{AB}{}^*$.

If the two loci are independent, the population will reach equilibrium after 12 generations of random mating (Table 8.1) if population size is sufficiently large. In general, the disequilibrium reduces approximately by half with each generation of random mating (except for the first generation in some situations). For some cases, one generation of random mating is sufficient to establish equilibrium, *e.g.*, the population starts with the double heterozygote individuals AB/ab or Ab/aB.

Table 8.2 shows a situation with the two loci linked with recombination fraction 0.2. By generation 34, the population reaches approximate equilibrium ($D_{AB} < 0.0001$). When the two loci are linked with recombination fraction 0.01, after 100 generations of random mating the population is still in disequilibrium with $D_{AB} = 0.0691$. This is about 40% of the original disequilibrium (Table 8.3). The inverse of Equation (8.11) gives the number of generations needed to reach equilibrium for large population under random mating, which is

$$t_{D_{AB}} = [\log (D_{AB}/D_{AB}^0)] / [\log (1 - r)] \qquad (8.13)$$

For example

$$t_{0.01} = [\log (0.01/0.1875)] / [\log (1 - 0.01)] = 292$$

generations are needed to reduce the disequilibrium coefficient from 0.1875 to 0.01 when two loci are linked with recombination fraction 0.01. It is common to estimate number of generations for the disequilibrium to be reduced by half, which is

$$t_{half} = [\log 0.5] / [\log (1 - r)] \qquad (8.14)$$

and is called median equilibrium time. Equations for the disequilibrium analysis may not work for the initial generation. The assumption for those formulae is that the population is under random mating and the population is sufficiently large.

8.1.4 DISEQUILIBRIUM-BASED ANALYSIS

Figure 8.2 shows the effect of linkage on the disequilibrium coefficient. It is important for the theory behind the use of linkage disequilibrium in genomic analysis. Linkage-based analyses are limited by the lack of the recombinant in mapping populations when linkage is tight. Linkage analyses have low statistical power when there is linkage heterogeneity at the population level. For example, when two genes are linked with a recombination fraction of 0.005 in a conventional F2 mapping population, the probability of no recombinant being observed in a sample size of 100 is 0.606. If we put a level of confidence (γ) on observing at least one recombinant, the sample size required is usually

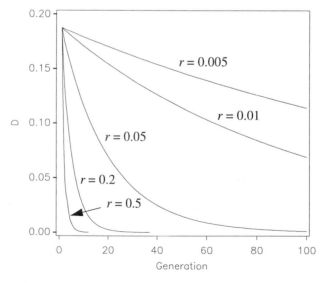

Figure 8.2 Effect of linkage on the disequilibrium coefficient for a two-locus model in generations of random mating. The population starts with half AABB and half AB/ab individuals.

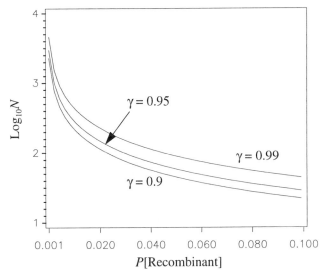

Figure 8.3 Sample size needed for observing at least one recombinant in log scale for different levels of confidences and the expected frequencies of recombinant in population.

large and can be computed using

$$N \geq \log (1 - \gamma) / \log (1 - p)$$

where p is the frequency for recombinant individuals in the population. For example, if $p = 0.001$ and $\gamma = 0.99$, then $N \geq 4603$. This means that a sample size of at least 4603 is needed if the probability of observing a recombinant is 0.001 and we want to be 99% sure of observing at least one recombinant individual. Figure 8.3 shows the sample size needed for some combinations of experimental conditions and confidence. Resolution of linkage-based analysis is usually low when sample size is small. When linkage heterogeneity exists in the population, analysis may fail to detect linkage.

Linkage disequilibrium-based analysis can increase the resolution of gene mapping. The probability of observing a recombinant individual in a population after t generations of random mating is

$$P_t[R] = P_{t-1}[R] + 0.5 \{ 1 - P_{t-1}[R] \} r \qquad (8.15)$$

where r is the recombination fraction between two loci. When r is small, it can be approximated as

$$P_t[R] = 0.5tr \qquad (8.16)$$

For example, when two loci are linked with a recombination fraction of 0.001 (Figure 8.5), the expected frequency of recombinant in the population is about 5% after 100 generations of random mating. For linkage-based analysis, 0.001 is an impractical level of resolution for most experiments. In natural populations, the recombinants have been accumulated from generations of random mating. Figure 8.4 shows the number of individuals needed to observe at least one recombinant after 10 generations of random mating for the same conditions described in Figure 8.3. The sample size needed is much smaller after 10 generations of random mating.

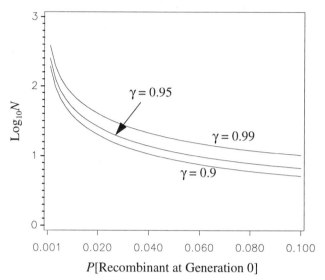

Figure 8.4 Sample size needed for observing at least one recombinant after 10 generations of random mating for different levels of confidences and the expected frequencies of a recombinant in an original population.

In practice, resolutions of 0.01 and 0.001 may essentially have no difference for linkage-based applications. However, they have a fundamental difference when the application is map based cloning. For example, in humans, 0.01 map units may correspond to 1Mb of DNA. A large effort is needed to obtain a clone of a gene in a 2 Mb genome segment. However, if the resolution can increase to 0.001, cloning the gene becomes more feasible.

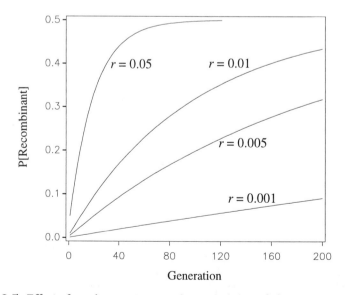

Figure 8.5 Effect of random mating on the probability of observing a recombinant. The population starts with AB/ab.

If the true recombination fraction between two loci is high, disequilibrium-based analysis is usually not effective because the linkage has been diluted in the generations of random mating (Figure 8.5). This is the reason that population association analysis is not suitable when it is necessary to scan a whole genome. A larger number of markers are needed to cover a reasonable portion of genome using population association analysis, compared to the number needed in linkage-based analysis. Another disadvantage of the disequilibrium-based analysis is that the linkage, disequilibrium, generations from the initial disequilibrium and other possible sources of disequilibrium at the population level, such as recent population admixture, are commonly confounded. A direct linkage relationship usually cannot be precisely quantified.

8.2 THE TRANSMISSION DISEQUILIBRIUM TEST (TDT)

TDT has been used for finding markers linked to disease genes in humans (Julier *et al.* 1991; Spielman *et al.* 1993). The basic concept of TDT is that marker alleles associated with disease have a high probability of being transmitted to affected offspring (Spielman *et al.* 1993; Ewens and Spielman 1995). The underlying genetics of TDT is linkage disequilibrium at the population level. The TDT has a similarity with the haplotype relative risk test (HRR, Falk and Rubinstein 1987; Ott 1989; Terwilliger and Ott 1992); both these studies rely on linkage disequilibrium in the population and compare marker alleles found in parents who transmit a disease gene to affected offsprings with a control parent who did not transmit a disease gene. Here, the basic framework of TDT will be given. Readers interested in a detailed discussion should refer to Ott (1989), Spielman *et al.* (1993) and Ewens and Spielman (1995).

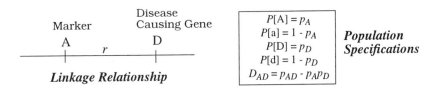

Figure 8.6 Hypothetical linkage relationship between a marker A with alleles A and a and a disease gene D with alleles D and d.

8.2.1 GENETIC MODEL

Let us assume the situation described in Figure 8.6. A disease gene D with alleles D and d is linked to a marker A with alleles A and a. The allelic frequencies corresponding to the four alleles in the population are p_A, $(1 - p_A)$, p_D and $(1 - p_D)$ for alleles A, a, D and d, respectively. The recombination fraction between the gene and the marker is r. The frequencies of the four possible haplotypes, AD, Ad, aD and ad in the population are

$$p_{AD} = p_A p_D + D_{AD}$$
$$p_{Ad} = p_A (1 - p_D) - D_{AD}$$
$$p_{aD} = (1 - p_A) p_D - D_{AD}$$
$$p_{ad} = (1 - p_A) (1 - p_D) + D_{AD}$$

where

$$D_{AD} = p_{AD} p_{ad} - p_{Ad} p_{aD} = p_{AD} - p_A p_D$$

is the disequilibrium coefficient between the marker and the disease gene. The genotypic distribution for the population is given in Table 8.4.

Table 8.4 Distribution of genotypes for a marker and a disease gene in a natural population assuming random union of gametes.

Genotype	Frequency
AADD	$f_1 = p_{AD}^2 = (p_A p_D + D_{AD})^2$
AADd	$f_2 = 2 p_{AD} p_{Ad} = 2 (p_A p_D + D_{AD}) [p_A (1 - p_D) - D_{AD}]$
AAdd	$f_3 = p_{Ad}^2 = [p_A (1 - p_D) - D_{AD}]^2$
AaDD	$f_4 = 2 p_{AD} p_{aD} = 2 (p_A p_D + D_{AD}) [(1 - p_A) p_D - D_{AD}]$
AD/ad	$f_5 = 2 p_{AD} p_{ad} = 2 (p_A p_D + D_{AD}) [(1 - p_A) (1 - p_D) + D_{AD}]$
Ad/aD	$f_6 = 2 p_{Ad} p_{aD} = 2 [p_A (1 - p_D) - D_{AD}] [(1 - p_A) p_D - D_{AD}]$
Aadd	$f_7 = 2 p_{Ad} p_{ad} = 2 [p_A (1 - p_D) - D_{AD}] [(1 - p_A) (1 - p_D) + D_{AD}]$
aaDD	$f_8 = p_{aD}^2 = [(1 - p_A) p_D - D_{AD}]^2$
aaDd	$f_9 = 2 p_{aD} p_{ad} = 2 [(1 - p_A) p_D - D_{AD}] [(1 - p_A) (1 - p_D) + D_{AD}]$
aadd	$f_{10} = p_{ad}^2 = [(1 - p_A) (1 - p_D) + D_{AD}]^2$

Table 8.5 Frequency table for transmitted and nontransmitted marker alleles A and a among 2N parents of N affected children.

Transmitted	Nontransmitted		Marginal
	A	a	
A	n_{11}	n_{12}	$n_{1\circ}$
a	n_{21}	n_{22}	$n_{2\circ}$
Marginal	$n_{\circ 1}$	$n_{\circ 2}$	$2N$

8.2.2 TRANSMISSION/DISEQUILIBRIUM TEST

TDT is based on a 2 x 2 table containing frequencies for the marker alleles transmitted or not transmitted from parents to affected children. Assume one affected child is sampled from each pair of parents. For example, if N affected children are sampled, then $2N$ corresponding parents that are heterozygous for the marker also are sampled. Data can be organized according to Table 8.5. Each of the $2N$ parents can be designated by which allele was transmitted or not transmitted to the affected child. For example, if the child is a homozygote AA, then both the alleles A and A in the two parents must have been transmitted to the child and the frequency of n_{12} increases by two. Similarly, frequency n_{21} increases by two if the child is a homozygote aa. The frequencies n_{11} and n_{22} contributed by the homozygous parents are irrelevant to the TDT. The statistic

$$\chi^2_{TDT} = (n_{12} - n_{21})^2 / (n_{12} + n_{21}) \tag{8.17}$$

following a chi-square distribution asymptotically with one degree of freedom can be used to test if there is an association between marker A and the disease gene, where $n_{12} + n_{21}$ is an estimate for the variance of $n_{12} - n_{21}$. The test is one kind of McNemar's test (Sokal and Rohlf 1981). Spielman *et al.* (1993) showed that this test can also be applied to the situation where there is more than one affected child in a family.

Table 8.6 Distribution of parental genotype for the parents that have an affected child.

Genotype	Frequency
AADD	$p_{AD}^2/p_D = (p_A p_D + D_{AD})^2/p_D$
AADd	$p_{AD}p_{Ad}/p_D = (p_A p_D + D_{AD})[p_A(1-p_D) - D_{AD}]/p_D$
AAdd	0
AaDD	$2p_{AD}p_{aD}/p_D = 2(p_A p_D + D_{AD})[(1-p_A)p_D - D_{AD}]/p_D$
AD/ad	$p_{AD}p_{ad}/p_D = (p_A p_D + D_{AD})[(1-p_A)(1-p_D) + D_{AD}]/p_D$
Ad/aD	$p_{Ad}p_{aD}/p_D = [p_A(1-p_D) - D_{AD}][(1-p_A)p_D - D_{AD}]/p_D$
Aadd	0
aaDD	$p_{aD}^2/p_D = [(1-p_A)p_D - D_{AD}]^2/p_D$
aaDd	$p_{aD}p_{ad}/p_D = [(1-p_A)p_D - D_{AD}][(1-p_A)(1-p_D) + D_{AD}]/p_D$
aadd	0

Table 8.7 Expected frequency for transmitted and nontransmitted marker alleles.

Transmitted	Nontransmitted		Marginal
	A	a	
A	$p_A^2 + D_{AD}\,p_A/p_D$	$p_A(1-p_A) + D_{AD}(1-r-p_A)/p_D$	$p_A + D_{AD}(1-r)/p_D$
a	$p_A(1-p_A) + D_{AD}(r-p_A)/p_D$	$(1-p_A)^2 - D_{AD}(1-p_A)/p_D$	$1 - p_A - D_{AD}(1-r)/p_D$
Marginal	$p_A + rD_{AD}/p_D$	$(1-p_A) - rD_{AD}/p_D$	1

8.2.3 GENETIC INTERPRETATION OF TDT

Let us assume DD is the genotype resulting in disease. Ott (1989) gave the distribution of parental genotypes based on the condition that the parents have an affected child, using the allelic frequencies and the disequilibrium coefficient specified in Figure 8.6 and Equation (8.17). From the conditional frequencies (Table 8.6), the expectations for the transmitted and nontransmitted marker alleles to an affected child can be obtained as listed in Table 8.7. The expectation of the difference between the two categories $(n_{12} - n_{21})$ is

$$
\begin{aligned}
E(n_{12} - n_{21}) &= 2N\{ [p_A(1-p_A) + D_{AD}(1-r-p_A)/p_D] \\
&\quad - [p_A(1-p_A) + D_{AD}(r-p_A)/p_D] \} \\
&= 2N(1-2r)D_{AD}/p_D
\end{aligned}
\tag{8.18}
$$

If A and D are independent $(r = 0.5)$, the expectation goes to zero. The test corresponds to that for $D_{AD} = 0$ or $r = 0.5$.

Example: Insulin-Dependent Diabetes Mellitus (IDDM)

Spielman *et al.* (1993) used TDT to analyze data from 94 families. Among these, families of 57 parents were heterozygous for a marker on chromosome 11p. A total of 62 children were affected by IDDM, with 124 alleles transmitted to the children. Among them there were 78 "1" alleles and 46 "X" alleles. A chi-square was obtained as $(78 - 46)^2 / (78 + 46) = 8.258$, which is significant at 0.004 level.

8.2.4 STATISTICAL POWER OF TDT

Using Equation (8.18) and Table 8.7, the expected chi-square for TDT can be obtained and is approximately

$$
E[\chi^2_{TDT}] = E\left[\frac{(n_{12} - n_{21})^2}{n_{12} + n_{21}}\right] \approx \frac{2N(1-2r)^2 D^2_{AD}}{2p_A(1-p_A)p_D^2 + D_{AD}(1-2p_A)p_D}
\tag{8.19}
$$

The statistical power is

$$
Power_{TDT} = Pr[\chi^2_{df, E[\chi^2_{TDT}]} \geq \chi^2_\alpha]
\tag{8.20}
$$

where χ^2_α is the critical value to reject a null hypothesis at significance level α and

$$
\chi^2_{df, E[\chi^2_{TDT}]}
$$

is a deviate from a non-central chi-square distribution with the degrees of freedom df and non-centrality parameter $E[\chi^2_{TDT}]$.

Figure 8.7 shows the statistical power using TDT to detect linkage disequilibrium between a marker and a disease gene. The statistical power increases as disequilibrium increases and the recombination fraction decreases. The statistical power is high when the disease gene allelic (D allele) frequency in the population is lower. The marker allelic frequency has a small effect on the statistical power.

In general, the statistical power using TDT is lower than using the simple disequilibrium detection (Figure 8.1). One of the reasons for low statistical power using TDT is that TDT only uses a portion of the data. Another reason is that informative genotypic classes for detecting disequilibrium between two loci are those classes with homozygous genotypes for the marker and the disease, such as AADD, AAdd, aaDD and aadd and TDT uses only individuals with genotypes AADD, AaDD and aaDD.

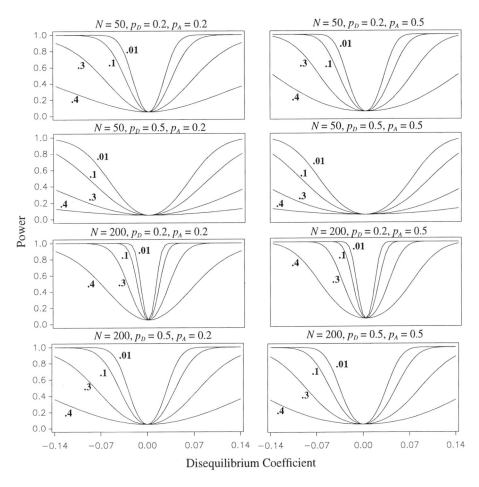

Figure 8.7 Statistical power using TDT to detect linkage disequilibrium between a marker and a disease gene for several settings of sample sizes (number of affected children), marker and disease gene allelic frequencies, true linkage between the marker and the gene and true disequilibrium between the loci. The recombination fraction is labeled on each of the curves. The significance level is 0.05.

8.2.5 WHY TDT?

TDT has lower statistical power than the simple disequilibrium detection. Why should TDT be used to detect disequilibrium? To answer this question, sources of disequilibrium should be considered. Disequilibrium can be generated by genetic drift, migration, selection, inbreeding and mutation (Hill and Robertson 1968; Hill 1975). Mutation is the initial source of the disequilibrium. As the population evolves, the other factors continuously generate or diminish disequilibrium. Random mating and recombination are the sources of decay of the disequilibrium. If there are no other sources of new disequilibrium, linkage reduces the speed of disequilibrium diminishing in the population. Simple disequilibrium detection using the log likelihood ratio test statistic is a test for detecting the combined effects of all the sources of disequilibrium. Among these factors, linkage is what we are interested in. A valid test should test only linkage. Ewens and Spielman (1995) studied effects of population history, subdivision and recent admixture on TDT. They showed that TDT remains a valid chi-square statistic for the linkage hypothesis, regardless of population history.

As shown by Table 8.1, even when the two loci are independent, the disequilibrium could be maintained in the population for a number of generations. This remaining disequilibrium could create a false positive when using the disequilibrium test to infer linkage. Let us assume a simple model for population admixture, large populations (no random drift) and no selection or inbreeding. Simulations were done to investigate effects of population admixture on disequilibrium-based genomic analysis (Table 8.8). Two types of admixture were simulated. In one case, a population is formed by admixture between two populations with equal contributions. The relationship of the population specification between the initial population and the founding populations are listed in Figure 8.8. Random mating is assumed after the admixture. In the other case, a population is formed by a base population and a donor population. A small portion of the population comes from the donor population of each generation. The migration continues for 10 generations. After the 10 generations, the population is under random mating (Figure 8.8).

In the first case, allelic frequencies remain constant after the initial admixture. The disequilibrium coefficient at generation zero is

$$D_{AD} = mD_{1AD} + (1 - m) D_{2AD} + m (1 - m) (p_{1A} - p_{2A}) (p_{1D} - p_{2D})$$
$$= m (1 - m) (p_{1A} - p_{2A}) (p_{1D} - p_{2D})$$

(8.21)

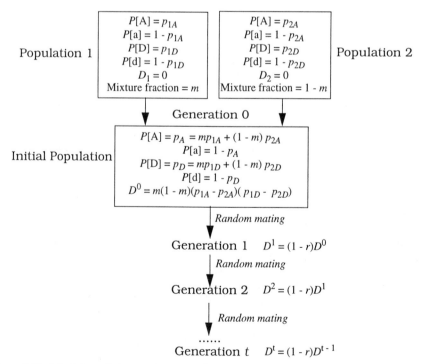

Figure 8.8 Admixture between two populations to generate an initial population (case one, one time mixture). The allelic frequencies do not change after the admixture. Random mating is assumed after the admixture.

Table 8.8 Conditions used in a simulation study on disequilibrium between a marker and a disease gene.

	Simulation Conditions
Population size	Infinite
Admixture process	1. (.5, .5) then random mating 2. (.98, .02), (.98, .02), ..., (.98, .02) for 10 generations and then random mating
Sample size	20, 50, 100, 200, and 1000
Marker allelic frequency (p_A)	0.01, 0.05, 0.1, 0.2, 0.3, 0.4, 0.5
Disease allelic frequency (p_D)	0.01, 0.05, 0.1, 0.2, 0.3, 0.4, 0.5, 0.6, 0.8, 1.0
Generation since admixture	0 to 100
Recombination fraction between A and D	0.01, 0.05, 0.1, 0.3, 0.5

where m and $1 - m$ are the mixture proportions for the two populations, D_{1AD} and D_{2AD} are disequilibrium coefficients for the two populations and are assumed to be zero and the ps are the corresponding allelic frequencies in the two populations. Because of random mating, the disequilibrium decays at an expected rate of $1 - r$ per generation. The allelic frequencies in the initial population are

$$p_A = mp_{1A} + (1 - m)p_{2A}$$
$$p_D = mp_{1D} + (1 - m)p_{2D}$$

(8.22)

The allelic frequencies remain the same in a population from generation to generation because an infinite population size is assumed.

Figure 8.9 shows simulation results for case one. In general, traditional disequilibrium detection using the log likelihood ratio test statistic has a higher statistical power than does TDT. However, in some situations, TDT has higher statistical power than does the traditional disequilibrium test. The most important finding from the simulation is that the traditional disequilibrium test shows a high probability of signicicance for disequlibrium when there is no linkage, which is shown by the curves for $r = 0.5$. The traditional disequilibrium test is not a valid test for linkage. The expected probability of false positive for TDT should be 0.05, because we used 0.05 as the significance level. TDT is a valid test even when the population involves a recent admixture.

Ewens and Spielman (1995) show this point using a multiple population admixture. In a population composed of k subpopulations, the original proportions for each of the k subpopulations are $m_1, m_2, ..., m_k$. The specifications for each of the subpopulations are

$$p_{A1}, p_{A2}, ..., p_{Ak} \qquad P[A]$$
$$p_{D1}, p_{D2}, ..., p_{Dk} \qquad P[D]$$
$$D_1, D_2, ..., D_k \qquad Disequilibrium$$

The population is assumed to be random mating after the initial admixture. The allelic frequencies in the new population are

$$p_A = \sum_{i=1}^{k} m_i p_{Ai} \qquad p_D = \sum_{i=1}^{k} m_i p_{Di}$$

The frequencies for the two cells relevant to TDT (n_{12}, n_{21}) in Table 8.5 are

$$E\left[\frac{n_{12}}{N}\right] = \sum_{i=1}^{k} \{m_i p_{Di} [p_{Di} p_{Ai} (1 - p_{Ai}) + D_i (1 - r - p_{Ai})]\} / \sum_{i=1}^{k} [m_i p_{Di}^2]$$

$$E\left[\frac{n_{21}}{N}\right] = \sum_{i=1}^{k} \{m_i p_{Di} [p_{Di} p_{Ai} (1 - p_{Ai}) + D_i (r - p_{Ai})]\} / \sum_{i=1}^{k} [m_i p_{Di}^2]$$

(8.23)

The expectation of $n_{12} - n_{21}$ is

$$E(n_{12} - n_{21}) = 2N(1 - 2r) \sum_{i=1}^{k} [m_i p_{Di} D_i] / \sum_{i=1}^{k} [m_i p_{Di}^2]$$

(8.24)

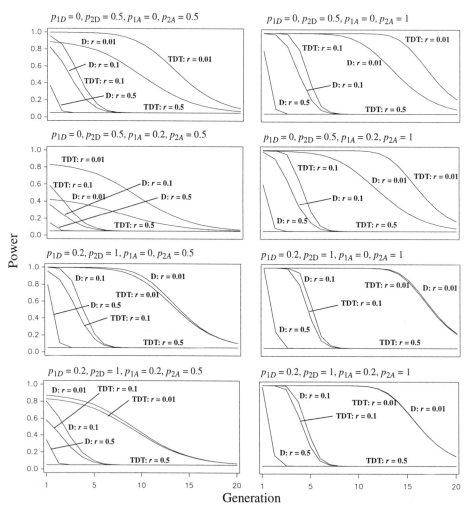

Figure 8.9 Statistical power for TDT and the simple disequilibrium test (D) under different settings of the allelic frequencies of the founding population and recombination frequencies between the marker A and the disease gene D. Sample size is 100 and the significance level is 0.05.

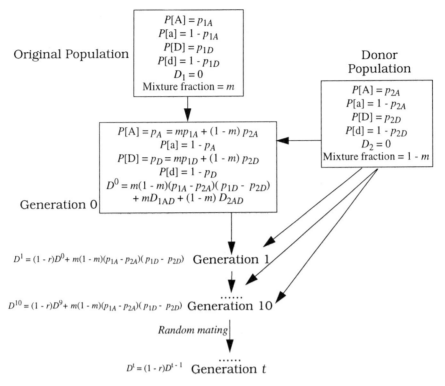

Figure 8.10 Admixture between two populations to generate an initial population (case two, slow mixture). The allelic frequencies change from generation to generation.

So, the TDT tests

$$r = 0.5 \text{ or}$$

$$\sum_{i=1}^{k} [m_i p_{Di} D_i] = 0$$

For the second case, the population is a result of continuing migration. The allelic frequencies change from generation to generation when the migration is effective. These frequencies for generation i can be estimated using

$$p_A^i = (1-m)\, p_A^{i-1} + m p_{2A} \tag{8.25}$$

$$p_D^i = (1-m)\, p_D^{i-1} + m p_{2D}$$

where p_A^{i-1} and p_D^{i-1} are the allelic frequencies for the previous generation in the base population, p_{2A} and p_{2D} are the allelic frequencies in the donor population and m is the proportion of the migrants in the new base population (Figure 8.10). The disequilibrium coefficient in generation i in the population is

$$D_{AD}^i = (1-m)\, D_{AD}^{i-1} + m\,(1-m)\,(p_A^{i-1} - p_{2A})\,(p_D^{i-1} - p_{2D}) \tag{8.26}$$

where D_{AD}^{i-1} is the disequilibrium coefficient for the previous generation in the base population and equilibrium is assumed in the donor population.

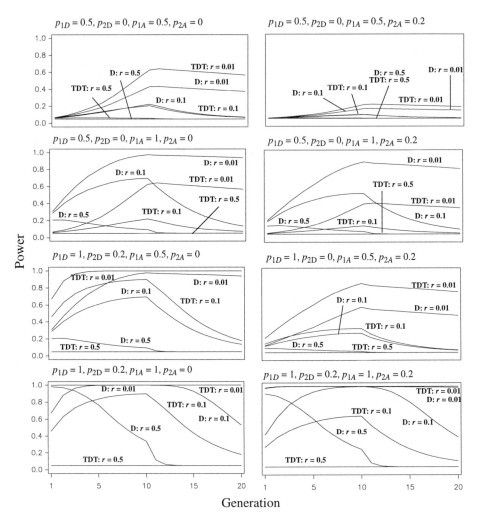

Figure 8.11 Statistical power for TDT and a simple disequilibrium test (D) under different allelic frequencies of a founding population and recombination frequencies between the marker A and the disease gene D. The population is a result of a base population and a migrant donor population (see Figure 8.10). Sample size is 100 and the significance level is 0.05.

Figure 8.11 shows some of simulation results for the second case. The migration rate was set at 2.5% for each of the first ten generations. The basic conclusions are similar to these in the first case. The simple disequilibrium analysis has a higher power to detect disequilibrium. TDT shows no significant influence from population admixture. TDT has higher statistical power than does the simple disequilibrium test for some of the conditions, while the simple disequilibrium test shows more power for the other conditions. Whenever possible, TDT is recommended over the simple disequilibrium test. Caution is needed in interpreting the results from simple disequilibrium tests. For example, precise information on map position will not be obtained from estimates of linkage disequilibrium (Hill and Weir 1994).

8.3 OTHER DISEQUILIBRIUM BASED ANALYSES

Besides TDT, several other types of analyses based on population disequilibrium are used to find genes controlling human diseases. The methods of relative risk (RR), haplotype relative risk (HRR) and linkage analysis using population admixture will be briefly introduced. TDT is closely related to RR and HRR.

Table 8.9 Frequency table of disease status and the marker classes.

| | Marker Group | | |
	A (AA and Aa)	a (aa)	Marginal
Affected	n_{11}	n_{12}	$n_{1\circ}$
Nonaffected	n_{21}	n_{22}	$n_{2\circ}$
Marginal	$n_{\circ 1}$	$n_{\circ 2}$	$2N$

8.3.1 RELATIVE RISK

We build the concept of relative risk using the genetic model shown in Figure 8.6. If we observe two independent groups of individuals where one group has marker genotype A and the other has a, then the relative risk is defined as

$$RR = p_{D|A}/p_{D|a} \tag{8.27}$$

where $p_{D|A}$ and $p_{D|a}$ are frequencies of affected children for the two marker groups, respectively. As shown by Knapp *et al.* (1993), Equation (8.27) can be written as the frequencies of markers conditional on disease status, which is

$$RR = \frac{p_{A|D}(1 - p_{A|\bar{D}})}{(1 - p_{A|D}) p_{A|\bar{D}}} = \frac{p_{A|D} p_{a|\bar{D}}}{p_{a|D} p_{A|\bar{D}}} \tag{8.28}$$

where subscript $\bar{D}$ denotes the nonaffected individual. If we observed the frequencies in Table 8.9, then RR can be estimated using

$$\hat{RR} = [n_{11}(n_{12} + n_{22})] / [n_{12}(n_{11} + n_{21})] \tag{8.29}$$

Table 8.10 Joint distribution of disease status and the marker classes.

| | Marker Group | | |
	A (AA and Aa)	a (aa)	Marginal
Affected (DD)	$p_D^2 - [p_D(1 - p_A) - D_{AD}]^2$	$[p_D(1 - p_A) - D_{AD}]^2$	p_D^2
Nonaffected (Dd and dd)	$p_A(2 - p_A) - p_D^2 + [p_D(1 - p_A) - D_{AD}]^2$	$(1 - p_A)^2 - [p_D(1 - p_A) - D_{AD}]^2$	$1 - p_D^2$
Marginal	$p_A(2 - p_A)$	$(1 - p_A)^2$	

Table 8.10 shows the joint distribution of disease status and the marker classes if we assume individuals DD have the disease. The expectation of RR under the null hypothesis $(D_{AD} = 0)$ is

$$E[RR]_{D_{AD}=0} = \frac{[p_D^2 - p_D^2(1-p_A)^2](1-p_A)^2}{p_D^2(1-p_A)^2 p_A(1-p_A)} = 1 \qquad (8.30)$$

Departure of RR from unity is evidence of association between the marker and the disease. The hypothesis for the departure can be implemented using a likelihood ratio test statistic or a simple test for independence using a chi-square. For the likelihood-based test, data in Table 8.10 has two degrees of freedom. As shown by Ott (1989), the maximum likelihood estimates of two estimable parameters are

$$\hat{p}_A = 1 - \sqrt{(n_{12} + n_{22})/(2N)}$$

$$\frac{\hat{D}_{AD}}{p_D} = \sqrt{(n_{12} + n_{22})/(2N)} - \sqrt{(n_{11} + n_{21})/(2N)} \qquad (8.31)$$

It is obvious that the disequilibrium coefficient and the disease allelic frequency are completely confounded.

The RR can also be estimated when individuals with genotypes DD and Dd have the disease (dominant disease model). Certainly, if the three possible genotypes for the disease gene and the three possible genotypes for the marker (two allele model) can be observed, the analysis can be extended as a simple disequilibrium analysis using the log likelihood ratio test. The RR-based test is a component of the simple disequilibrium test. The RR test has the same limitations as does the simple disequilibrium test and also a low statistical power.

Table 8.11 Frequency table for transmitted and nontransmitted marker alleles A and a among 2N parents of N affected children using marker haplotype.

	Allele		
	A	a	Marginal
Transmitted allele	$n_{11}{}^*$	$n_{12}{}^*$	$n_{1\circ}{}^*$
Nontransmitted allele	$n_{21}{}^*$	$n_{22}{}^*$	$n_{2\circ}{}^*$
Marginal	$n_{\circ 1}{}^*$	$n_{\circ 2}{}^*$	$4N$

8.3.2 GENOTYPE AND HAPLOTYPE RELATIVE RISK (GRR AND HRR)

Haplotype relative risk (HRR) was proposed by Rubinstein *et al.* (1981) and Falk and Rubinstein (1987). Ott (1989) and Terwilliger and Ott (1992) provided statistical properties for HRR. Table 8.5 can be rewritten into Table 8.11. The relationship between the two tables is

$$n_{11}{}^* = n_{11} + n_{12}$$

$$n_{12}{}^* = n_{21} + n_{12}$$

$$n_{21}{}^* = n_{11} + n_{21}$$

$$n_{22}{}^* = n_{12} + n_{22}$$

Terwilliger and Ott (1992) defined a haplotype-based HRR (HHRR) chi-square as

$$\chi^2_{HHRR} = \frac{4N(n_{11}{}^* n_{22}{}^* - n_{12}{}^* n_{21}{}^*)^2}{(n_{11}{}^* + n_{12}{}^*)(n_{21}{}^* + n_{22}{}^*)(n_{11}{}^* + n_{21}{}^*)(n_{12}{}^* + n_{22}{}^*)}$$

$$= \frac{(n_{12} - n_{21})^2}{(2n_{11} + n_{12} + n_{21})[1 - (2n_{11} + n_{12} + n_{21})]/(4N)} \qquad (8.32)$$

The difference between Equation (8.32) and the TDT chi-square is the variance for $n_{12} - n_{21}$. The TDT chi-square uses a variance estimated from only the heterozygous parents. The variance in Equation (8.32) is estimated using data from both heterozygous and homozygous parents. Terwilliger and Ott (1992) showed that the HHRR approach has more statistical power than does the TDT. Spielman *et al.* (1993) showed that the test statistic of Equation (8.32) is a valid test for $D_{AD} = 0$ and $r = 0$, but we are interested in testing $r = 0.5$.

The test statistic of Equation (8.32) also can be implemented as the log likelihood ratio test statistic using

$$G_{HHRR} = 2\left\{ \sum_{i=1}^{2}\sum_{j=1}^{2} n_{ij}{}^* \log n_{ij}{}^* - \sum_{i=1}^{2} n_{i\circ}{}^* \log n_{i\circ}{}^* - \sum_{j=1}^{2} n_{\circ j}{}^* \log n_{\circ j}{}^* + 4N\log(4N) \right\} \qquad (8.33)$$

This may be slightly different from Equation (8.32). There is usually no difference in terms of biological inference.

Table 8.12 Frequency table for transmitted and nontransmitted marker alleles A and a among 2N parents of N affected children using marker genotypes.

	Marker Class		Marginal
	A (AA and Aa)	a (aa)	
Transmitted	a	b	$a + b$
Nontransmitted	c	d	$c + d$
Marginal	$a + c$	$b + d$	$2N$

Cells in Table 8.11 can be filled using genotypic data instead of haplotype data. The columns in the table will be the same as in Table 8.5. Table 8.11 becomes Table 8.12. In Table 8.11, the cells correspond to transmitted marker alleles. In Table 8.12, the cells correspond to marker classes which are genotypic groups instead of haplotypes. The genotype-based HRR (GHRR) statistic is

$$\chi^2_{GHRR} = \frac{2N(ad - bc)^2}{(a+b)(c+d)(a+c)(b+d)} \qquad (8.34)$$

This statistic also can be implemented as the log likelihood ratio test statistic, which is

$$\begin{aligned} G_{GHRR} = 2\,[\,&a\log a + b\log b + c\log c + d\log d - (a+b)\log(a+b) \\ &- (c+d)\log(c+d) - (a+c)\log(a+c) + 2N\log(2N)\,] \end{aligned} \qquad (8.35)$$

8.3.3 LINKAGE ANALYSIS USING POPULATION ADMIXTURE

We have discussed how population admixture affects the inference of linkage from association analysis. Approaches have been proposed to solve the problem. However, natural population admixture can be used to study linkage by the careful selection of loci and populations (Chakraborty and Smouse 1988; Chakraborty and Weiss 1988; Briscoe *et al.* 1994).

Briscoe *et al.* (1994) list four conditions for using population admixture for gene mapping:

(1) A large number of markers are available to cover the entire genome.

(2) The linkage disequilibrium in the admixed population is large.

(3) Times of sampling from the founding population and the admixed population should be close.

(4) Noise disequilibrium can be discriminated from the true linkage disequilibrium.

If we carefully study two populations and obtain allelic frequencies for them and for a population recently admixed with the individuals from the two populations, then we can make inferences about linkage disequilibrium. If two loci are in equilibrium in the two founding populations where δ_A and δ_D are the differences for the allelic frequencies between the two populations, then the disequilibrium in the initial admixed population is expected to be

$$D^0 = m\,(1-m)\,\delta_A\delta_D \qquad (8.36)$$

where m and $1-m$ are the proportions of the founding populations in the initial admixed population. If we can also estimate the disequilibrium in advanced generations, then the linkage between the two loci can be inferred from the differences among the disequilibrium estimates for different generations.

The admixture population approach can be used for many animal and plant species. The approach can be used for several human ethnic groups in North America, such as Caucasian-American, African-American, Asian-American and American Indian (Briscoe *et al.* 1994). Briscoe *et al.* (1994) also listed some potential animal species with known dates of population origin that can be used in the admixture analysis. For some plant species, such as maize and some outbred tree species, the population admixture may be artificially developed.

The key for analysis using population admixture is to pick the loci and to establish random mating after the admixture. As new molecular markers are developed and the genome information is accumulated for many plant and animal species, there will be no shortage of the loci to meet the conditions. However, the random mating assumption may be hard to meet, especially for a finite population.

8.4 ESTIMATION OF RECOMBINATION FRACTION

TDT and the other types of disequilibrium-based analyses can detect association between known genetic markers and a disease-causing gene by comparing marker allelic frequencies between normal and diseased samples. The normal samples can be true independent normal individuals or diseased individuals as internal controls (certain marker alleles do not pass to the individuals from parents). However, these types of analyses do not provide information on how close the marker is to the disease gene. In other words, the disequilibrium estimated from samples of the current generation is confounded with the true linkage between the marker and the disease gene, the number of generations since the initial disequilibrium was created and other factors having an impact on disequilibrium, such as mutation rate, non-random mating and recent population admixture. In this section, methods for partitioning the confounded factor in order to estimate recombination will be discussed. For some open-pollinated plant species, the relationship between the disequilibrium coefficient and recombination fraction may be simple if it is assumed that the population is sufficiently large, under random mating and the number of generations since the initial disequilibrium is known. Both situations will be discussed in this section.

8.4.1 FIXED LARGE POPULATION SIZE

Let us first assume that a population started with AB/ab for two loci A and B and the population is large (Figure 8.12). If A and B are r recombination units apart, then the initial disequilibrium coefficient for the two loci in the first generation of random mating is

$$D^0 = p_{AB}p_{ab} - p_{Ab}p_{aB} = 0.25\left[(1-r)^2 - r^2\right] = 0.25(1-2r) \qquad (8.37)$$

and the disequilibrium coefficient at generation t is expected to be

$$D_{AB}^t = 0.25(1-2r)(1-r)^t \qquad (8.38)$$

If r is small, we can write

$$D_{AB}^t \approx 0.25(1-r)^{t+2} \qquad (8.39)$$

If we know the approximate number of generations of random mating since the initial disequilibrium was created, then the recombination fraction can be

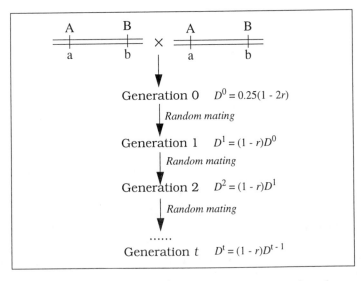

Figure 8.12 Two loci in a large population under t generations of random mating.

obtained using the disequilibrium estimate at the current generation, D_{AB}^t, by

$$r = 1 - \exp\left[\log\left(4D_{AB}^t\right) / (t+2)\right] \qquad (8.40)$$

Another way to infer linkage from disequilibrium is by using the disequilibrium estimates from multiple generations. For example, if we get the disequilibrium coefficients for generations t and $t+1$, then the recombination fraction between the two loci can be obtained by

$$r = 1 - \hat{D}_{AB}^{t+1} / \hat{D}_{AB}^t \qquad (8.41)$$

If we know the recombination fraction, then the age of the gene (generations from the initial disequilibrium) can be approximately estimated using

$$t = \left[\log\left(4D_{AB}^t\right)\right] / \left[\log\left(1-r\right)\right] - 2 \qquad (8.42)$$

Another way to approach this problem is to estimate the frequency of the ancestral haplotype in the current generation (p_{Adg}). This is the probability that a disease gene allele and a marker allele remain together without crossover between them since the disease mutation arose. p_{Adg} has the expectation

$$E[p_{Adg}] = (1-r)^g + \left[1 - (1-r)^g\right]p_{ANg} \approx e^{-rg} + (1 - e^{-rg})p_{ANg} \qquad (8.43)$$

where r is the recombination fraction between the disease gene and the marker, g is the number of generations since the disease gene arose (assuming that the marker is older than the disease) and p_{ANg} is the frequency of the A allele in the normal population. So the estimate of r is

$$r = 1 - \left(\frac{\hat{p}_{Adg} - \hat{p}_{ANg}}{1 - \hat{p}_{ANg}}\right)^{1/g} \approx -\frac{1}{g}\log\left(\frac{\hat{p}_{Adg} - \hat{p}_{ANg}}{1 - \hat{p}_{ANg}}\right) \qquad (8.44)$$

if the number of generations (g) since the disease arose is known. The term inside the logarithm has been defined as

$$p_{excess} = \frac{\hat{p}_{Adg} - \hat{p}_{ANg}}{1 - \hat{p}_{ANg}} \qquad (8.45)$$

In some situations, p_{ANg} can be ignored (Hästbacka et al. 1992). The equations can be simplified as

$$E[p_{Adg}] = e^{-rg}$$
$$\hat{r} = -\frac{1}{g}\log\left(\hat{p}_{Adg}\right) \qquad (8.46)$$

The above approaches are best applied to situations where the evolutionary process of the populations is known, such as for some of the field crop species. For most disequilibrium-based analyses, these approaches may not be adequate. The stochastic nature of the evolutionary history of a disease gene and markers linked to the gene should be taken into consideration in the estimation procedures. It is important to build a confidence interval for the estimated recombination fraction between the disease gene and the marker. Two approaches to estimate the confidence interval for recombination fraction based on disequilibrium will be introduced. One approach is to use the Luria-Delbrück algorithm proposed by Hästbacka et al. (1992) and the other approach is to use the simulated likelihood method proposed by Kaplan and Weir (1995a and 1995b).

Table 8.13 Notations used for deriving methodology for estimating recombination fraction between a marker A and a disease gene D using a dynamic population. g is the number of generations from the initial mutation, occurring before the generation in which the samples were taken.

	Generation		
	0	t	g
Population size (normal and affected)	N_0	N_t	N_g
Affected individuals	N_{dd0}	N_{ddt}	N_{ddg}
Number of d allele carriers	N_{d0}	N_{dt}	N_{dg}
Number of Ad (Affected with allele A)	$N_{Ad0} = 1$	N_{Adt}	N_{Adg}
Number of ad (Affected with allele a)	$N_{ad0} = 0$	N_{adt}	N_{adg}
A allele frequency in normal population	p_A	p_A	p_A
A allele frequency in disease population	1	p_{At}	p_{Ag}

8.4.2 MODEL

A disease gene may start as a mutation. Assume that a new mutation from D to d creates linkage disequilibrium between marker locus A and disease gene locus d. A and D are r recombination units apart. This gene may spread in the population and grow with it. New mutations may come along with the gene in the process of the disease spreading. The generation in which the disequilibrium was created is labeled as generation zero. We assume that the homozygous genotype dd is a diseased individual that does not transmit the disease alleles to the next generation. Individuals with genotype Dd are disease gene carriers that can pass the disease gene to the next generation. Table 8.13 shows the notation that will be used to derive the methodology for estimating recombination fraction between a marker and a disease gene in a dynamic population.

When the original disease mutation arises, the disease population can be modelled as a Poisson branching process (Ewens 1979; Kaplan and Weir 1995). Using the notation in Table 8.13 and Figure 8.13, we can derive the general

Generation 0 $N_{Ad0} = 1, N_{ad0} = 0, N_{dd0} = 1$

Generation 1 $N_{Ad1} \sim \text{Poisson} \{(1 + \lambda)[(1 - r) + r p_A]\}$

           $N_{ad1} \sim \text{Poisson} \{(1 + \lambda)[r(1 - p_A)]\}$

       ......

Generation t $N_{Adt}, N_{adt}, N_{ddt} = N_{Adt} + N_{adt}$

Generation $t + 1$ $N_{Ad,t+1} \sim \text{Poisson} \{(1 + \lambda)[(1 - r)N_{Adt} + r N_{ddt} p_A]\}$

                 $N_{ad,t+1} \sim \text{Poisson} \{(1 + \lambda)[(1 - r)N_{adt} + r N_{ddt}(1 - p_A)]\}$

                 $N_{T,t+1} = N_{Ad,t+1} + N_{ad,t+1}$

       ......

Generation g $N_{Adg}, N_{adg}, N_{ddg} = N_{Adg} + N_{adg}$

Figure 8.13 Poisson branching process to model disease population.

procedures to estimate the recombination fraction between the marker and the disease gene. In the original population, the A allele at a locus had a frequency of p_A. A single mutation from D to d arises in generation 0 and the disease locus is r recombination units away from a marker A. There was one Ad haplotype and zero ad haplotypes in generation 0. The total number of disease haplotypes is $1 + 0 = 1$. If we assume that the population growth rate on average is λ, then the numbers of Ad haplotypes and ad haplotypes in the next generation (generation 1) are independent Poisson variables with means

$$(1 + \lambda) \, [\, (1 - r) + r p_A] \qquad Ad$$
$$(1 + \lambda) \, [\, (1 - r) + r (1 - p_A) \,] \qquad ad$$

In general, the numbers of Ad and ad haplotypes in generation $t + 1$, conditional on the numbers of Ad and ad haplotypes in generation t, are Poisson variables with means

$$(1 + \lambda) \, [\, (1 - r) \, N_{Adt} + r N_{ddt} p_A] \qquad Ad$$
$$(1 + \lambda) \, [\, (1 - r) \, N_{adt} + r N_{ddt} (1 - p_A) \,] \qquad ad$$

Kaplan and Weir (1995) give the expectation of the frequency of the Ad haplotype in the disease population of the current generation (g) as

$$E\left(p_{Adg}\right) \, = \, E\left(\frac{N_{Adg}}{N_{ddg}}\right) \, = \, (1 - r)^g + [1 - (1 - r)^g] \, p_A \approx e^{-rg} + (1 - e^{-rg}) \, p_A$$

If we estimate the expectation using the observed haplotype frequency in the current generation ($\hat{p}_{Adg}$), then an estimate of the recombination fraction is

$$\hat{r} \, = \, 1 - \left(\frac{\hat{p}_{Adg} - \hat{p}_A}{1 - \hat{p}_A}\right)^{1/g} \approx -\frac{1}{g} \log\left(\frac{\hat{p}_{Adg} - \hat{p}_A}{1 - \hat{p}_A}\right)$$

which is the same as Equation (8.44).

8.4.3 THE LURIA-DELBRÜCK ALGORITHM

Hästbacka et al. (1992) proposed a method for building a confidence interval for the moment estimate of the recombination fraction between a disease gene and a marker based on the Luria-Delbrück algorithm. If a point estimate is obtained using Equation (8.44), the 95% lower and upper bounds can be obtained by solving

$$\frac{\hat{r}^U}{\lambda}\left[\log\left(\frac{\hat{r}^U N_{ddg}}{\lambda}\right) - 2\right] \, = \, 1 - \hat{p}_{Adg}$$
$$\frac{\hat{r}^L}{\lambda}\left[\log\left(\frac{\hat{r}^L N_{ddg}}{\lambda}\right) + 2\right] \, = \, 1 - \hat{p}_{Adg} \tag{8.47}$$

Simple solutions cannot be obtained. An iterative approach is needed to solve the equations.

8.4.4 MAXIMUM LIKELIHOOD APPROACH

Kaplan et al. (1995) and Kaplan and Weir (1995) used a Poisson branching process to model the population dynamic and the relationship between the disease gene and a normal genetic marker (Figure 8.13). They proposed to simulate the disease population for different recombination fraction values. An example would be the simulated data at every 0.001 recombination units. For each recombination fraction point, a number of replications are simulated. The

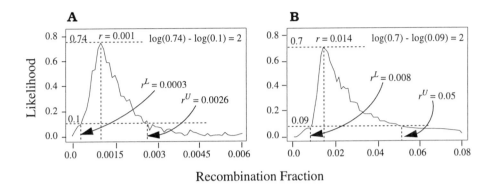

Figure 8.14 Likelihood plot for two simulated samples A and B. The number of simulation replications is 50.

likelihood at a point r is

$$L = \frac{1}{rep}\sum_{rep} \{ \left(\frac{p_{Ad}}{\hat{f}_{Ad}}\right)^{n_{Ad}} \left[\frac{1 - (p_{Ad})^{n_{Ad}}}{1 - (\hat{f}_{Ad})^{n_{Ad}}}\right]^{n_{ad}} \}$$

where $\hat{f}_{Ad}$ is an observed proportion of Ad haplotypes in a sample, p_{Ad} and n_{Ad} are the proportion and the count of Ad haplotypes in one simulation and rep is the number of replications of the simulation. Figure 8.14 shows two likelihood plots. The maximum likelihood point estimate of the recombination fraction corresponds to the peaks of the likelihood curves. The 95% lower and upper bounds are the two points corresponding to approximately 2 units below the likelihood peak on the log scale. For example, in the plot of Figure 8.14A, the maximum likelihood estimate is 0.001 with a likelihood value of 0.74. The 95% lower and upper bounds are 0.0003 and 0.0026, which correspond to a likelihood value of 0.1 because the difference from the peak is 2 on the log scale

$$\log (0.74) - \log (0.1) = 2$$

For Figure 8.14B, the approximate 95% confidence interval is (0.009 - 0.05), because

$$\log (0.7) - \log (0.09) = 2$$

8.5 EXERCISES

Exercise 8.1

A sample with 200 individuals was obtained from a randomly mating population. Two loci were genotyped for the 200 individuals. There are two alleles for each of the two loci in the population. The observed frequencies for the nine genotypic classes are:

Genotype	Frequency
AABB	5
AABb	8
AAbb	11
AaBB	10
AaBb	62
Aabb	19
aaBB	9
aaBb	15
aabb	61

(1) Estimate the allelic frequencies in the population.

(2) Estimate the disequilibrium coefficients for each of the loci and for the combination of the two loci.

(3) Predict the genotypic frequency in the next generation assuming random mating and independence between the two loci.

(4) Predict the genotypic frequency in the next generation assuming random mating and 10% recombination between loci A and B.

(5) How many generations of random mating are needed to reduce the disequilibrium coefficient to 0.0001 if loci A and B are linked with a 10% recombination fraction?

Exercise 8.2

Explain the relationship between linkage and disequilibrium using a two-locus model. (Review the sources of the original disequilibrium and the forces to reduce or to maintain the disequilibrium.)

Exercise 8.3

In an experiment to locate a gene partially responsible for a disease, 80 families were sampled. At least one child was affected by the disease in each of the 80 families. An RFLP marker which was previously determined to be linked with the disease gene by a linkage-based study was genotyped for all the parents and the affected children. All parents are heterozygous for the marker locus. Among the 160 alleles transmitted to the affected children from the parents, 100 were alleles denoted by 'G' and the remaining 60 alleles were alleles denoted by 'X'.

(1) Carry out a TDT (transmission/disequilibrium test) for the data.

(2) Give a genetic interpretation of the result of TDT. (Hint: Consider a genetic model of the disease and its relationship with the marker.)

(3) A clone of the gene that causes the disease and a complete physical map in the region of the disease gene are obtained in a recent experiment. The gene is 5 kb away from the restriction site which creates the polymorphic RFLP marker. Can you speculate on the connection between the physical distance and the TDT result?

(4) If the population from which the 80 families were drawn is stable and relatively isolated, can you find out the approximate age of the disease gene? (Hint: Combine the results of TDT and the physical distance.)

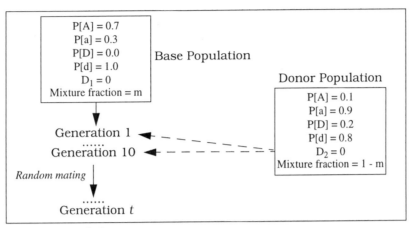

For Exercise 8.4

Exercise 8.4

A genetic marker A with alleles A and a and a disease gene D with alleles D and d were investigated in two populations (see the following illustration). The two populations were under 10 generations of population mixture. After these 10 generations, the population underwent a number of generations of random matings. The two loci were in equilibrium in each of the original populations. Assume that the populations are large.

(1) Assuming that loci A and D are not linked, compute the population parameters for a) $m = 0.95$ and $t = 20$, b) $m = 0.95$ and $t = 100$ and c) $m = 0.5$ and $t = 20$ (Note: m is a mixture fraction of the base population, $1 - m$ is the mixture fraction of the donor population and t is the number of generations from the initial population mixture. Hint: Compute the allelic frequencies for the two loci and disequilibrium parameters for each of the two loci and the two-locus model.).

(2) Assuming that loci A and D are linked, compute the population parameters for a) $r = 0.1$, $m = 0.95$ and $t = 20$, b) $r = 0.01$, $m = 0.95$ and $t = 20$ and c) $r = 0.01$, $m = 0.95$ and $t = 100$ (Note: r is the recombination fraction between loci A and D.).

(3) How many generations of random mating are needed to reduce the disequilibrium coefficient to 0.01 if the recombination fraction between the two loci is 0.001?

(4) A disequilibrium coefficient 0.01 is estimated from a sample drawn after 90 generations of random mating ($t = 100$). Review the procedures to estimate the recombination fraction between the two loci. Compute the recombination fraction between the two loci using any procedure you choose.

(5) Review the problems in inferring recombination fraction from the population disequilibrium data.

Exercise 8.5

Several factors, such as genetic linkage, mutation, selection, random drift, population admixture and random mating contribute to create, maintain or reduce two-locus disequilibrium at a population level. Ignoring mutation, selection, random drift and population admixture, consider linkage as the major mechanism to reduce disequilibrium caused by random mating. Linkage and the number of generations of random mating are important in studies of the genetic structure of populations. However, linkage and the number of generations of random mating are confounded in the population-based data. Even when the number of generations is known, the relationship between population-based disequilibrium and genetic linkage is still complex. Explain why and outline your strategies for solving the problem.

CHAPTER 9

LINKAGE GROUPING AND
LOCUS ORDERING

In this chapter, linkage grouping and locus ordering will be discussed. Linkage grouping is placing loci into linkage groups based on their linkage relationships. Locus ordering is to infer the relative order of the loci in a linkage goup or locus locations on chromosomes. Different criteria for linkage grouping and locus ordering will be introduced. Special attention will be paid to computation problems and uncertainty measurement for locus ordering. Linkage grouping and locus ordering in this chapter are different from the traditional cytogenetic approaches. The approaches in this chapter are solely for polymorphic markers or genes segregating in mapping populations.

9.1 LINKAGE GROUPING

A linkage group is defined biologically as a group of genes with their loci located on the same chromosome. Linkage group in this chapter is defined statistically as a group of loci inherited together according to certain statistical criteria. The statistically defined linkage group may differ from the biologically defined linkage group. In theory, if no false linkage is assumed, then a linkage group obtained based on statistics should also be a biological linkage group. However, genes on the same chromosome may be grouped into different linkage groups based on statistics because genes on a large segment of chromosome may not be observed. In this section, the criteria of linkage statistics and the ways to use these criteria to group loci will be introduced.

9.1.1 LINKAGE GROUPING CRITERIA

Let θ_{ij}, z_{ij} and p_{ij} denote a two-point recombination fraction, a lod score and a significant P-value, respectively, for a pair of loci i and j. Linkage grouping is usually based on one of the combinations of θ_{ij}, z_{ij} and p_{ij}. Recombination fractions and significant P-values may be used as criteria to infer whether loci i and j belong to same linkage group or not:

- if $\{ [\theta_{ij} \leq c] \, and \, [p_{ij} \leq b] \}$,
- then loci i and j belong to same linkage group

or

- if $\{ [\theta_{ij} \leq c] \, or \, [p_{ij} \leq b] \}$,
- then loci i and j belong to same linkage group

c is the maximum recombination fraction value to to be declared a linkage and b is the maximum significant P-value for declaring a linkage. Lod score and recombination fraction have also been used in combination for linkage grouping. The criteria are

- $\{ [\theta_{ij} \le c] \, and \, [z_{ij} \ge a] \}$ or
- $\{ [\theta_{ij} \le c] \, or \, [z_{ij} \ge a] \}$

where c is the maximum recombination value and a is the minimum lod score value to declare a linkage.

9.1.2 PROCEDURES

In practical genomic mapping, linkage grouping is done as a double-iteration process:

(1) iterate grouping criteria to find an acceptable grouping pattern
(2) iterative searching to determine grouping pattern, given the criteria

. Given the criteria, there is a unique grouping pattern. The following has been used in practical genomic map construction as an empirical guideline:

(1) If a grouping pattern does not change over a wide range of grouping criteria, the pattern has high confidence.
(2) If a grouping pattern meets a biological expectation, the grouping pattern has high confidence.
(3) Unexpectedly large linkage groups are a sign of false linkage and the grouping criteria needs to be more restrictive.
(4) A large number of unlinked genetic markers (or so-called single marker linkage groups) is a sign of low quality data, small population size or a small number of genetic markers.

For example, using criteria

$$\{ [\theta_{ij} \le 0.3] \, or \, [p_{ij} \le 0.00001] \}$$

a grouping pattern could be

```
Group#1:  1  2  5  6  9  10  13  14
Group#2:  3  4  7  8 11  12  15  16
```

where the numbers are orders of the genetic markers in the data set.

If a genomic map already exists, it is another type of linkage grouping problem to determine which linkage group a new genetic marker belongs to. Approaches using whole map log likelihood have been proposed. This problem will be discussed after locus ordering because the new loci will be analyzed and located in a genome with all markers simultaneously.

In practice, balancing c and b or a, biological information, such as number of chromosomes, can be used in linkage grouping.

9.2 THREE-LOCUS ORDER

9.2.1 INTRODUCTION TO LOCUS ORDERING

Locus order is defined as the linear arrangement of genes or genetic markers in a linkage group. Let

$$a_1, a_2, \ldots, a_l$$

denote l loci in a linkage group, whose map distances are

$$D = (d_1, d_2, ..., d_{l-1})$$

Map distance will be discussed in Chapter 10. For now, the "map distance" is the distance related to recombination fraction and crossover interference. An order of markers of genes on a linkage group is a locus order. For example, $1, 2, 3, ..., i, ..., l-1, l$ is a locus order.

For l genes, there are $l!/2$ possible gene orders if the orientation of the orders can be ignored. Locus ordering searches for the best locus order among the possible orders and uses specific algorithms or computational procedures to obtain the order. The gene ordering problem can be solved using algorithms used for solving the traveling salesman problem (TSP). For the method of the sum of adjacent recombination fractions, the two-point recombination fraction matrix for a gene ordering problem is equivalent to the distance matrix for the TSP. Locus ordering presents a computation problem because there is a large number of possible gene orders when a linkage group contains a large number of loci (Table 9.1). The computational methods or algorithms for precise gene ordering are limited to a small number of loci in a linkage group.

Table 9.1 Numbers of possible locus orders and three-locus combinations.

Number of Loci l	Possible Gene Orders $l!/2$	Number of Triplets $l(l-1)(l-2)/6$
2	1	0
3	3	1
5	60	10
10	$1,814,400$	120
20	1.22×10^{18}	1,140
40	4.08×10^{47}	9,880

Obtaining high or medium density genomic maps is one of the goals of many programs of research. Locus ordering is a key step for genomic mapping. Many of the basic inferences and practical applications of genome research rely on accurate gene orders. Several methods for gene ordering have been developed, such as seriation (Buetow and Chakravarti 1987; Buetow 1987), minimum sum of adjacent recombination fractions (SARF) (Falk 1989), minimum product of adjacent recombination fractions (PARF) (Wilson 1988), maximum sum of adjacent lod scores (SALOD) (Weeks and Lange 1987), minimum sum of the probability of double recombinants (PDR) (Knapp et al. 1989) and maximum likelihood (ML) (Lander and Green 1987). Thompson (1987) proposed a method for finding gene order with minimum obligatory crossovers (MOC) using a branch-and-bound method. For certain situations, MOC is the same as minimum SARF (Falk 1989). Among the above methods, SARF, PARF, SALOD, ML and MOC are criteria for defining the best locus order and seriation is one of the algorithms for obtaining minimum SARF. Some comparisons among those methods have been given in a small scale (6 or 7 loci in a linkage group) using computer simulations (Olson and Boehnke 1990, Kammerer and MacCluer 1988).

There is only one possible locus order for two loci. Three is the minimum number of loci to be ordered. Methods for two-locus models have been described. As we will see, results of the two-locus analysis are the basis for all of the advanced analyses. The next extension of the two-locus model is the three-locus model. For a three-locus model, there are three possible locus orders, three two-locus combinations and a parameter for crossover interference. Locus ordering for more than three loci can be an extension of three-locus ordering.

Table 9.2 Expected genotypic frequency for a triple backcross (ABC/abc × abc/abc) progeny set when ABC is the order of the loci. C is the coefficient of coincidence.

Genotype	Observed Count	Expected Frequency	
		Without Interference (p_i)	Interference Model (p_{Ci})
AaBbCc	f_1	$0.5(1-\theta_1)(1-\theta_2)$	$0.5(1-\theta_1-\theta_2+C\theta_1\theta_2)$
AaBbcc	f_2	$0.5(1-\theta_1)\theta_2$	$0.5(\theta_2-C\theta_1\theta_2)$
AabbCc	f_3	$0.5\theta_1\theta_2$	$0.5C\theta_1\theta_2$
Aabbcc	f_4	$0.5(1-\theta_2)\theta_1$	$0.5(\theta_1-C\theta_1\theta_2)$
aaBbCc	f_5	$0.5(1-\theta_2)\theta_1$	$0.5(\theta_1-C\theta_1\theta_2)$
aaBbcc	f_6	$0.5\theta_1\theta_2$	$0.5C\theta_1\theta_2$
aabbCc	f_7	$0.5(1-\theta_1)\theta_2$	$0.5(\theta_2-C\theta_1\theta_2)$
aabbcc	f_8	$0.5(1-\theta_1)(1-\theta_2)$	$0.5(1-\theta_1-\theta_2+C\theta_1\theta_2)$

9.2.2 THREE-LOCUS LIKELIHOOD AND THE CONCEPT OF INTERFERENCE

Let us first assume a situation which has three linked loci, A, B and C. There are three possible locus orders, ABC, ACB and BAC. For a backcross model, there are 8 possible genotypes in the progeny. The expected genotypic frequencies are not defined if the order of the loci on the genome is not certain. However, the three two-locus pairwise recombination fractions, θ_{ab}, θ_{ac} and θ_{bc}, can be obtained without knowing the genome positions of the loci.

Given that the order of three loci is ABC, expected frequencies for the 8 genotypes are listed in Table 9.2. In the table, θ_1 and θ_2 are recombination fractions between A and B and between B and C, respectively. C is defined as the coefficient of coincidence and $1-C$ is defined as crossover interference. If there is no interference and crossovers occur independently, then the expected double recombinant frequency will be $\theta_1\theta_2$ and

$$Interference = 1 - C = 0$$

If crossovers in intervals AB and BC are not independent, the observed double recombinant frequency may not equal the expectation. If we use θ_{12} to denote the true double recombinant frequency, the coefficient of coincidence is

$$C = \frac{\theta_{12}}{\theta_1\theta_2}$$

Interference is negative if

$$C > 1 \text{ and } Interference = 1 - C < 0$$

positive if

$$C < 1 \text{ and } Interference = 1 - C > 0$$

absent if

$$C = 1 \text{ and } Interference = 1 - C = 0$$

and complete if

$$C = 0 \text{ and } Interference = 1 - C = 1$$

A significant negative interference means that the observed double crossovers are more than expected, a significant positive interference means that the observed double crossovers are less than expected, absence of interference means that they are equal and complete interference means that no double crossovers were observed. From Table 9.2, the likelihood function for a three-locus model with locus order ABC is

$$
\begin{aligned}
L_{ABC}(\theta_1, \theta_2, C) &= \sum_i f_i \log p_{Ci} \\
&= (f_1 + f_8) \log (1 - \theta_1 - \theta_2 + C\theta_1\theta_2) + (f_3 + f_6) \log (C\theta_1\theta_2) \\
&\quad + (f_2 + f_7) \log (\theta_2 - C\theta_1\theta_2) + (f_4 + f_5) \log (\theta_1 - C\theta_1\theta_2)
\end{aligned}
\tag{9.1}
$$

The maximum likelihood estimates of the two-locus recombination fractions for θ_{ab}, θ_{ac} and θ_{bc} are

$$\hat{\theta}_{ab} = (f_3 + f_4 + f_5 + f_6) / N \tag{9.2}$$

$$\hat{\theta}_{ac} = (f_2 + f_4 + f_5 + f_7) / N$$

$$\hat{\theta}_{bc} = (f_2 + f_3 + f_6 + f_7) / N$$

where N is the population size. These estimates do not depend on specific locus order. However, the relationships between the estimates do depend on this order. For locus order ABC, the relationships are

$$
\begin{aligned}
\theta_{ab} &= \theta_1 \\
\theta_{ac} &= \theta_1 + \theta_2 - 2C\theta_1\theta_2 \\
\theta_{bc} &= \theta_2
\end{aligned}
\tag{9.3}
$$

The maximum likelihood estimate of the coefficient of coincidence is

$$
\begin{aligned}
\hat{C} &= \frac{2N(f_3 + f_6)}{(f_3 + f_4 + f_5 + f_6)(f_2 + f_3 + f_6 + f_7)} \\
&= \frac{\hat{\theta}_{ab} + \hat{\theta}_{bc} - \hat{\theta}_{ac}}{2\hat{\theta}_{ab}\hat{\theta}_{bc}}
\end{aligned}
\tag{9.4}
$$

which is a function of locus order. The large-sample variance is (see Bailey 1961)

$$Var(\hat{C}) = \frac{\hat{\theta}_{bc}^2 (\hat{\theta}_{ac} - \hat{\theta}_{bc})^2 Var(\hat{\theta}_{ab}) + \hat{\theta}_{ab}^2 (\hat{\theta}_{ac} - \hat{\theta}_{ab})^2 Var(\hat{\theta}_{bc}) + \hat{\theta}_{ab}\hat{\theta}_{bc} Var(\hat{\theta}_{ac})}{4\hat{\theta}_{ab}^4 \hat{\theta}_{bc}^4} \tag{9.5}$$

where $Var(\hat{\theta}_{ab})$, $Var(\hat{\theta}_{bc})$ and $Var(\hat{\theta}_{ac})$ are the large-sample variances for $\hat{\theta}_{ab}$, $\hat{\theta}_{bc}$ and $\hat{\theta}_{ac}$, respectively (see Chapter 6).

For the coefficient of coincidence, the log likelihood ratio test statistics of the log likelihood under the maximum likelihood estimate of the coefficient over the log likelihood under no interference ($C = 1$) can be used, which is

$$G = 2 \left[L_{ABC}(\hat{\theta}_1, \hat{\theta}_2, \hat{C}) - L_{ABC}(\hat{\theta}_1, \hat{\theta}_2, 1) \right] \tag{9.6}$$

being distributed as a chi-square variable with one degree of freedom. When $C = 0$, the log likelihood function of Equation (9.1) becomes a complete interference model

$$\begin{aligned} L_{ABC}(\theta_1, \theta_2) &= \sum_{i=1}^{8} f_i \log p_i \\ &= (f_1 + f_8) \log (1 - \theta_1 - \theta_2) + (f_4 + f_5) \log \theta_1 + (f_2 + f_7) \log \theta_2 \end{aligned} \tag{9.7}$$

Table 9.3 Observed genotypic frequency for a triple backcross (ABC/abc × abc/abc).

Genotype	Notation	Observed Count
AaBbCc	f_1	34
AaBbcc	f_2	5
AabbCc	f_3	11
Aabbcc	f_4	0
aaBbCc	f_5	1
aaBbcc	f_6	10
aabbCc	f_7	4
aabbcc	f_8	35

Table 9.4 Maternal gametic frequency for data in Table 9.3.

Gametes	Notation	Count
ABC + abc	$f_1 + f_8$	69
ABc + abC	$f_3 + f_6$	21
AbC + aBc	$f_2 + f_7$	9
Abc + aBC	$f_4 + f_5$	1

In practice, the order of the three loci in the genome can be determined in several ways, such as the double-crossover approach, the two-locus recombination fraction approach and the likelihood approaches. An example, shown in Table 9.3, will be used to illustrate these methods to determine a three-locus gene order and shows a likelihood analysis for a three-locus model. Data was obtained from a triple backcross (ABC/abc × abc/abc) progeny set. The progeny size is 100. No segregation distortion is significant for the three loci.

9.2.3 DOUBLE CROSSOVER APPROACH

The order of the three loci can be determined by finding the rarest genotypes and assigning the genotypes as double recombinants. Once the double recombinant classes are identified, the order of the loci is determined. For data in Table 9.3, maternal gametes can be classified into four classes: one non-recombinant class (ABC and abc) and three recombinant classes (ABc and abC), (AbC and aBc) and (Abc and aBC). Paternal gametes do not segregate and do not contribute to this linkage analysis. Depending on the order of the three loci, the three recombinant classes could be single recombinant or double recombinant classes. Maternal gametic frequency for the data is listed in Table 9.4. It is easy to identify Abc and aBC as the rarest recombinant classes. If Abc and aBC are assigned as the double recombinant class, the order of the three loci has to be BAC.

9.2.4 TWO-LOCUS RECOMBINATION FRACTION APPROACH

It is common to determine the order of the three loci by the magnitude of the estimates of the three recombination fractions. For the data in Table 9.3, the estimated recombination fractions are

$$\hat{\theta}_{ab} = \frac{9+1}{100} = 0.1$$

$$\hat{\theta}_{ac} = \frac{21+1}{100} = 0.22$$

$$\hat{\theta}_{bc} = \frac{21+9}{100} = 0.3$$

The estimated recombination fraction is larger when the loci are located further away. So, the loci located at either end will show the largest recombination fraction between them. For the example data, B and C are located at the ends and A is located in middle. The estimated locus order based on the magnitude of the two-point recombination fractions is BAC.

Another method for determining the possible three-locus order using the magnitude of the two-locus recombination fractions is to pick the tightest linked loci and put them close together. Then the next tightest locus to the two loci is placed flanking those two loci. For example, A and B are the tightest linked loci and are put together. The next tightest linkage is between A and C, so C is placed beside A. The order is BAC.

Table 9.5 Estimated coefficients of coincidence and log likelihood for data in Table 9.3.

	Order		
	ABC	ACB	BAC
Coefficient of Coincidence	3	3.1818	0.455
Log Likelihood, Equation (9.1)	-84.65	-84.65	-84.65
Log Likelihood ($C = 0$)	-86.82	-91.81	-87.20
Log Likelihood ($C = 1$)	-93.60	-113.78	-87.20
Log Likelihood ($C \leq 1$)	-93.60	-113.78	-84.65

9.2.5 LOG LIKELIHOOD APPROACH

The other commonly used method for three-locus ordering is to compare the likelihoods for the three possible orders. The order with the highest likelihood is the most possible order. For the example data, the log likelihoods for the three possible orders are listed in Table 9.5. For a locus order ABC, the log likelihood is

$$L_{ABC}(0.1, 0.3, 3) = 69\log(1 - 0.1 - 0.3 + 3 \times 0.1 \times 0.3) + 9\log(3 \times 0.1 \times 0.3)$$

$$+ 21\log(0.3 - 3 \times 0.1 \times 0.3) + \log(0.1 - 3 \times 0.1 \times 0.3)$$

$$= -84.65$$

For a locus order of BAC, the log likelihood is

$$L_{BAC}(0.1, 0.22, 0.455) = 69\log(1 - 0.1 - 0.22 + 0.455 \times 0.1 \times 0.22)$$
$$+\log(0.455 \times 0.1 \times 0.22)$$
$$+21\log(0.22 - 0.455 \times 0.1 \times 0.22)$$
$$+9\log(0.1 - 0.455 \times 0.1 \times 0.22)$$
$$= -84.65$$

The log likelihoods are the same for the three possible orders when the likelihood function is fully parameterized, therefore, the most likely order cannot be determined this way. When the coefficient of coincidence is set to 1.0 (without interference) or 0.0 (complete interference), the order of BAC gives the highest log likelihoods. In the last row in the table, log likelihoods were obtained for the three possible locus orders under assumption of absence of negative interference. Negative interference has been observed for many species and it is most likely the result of gene conversion over very short DNA sequences instead of multiple crossovers at the chromosome level.

If a biological assumption, which is that negative interference ($C > 1$) does not occur, is applied to the statistical model, then the likelihood approach can be used to find the most likely locus order. For example, BAC is identified as the most likely locus order for the three loci using the restrained likelihood functions ($C \leq 1$). The likelihood approach has been said to be the method of choice for many statistical problems; however, questions remain about the use of the likelihood approach in locus ordering.

When a perfect model (no constraints on the coefficient of coincidence) is used for a mapping data set, the log likelihoods should be the same for all possible locus orders for the same data set. The example data supports this point. The general argument to support this point is that parameters for a fully parameterized model are sufficient statistics to explain all the variation in the data. For the gene ordering problem, if the parameters for the data are sufficient and the maximum likelihood estimates of the parameters are unbiased for different orders, the log likelihoods evaluated using the maximum likelihood estimate should be the same for all possible gene orders.

Another argument against the likelihood approach is that the likelihood comparison cannot be implemented as the former log likelihood ratio test because there is no degree of freedom difference among the possible locus orders. The usual log likelihood ratio test statistic involves two or more likelihood functions which contain different numbers of parameters. For example, the odds ratio between an order of BAC and an order of ABC for the constrained model of $C \leq 1$ is

$$\frac{Pr[BAC]}{Pr[ABC]} = \exp[-84.65 - (-93.60)] = 7708$$

The interpretation of this odds ratio, which is that the order of BAC is 7708 times more likely than the order of ABC, is questionable. If we constrain the model without consideration of interference ($C = 1$), the odds ratio is

$$\frac{Pr[BAC]}{Pr[ABC]} = \exp[-87.20 - (-93.60)] = 602$$

Under this biological assumption, the order of BAC is 602 times more likely than the order of ABC.

9.3 MULTIPLE-LOCUS ORDERING

The restrained likelihoods ($C \leq 1$, $C = 1$ and $C = 0$) have been useful for determining the most probable locus order for three-locus situations (Table 9.5). As we will see in the next section, no interference models have been commonly used for locus ordering involving a large number of loci and many other criteria for locus ordering, such as a minimum SARF, PARF, SALOD and MOC.

Lathrop *et al.* (1987) show several exact tests for gene order for three loci using backcross matings. They suggest that the tests should be performed assuming either positive interference or lack of interference. If lack of interference is assumed, computation is simplified and the tests are robust even when interference is present. It is necessary to constrain the coefficient of coincidence ($C \leq 1$) in order to use the likelihood approach for locus ordering. Edwards (1987) shows coding procedures for likelihood calculation used in locus ordering. He shows how haplotypes can be deduced according to phenotypes and family structures and then the likelihoods for different orders can be calculated.

9.3.1 MULTIPLE-LOCUS ORDERING STATISTIC

Historically, locus ordering for multiple loci has been a process of minimizing the number of crossovers (Sturtevant 1913). The criteria for gene ordering commonly used in recent gene mapping, such as a minimum SARF, PARF and MOC, also have a strong association with the idea of minimizing crossovers.

In Section 9.3, methods for determining locus order for three loci were described. Multiple-locus ordering statistics are usually extensions of the three-locus approaches. For a linkage group with a large number of loci, it is impossible to fit a full parameterized model because the number of genotypic classes needed for the model is too large. For example, for a 10-locus full model using backcross progeny, there are $2^{10} = 1,024$ possible genotypic classes and the number is $3^{10} = 59,049$ for an F2 codominant marker model. For number of loci greater than 6, it is reasonable to assume an absence of interference model for locus ordering. In Chapter 10, when mapping functions are introduced, there will be more discussion of interference models.

Notation

Let $a_1, a_2, ..., a_l$ denote l loci in a linkage group. The following matrices are used to denote the statistics from the two-locus analysis

$$R = \left[\hat{\theta}_{a_i a_j} \right]_{l \times l}$$

$$Z = \left[\hat{z}_{a_i a_j} \right]_{l \times l} \qquad (9.8)$$

$$n = \left[n_{a_i a_j} \right]_{l \times l}$$

where

$$\hat{\theta}_{a_i a_j}, \hat{z}_{a_i a_j} \text{ and } n_{a_i a_j}$$

are the estimated recombination fraction, lod score and number of informative observations, respectively, between loci a_i and a_j and i and j are locus indicates for a certain locus order. i and j change as locus orders are different.

For example, 1 is for locus a_1 and 2 is for locus a_2 for a locus order $1, 2, 3, ..., l-1, l$. However, 1 is for locus a_2 and 2 is for locus a_1 for a locus order $2, 1, 3, ..., l-1, l$. The indicates variables in Equation (9.8) are genome position indicators instead of locus indicators.

Three-Locus Approach

For a linkage group with more than three loci, the locus ordering problem can be partitioned into a number of three-locus ordering problems. For example, 4 linked loci can be organized into 4 possible three-locus combinations. For each of the triplets, there are three possible locus orders and the best order can be determined using the approaches described in Section 9.3. In general, for l loci, there are

$$\binom{l}{3} = \frac{l!}{3! \, (l-3)!}$$
$$= \frac{1}{6} l (l-1) (l-2)$$

(9.9)

possible three-locus combinations (Table 9.1). Among them, some of the triplets may have no meaning because they are not linked to each other. For the triplets with all three linked loci, the best order can be determined.

In practice, the tightest triplet can be picked first. Then, based on that triplet, it is possible to add the next tightest locus and do a three-locus analysis for combinations of the new locus and the possible existing two-locus combinations. If there is a contradiction regarding the new locus in terms of locus order, the locus cannot be placed on the map. This procedure is repeated until all loci are tried. The disadvantages of the three-locus approach are:

(1) some of loci cannot be placed on the map due to contradictory results among different triplets

(2) a local optimal locus order may be obtained instead of a global optimal locus order

(3) information on locus order may be lost by subdividing the multiple loci into the triplets

Maximum Likelihood Approach

Consider a linear sequence of l loci, $a_1, a_2, ..., a_l$. The likelihood function for locus order $1, 2, 3, ..., l-1, l$ is

$$L(R) = \sum_{i=1}^{l-1} n_{a_i a_j} [\theta_{a_i a_j} \log \theta_{a_i a_j} + (1 - \theta_{a_i a_j}) \log (1 - \theta_{a_i a_j})]$$

(9.10)

where $j = i + 1$ and n_{ij} is number of informative sample sizes for locus combination i and j. n_{ij} may be different from locus to locus due to missing data or missing linkage information for the two adjacent loci. The order with the maximum likelihood is defined as the maximum likelihood locus order.

Minimum Sum or Product of Adjacent Recombination Fractions (SARF and PARF)

The SARF for a given order $1, 2, 3, ..., l-1, l$ is

$$SARF = \sum_{i=1}^{l-1} \hat{\theta}_{a_i a_j}$$

(9.11)

An order with minimum SARF is defined as a minimum SARF locus order. If there is no missing data and no segregation ratio distortion, there is a high correlation between the minimum SARF and maximum likelihood locus orders. The PARF is

$$SARF = \prod_{i=1}^{l-1} \hat{\theta}_{a_i a_j} \qquad (9.12)$$

An order with minimum PARF is defined as a minimum PARF locus order. To avoid PARF zero, a small tolerance, such as 0.00001, can replace zero recombination fraction.

Maximum Sum of Adjacent Lod Score (SALOD)

SALOD for an specific locus order is

$$SALOD = \sum_{i=1}^{l-1} z_{a_i a_j} \qquad (9.13)$$

An order with maximum SALOD is the maximum SALOD locus order. As mentioned by Olson and Boehnke (1990), these criteria are sensitive to locus informativity and tend to place highly informative loci together. For this reason, they proposed two modifications for SALOD. The first is

$$SALOD^{PIC} = \sum_{i=1}^{l-1} \frac{z_{a_i a_j}}{PIC_{a_i} PIC_{a_j}} \qquad (9.14)$$

where PIC is polymorphism information content, defined by Botstein $et\ al.$ (1980) as

$$PIC = 2 \sum_{i=1}^{a-1} \sum_{j=i+1}^{a} u_i u_j (1 - u_i u_j) \qquad (9.15)$$

for a single locus, where a is the number of alleles of the locus in populations and u_i is the frequency of i allele in the populations (see Chapter 5). The other modification is

$$SALOD^{EN} = \sum_{i=1}^{l-1} \frac{z_{a_i a_j}}{EN_{a_i a_j}} \qquad (9.16)$$

where EN is defined as the equivalent number of fully informative meioses, which can be estimated by

$$EN = Log_{10} 2 + \hat{\theta} Log_{10} \hat{\theta} + (1 - \hat{\theta}) Log_{10} (1 - \hat{\theta}) \qquad (9.17)$$

where $\hat{\theta}$ is the estimated recombination fraction between the two loci.

Olson and Boehnke (1990) carried out a simulation study to compare the power of methods for locus ordering. They used the methods described in previous paragraphs, except for the maximum likelihood approach. They concluded that the SARF and a modified SALOD^{EN} are the best overall locus ordering methods. However, they also mentioned that performance of the locus ordering methods may vary according to different genetic settings.

Least Square Method

Given a locus order, the interlocus map distances can be estimated from the pairwise distances using least squares (Jensen 1971; Buetow and Chakravarti 1987; Weeks and Lange 1987). The estimation procedures will be described in Chapter 10 because the least square approach is a way to obtain least square multiple-point map distances. The order with the minimum least square is defined as least square locus order.

9.3.2 THE TRAVELING SALESMAN PROBLEM

The definitions of the locus ordering statistics are easy to understand. However, the problem is how to achieve the global maximum or minimum when the number of loci in a linkage group is large. This has been recognized as a special case of the traveling salesman problem (TSP).

Problem

TSP is a classical example of NP (non-deterministic polynomial time) complete problems that have gained the attention of mathematicians, computer scientists and practitioners of many other branches of science for many decades because of their theoretical and applied importance (Miller and Pekny 1991). A general description of the symmetric TSP is: given $C_{l \times l}$, a symmetric matrix of distances among a set of l cities denoted by the set of $V = \{1, 2, ..., l\}$, find the shortest cyclical itinerary for a traveling salesman who must visit each of l cities optimally once. Let c_{ij} be the distance between cities i and j, then the problem becomes that of finding an order with

$$\sum_{i \in V} \sum_{j \in V} c_{ij} x_{ij}$$

maximized, subject to

$$\begin{cases} \sum_{i \in V} x_{ij} = 1, j \in V \\ \sum_{j \in V} x_{ij} = 1, i \in V \\ \sum_{i \in V} \sum_{j \in V} x_{ij} \leq |S| - 1 \quad for \quad S \subset V, S \neq 0 \\ x_{ij} \in \{0, 1\}, i, j \in V \end{cases}$$

where $x_{ij} = 1$ if (i, j) is in the solution and $x_{ij} = 0$ otherwise. For the gene ordering problem, constraint

$$\sum_{i \in V} \sum_{j \in V} x_{ij} \leq |S| - 1$$

is designed to guarantee a linear order instead of a circle. In the general case, there are $l!$ possible routes for l cities. When l is large, it is not practical to find the shortest path by evaluating every possible route (at current computer speed). For example, there are 8.16×10^{47} possible routes for 40 cities.

Algorithms

Algorithms are procedures for solving computation problems. Algorithms to solve the TSP can be put into two classes: approximation or exhaustive search algorithms. The approximation algorithms usually do not guarantee optimal solutions. However, the results are usually good enough for practical applications. For locus ordering, seriation, which is a modified version of a greedy algorithm for obtaining minimum SARF, and a simulated annealing (SA) algorithm have been used as approximation algorithms for obtaining minimum SALOD and least square criterion (Weeks and Lange 1987; Liu *et al.* 1992). The exhaustive search does provide optimal solutions, but it usually can be applied only to a specific problem when the searching space is large. The branching and bound (BB) algorithm was proposed for locus ordering (Thompson 1987).

Algorithms such as simulated annealing (SA) (Kirkpatrick 1983; Aarts and Korst 1989) and a genetic algorithm (Holland 1975; Goldberg 1989) have been used to reach approximate solutions for the TSP. These algorithms cannot guarantee the optimal solution. For some specific problems, optimal solutions were obtained by taking advantage of specific data features, such as the shortest path among 532 US cities using heavy branching and cutting (Padberg and Rinaldi 1987). Search space can be greatly reduced by starting with an approximate solution and using branching and bounding constraints, such as only 1,278 nodes for the 532 US city problem. For the general TSP, optimal solutions can be obtained by using exclusive searching algorithms, such as BB (Miller and Pekny 1991). Effectiveness of BB depends on the features of the distance matrix. For general cases, it is still unpractical to solve the TSP for a large number of cities even if BB is used (Miller and Pekny 1991).

Seriation

Seriation for locus ordering was first introduced by Buetow and Chakravarti (1987). Seriation is designed to obtain minimum SARF. Seriation is a set of rules for placing each of the l loci based on the two-locus recombination fractions. Here are the rules described by Buetow and Chakravarti:

(1) Starts: Start with each of the l loci $a_1, a_2, ..., a_l$. When a_i is a starter, place a_j to the right of a_i if the recombination fraction is the smallest among the recombination fractions between a_i with all other $l-1$ loci. After this step, $l-2$ loci are to be placed.

(2) Remaining $l-2$ Loci: Choose the locus a_k from the unplaced loci with the smallest recombination fraction to a_i. Compare the recombination fraction of a_k with the two loci which are currently external in the cluster of placed loci, a_{left} (the locus on the left side) and a_{right} (the locus at the right side). If

$$\hat{\theta}_{a_k a_{right}} > \hat{\theta}_{a_k a_{left}}$$

place a_k to the left of the cluster of currently placed loci. If

$$\hat{\theta}_{a_k a_{right}} < \hat{\theta}_{a_k a_{left}}$$

place a_k to the right. Repeat the above procedure until all loci are placed.

(3) Resolving Ties: When recombination fractions involved in the placing process are identical, here are some of the rules for resolving the ties. The ties need to be considered only when the loci are to be placed on the same side, for example

$$a_{left}, ..., a_{right}, (a_j, a_k) \text{ and } \hat{\theta}_{a_j a_{right}} = \hat{\theta}_{a_k a_{right}}$$

The two loci, a_j, a_k, should be ordered with respect to the locus most external on that side in the ordered locus cluster. If this approach fails to resolve the tie, the next two internal loci in the ordered cluster should be considered.

(4) Determination of Locus Order: For l loci, l locus orders will be derived. When the recombination fraction matrix is monotonic, only one unique order can be derived if orientation of the orders is ignored. If the matrix is not monotonic, a continuity index (CI) can be used to measure the goodness-of-fit to monotonicity. For each of the l orders, the CI is estimated using

$$CI = \sum_{i<j} \hat{\theta}_{ij} / (j-i)^2$$

where i and j are indicators for the genome positions and $\hat{\theta}_{ij}$ is the recombination fraction between the loci located at the two genome positions. The best order is the one with the lowest CI.

The seriation method is simple and needs the least computing among all the algorithms. It is good for hand calculation of locus ordering. However, the disadvantage of the method is that it frequently fails to reach a global optimal solution when the recombination fraction matrix is not monotonic.

Simulated Annealing Algorithm

Simulated annealing is a method developed through analogy to the thermodynamics of liquid crystallization. At high temperatures, molecules of a liquid move freely. As the liquid is cooled slowly, thermal mobility is lost. The atoms are often able to line themselves up and form a pure crystal. This crystal is the state of minimum energy for this system. The essential basis of the process is slow cooling, allowing ample time for redistribution of the atoms as they lose mobility. This is the definition of annealing and annealing is essential for ensuring that a low energy state will be achieved.

The annealing process can be modeled using computer simulation (Metropolis *et al.* 1953). The algorithm is based on Monte Carlo techniques and generates a sequence of states of the solid in the following way: Given a current state i of the solid with energy E_i, then a subsequent state j is generated by applying a perturbation mechanism which transforms the current state into the next state by a small distortion, for instance, by displacement of a particle. The energy of the next state is E_j. If the energy difference $(E_i - E_j) \geq 0$, the state j is accepted as the current state. If $(E_i - E_j) < 0$, the state j is accepted with a probability given by

$$\frac{1}{k_b T} \exp(E_i - E_j) \qquad (9.18)$$

where T is the temperature and k_b is a physical constant called the Boltzman constant. This process is known as the Metropolis algorithm (MA) (Metropolis 1953; Kirkpatrick *et al.* 1983). If the lowering of T is sufficiently slow, the solid can reach thermal equilibrium at each T. In the MA, this is achieved by generating a large number of transitions at a given T value. Thermal equilibrium is characterized by the Boltzman distribution. This distribution gives the probability of a solid being in a state i with energy E_i at T and is

$$P_T(x = i) = \frac{1}{Z(T) k_b T} \exp(E_i - E_j) \qquad (9.19)$$

where x is a stochastic variable denoting the current state of the solid. $Z(T)$ is the partition function, which is defined as

$$Z(T) = \sum \exp\left[-\frac{E_j}{k_b T}\right] \tag{9.20}$$

where the summation extends over all possible states.

The simulated annealing algorithm is an iteration of the MA. In combinatorial optimization, a similar process can be defined. This process can be formulated as the problem of finding a solution with optimum distance, cost or other functions.

The simulated annealing algorithm can be implemented for a gene ordering problem using the statistics explained in the previous section. Here, SARF is used to illustrate the procedures:

(1) Configuration: The l loci in a linkage group are numbered as $a_1, a_2, ..., a_i, ..., a_l$ and the recombination fraction between locus a_i and a_j is $\theta_{a_i a_j}$. A configuration is the permutation of $a_1, a_2, ..., a_l$, interpreted as the order of the loci.

(2) Rearrangements: An efficient set of changes was suggested by Lin for the TSP (Lin 1965). For the gene ordering problem using SARF, similar approaches can be used. A segment is removed and then replaced with the same segment in the opposite order, or a locus is replaced by a randomly chosen one.

(3) Objective function: E_i is the SARF for intermediate gene order of the stage i.

(4) Annealing schedule: T is usually chosen to be larger than the largest changes of E_i from stage to stage. For gene ordering, using SARF $T = 50$ cM will be sufficient. The 'cooling' speed 0.9 has been adequate in a simulation study.

Branch-and-Bound (BB)

BB is a back track algorithm used to reduce search space for the TSP problem. Assume that $a_1, a_2, ...a_{k-1}$ is a partial solution for a gene ordering problem. For a given order, we have

$$SARF(a_1, a_2, ..., a_{k-1}) \leq SARF(a_1, a_2, ..., a_{k-1}, a_k)$$

so $(a_1, a_2, ...a_{k-1})$ can be discarded if its SARF is greater than or equal to the SARF of that previously computed. An H (heuristic function) can be used to further discard partial solutions when H is greater than or equal to the SARF of the previously lowest SARF. The heuristic function is

$$H(a_1, a_2, ..., a_{k-1}) = SARF(a_1, a_2, ..., a_{k-1}) + H(a_k, a_{k+1}, ..., a_l)$$

with

$$H(a_k, a_{k+1}, ..., a_l) \leq min\{c(a_k, a_{k+1}, ..., a_l) | a_i \in S, i = k, ..., l\}$$

where S is the set of unvisited nodes. It is difficult to obtain an optimal value for $H(a_k, a_{k+1}, ..., a_l)$, but a conservative estimate can be computed.

For the minimum of example data as shown in Table 9.6, a set of nodes $(a_1, a_2, ...a_{k-1})$ is visited. SARF for the set of unvisited nodes $(a_k, a_{k+1}, ..., a_l)$ can be estimated based on the distribution listed in Table 9.7. For the gene ordering problem, a maximum of two distance values can be used to estimate $H(a_k, a_{k+1}, ..., a_l)$ for a single locus and this approach can be defined as a "two minimum rule" for the gene ordering problem.

Table 9.6 Recombination fraction matrix for 26 loci of barley chromosome I in %. Data was generated by the North American Barley Genome Mapping Project.

		1	2	3	4	5	6	7
2	ABG313B	11.18						
3	CDO669	16.28	4.58					
4	BCD351D	29.49	19.14	13.34				
5	BCD402B	30.70	20.92	14.03	3.79			
6	TubA1	42.75	31.11	24.23	10.86	10.31		
7	BCD265B	40.61	28.77	22.30	10.36	11.47	0.75	
8	Dhn6	44.20	34.27	27.40	14.48	17.27	7.11	7.97
9	ABG3	44.08	34.61	28.44	14.27	17.96	4.80	7.69
10	WG1026B	46.04	37.46	28.88	16.32	16.79	7.24	8.14
11	Adh4	48.86	36.76	29.83	19.28	18.25	8.02	9.62
12	ABA3	47.58	37.75	30.52	19.01	19.68	8.95	10.68
13	ABG484	48.91	37.85	32.08	20.69	19.83	13.12	14.17
14	Pgk1	47.43	34.55	29.99	17.98	19.37	9.84	11.45
15	ABR315	43.50	33.07	28.33	17.51	19.34	12.21	13.49
16	WG464	41.34	33.08	31.53	27.36	24.18	17.54	19.37
17	BCD453B	44.60	39.98	36.57	32.40	32.80	26.27	27.62
18	ABG472	47.72	42.54	41.34	36.41	37.89	30.59	31.05
19	iAco2	51.05	48.24	49.64	47.92	48.85	47.00	47.98
20	ABG500B	52.15	48.90	48.51	47.82	47.29	43.38	44.35
21	WG114	52.19	48.50	49.21	48.58	49.21	47.78	46.86
22	ABG498	52.19	48.20	49.25	48.95	49.59	46.34	47.36
23	ABG54	52.19	47.10	48.47	47.55	49.23	47.18	46.22
24	ABG394	54.47	50.35	51.14	49.98	52.35	47.71	48.03
25	ABG397	54.00	49.64	49.60	46.52	48.46	47.03	48.10
26	ABG366	57.02	51.90	53.22	50.74	53.27	49.97	48.80

		8	9	10	11	12	13	14
9	ABG3	0.76						
10	WG1026B	2.07	1.50					
11	Adh4	7.03	4.65	2.85				
12	ABA3	6.37	7.38	4.22	2.20			
13	ABG484	9.03	8.33	6.89	4.31	4.96		
14	Pgk1	6.46	6.25	4.31	2.25	2.22	4.31	
15	ABR315	8.08	7.93	7.84	7.35	4.50	7.30	3.01
16	WG464	13.13	14.28	12.30	10.45	11.18	13.87	9.84
17	BCD453B	21.52	23.48	21.37	21.56	21.27	23.12	20.86
18	ABG472	28.06	27.54	26.42	26.46	27.92	26.07	24.05
19	iAco2	42.17	41.46	40.55	38.73	38.18	36.72	39.00
20	ABG500B	41.25	40.46	40.27	36.94	37.11	36.09	38.39
21	WG114	43.88	44.09	42.84	39.54	40.00	39.27	41.04
22	ABG498	44.74	44.27	43.43	39.84	40.70	39.57	41.60
23	ABG54	43.65	43.05	43.36	40.15	41.01	39.43	41.90
24	ABG394	47.31	44.87	43.56	40.29	41.16	39.55	41.80
25	ABG397	44.74	44.26	43.04	40.56	42.13	39.16	41.28
26	ABG366	47.38	46.39	47.51	42.63	43.07	39.84	43.40

		15	16	17	18	19	20	21
16	WG464	7.75						
17	BCD453B	16.78	11.67					
18	ABG472	19.99	17.78	6.52				
19	iAco2	38.11	32.86	23.12	17.73			
20	ABG500B	37.49	30.66	22.90	16.06	3.42		
21	WG114	40.47	33.58	24.28	18.03	7.67	2.16	
22	ABG498	40.72	33.56	27.00	18.83	7.47	2.09	0.00
23	ABG54	41.03	37.54	28.16	22.05	6.89	4.25	2.20
24	ABG394	41.22	37.08	27.16	19.54	4.93	2.88	0.74
25	ABG397	41.47	37.95	27.96	21.88	8.90	7.03	7.11
26	ABG366	42.95	37.80	30.07	24.40	11.02	10.59	8.58

		22	23	24	25
23	ABG54	2.12			
24	ABG394	0.72	1.46		
25	ABG397	4.93	7.67	3.57	
26	ABG366	8.33	9.15	6.97	7.57

Table 9.7 Frequency of map distances used to compute H (heuristic function) for the barley data.

Mid-point (cM)	Frequency
0.5	7
1.5	11
2.5	3
3.5	4
4.5	8
7.5	2
6.5	1
7.5	3
8.5	1
9.5	1
>10	12

For the barley data, for each locus, the two lowest map distances were counted ("two minimum rule"). If a partial solution $(a_1, a_2, ...a_7)$ with a cost of 128 cM and two 1s used in the cost calculation is assumed, then for the 19 unvisited nodes $(a_9, a_{10}, ..., a_{26})$, heuristic function can be estimated as

$$128 + 5 \times 0.5 + 11 \times 1.5 + 3 \times 2.5 = 152$$

Because the estimated lowest cost from simulated annealing was 148, the partial solution was discarded.

A Combination of SA and BB

This strategy includes finding an initial order and minimum SARF by SA and then computing the heuristic function using the features of the recombination matrix and a "two minimum rule". Extensive computer simulations have been done for a number of genes in one linkage group $(l \geq 15)$ to evaluate the effectiveness of the algorithm using SARF as the objective function. For SA alone, the number of genes in one linkage group is practically unlimited, but SA does not guarantee the best solution.

A combination of SA and BB does guarantee the best solution. A gene ordering problem with 500 genes has been solved using the algorithm on a SUN SPARC 10-52 workstation. SA yielded a minimum SARF of 27.09 for a true SARF of 24.98. By sending the results to BB, the optimal solution with a SARF of 24.98 was obtained. The number of genes can be increased by carefully picking an initial order and semi-minimum SARF and BB criteria.

9.4 PROBABILITY OF ESTIMATED LOCUS ORDERS

Sampling and inadequate criteria and algorithms are error sources for determining locus order. Some genome researchers have believed that locus order is unique and not subject to sampling error. They believe that the locus order should be determined by means of molecular biology, such as physical mapping. However, locus order is determined using statistics and data, so locus ordering is subject to sampling error. In addition, locus ordering is subject also to the ordering criteria and the algorithms, due to properties of the statistics used for locus ordering and the computational problems. The question is how to quantify the locus ordering errors or the confidence of locus order. In this section, the likelihood approaches and the nonparametric approaches for measuring the confidence of locus ordering will be discussed.

9.4.1 LIKELIHOOD APPROACH

Likelihood approaches have been used for locus ordering. Intuitively, likelihood ratios might be easily derived and used to measure the relative confidence of an estimated locus order. However, as mentioned by Lathrop *et al.* (1987), interpretation of the likelihood ratio statistic is difficult, due to the lack of a distribution theory that could lead to the calculation of a significance level. Lack of degrees of freedom difference among different locus orders could prevent use of the standard theory of likelihood ratio test. Nevertheless, approaches using likelihood to quantify confidence of locus order have been proposed and used (Edwards 1987; Lathrop *et al.* 1987; Smith 1990; Keats *et al.* 1991).

Log likelihood functions have been defined for three-locus and multiple-locus models. If absence of interference is assumed, the multiple-locus log likelihood ratio test statistic is

$$G = 2 \sum_{i=1}^{l-1} n_{a_i a_j} \{ \hat{\theta}_{a_i a_j} Log [2\hat{\theta}_{a_i a_j}] + [1 - \hat{\theta}_{a_i a_j}] Log [2 - 2\hat{\theta}_{a_i a_j}] \} \qquad (9.21)$$

for locus order $1, 2, 3, ..., l-1, l$ for a set of l loci, $a_1, a_2, ..., a_l$, where $j = i+1$ and n_{ij} is the number of informative sample sizes for locus combination i and j, which may be different from locus to locus due to missing data or missing linkage information for the two adjacent loci. If the log likelihood ratio test statistic for an alternative locus order is G', the log likelihood ratio for the two locus orders is $e^{G-G'}:1$. For example, the likelihood ratio for the three locus orders BAC, ABC and ACB is $2.58 \times 10^{12} : 4447 : 1$ (Table 9.5).

For human genetic linkage analysis, the multiple-locus lod score has been commonly used, which is

$$Z = \sum_{i=1}^{l-1} n_{a_i a_j} \{ \hat{\theta}_{a_i a_j} Log_{10} [2\hat{\theta}_{a_i a_j}] + [1 - \hat{\theta}_{a_i a_j}] Log_{10} [2 - 2\hat{\theta}_{a_i a_j}] \} \qquad (9.22)$$

for the locus order. If lod score for an alternative locus order is Z', the likelihood ratio for the two locus orders is $10^{Z-Z'}:1$. It is said that the order is $10^{Z-Z'}$ times better or worse than the alternative order. The concept of framework mapping, which is based on the principles explained in this section, will be discussed in section 9.5.3.

The drawback of the likelihood approaches is due to the relativeness of the likelihood and the unknown distribution of the log likelihood difference or lod score differences. The relative likelihood ratio can mean only that one order is better than an alternative order. It cannot determine if confidence of the order is high. For example, the likelihood ratio $2.58 \times 10^{12} : 4447 : 1$ for the three locus orders BAC, ABC and ACB means that BAC is the best possible order. However, it does not indicate high confidence for the specific order BAC itself. As with commonly used estimation procedures, variance or probability need to be attached to show the confidence of the estimation even when the estimator is the best and unbiased. Furthermore, the distribution of the likelihood ratio is unknown and the ratio may not mean that the order is better than the alternative in the quantity of the ratio. For a locus ordering problem, the estimate of the locus order may or may not be unbiased due to the criteria and the nature of the algorithms.

9.4.2 BOOTSTRAP APPROACH

Percentage of Correct Gene Order

For a group of genes or genetic markers, there is a linear order for these loci in the genome. If there are 5 genes, G1, G2, G3, G4 and G5 in one linkage group, they should correspond to 5 loci, L1, L2, L3, L4 and L5 in a genome. The objective of gene ordering is to estimate the unknown order. Let us assume that we repeat the experiment to estimate the gene order a number of times. For each of the repeated experiments, we will get an estimated gene order. If we use genome positions or true locus order as column indicators and use the genes as row indicators, we can build a frequency matrix. If we have 100% confidence in the estimated gene order (or we obtain the same gene order for every repeated experiment), then we would observe a matrix with ones on the diagonal and zeroes off the diagonal (Figure 9.1). In other words, all genes are located on their corresponding genome positions in all the repeated experiments. On the other hand, if we do not have any confidence in the estimated gene orders (or we randomly assign the genes to a genome position), then the elements of the matrix are expected to be 0.2 for 5 loci (Figure 9.1). In reality, the matrix may be observed to be between the matrices for 100% confidence and for zero confidence (Figure 9.1).

<table>
<tr><th colspan="6">100% Confidence</th></tr>
<tr><th></th><th>L1</th><th>L2</th><th>L3</th><th>L4</th><th>L5</th></tr>
<tr><td>G1</td><td>1</td><td>0</td><td>0</td><td>0</td><td>0</td></tr>
<tr><td>G2</td><td>0</td><td>1</td><td>0</td><td>0</td><td>0</td></tr>
<tr><td>G3</td><td>0</td><td>0</td><td>1</td><td>0</td><td>0</td></tr>
<tr><td>G4</td><td>0</td><td>0</td><td>0</td><td>1</td><td>0</td></tr>
<tr><td>G5</td><td>0</td><td>0</td><td>0</td><td>0</td><td>1</td></tr>
</table>

<table>
<tr><th colspan="6">Zero Confidence</th></tr>
<tr><th></th><th>L1</th><th>L2</th><th>L3</th><th>L4</th><th>L5</th></tr>
<tr><td>G1</td><td>.2</td><td>.2</td><td>.2</td><td>.2</td><td>.2</td></tr>
<tr><td>G2</td><td>.2</td><td>.2</td><td>.2</td><td>.2</td><td>.2</td></tr>
<tr><td>G3</td><td>.2</td><td>.2</td><td>.2</td><td>.2</td><td>.2</td></tr>
<tr><td>G4</td><td>.2</td><td>.2</td><td>.2</td><td>.2</td><td>.2</td></tr>
<tr><td>G5</td><td>.2</td><td>.2</td><td>.2</td><td>.2</td><td>.2</td></tr>
</table>

$$PCO = (P_{11} + P_{22} + P_{33} + P_{44} + P_{55}) / 5$$

<table>
<tr><th></th><th>L1</th><th>L2</th><th>L3</th><th>L4</th><th>L5</th></tr>
<tr><td>G1</td><td>P_{11}</td><td>P_{21}</td><td>P_{31}</td><td>P_{41}</td><td>P_{51}</td></tr>
<tr><td>G2</td><td>P_{12}</td><td>P_{22}</td><td>P_{32}</td><td>P_{42}</td><td>P_{52}</td></tr>
<tr><td>G3</td><td>P_{13}</td><td>P_{23}</td><td>P_{33}</td><td>P_{43}</td><td>P_{53}</td></tr>
<tr><td>G4</td><td>P_{14}</td><td>P_{24}</td><td>P_{34}</td><td>P_{44}</td><td>P_{54}</td></tr>
<tr><td>G5</td><td>P_{15}</td><td>P_{25}</td><td>P_{35}</td><td>P_{45}</td><td>P_{55}</td></tr>
</table>

Figure 9.1 Frequency matrices using genome positions as column indicators and genes or markers as row indicators.

The mean probability that a gene or marker is located at its corresponding genome position is defined as the percentage of the correct order (PCO). For l loci, the PCO is

$$PCO = \frac{1}{l}\sum P_{ii} \qquad (9.23)$$

where P_{ii} is the probability on the diagonal of the frequency matrix. In practical genomic experiments, repeated experiments are rarely done. The frequency matrix can be estimated using a resampling technique, such as bootstrapping (see Chapter 19).

Repeated bootstrap samples can be drawn from the experiment data for a linkage group. A locus order can be estimated for each bootstrap sample. For t bootstrap gene orders, the probability of locus i being located at position j is P_{ij}. A frequency matrix can be obtained and used to estimate the PCO.

An Example

A linkage group which has 8 loci is used to illustrate the methods of obtaining PCO using a bootstrapping approach. Table 9.8 shows results of gene ordering and multi-point analysis (see Chapter 10 for multi-point analysis). The best order estimated from the data is

LOCUS 1, LOCUS 2, ..., LOCUS 8

The log likelihood for the order is -667.22.

Table 9.8 A linkage map for a linkage group with 8 loci. (PGRI output)

Locus (code)	2-R	ML-R		H-D		K-D	
		Point	Sum	Point	Sum	Point	Sum
LOCUS 1 (1)	0.00	0.000	0.000	0.000	0.000	0.000	0.000
LOCUS 2 (9)	0.17	0.167	0.167	0.204	0.204	0.174	0.174
LOCUS 3 (10)	0.17	0.155	0.322	0.185	0.389	0.160	0.334
LOCUS 4 (2)	0.17	0.173	0.496	0.213	0.602	0.181	0.515
LOCUS 5 (5)	0.26	0.266	0.762	0.380	0.982	0.296	0.812
LOCUS 6 (6)	0.23	0.224	0.986	0.297	1.279	0.241	1.053
LOCUS 7 (13)	0.16	0.160	1.146	0.193	1.473	0.166	1.219
LOCUS 8 (14)	0.15	0.154	1.301	0.185	1.657	0.160	1.379

code = the rank of the locus in the original data set
2-R = two-point recombination fraction to the next adjacent locus
ML-R = maximum likelihood multipoint recombination fraction
H-D = Haldane map distance
K-D = Kosambi map distance
Point = point estimator
Sum = summation start at locus one
Output from PGRI computer software

Table 9.9 Ordered recombination fractions for gene order in Table 9.8. (PGRI output)

Locus	1	2	3	4	5	6	7
LOCUS 2	0.17						
LOCUS 3	0.25	0.17					
LOCUS 4	0.33	0.25	0.17				
LOCUS 5	0.48	0.44	0.38	0.26			
LOCUS 6	0.49	0.50	0.46	0.35	0.23		
LOCUS 7	0.52	0.49	0.51	0.43	0.31	0.16	
LOCUS 8	0.51	0.50	0.55	0.49	0.37	0.28	0.15

Table 9.9 shows the pairwise recombination fraction matrix. The values on the diagonal are the recombination fractions between two adjacent loci. For example, the recombination fraction between LOCUS 1 and LOCUS 2 is 17% and the loci are adjacent to each other. If the pairwise recombination fraction matrix is monotonic, then the recombination fraction on the diagonal should be the smallest value among the values on the same row or the same column and should increase as it moves away from the diagonal. If the recombination fraction matrix follows the triangle inequality, then the relationship among the

recombination fractions on any triangle component in the matrix should not be significantly different from

$$r_{AC} = r_{AB} + r_{BC} - 2r_{AB}r_{BC}$$

where loci A, B and C are in the order of ABC, recombination fractions between A and B and between B and C are on the diagonal and recombination fraction between A and C is off the diagonal. For example, recombination fraction between LOCUS 1 and LOCUS 2 and between LOCUS 2 and LOCUS 3 is 0.17 and between LOCUS 1 and LOCUS 3 is 0.25. The predicted recombination fraction between LOCUS 1 and LOCUS 3 based on the triangle inequality and the recombination fractions between LOCUS 1 and LOCUS 2 and between LOCUS 2 and LOCUS 3 is $0.17 + 0.17 - 2 \times 0.17 \times 0.17 = 0.28$, which differs from the observed value 0.25. Hypothesis testing on the difference can be implemented using the log likelihood ratio test statistics.

Table 9.10 Bootstrapping results for the linkage group with 8 loci. The frequencies were obtained from 100 bootstrap replications. If the confidence of the gene order is 100%, then the count should be located on the diagonal (the locus is always located at the same genome position). (The result was obtained using PGRI.)

Locus (Code)	Genome Position							
	1	2	3	4	5	6	7	8
1	97	0	0	3	0	0	0	0
9	0	97	3	0	0	0	0	0
10	0	3	96	1	0	0	0	0
2	3	0	1	96	0	0	0	0
5	0	0	0	0	96	0	1	3
6	0	0	0	0	0	96	3	1
13	0	0	0	0	0	4	96	0
14	0	0	0	0	4	0	0	96

Table 9.11 Bootstrapping results for simulated data according to different sample sizes. The original sample size for the data set is 150. As the sample size decreases, the confidence on estimated gene order decreases. (Output from PGRI computer software)

Sample Size	Genome Position	Individual Locus							
		1	2	3	4	5	6	7	8
100	1	89			10	1			
	2		89	10	1				
	3	1	10	88	1				
	4	10	1	1	88				
	5				1	90	1		8
	6						90	9	1
	7						8	91	1
	8						9	1	90
50	1	73	10	1	14	2			
	2	13	73	10	4				
	3	3	14	74	8		1		
	4	11	2	13	65			1	8
	5		1	1		52	14	13	19
	6			1		4	53	29	13
	7					2	9	29	42	18
	8					7	33	3	15	42

Table 9.10 shows the results from the bootstrap analysis for the data shown in Tables 9.8 and 9.9. If the data fits the gene order (1, 9, 10, 2, 5, 6, 13 and 14) perfectly and the sample size is sufficiently large, then the same gene order should be obtained for every bootstrap replication. The confidence on the gene order is 100% when the percentile scores are all located on the diagonal (the locus is always located at same genome position). The scores that are off-diagonal represent the frequencies of those markers placed at wrong genome positions due to sampling error. For example, locus 1 was located at the same genome position 97 times out of 100 replications and 3 times at a different genome position.

Sample Size and PCO

It is intuitive that the larger the sample size, the higher the PCO. To see how sample size affects the confidence of estimated gene order, data for the 8 loci were simulated according to the gene order for two sample sizes, 100 and 50 (Table 9.11). The simulation was done for 100 independent replications. For a sample size of 100, the confidence on the estimated gene order is 90% and it is 60% for a sample size of 50. The confidence for the estimated gene order decreases quickly as the sample size decreases.

The confidence for an estimated gene order can be shown graphically by plotting the frequency matrix against the genome position and individual locus number (Figure 9.2). Probability surfaces between the gene and its

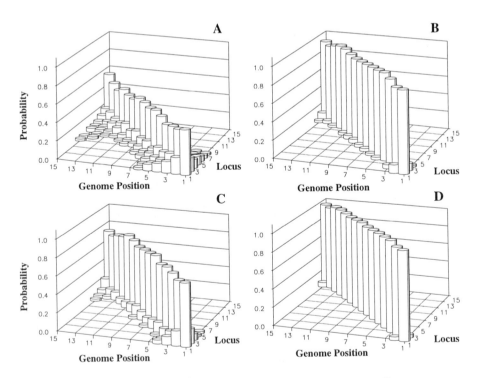

Figure 9.2 Probability surfaces between the gene and its corresponding genome locations from the 1,000 simulated gene orders. The number of genes in a linkage group is 15. A. Gene spacing in cM 1, 1, 1, ... and population size is 100; B. Gene spacing in cM 1, 1, 1, ... and population size is 500; C. Gene spacing in cM 5, 5, 5, ... and population size is 50; and D. Gene spacing in cM 5, 5, 5, ... and population size is 200.

corresponding genome locations from the 1,000 simulated gene orders can be plotted for some combinations of gene spacing and population size. The higher the cylinders on diagonals, the more confidence the gene order has. Plots A and C show relative low confidence and plots B and D show high confidence. When genes are closely linked (plots A and B, 1,1,1,..., 1cM apart), a large sample size is needed to obtain a gene order with high confidence.

9.4.3 INTERVAL SUPPORT FOR LOCUS ORDER

For some applications of genomic mapping, precise gene order and high resolution of the linkage map are not necessary. However, for some applications, such as the cloning of human disease genes, both precise gene order and high resolution of the linkage map are needed (Boehnke 1994). To increase the precision of an inferred gene order, a framework map approach was suggested by Keats *et al.* (1991) and has been used as a guideline for evaluating the degree of certainty of the estimated gene orders. The framework map approach is useful in practice and its intuitive interpretation is simple. However, precise interpretation of a framework can be limited if the nature of the likelihood of the gene order is considered. Interpretation of a nonparametric confidence interval of gene order (precise possible genome positions of an individual locus) can be simple and the approach can be effective.

Framework Map

Loci with interval support of at least lod 3 were defined as framework loci and the linkage map was defined as a framework map. The interval support is defined as a base 10 logarithm of a likelihood ratio

$$Log_{10}(L_1/L_2)$$

where L_1 is the likelihood with the locus in a particular interval of the linkage map and L_2 is the highest likelihood, placing the locus in any other interval. It is said that the gene order for the framework map is 1,000 times more likely than is any other order.

In practice, framework maps have been achieved by dropping loci which do not meet the lod score 3 support interval. Problems associated with this approach are:

(1) If the starting gene order is a local optimum, then there is a small chance of recovering a gene order with global optimum when a large number of genes are in the linkage group.

(2) For the same data set, different framework maps may be produced by using different locus dropping strategies.

(3) The argument that the lod score support 3 means that the gene order of the framework map is 1,000 times more likely than any other gene order may need further study, because the structure of the likelihood ratio is not nested and has 0 degrees of freedom.

Even though the framework map approach has not been rigorously studied, it may be adequate for many practical applications. But it should be noted that the confidence (or lod score support) for the gene order is for the framework loci only and not for the complete data set. The framework approach will reduce the resolution of the linkage map. A large sample size is needed to ensure a precise gene order and a high resolution of a linkage map. New methods of gene ordering may be needed to achieve the most precise gene order for a given amount of data.

Confidence Interval for Gene Order

For t bootstrap gene orders, the probability that the i locus is located at the j position of the original gene order is P_{ij}. There is also a vector with l elements corresponding to each genome position for each of the l genes and this can be represented by a $[l \times l]$ matrix. The most likely genome position (or linkage map location) for the i gene will be the position with the highest probability. For the whole genome, the order (or map location for every gene) corresponding to a rank of the matrix with maximum trace, which is the summation of the diagonal elements at maximum, is the most likely gene order for the data. In a graphic representation, the most likely gene order corresponds to the peak of a three dimensional surface with axes of gene identifications, map locations and probabilities. A bootstrap confidence interval can be constructed by counting a symmetric percentile from the diagonals. For example, a 95% confidence interval of possible map locations for gene i is the map location corresponding to the 2.5 percentile and the 97.5 percentile for the probability vector.

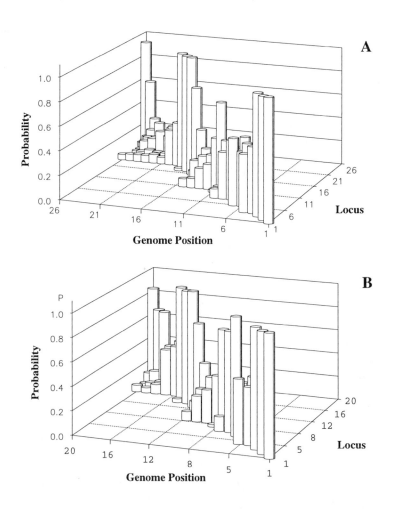

Figure 9.3 Probability graphs of specific loci and their corresponding linkage map locations from the 1,000 bootstrap gene orders for the barley chromosome IV. (A). 26 loci and (B). 20 loci.

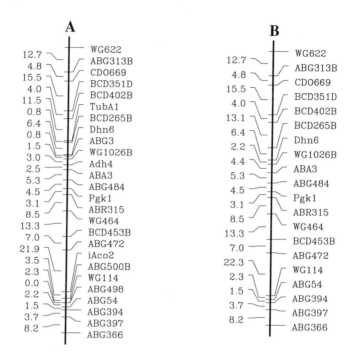

Figure 9.4 Linkage map of barley chromosome IV with 26 (A) and 20 (B) loci generated from data provided by the North American Barley Genome Mapping Project (NABGMP). For this data set, 1,000 bootstrap samples were drawn.

An Example

Figure 9.3 shows the probabilities for specific marker loci and their corresponding linkage map locations from the 1,000 bootstrap locus orders for the barley chromosome IV, using data provided by the North American Barley Genome Mapping Project (NABGMP). In the figure, A has 26 loci on the chromosome and B has 20 loci, which is the result of discarding 6 loci with unfavorable effects on gene order. The graph shows the order with the highest agreement among the bootstrapped orders. For the 26 loci, the mean confidence of correct gene order was 57%. By dropping 6 loci with low confidence, the mean confidence for the 20 loci increased to 72%. The discarded loci are TubA1, ABG3, Adh4, iAco2, ABG500B and ABG498. The relative positions for the remaining loci are the same for the both linkage maps (Figure 9.4).

Table 9.12 shows the percentage of correct map positions and confidence intervals of the map locations for the barley chromosome IV with 26 loci and 20 loci. In this table, P_{ii} is the probability of locus i located at map position i, and 95% interval means that among the 1,000 bootstrap gene orders, the locus was located at the map position interval 95% of the time. For example, locus WG622 is located at the genome position one 99% of the time in the 26 loci map. So, the 95% confidence interval for the locus is genome position one. Locus BCD265B is located at the genome position seven 54% of the time and at the genome position six 44%. Therefore, the 95% confidence interval for BCD265B is from genome positions six to seven. When locus TubA1 is discarded from the data, BCD265B is located at the same genome position 98% of the time. As the confidence of the estimated gene order increases, the range of the confidence interval decreases.

Table 9.12 Percentage of correct map position (P_{ii}) and confidence intervals of the possible map locations for the barley chromosome IV with 26 loci and 20 loci.

Locus Name	26 Loci Linkage Map		20 Loci Linkage Map	
	P_{ii}	95% Interval	P_{ii}	95% Interval
WG622	0.99	1 - 1	1.0	1 - 1
ABG313B	0.99	2 - 2	0.98	2 - 2
CDO669	0.98	3 - 3	0.99	3 - 3
BCD351D	0.53	4 - 5	0.53	4 - 5
BCD402B	0.53	4 - 5	0.53	4 - 5
TubA1	0.54	6 - 7		
BCD265B	0.54	6 - 7	0.98	6 - 6
Dhn6	0.48	8 - 11	0.83	7 - 8
ABG3	0.46	8 - 11		
WG1026B	0.75	8 - 13	0.81	7 - 9
Adh4	0.43	11 - 15		
ABA3	0.27	11 - 15	0.40	9 - 12
ABG484	0.29	11 - 15	0.36	9 - 12
Pgk1	0.42	11 - 15	0.44	9 - 12
ABR315	0.75	11 - 15	0.75	9 - 12
WG464	0.99	16 - 16	0.99	13 - 13
BCD453B	0.98	17 - 17	0.97	14 - 14
ABG472	0.98	18 - 18	0.97	15 - 15
iAco2	0.36	19 - 24		
ABG500B	0.31	19 - 25		
WG114	0.20	19 - 25	0.43	16 - 19
ABG498	0.21	19 - 25	0.40	16 - 19
ABG54	0.27	19 - 25		
ABG394	0.28	19 - 25	0.67	16 - 19
ABG397	0.58	19 - 25	0.66	17 - 20
ABG366	0.91	25 - 26	0.82	18 - 20

Combination of Jackknife and Bootstrap

In practical linkage map construction, strategies such as varying locus positions in a small genome region to obtain gene order with local maximum likelihood and examining pairwise recombination fractions have been used to obtain gene order with high confidence support. These strategies are effective for limited situations because a low confidence support of gene order for a locus defined by those approaches may not be caused by the locus itself. Order is a complex statistic. Many loci have effects on others to some extent and can influence results of locus ordering. A combination of the jackknife and the bootstrap approaches can be used to calculate locus order.

Table 9.13 Jackknifing results for the linkage group with 8 loci (see Table 9.8).

Locus (code)	Chi	Missings	PCO
LOCUS 1 (1)	7.120	0	1.000
LOCUS 2 (9)	7.780	0	1.000
LOCUS 3 (10)	3.380	0	0.914
LOCUS 4 (2)	2.000	0	0.943
LOCUS 5 (5)	1.280	0	0.886
LOCUS 6 (6)	0.320	0	0.943
LOCUS 7 (13)	0.020	0	1.000

"Chi" = chi-square for single-locus segregation ratio test
"Missings" = number of missing values
"PCO(%)" = average percentage of correct gene order after the locus is discarded
(Output from PGRI computer software)

Table 9.13 shows results from jackknifing analysis of the linkage group with 8 loci (see Table 9.8). The jackknife procedure combined with bootstrapping is used for evaluating the impact of a locus on overall gene order estimation. PCO in Table 9.13 is estimated for each locus using a bootstrap procedure when the locus is discarded from the data (jackknifing). If the PCO of the estimated gene order increases significantly when a locus is discarded from the data, then the locus has a "bad" effect on the confidence of estimated gene order. On the other hand, if the PCO of the estimated gene order decreases significantly when a locus is discarded from the data, then the locus is essential for obtaining a high confidence gene order.

Table 9.14 Two-Locus Model (P = p-value, 0.00 = <0.005, R = recombination fraction).

Locus		MWG836	MWG8511	RISIC101	RISP103	Wx	iEst5	ABC1511
MWG8511	P	0.47						
	R	0.46						
RISIC101	P	0.62	0.00					
	R	0.47	0.06					
RISP103	P	0.13	0.71	1.00				
	R	0.40	0.48	0.53				
Wx	P	0.01	0.00	0.00	1.00			
	R	0.35	0.19	0.11	0.53			
iEst5	P	0.03	0.00	0.00	1.00	0.00		
	R	0.37	0.17	0.09	0.55	0.01		
ABC1511	P	0.00	0.01	0.00	1.00	0.00	0.00	
	R	0.26	0.33	0.25	0.51	0.14	0.16	
ABC156D	P	0.00	0.40	0.90	0.01	0.09	0.15	0.04
	R	0.18	0.45	0.49	0.34	0.40	0.41	0.38
ABC1671	P	0.00	0.18	0.06	0.70	0.00	0.00	0.00
	R	0.17	0.42	0.38	0.48	0.27	0.28	0.12
ABC3102	P	0.04	0.47	0.80	0.00	0.81	1.00	0.81
	R	0.37	0.46	0.48	0.05	0.49	0.50	0.49
ABC455	P	0.01	0.07	0.26	0.00	0.05	0.09	0.90
	R	0.34	0.39	0.43	0.22	0.39	0.40	0.49
ABG380	P	0.00	0.22	0.04	0.61	0.00	0.00	0.00
	R	0.18	0.43	0.38	0.47	0.26	0.28	0.12
ABG461	P	0.39	0.33	0.80	0.00	0.81	0.63	1.00
	R	0.45	0.44	0.48	0.18	0.49	0.47	0.51
ABG476	P	0.01	0.04	0.26	0.00	0.07	0.12	1.00
	R	0.35	0.38	0.43	0.24	0.39	0.41	0.51
Brz	P	0.00	0.22	0.80	0.13	0.05	0.09	0.01
	R	0.12	0.43	0.48	0.40	0.38	0.40	0.35
Locus		ABC156D	ABC1671	ABC3102	ABC455	ABG380	ABG461	ABG476
ABC1671	P	0.00						
	R	0.31						
ABC3102	P	0.00	0.54					
	R	0.29	0.46					
ABC455	P	0.00	0.54	0.00				
	R	0.19	0.46	0.18				
ABG380	P	0.00	0.00	0.39	0.63			
	R	0.32	0.01	0.45	0.47			
ABG461	P	0.05	0.46	0.00	0.01	0.39		
	R	0.38	0.45	0.21	0.35	0.45		
ABG476	P	0.00	0.71	0.00	0.00	0.81	0.03	
	R	0.19	0.48	0.19	0.01	0.49	0.37	
Brz	P	0.00	0.00	0.02	0.00	0.00	0.18	0.00
	R	0.09	0.26	0.36	0.26	0.27	0.42	0.25

EXERCISES

Exercise 9.1

The following table lists pairwise two-point recombination fraction (R) and the significant p-value (P) among 15 loci (Table 9.14), using this conditional statement:

"If $\{\,[R_{ij} \le c\,]\,and\,[P_{ij} \le b]\,\}$ then loci i and j belong to same linkage group"

as criteria to group the 15 loci. Pay attention to be number of linkage groups, the sizes of the linkage groups and possible false linkages. If a single significant two-point recombination fraction joins two distinct linkage groups, then it is possible that the significant two-point recombination fraction is a false linkage.

(1) c = 0.2 and b = 1.0

(2) c = 0.2 and b = 0.005

(3) c = 0.5 and b = 0.005

(4) c = 0.3 and b = 0.001

(5) Discuss the relationship between recombination fraction and the significant p-value.

Exercise 9.2

Observed genotypic frequency in progeny of a triple backcross is listed in Table 9.15. Of three loci A, B and C, each has two alleles, denoted by the upper and lower cases of the corresponding letters. Genotypes of the two parents are $AbC/aBc \times abc/abc$. Do the following using the data in the table.

(1) Can you identify the most possible locus order by looking only at the frequencies?

(2) Estimate pairwise recombination fraction among the three loci.

Table 9.15 Genotypic frequencies in a triple backcross progeny set.

Genotype	Observed Frequency
ABC/abc	1
ABc/abc	2
AbC/abc	48
Abc/abc	9
aBC/abc	1
aBc/abc	35
abC/abc	4

(3) Determine the most probable locus order using the pairwise recombination fractions.

(4) Estimate the coefficient of coincidence for the three possible locus orders.

(5) Estimate the log likelihood for the three possible locus orders using

        a) a complete crossover interference model ($c = 0$)

        b) absence of crossover interference ($c = 1$)

c) positive crossover interference ($c \leq 1$)

d) no restriction on crossover interference ($c = C$),

where c is the coefficient of coincidence.

(6) Compute the probability of double-crossovers, assuming absence of crossover interference for the true locus order ABC. Is one observed double-crossover event in the 100 observations expected or unexpected?

Exercise 9.3

In this chapter, we have discussed confidence for the estimated locus order. Lod score (which is the base ten log of the likelihood ratio) difference between two locus orders has been commonly used to determine the confidence of the estimated locus order. The lod score difference between two locus orders can be estimated using

$$LodD = Log_{10}(L_1/L_2)$$

where L_1 and L_2 are the likelihoods for the two alternative locus orders. The interpretation of the lod score difference is that the locus order 1 is 1,000 times more likely than the locus order 2 if the lod score difference is 3 (if $LodD = 3$ then $L_1/L_2 = 1000$). If the lod score difference between the most likely locus order and any other locus orders is equal to or greater than 3 by altering the position of a single locus, the locus is defined as a framework locus. If every locus is a framework locus, then the locus order has a lod score support of 3 and the linkage map with the locus order is a framework map. Another way to estimate the confidence of a locus order is using the nonparametric approach, specifically the bootstrap approach. Bootstrap samples can be drawn from the observed data. A locus order can be estimated from each of the bootstrap samples. If there is a high probability that every locus is optimally located based on bootstrapping, then the entire locus order has high confidence. Do the following problems using the data in Exercise 9.2.

(1) Estimate the lod score differences between order ABC and orders ACB and BAC using

a) a complete crossover interference model ($c = 0$)

b) absence of crossover interference ($c = 1$)

c) positive crossover interference ($c \leq 1$)

d) no restriction on crossover interference ($c = C$),

where c is the coefficient of coincidence).

Interpret the results in terms of confidence for locus order ABC.

(2) Draw 10 bootstrap samples from the original data using the following hints. Estimate the most likely locus order for each of the bootstrap samples using the simple double-crossover approach. How many times is the estimated locus order ABC? What do you expect if you draw 1,000 bootstrap samples?

Hints on Bootstrapping: The cumulative probability (p_i) for the genotypes is listed in Table 9.19. Define f_i as the frequency of genotype i in a bootstrap sample. Draw 100 random numbers from uniform (0, 1). Comparing the random numbers (x) with the cumulative probability, if $p_{i-1} < x \leq p_i$, then a genotype i is drawn and one is added to the frequency of f_i. The 100 random numbers correspond to 100 genotypes and provide one bootstrap sample.

Table 9.16 Cumulative probability for the bootstrap sample.

Genotype	Observed Frequency	Cumulative Probability (Pi)
ABC/abc	1	0.01
ABc/abc	2	0.03
AbC/abc	48	0.51
Abc/abc	9	0.60
aBC/abc	1	0.61
aBc/abc	35	0.96
abC/abc	4	1.00

(3) How many genotypes resulting from the double-crossovers are sampled in the 10 bootstrap samples? Are those double-crossover events expected or unexpected?

(4) Compare the advantages and disadvantages of the bootstrap approach with the lod score approach to infer confidence of the estimated locus order using the results of this exercise and of the Exercise 9.2.

Exercise 9.4

Table 9.17 lists pairwise recombination fractions (R), the significant p-value (P), number of recombinants (Nr) and number of non-recombinants (Nn) among 6 loci in one linkage group.

Table 9.17 Two-locus analysis.

Locus		MWG836	ABC1511	ABC156D	ABC1671	ABG380
ABC1511	P	0.00				
	R	0.26				
	Nr,Nn	26,74				
ABC156D	P	0.00	0.04			
	R	0.18	0.38			
	Nr,Nn	18,82	42,68			
ABC1671	P	0.00	0.00	0.00		
	R	0.17	0.12	0.31		
	Nr,Nn	17,83	15,110	35,78		
ABG380	P	0.00	0.00	0.00	0.00	
	R	0.18	0.12	0.32	0.01	
	Nr,Nn	18,82	13,95	32,68	1,99	
Brz	P	0.00	0.01	0.00	0.00	0.00
	R	0.12	0.35	0.09	0.26	0.27
	Nr,Nn	12,88	40,74	9,91	32,91	34,92

P = p-value
0.00 = <0.005
R = recombination fraction
Nr = number of recombinants
Nn = number of non-recombinants
N = Nr + Nn = total number of informative samples

(1) Obtain the most probable locus order using the seriation approach.

(2) Compute the base 10 log likelihood, sum of the adjacent recombination

fractions, and sum of the adjacent lod scores for the following orders:

```
ABC1511 - ABG380 -  ABC1671 - MWG836 - Brz -      ABC156D
ABG380 -  ABC1511 - ABC1671 - MWG836 - Brz -      ABC156D
ABC1511 - ABC1671 - ABG380 - MWG836 - Brz -       ABC156D
ABC1511 - ABG380 -  MWG836 -  ABC1671 - Brz -     ABC156D
ABC1511 - ABG380 -  ABC1671 - Brz -    MWG836 -   ABC156D
ABC1511 - ABG380 -  ABC1671 - MWG836 - ABC156D - Brz
```

(3) If ABC1511-ABG380-ABC1671-MWG836-Brz-ABC156D is the most probable locus order, can you infer the confidence of the order from the likelihoods computed in (2)?

(4) The following illustration shows a summary result of an estimated best locus order from 1000 replications of bootstrapping. The frequency is the number of times among the 1000 replications that the locus is placed at that position. For example, locus ABC1511 is at the first position in all 1000 bootstrap replications and locus ABG380 is placed at the second position 750 times and at the third position 250 times. Estimate the average percentage of the correct locus order and the approximate 99% confidence intervals for the location of the 6 loci.

```
                  .12  .01    .17         .12      .09

                  ABC1511  ABG380  ABC1671  MWG836  Brz  ABC156D

        ABC1511   1000     0       0        0       0      0
        ABG380    0        750     250      0       0      0
        ABC1671   0        250     750      0       0      0
        MWG836    0        0       0        980     15     5
        Brz       0        0       0        12      982    6
        ABC156D   0        0       0        8       3      989
```

Ex. 9.4 (4)

(5) Compare the advantages and disadvantages of the bootstrap approach with the lod score approach to infer the confidence of the estimated locus order.

(6) After the best locus order is obtained, the pairwise recombination fraction can be rearranged according to the locus order. For the 6 loci, the ordered recombination fraction matrix is shown in Table 9.18. Can you find evidence of the crossover interference from the data?

Table 9.18 Two-Locus Model (Pairwise recombination fractions).

Locus	ABC1511	ABG380	ABC1671	MWG836	Brz
ABG380	0.12				
ABC1671	0.12	0.01			
MWG836	0.26	0.18	0.17		
Brz	0.35	0.27	0.26	0.12	
ABC156D	0.38	0.32	0.31	0.18	0.09

(7) Explain why the most likely locus orders obtained using maximum likeli-hood, minimum sum of adjacent recombination fractions and the seria-tion approach are the same. Do you expect that this result is always true? Why?

Exercise 9.5

Obtain the most probable locus order using a seriation approach and the pairwise recombination fractions in Table 9.19. Use the first three loci and the last three loci as starting points for the algorithm. Answer the following questions.

(1) If you find that all locus orders obtained using different starting points are the same, explain why. If you find that some of the locus orders are obtained using different starting points are different, explain why.

(2) What are the advantages and disadvantages of the seriation algorithm for locus ordering?

(3) What factors influence the quality of an estimated locus order?

Table 9.19 Two-point recombination fraction.

Locus	MWG934	Nir	Amy1	MWG820	ABC163	ksuD17
Nir	0.09					
Amy1	0.16	0.08				
MWG820	0.25	0.20	0.11			
ABC163	0.28	0.24	0.15	0.04		
ksuD17	0.31	0.25	0.19	0.05	0.02	
ABG458	0.33	0.30	0.25	0.17	0.13	0.12
MWG916	0.40	0.38	0.30	0.24	0.22	0.21
ABG387B	0.37	0.31	0.31	0.25	0.22	0.19
cMWG652A	0.40	0.41	0.38	0.35	0.33	0.32
ABG378	0.49	0.49	0.49	0.45	0.42	0.43
Nar1	0.49	0.48	0.48	0.46	0.45	0.46
MWG663B	0.51	0.48	0.50	0.47	0.47	0.47
ABG466	0.49	0.48	0.49	0.47	0.45	0.47
PSR167	0.50	0.50	0.52	0.51	0.51	0.53
Locus	ABG458	MWG916	ABG387B	cMWG652A	ABG378	MWG663B
MWG916	0.14					
ABG387B	0.16	0.07				
cMWG652A	0.30	0.20	0.16			
ABG378	0.40	0.34	0.29	0.13		
MWG663B	0.45	0.41	0.34	0.17	0.05	
ABG466	0.46	0.37	0.33	0.18	0.06	0.01
PSR167	0.51	0.47	0.40	0.26	0.12	0.07
Locus	ABG466					
PSR167	0.06					

CHAPTER 10

MULTI-LOCUS MODELS

In previous chapters, methods for detecting and estimating recombination fraction in a two-locus model have been discussed. One of the most important objectives of estimating recombination fractions is to make a genetic map. Ideally, a genetic map of each linkage group will correspond to one of the chromosomes. Traditionally, a linkage relationship among the genes is quantified using the recombination fraction between a pair of loci. However, the two-locus recombination fraction may not be efficient for more than two loci. The two-locus recombination fraction is not usually additive, even when the loci are linearly arranged on a genetic map.

Multi-locus models, discussed in this chapter, differ from two-locus models because information other than the interval between two loci is taken into consideration. For example, information on intervals between loci A and B and between B and C will be included in estimating map distance between A and C if the locus order for the three loci is ABC.

10.1 INTERPRETATION OF MAP DISTANCE

10.1.1 MAP AND PHYSICAL DISTANCES

The relationship between a multi-point map distance and physical distance is a confused one. The physical distance between two genes can be quantified using the number of DNA base pairs between the two genes. The multi-point map distance is based on a statistical estimate of crossover and has more in common with recombination fraction than with the physical distance. For most cases, map distance is a transformation of the recombination fraction. If map distance is additive among multiple segments, this additivity is rarely true for the commonly used mapping functions. On the average, any specific organism has an average ratio between physical and map distances. For example, one cM (centi-Morgan) of map distance corresponds to approximately 1,100 kb of nucleotides in human males. The ratio differs from species to species and can be different in the sexes of the same species. Even within a single species, the ratio could differ greatly according to genome location. For example, the observation of crossover hot spots means that crossover events do not occur randomly in the genome according to physical distance. It is well known that the frequency of crossover in the centromere region may differ greatly from that in the other parts of the chromosome.

The relationship between map distance or recombination fraction and physical distance may vary greatly at the genome segment level. This lack of clarity has created problems for physical distance-based applications which rely on information from linkage-based analysis. For example, gene cloning based on map distance between the target gene and a known marker or sequence (called map-based cloning) may fail when the ratio between physical distance and map distance is incorrectly estimated or when the estimate of the map distance is not accurate.

0-8493-3166-8/97/$0.00+$.50

Despite the lack of direct correspondence between physical and map distance, many applications of the physical and map distances are independent of the relationship. Map distance and other crossover based statistics are used mainly in statistics-based applications, such as individual identification or prediction of performance based on linkage relationships. These applications have no relationship with the physical distances among genes. For some physical distance-based applications, such as gene cloning or physical map construction, recombination fraction may be irrelevant. In cases like these, establishing the relationship between map distance and physical distance is not necessary. However, as genomic information accumulates, such as the complete DNA sequence of an organism and physical and genetic maps, the relationship between map distance and physical distance will be readily calculated for each segment of a genome. Many statistical problems related to genomics, such as gene ordering, discussed in Chapter 9, and the multi-locus analysis, will be solved or partially solved as genomic information accumulates.

The inconsistent relationships between genetic distance and physical distance do not create a problem for linkage analysis and genetic map construction. For example, 7 genetic markers are shown that correspond to 7 restriction sites on a genome segment (Figure 10.1). The seven markers may be evenly spread on the genetic map but not on the physical map (Figure 10.1A). They also could be spread evenly on a physical map but not on the genetic map (Figure 10.1B). Those relationships do not influence the analysis based only on physical distance or crossover. However, we do expect the order to be colinear in both types of maps.

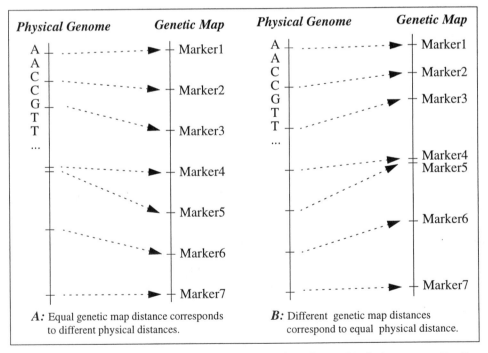

Figure 10.1 Two hypothetical situations to show the relationship between genetic distance and physical genome location.

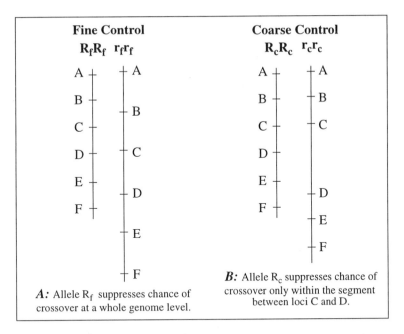

Fine Control
R_fR_f r_fr_f

A ┼   ┼ A

B ┼
          ┼ B
C ┼

D ┼   ┼ C

E ┼
        ┼ D
F ┼

        ┼ E

        ┼ F

A: Allele R_f suppresses chance of crossover at a whole genome level.

Coarse Control
R_cR_c r_cr_c

A ┼   ┼ A

B ┼   ┼ B

C ┼   ┼ C

D ┼

E ┼
        ┼ D
F ┼
        ┼ E

        ┼ F

B: Allele R_c suppresses chance of crossover only within the segment between loci C and D.

Figure 10.2 Two hypothetical situations of genetic control of crossover frequency.

10.1.2 POSSIBLE GENETIC CONTROL OF CROSSOVER

The greatest challenge to the commonly used crossover-based methods comes from the factors influencing the frequency of crossover. Crossover is influenced by both genetics and environment. This challenge is especially true when the parents used to generate progeny for linkage analysis are heterogeneous for the factors influencing chance of crossover.

Genetic recombination was found to be affected by several environmental agents, such as temperature (Bridges 1915; Levine 1955; Suzuki 1965; Simchen and Stamberg 1969). Gowen (1919) first found that recombination in *Drosophila* was influenced by the genetic background. Great variability of genetic recombination fraction was found in maize (Stadler 1925). The recombination fractions between the Hor 1 and Hor 2 loci on the short arm of chromosome 5 were found to be significantly different among three varieties of barley (Sall 1990 and 1991). In *Petunia hybrida,* a recombination modulator gene (Rm1), which acts on the seven chromosomes of the species, was found on chromosome II. The Rm1 gene often increases recombination, especially for strongly linked genes, but its effects vary according to the chromosome segments studied. In *E. coli,* many mutations have been found to have effects on recombination (Hastings 1992; Lanzov *et al.* 1991; Takahashi 1992). In *C. elegans,* a recessive mutation, rec-1, was found to increase recombination fraction at least threefold higher than that found in the wild type. Experiments of two-direction selection on recombination fraction have been done on lima bean (Allard 1963) and *Drosophila* (Chinnici 1971; Kidwell 1972a and b; Cederberg 1985; Charlesworth and Charlesworth 1985a and b). Recombination fraction changes significantly under bidirectional selection and the results further suggest that the system of genetic control of recombination is polygenic in nature.

Figure 10.2 shows two hypothetical situations of genetic control of crossover. Fine control of crossover is defined as the frequency of crossover at a whole genome level being affected by a gene or a number of genes (Figure 10.2A). The genetic effect on crossover rate is evenly distributed on the genome. For example, gene R_f, with alleles R_f and r_f, has an effect on crossover rate. If parental genotype is $R_f R_f$, then the frequency of a recombination event being in the progeny is significantly reduced at a whole genome level. Another form of genetic control of crossover rate is coarse control, which is defined as the frequency of crossover in a specific segment or several segments being affected by a gene or a number of genes (Figure 10.2B). For example, gene R_c, with alleles R_c and r_c, has an effect on the frequency of crossover in a segment between markers C and D (Figure 10.2B). If parental genotype is $r_c r_c$, then the frequency of a recombination event between markers C and D is greatly increased.

When two parents have different genotypes for the genes controlling crossover rate, the methodologies currently used in linkage analysis and map construction are not adequate. New methods taking the genotypes into the model of analysis are needed. When two parents have the same genotype, the methodologies are adequate. However, the interpretation of the estimated recombination fraction is not straightforward. In any case, caution is needed when there is evidence of genetic control of crossover.

Table 10.1 Expected genotypic frequency for loci A and B in progeny of two heterozygous parents (AaBb X AaBb). The recombination fraction between loci A and B in one parent (r_1) may differ from the other parent (r_2).

Genotype	Expected Frequency	
	$r_1 \neq r_2$	$r_1 = r_2 = r$
AABB	$0.25\,(1 - r_1 - r_2 + r_1 r_2)$	$0.25\,(1 - r)^2$
AABb	$0.5\,(r_1 + r_2 - r_1 r_2)$	$0.5r\,(1 - r)$
AAbb	$0.25 r_1 r_2$	$0.25 r^2$
AaBB	$0.5\,(r_1 + r_2 - r_1 r_2)$	$0.5r\,(1 - r)$
AaBb	$0.5\,(1 - r_1 - r_2 + 2r_1 r_2)$	$0.5\,(1 - 2r + 2r^2)$
Aabb	$0.5\,(r_1 + r_2 - r_1 r_2)$	$0.5r\,(1 - r)$
aaBB	$0.25 r_1 r_2$	$0.25 r^2$
aaBb	$0.5\,(r_1 + r_2 - r_1 r_2)$	$0.5r\,(1 - r)$
aabb	$0.5\,(1 - r_1 - r_2 + r_1 r_2)$	$0.25\,(1 - r)^2$

10.1.3 GENOME STRUCTURE VARIATION AMONG PARENTS

Another complication of crossover-based analysis is caused by genome rearrangements such as insertions, deletions, inversions and translocations (Figure 10.3). If any of those genome rearrangements are present in one of the parents, the expected genotypic frequency (as a function of recombination fraction and parental genotypes) in their progeny, for some marker combinations, will differ from expectation. The methodologies currently used in linkage analysis and genetic map construction often ignore genome rearrangement and may not be adequate for use in mapping populations involving parents with genome rearrangements. In cases of parents involving genome rearrangement, the pairwise two-locus recombination fractions could be the pooled estimates between two parents. However, in a multiple-locus context, a meaningful analysis may be difficult to obtain. In practical data analysis, genome rearrangements may have lethal effects and exist in a low frequency in nature.

At a two-locus model level, the effect of the different genotypes of the parents for genes controlling crossover and genome rearrangement can be summarized as heterogeneity of recombination fraction between two parents. This differs from recombination fraction heterogeneity among different crosses (or populations). For the latter case, recombination fraction homogeneity is generally assumed between the two parents. The linkage map merging problem deals with the latter case (see Chapter 11). For example, in the case where one of the parents has a genotype with a genome segment inversion (Figure 10.3C), the expected genotypic frequency for loci A and B in the progeny of two heterozygous parents (AaBb X AaBb) is listed in Table 10.1. It has been generally assumed that recombination fraction between loci A and B in one parent (r_1) is the same as that in the other parent (r_2). When this assumption is violated, the analysis should be derived from the expected frequency of column 2 in Table 10.1 for $r_1 \neq r_2$. Two separate recombination fractions can be estimated for each of the parents. Interpretation of the results and dealing with multiple-locus models are complex when a large number of loci are considered. This difficulty is especially true when the heterogeneity is not uni-directional.

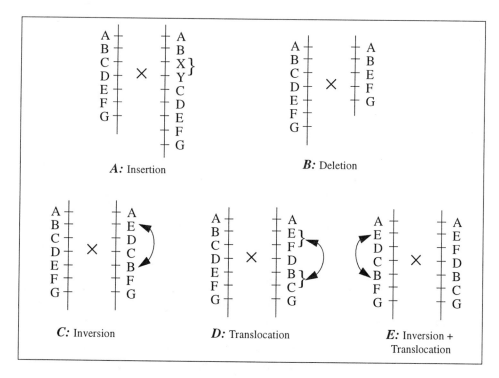

Figure 10.3 Crosses involving parents with genotypes resulting from genome rearrangements. A: A segment labeled X and Y is inserted into the genome of one of the parents. B: A segment labeled C and D is deleted from the genome of one of the parents. C: One parent has a genotype with a genome segment inversion. D: One parent has a genotype with a genome segment translocation. E: One parent has a genotype with genome segment inversion and the other with translocation.

The heterogeneity is not uni-directional for most cases of genome rearrangements and genetic coarse control of crossover.

10.2 THREE-LOCUS MODELS

A three-locus model is the simplest multi-locus model. Concepts such as double crossover, crossover interference and mapping functions were all based on a three-locus model (Chapter 9). As we will see in discussion of QTL mapping, a three-locus model is essential for the development of advanced methods to search for genes controlling quantitative traits.

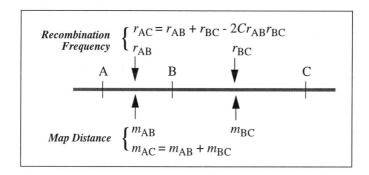

Figure 10.4 Linkage relationships among three loci A, B and C are quantified using both pairwise recombination fraction and map distance.

10.2.1 THREE-LOCUS MODEL

For three linked loci, A, B and C, there are three possible recombination fractions, r_{AB}, r_{BC} and r_{AC} (Figure 10.4). If the three loci are in order ABC on the chromosome and crossovers are completely at random, then the relationship of the three recombination fractions is

$$r_{AC} = r_{AB} + r_{BC} - 2r_{AB}r_{BC} \qquad (10.1)$$

where $2r_{AB}r_{BC}$ is expected to be the double crossover frequency (crossover between A and B and B and C simultaneously). However, departure from this expectation has been observed. This departure is defined as crossover interference between the two segments. This phenomenon is quantified by adding a coefficient in Equation (10.1)

$$r_{AC} = r_{AB} + r_{BC} - 2Cr_{AB}r_{BC} \qquad (10.2)$$

where C is defined as coefficient of coincidence and $1 - C$ is defined as interference.

If there is no interference and crossover occurs randomly, then the expected double recombinant frequency will be $2r_{AB}r_{BC}$, $C = 1$ and *Interference* = $1 - C = 0$. If crossovers in interval AB and BC are not independent, the observed double recombinant frequency may not equal the expectation. If we use r_{12} to denote the true double recombinant frequency, the coefficient of coincidence is

$$C = \frac{r_{12}}{2r_{AB}r_{BC}} \qquad (10.3)$$

Interference is

- negative for $C > 1$ and $Interference = 1 - C < 0$

- positive for $C < 1$ and $Interference = 1 - C > 0$

- absent for $C = 1$ and $Interference = 1 - C = 0$

- complete for for $C = 0$ and $Interference = 1 - C = 1$

When there is absence of the crossover interference, Equation (10.2) reduces to Equation (10.1).

For the meanings of different types of interference, see Section 9.2.2. High level interference has been a limiting factor in developing complete multiple-locus-models. Many models have been developed assuming no interference or no high level of interference.

The recombination fraction for a large genome segment is not the summation of the small intervals within the large segment. However, if the expected number of crossovers in each of the intervals can be estimated, the expected number of crossovers within the large segment should be the summation over the intervals. The expected number of crossovers within a genome segment is defined as a map distance. For example

$$m_{AC} = m_{AB} + m_{BC}$$

can be used to model the three loci in Figure 10.4, where m_{AB}, m_{BC} and m_{AC} are map distances between A and B, B and C and A and C, respectively. Mapping functions have been developed to convert recombination fraction into map distance.

10.2.2 CROSSOVER IN A THREE-LOCUS MODEL

Configurations

There are four possible crossover configurations for three loci during meiosis. If a parent is heterozygous (ABC/abc) for three loci, A, B and C, then one of the following will happen during meiosis (Figure 10.5):

(1) no crossover happens between A and C
(2) crossover happens between A and B, but not between B and C
(3) crossover happens between B and C, but not A and B
(4) crossover happens in both between A and B and B and C

If the parent is heterozygous for the three loci, then the frequency of the four possible crossover configurations can be observed by genotyping the three loci for a number of individuals in the progeny. Figure 10.5 shows the gametes produced by a heterozygous parent in association with the four configurations of crossover. Gametes ABC and abc are produced when no crossover happens during meiosis. A single crossover between A and B will produce gametes Abc and aBC and between B and C will produce ABc and acC. The double crossover will result in gametes AbC and aBc.

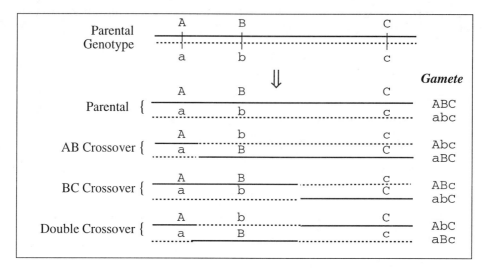

Figure 10.5 Gametes produced by a heterozygous parent (ABC/abc) for three loci A, B and C in order ABC. The expected frequency for the gametes is listed in Table 10.2.

Table 10.2 lists the gametes and their expected frequencies for the three loci. If we assume no crossover interference, then the probabilities of no crossover, BC crossover, AB crossover and double crossover are

$$(1 - r_{AB})\,(1 - r_{BC}) \;=\; 1 - r_{AB} - r_{BC} + r_{AB}r_{BC}$$

$$(1 - r_{AB})\,r_{BC} \;=\; r_{BC} - r_{AB}r_{BC}$$

$$r_{AB}\,(1 - r_{BC}) \;=\; r_{AB} - r_{AB}r_{BC} \tag{10.4}$$

$$r_{AB}r_{BC}$$

Table 10.2 Expected frequencies of gametes produced by a heterozygous parent (ABC/abc) for three loci, A, B and C in the order ABC. C is the coefficient of coincidence.

Gamete	Notation	Expected Frequency Double Crossover	No Double Crossover
ABC	p_1	$0.5\,(1 - r_{AB} - r_{BC} + Cr_{AB}r_{BC})$	$0.5\,(1 - r_{AB} - r_{BC}) = 0.5\,(1 - r_{AC})$
ABc	p_2	$0.5\,(r_{BC} - Cr_{AB}r_{BC})$	$0.5\,r_{BC}$
AbC	p_3	$0.5\,Cr_{AB}r_{BC}$	0
Abc	p_4	$0.5\,(r_{AB} - Cr_{AB}r_{BC})$	$0.5\,r_{AB}$
aBC	p_4	$0.5\,(r_{AB} - Cr_{AB}r_{BC})$	$0.5\,r_{AB}$
aBc	p_3	$0.5\,Cr_{AB}r_{BC}$	0
abC	p_2	$0.5\,(r_{BC} - Cr_{AB}r_{BC})$	$0.5\,r_{BC}$
abc	p_1	$0.5\,(1 - r_{AB} - r_{BC} + Cr_{AB}r_{BC})$	$0.5\,(1 - r_{AB} - r_{BC}) = 0.5\,(1 - r_{AC})$

respectively. When the crossover interference is taken into consideration, Equation (10.4) becomes

$$(1 - r_{AB})\,(1 - r_{BC}) \; = \; 1 - r_{AB} - r_{BC} + Cr_{AB}r_{BC}$$

$$(1 - r_{AB})\,r_{BC} \; = \; r_{BC} - Cr_{AB}r_{BC}$$

$$r_{AB}\,(1 - r_{BC}) \; = \; r_{AB} - Cr_{AB}r_{BC} \tag{10.5}$$

$$Cr_{AB}r_{BC}$$

Each of the crossover configurations creates two equally frequent gametes (assuming no or equal selection pressure on the two gametes). The gamete frequency is obtained using the corresponding terms for Equation (10.5), dividing by 2.

Double Crossover Issue

In some applications of genomic mapping, double crossover is ignored in a three-locus model. For example, some of the interval mapping procedures for QTLs are derived under the assumption that double crossover between the flanking markers and the tentative QTL is absent (see Chapter 13). This assumption may be valid when the two segments are closely linked, because the probability of double crossover is low (Figure 10.6). This assumption may be violated when the segment is relatively large, especially when the locus is located in the middle of the segment (Figure 10.6). For example, if a locus is in the middle of a 30cM segment, then the expected double crossover rate is 2.25%. On the other hand, if a locus is located 5cM from the end of the segment, then the expected double crossover rate is 1.25%. The increase of the double crossover rate has a nonlinear relationship with the size of the segment. If map distance for the two adjacent segments is 30cM (a locus in the middle of a 60cM segment), then the expected double crossover rate is 9%, which is almost 4 times the rate for the segment half the size. Table 10.2 also shows the expected gametic frequency, ignoring double crossovers.

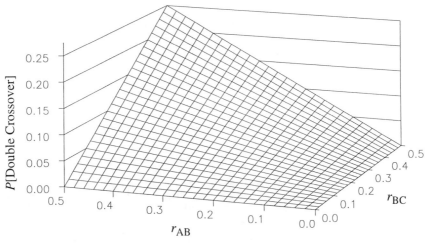

Figure 10.6 Expected double crossover rate for a three-locus model. The three loci are A, B and C and are in order of ABC.

In practice, gametes produced by double crossovers have often been treated as potential experimental errors. In many cases, this is true. These double crossovers are either corrected after checking the data or discarded. This has been an efficient way to achieve a "clean genetic map" empirically. The genetic map built in this way is usually shorter in length and has a high lod score support for the locus order. However, the issue of the treatment is complex because:

- Double crossovers are usually ignored in the likelihood function for the genetic map on which the lod score support is based.

- The double crossover frequency strictly depends on locus order. So, the double crossovers identified using one locus order may entirely differ from those identified using another locus order.

- The expected double crossovers are usually treated the same as the unexpected double crossovers.

One should be cautious in using the double crossovers as criteria to identify experimental error. Any wrong doing may result in a biased genetic map. The biases could be an incorrect locus order, an underestimated map distance, or an unrealistic lod score support for the locus order.

10.2.3 LIKELIHOOD FUNCTION

Likelihood function for the three-locus model is the first step toward multi-locus likelihood. In this section, traditionally defined backcross and F2 populations are used as examples to illustrate the procedures to establish likelihood function for the three-locus model.

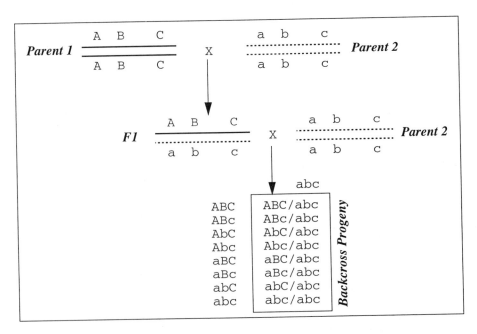

Figure 10.7 Backcross progeny for a three-locus model.

Triple-Backcross

Figure 10.7 shows progeny of a backcross mating for three loci, A, B and C. The F1 offspring of a cross between two different homozygous individuals are heterozygous for the three loci. The F1 produces eight types of gametes. Parent 2 is the recurrent parent and has been crossed with the F1. There are eight possible genotypes in the progeny and their expected frequencies with respect to the recombination fractions among the three loci are listed in Table 10.3.

Table 10.3 Expected genotypic frequency in triple backcross progeny when ABC is the order of the loci. C is the coefficient of coincidence.

Genotype	Observed Count	Notation	Expected Frequency	
			With Double Crossover	Ignoring Double Crossover
AaBbCc	f_1	p_1	$0.5 (1 - r_{AB} - r_{BC} + C r_{AB} r_{BC})$	$0.5 (1 - r_{AB} - r_{BC})$
AaBbcc	f_2	p_2	$0.5 (r_{BC} - C r_{AB} r_{BC})$	$0.5 r_{BC}$
AabbCc	f_3	p_3	$0.5 C r_{AB} r_{BC}$	0
Aabbcc	f_4	p_4	$0.5 (r_{AB} - C r_{AB} r_{BC})$	$0.5 r_{AB}$
aaBbCc	f_5	p_4	$0.5 (r_{AB} - C r_{AB} r_{BC})$	$0.5 r_{AB}$
aaBbcc	f_6	p_3	$0.5 C r_{AB} r_{BC}$	0
aabbCc	f_7	p_2	$0.5 (r_{BC} - C r_{AB} r_{BC})$	$0.5 r_{BC}$
aabbcc	f_8	p_1	$0.5 (1 - r_{AB} - r_{BC} + C r_{AB} r_{BC})$	$0.5 (1 - r_{AB} - r_{BC})$

In genomic analysis, two situations have been considered. When the three loci are located in a relatively small genome region, double-crossover has been ignored in some applications. For example, some interval mapping approaches to study genes controlling quantitative traits are derived under the model of ignoring double-crossover in a region flanked by two adjacent markers. This approach of ignoring double-crossover certainly has potential problems when the target region is not small; for example, the expected double-crossover rate is 4% when a marker is located in the middle of a 40cM genome region. Caution is needed when procedures derived under the assumption of absence of double-crossover are used for a relatively large genome region.

A full three-locus model should include double-crossover and crossover interference. Column 4 of Table 10.3 shows the expected genotypic frequencies considering double-crossover and crossover interference. If the coefficient of coincidence is set to one, then the events of crossover which occur in the two segments are considered independent.

A log likelihood function for a three-locus model with locus order ABC is

$$
\begin{aligned}
L(r_{AB}, r_{BC}) &= (f_1 + f_8) \log p_1 + (f_2 + f_7) \log p_2 + (f_3 + f_6) \log p_3 + (f_4 + f_5) \log p_4 \\
&= (f_1 + f_8) \log (1 - r_{AB} - r_{BC} + C r_{AB} r_{BC}) + (f_2 + f_7) \log (r_{BC} - C r_{AB} r_{BC}) \\
&\quad + (f_3 + f_6) \log (C r_{AB} r_{BC}) + (f_4 + f_5) \log (r_{AB} - C r_{AB} r_{BC}) \qquad (10.6) \\
&= (f_1 + f_8) \log (1 - r_{AC}) + (f_2 + f_7) \log (r_{AC} - r_{AB}) \\
&\quad + (f_3 + f_6) \log (C r_{AB} r_{BC}) + (f_4 + f_5) \log (r_{AC} - r_{BC})
\end{aligned}
$$

where f_i is the observed count, p_i is the expected genotypic frequency (defined in Table 10.3), C is the coefficient of coincidence, $\hat{r}_{AB}$ is the estimated recombination fraction between A and B and $\hat{r}_{BC}$ between B and C (Table 10.3). The maximum likelihood estimates of pairwise two-locus recombination fractions for r_{ab}, r_{ac} and r_{bc} are

$$\hat{r}_{AB} = (f_3 + f_4 + f_5 + f_6)/N$$
$$\hat{r}_{AC} = (f_2 + f_4 + f_5 + f_7)/N \qquad (10.7)$$
$$\hat{r}_{BC} = (f_2 + f_3 + f_6 + f_7)/N$$

where N is population size. These estimates do not depend on specific locus order. However, the expectations for the estimates depend on locus order. For locus order ABC, the expectations are

$$E[\hat{r}_{AB}] = r_{AB}$$
$$E[\hat{r}_{AC}] = r_{AC} = r_{AB} + r_{BC} - 2Cr_{AB}r_{BC} \qquad (10.8)$$
$$E[\hat{r}_{BC}] = r_{BC}$$

However, if locus order is BAC, the expectations are

$$E[\hat{r}_{AB}] = r_{AB}$$
$$E[\hat{r}_{AC}] = r_{AC} \qquad (10.9)$$
$$E[\hat{r}_{BC}] = r_{BC} = r_{AB} + r_{AC} - 2Cr_{AB}r_{AC}$$

The coefficients of coincidence in Equation (10.8) and Equation (10.9) are different. For locus order ABC, the maximum likelihood estimate of the coefficient of coincidence is

$$\hat{C} = \frac{2N(f_3 + f_6)}{(f_3 + f_4 + f_5 + f_6)(f_2 + f_3 + f_6 + f_7)}$$
$$= \frac{\hat{r}_{ab} + \hat{r}_{bc} - \hat{r}_{ac}}{2\hat{r}_{ab}\hat{r}_{bc}}$$

The estimate of coefficient of coincidence is order specific. It is defined under a certain locus order. Ignoring double-crossover, then it is equivalent to set $C = 0$. The likelihood function of Equation (10.6), the two-point recombination fraction estimation of Equation (10.7) and the expectations of estimates of the two-point recombination fractions for models considering double-crossover are applied to models ignoring double-crossover by simply setting $C = 0$.

F2 Progeny

Figure 10.8 shows an F2 progeny set produced by selfing F1 individuals from a cross between two homozygous parents with genotypes AABBCC and aabbcc for the three loci. The F1 individuals are therefore heterozygous for the three loci (AaBbCc) and produce 8 possible gametes (Figure 10.5). There are 27 possible genotypes in the progeny and their expected frequencies with respect to the recombination fractions among the three loci are listed in Table 10.4 (considering double-crossover) and Table 10.5 (ignoring double-crossover).

A log likelihood function for a three-locus model with locus order ABC is

$$L_{ABC}(r_{AB}, r_{BC}, C) = \sum_{i=1}^{27} f_i \log P_i \qquad (10.10)$$

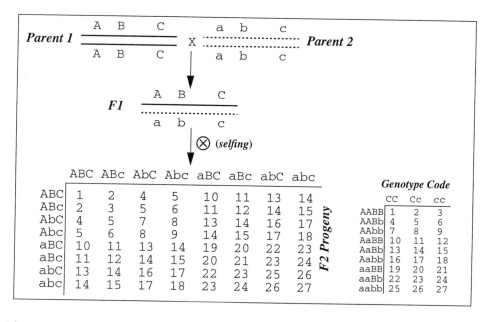

Figure 10.8 F2 progeny for a three-locus model.

where f_i and P_i are the observed and expected genotypic frequencies corresponding to the genotypes in Table 10.4 and Table 10.5, respectively, C is the coefficient of coincidence, $\hat{r}_{AB}$ is the estimated recombination fraction between A and B and $\hat{r}_{BC}$ is between B and C (Table 10.4 and Table 10.5). The maximum likelihood estimates of pairwise two-locus recombination fractions for r_{ab}, r_{ac} and r_{bc} cannot be obtained using a simple equation, as can be done for the backcross model. See Chapter 6 for methodology of estimation of recombination fraction using F2 progeny.

Table 10.4 Expected genotypic frequency in F2 progeny for three loci in order ABC assuming double crossovers occur. See Table 10.3 for definitions of the p_i in this table.

Loci A and B	Locus C		
	CC	Cc	cc
AABB	p_1^2	$2p_1p_2$	p_2^2
AABb	$2p_1p_3$	$2(p_1p_4 + p_2p_3)$	$2p_2p_4$
AAbb	p_3^2	$2p_3p_4$	p_4^2
AaBB	$2p_1p_4$	$2(p_2p_4 + p_1p_3)$	$2p_2p_3$
AaBb	$2(p_1p_2 + p_2p_3)$	$2(p_1^2 + p_2^2 + p_3^2 + p_4^2)$	$2(p_1p_2 + p_3p_4)$
Aabb	$2p_2p_3$	$2(p_2p_4 + p_1p_3)$	$2p_1p_4$
aaBB	p_4^2	$2p_3p_4$	p_3^2
aaBb	$2p_2p_4$	$2(p_1p_4 + p_2p_3)$	$2p_1p_3$
aacc	p_2^2	$2p_1p_2$	p_1^2

Table 10.5 Expected genotypic frequency in F2 progeny for three loci in order ABC assuming no double crossovers. See Table 10.3 for definitions of the p_i in this table.

Loci A and B	Locus C		
	CC	Cc	cc
AABB	$0.25(1-r_{AC})^2$	$0.5r_{BC}(1-r_{AC})$	$0.25r_{BC}^2$
AABb	0	$0.5r_{AB}(1-r_{AC})$	$0.5r_{AB}r_{BC}$
AAbb	0	0	$0.25r_{AB}^2$
AaBB	$0.5r_{AB}(1-r_{AC})$	$0.5r_{AB}r_{BC}$	0
AaBb	$0.5r_{BC}(1-r_{AC})$	$0.5[(1-r_{AC})^2+r_{AB}^2+r_{BC}^2]$	$0.5r_{BC}(1-r_{AC})$
Aabb	0	$0.5r_{AB}r_{BC}$	$0.5r_{AB}(1-r_{AC})$
aaBB	$0.25r_{AB}^2$	0	0
aaBb	$0.5r_{AB}r_{BC}$	$0.5r_{AB}(1-r_{AC})$	0
aabb	$0.25r_{BC}^2$	$0.5r_{BC}(1-r_{AC})$	$0.25(1-r_{AC})^2$

10.3 MAPPING FUNCTIONS

Mapping functions have been commonly used in genetic mapping. The recombination fractions are not additive. Mapping functions have been designed to solve this problem. Over several decades, a number of mapping functions have been developed. These functions may apply to general or specific situations. In this section, the basic foundation for deriving mapping functions will be introduced and some of them will be compared using general and specific cases.

10.3.1 DEFINITIONS

Ideally, genes or genetic markers are organized linearly on a map and their relative positions on the map can be quantified in an additive fashion. For example, regarding 6 loci A, B, C, D, E and F in order ABCDEF, the relationship can be quantified using

$$
\begin{aligned}
m_{AF} &= m_{AE} + m_{EF} \\
&= m_{AD} + m_{DE} + m_{EF} \\
&= m_{AC} + m_{CD} + m_{DE} + m_{EF} \\
&= m_{AB} + m_{BC} + m_{CD} + m_{DE} + m_{EF}
\end{aligned}
\tag{10.11}
$$

where m_{ij} is defined as the map distance between loci i and j and is derived from the expected number of crossovers between the two loci. If the expected number of crossovers is one in a genome segment, then the map distance between two genes or genetic markers flanking the segment is 1 Morgan (M) or 100 centiMorgans (cM). The Morgan is usually given in the smaller subunit, centiMorgan.

In theory, map distance can be obtained in three ways. For a long chromosome segment, the number of crossovers may be observed cytologically, if cytological markers can be visualized during meiosis. The cytological approach has been used to estimate genetic map length for chromosomes. It has rarely been used for building a detailed genetic map because it has low resolution. Map distance can also be obtained by summing recombination fractions over very short adjacent intervals. For genetic markers or genes located in a short interval, the chance of double and multiple-crossover is low. However, in practical genetic mapping, map distance has been estimated by converting recombination fraction to map distance using mapping functions.

As we found out in the simple three-locus model, the recombination fractions based on two-locus models are not additive. As the number of loci increases, the complexity of the relationships among locus positions and recombination fractions increases. For recombination fraction, r_{ij}, between two loci i and j, if a function

$$m_{ij} = F(r_{ij}) \tag{10.12}$$

exists for all pairs of genes or markers and is a continuous function, then $F(r_{ij})$ is defined as a mapping function. Map distance becomes a one-to-one function of recombination fraction. Equation (10.12) has been commonly used to convert recombination fraction into map distance. For some mapping functions, the inverse of the function is

$$r_{ij} = F^{-1}(m_{ij}) \tag{10.13}$$

and it is used to convert map distances to recombination fractions.

Table 10.6 A list of commonly used mapping functions and their inverses.

	Map Function ($m = F(r)$)	Inverse ($r = F^{-1}(m)$)
Morgan (1928)	r	m
Haldane (1919)	$-0.5\log(1 - 2r)$	$0.5(1 - e^{-2|m|})$
Kosambi (1944)	$\frac{1}{2}\tanh^{-1}2r = \frac{1}{4}\log\frac{1 + 2r}{1 - 2r}$	$\frac{1}{2}\tanh(2m) = \frac{1}{2}\frac{e^{4m} - 1}{e^{4m} + 1}$
Carter & Falconer (1951)	$0.5(\tan^{-1}2r + \tanh^{-1}2r)$	Not applicable
Rao et al. (1977)	Equation (10.30)	Not applicable
Sturt (1976)	Not applicable	$0.5\left[1 - \left(1 - \frac{m}{L}\right)e^{\frac{m}{L}(1 - 2L)}\right]$
Felsenstein (1979)	$\frac{1}{2(K - 2)}\log\frac{1 - 2r}{1 - 2(K - 1)r}$	$\frac{1 - e^{2(K - 2)m}}{2[1 - (K - 1)e^{2(K - 2)m}]}$
Karlin (1984) (binomial)	$0.5N[1 - (1 - 2r)^{1/N}]$	$0.5\left[1 - \left(1 - \frac{2m}{N}\right)^N\right]$

10.3.2 COMMONLY USED MAP FUNCTIONS

Table 10.6 lists some commonly used mapping functions and their inverses. The theoretical derivation for Haldane's and Kosambi's mapping functions will be shown. For the other mapping functions, a simple description will be given. Refer to the original papers for the mathematical perspectives underlying those mapping functions.

Morgan's Map Function

Sturtevant (1913) and Morgan (1928), in their early work on gene mapping in *Drosophila*, used the estimated recombination fraction as map distance, which is

$$m = r$$

This is known as Morgan's mapping function. When a small genome segment is considered, the chance that double or multiple-crossovers occur in the segment is low. In such cases, the estimated recombination fraction has the same expectation as the expected number of crossovers. For example, when three loci A, B and C are in a small segment and the estimated recombination fraction between A and B is 0.03 and between B and C is 0.01, the probability of double and multiple-crossovers is low. In this case, the number of expected crossovers is the same as that of the recombination fraction. The map distance between A and B is approximately 3cM and between B and C is 1cM. By adding the map distances in the two adjacent segments, we have the map distance of 4cM between loci A and C. Morgan's mapping function can be applied only over a small genome segment. When a segment is large, the expected number of multiple-crossovers is large and the Morgan's mapping function is not appropriately applied (see Figure 10.6).

Haldane's Map Function

If crossover interference is ignored, the relationship among the pairwise recombination fractions among three loci A, B and C with an order of ABC can be quantified as

$$r_{AC} = r_{AB} + r_{BC} - 2r_{AB}r_{BC} \qquad (10.14)$$

It is obvious that the recombination fraction is not additive. Equation (10.14) can be rewritten as

$$1 - 2r_{AC} = 1 - 2(r_{AB} + r_{BC} - 2r_{AB}r_{BC})$$
$$= (1 - 2r_{AB})(1 - 2r_{BC})$$

If more than three loci are considered, then we can write

$$1 - 2r_l = \prod_{i=1}^{l-1}(1 - 2r_i) \qquad (10.15)$$

where l is the number of loci on a genome segment, r_l is the recombination fraction between two genes or genetic markers flanking the whole segment and r_i is the recombination fraction between two markers flanking a sub-segment i. It is clear that an additive function for Equation (10.15) is

$$F(r) = c \log(1 - 2r) \qquad (10.16)$$

where c is a constant. Haldane (1919) derived his mapping function from Equation (10.16) by setting $c = -1/2$, which is

$$m = F(r) = \begin{cases} -\dfrac{1}{2}\log(1 - 2r) & \text{for } 0 \le r < 0.5 \\ \infty & \text{for } r \ge 0.5 \end{cases} \qquad (10.17)$$

with an inverse

$$r = \frac{1}{2}(1 - e^{-2m}) \tag{10.18}$$

The original derivation of Haldane's mapping function was purely mathematical. However, Haldane's mapping function has been extensively used and studied in the last several decades. A more biologically based derivation of Haldane's mapping function has been developed.

If we assume that crossovers occur uniformly (randomly) along the length of the chromosome (in the absence of crossover interference), then the events of crossover that occur on the chromosome can be modeled as a Poisson process. If we assume again that the average number of crossovers is m in a genome segment flanked by loci A and B, then the the probability of x crossovers occurring in an interval is

$$P[cros\sin gover = x] = \frac{e^{-m}m^x}{x!} \tag{10.19}$$

We know that the recombinant classes for loci A and B are observed only when an odd number of crossovers occur in the interval. We have

x	Probability	Chromosome	
0	e^{-m}		AB / ab
1	me^{-m}		Ab / aB
2	$m^2e^{-m}/2$		AB / ab
3	$m^3e^{-m}/6$		Ab / aB
...	...		
x	$e^{-m}m^x/x!$		

The expected recombination fraction (which is defined as the probability of recombinant genotypes in the progeny) in terms of the expected number of crossovers (map distance) is

$$r = \sum_{i=0}^{\infty} \frac{e^{-m}m^{(2i+1)}}{(2i+1)!} = e^{-m} \sum_{i=0}^{\infty} \frac{m^{(2i+1)}}{(2i+1)!}$$
$$= me^{-m} + \frac{m^3 e^{-m}}{6} + ... = \frac{1}{2}(1 - e^{-2m}) \tag{10.20}$$

which is Haldane's mapping function. When the recombination fraction is small, the map distances using Haldane's mapping function and recombination fraction are approximately equal. As the size of the segment increases, the expected number of multiple-crossovers increases and the map distance is adjusted for the multiple-crossovers through Haldane's mapping function. The recombination fraction approaches 0.5 and is independent of map distance when map distance is large.

To show how Haldane's mapping function works for a simple three-locus model, we write

$$m_{AC} = m_{AB} + m_{BC} \tag{10.21}$$

for a three-locus model with locus order ABC. Because of

$$m_{AB} = -1/2\log(1 - 2r_{AB})$$
$$m_{BC} = -1/2\log(1 - 2r_{BC})$$
$$m_{AC} = -1/2\log(1 - 2r_{AC})$$

Equation (10.21) can be rewritten as

$$-1/2\log(1 - 2r_{AC}) = -1/2\log(1 - 2r_{AB}) - 1/2\log(1 - 2r_{BC})$$

which reduces to $r_{AC} = r_{AB} + r_{BC} - 2r_{AB}r_{BC}$, and this is the form of Equation (10.1). This derivation verifies that the mapping function of Equation (10.17) precisely converts recombination fractions to additive map distance for the simple three loci case. Figure 10.9 shows a nonlinear relationship among the three recombination fractions and a linear additive relationship among the Haldane's mapping functions. Figure 10.9B is produced by simple conversion from recombination fractions to map distances using Equation (10.17).

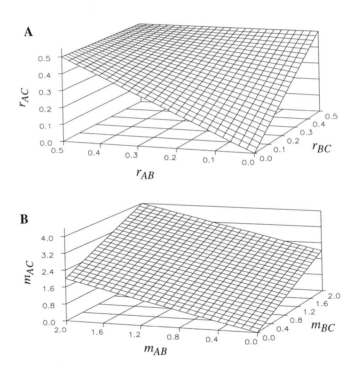

Figure 10.9 A: Relationships among the three recombination fractions (nonlinear and nonadditive) for the three loci ABC and B: relationships after converting the recombination fractions using Haldane's function (linear and additive).

Kosambi's Map Function

Haldane's mapping function works for situations with absence of crossover interference. However, experimental evidence has been found to support that crossover interference exists and crossovers occur non-randomly in genomes (Muller 1916). Taking into consideration crossover interference, the relationship among pairwise recombination fractions for three loci, A, B and C in order ABC, can be quantified using

$$r_{AC} = r_{AB} + r_{BC} - 2Cr_{AB}r_{BC} \qquad (10.22)$$

where C has been defined as the coefficient of coincidence and $1 - C$ as interference. As we have discussed, recombination fraction can be considered a function of the expected number of crossovers or map distance, $F(m)$, if we assume that the segment flanked by B and C is small. We can write

$$r_{AB} = F^{-1}(m)$$

$$r_{BC} = \Delta m$$

$$r_{AC} = F^{-1}(m + \Delta m)$$

because recombination fraction and map distance are approximately equal for a short segment. Equation (10.22) can be rewritten. Taking $\Delta m \to -\infty$, we have

$$F^{-1}(m + \Delta m) = F^{-1}(m) + \Delta m - 2CF^{-1}(m)(\Delta m)$$
$$\downarrow$$
$$[F^{-1}(m + \Delta m) - F^{-1}(m)]/\Delta m = 1 - 2CF^{-1}(m)$$
$$\downarrow$$
$$dF^{-1}(m)/dm = 1 - 2Cr$$
$$\downarrow$$
$$dF(r)/dr = \frac{1}{1 - 2Cr}$$

Now, we have the mapping function

$$F(r) = \int_0^r \frac{1}{1 - 2Cu} du \qquad (10.23)$$

If we set $C = 1$, then we have Haldane's mapping function

$$F_{Haldane}(r) = \int_0^r \frac{1}{1 - 2u} du$$

$$= \begin{cases} -\frac{1}{2}\log(1 - 2r) & \text{for } 0 \le r < 0.5 \\ \\ \infty & \text{for } r \ge 0.5 \end{cases} \qquad (10.24)$$

Kosambi (1944) took $C = 2r$, instead of $C = 1$ and gave his mapping function as

$$F(r) = m$$

$$= \int_0^r \frac{1}{1 - 4u^2} du$$

$$= \begin{cases} \frac{1}{2}\tanh^{-1} 2r = \frac{1}{4}\log\frac{1 + 2r}{1 - 2r} & \text{for } 0 \le r < 0.5 \\ \\ \infty & \text{for } r \ge 0.5 \end{cases} \qquad (10.25)$$

with an inverse function

$$r = F^{-1}(m)$$

$$= \frac{1}{2}\tanh(2m)$$

$$= \frac{1}{2}\frac{(e^{4m}-1)}{(e^{4m}+1)}$$

(10.26)

The rationale behind Kosambi's mapping function is that the crossover interference depends on the size of a genome segment. The interference is absent when a segment is sufficiently large (e.g., $C \to 1$ when $r \to 0.5$). The interference increases as the segment decreases (e.g., $C \to 0$ when $r \to 0.0$). The relationship between the size of the segment and the crossover interference is $C = 2r$.

Other Map Functions

Carter and Falconer (1951) and Felsenstein (1979) presented mapping functions that were also derived from Equation (10.23) by setting $C = 8r^3$ and $C = K - 2(K-1)r$, respectively (see Table 10.7 for definition of K). The Carter and Falconer mapping function is

$$F(r) = m$$

$$= \int_0^r \frac{1}{1 - 16u^4} du$$

(10.27)

$$= \frac{1}{4}(\tan^{-1}2r + \tanh^{-1}2r)$$

which has no simple inverse. In practice, recombination fraction can be obtained from the Carter and Falconer map distance using the numerical method. The Carter and Falconer mapping function is commonly used when there is evidence of strong crossover interference. Felsenstein's mapping function is

$$F(r) = m$$

$$= \int_0^r \frac{1}{1 - 2[K - 2(K-1)r]u} du$$

(10.28)

$$= \frac{1}{2(K-2)}\log\frac{1-2r}{1-2(K-1)r}$$

with an inverse

$$F^{-1}(r) = r$$

$$= \frac{1 - e^{2(K-2)m}}{2[1 - (K-1)e^{2(K-2)m}]}$$

(10.29)

where $-\infty < K < \infty$ is a parameter for crossover interference. If $K = 0$, Equation (10.28) is the the same as the Kosambi mapping function and if $K = 1$, Equation (10.28) is the the same as Haldane's mapping function.

Rao et al. (1977) proposed a mapping function as a weighted mean of the Morgan, Haldane, Kosambi and Carter and Falconer mapping functions. The Rao's mapping function is formulated as:

$$m = [p(2p-1)(1-4p)\log(1-2r)]/6$$
$$+ [8p(p-1)(2p-1)\tan^{-1}2r]/3$$
$$+ ([2p(1-p)(4p+1)\tanh^{-1}2r]/3)$$
$$+ (1-p)(1-2p)(1-4p)r$$

$$(10.30)$$

When $p = 0$, $p = 0.25$, $p = 0.5$ and $p = 1$, Equation (10.29) reduces to Morgan, Carter and Falconer, Kosambi and Haldane functions, respectively.

Table 10.7 Mapping functions based on Equation (10.23).

	C	Comments
Morgan (1928)	0	Complete interference, absence of multiple-crossover
Haldane (1919)	1	Absence of crossover interference
Kosambi (1944)	2r	Crossover interference is a function of recombination fraction
Carter & Falconer (1951)	$8r^3$	Strong crossover interference
Felsenstein (1979)†	K - (K - 1)2r	K = 1: Absence of crossover interference K < 1: Positive interference K > 1: Negative interference

The mapping functions discussed previously are all based on Equation (10.23). Those functions were obtained by setting different values for the coefficient of coincidence C. Table 10.7 lists their C values and comments on the mapping functions. As mentioned by Karlin (1984), there are two difficulties with deriving mapping functions using Equation (10.23):

- First, the mapping functions take into consideration only the map distance between genes or markers regardless of the location of these genes or markers on the chromosome. This consideration is not realistic when the mapping functions are applied to a large number of loci.
- Second, Karlin questions the existence of a global relationship between the marginal pairwise recombination fractions and the map distances. He suggests that a mapping function should include a map distance and a parameter for the location of the segment on the genome.

Karlin concludes that the mapping functions of Kosambi, Carter and Falconer and Felsenstein (for K not in the range between 1 and 2) are not valid mapping functions for a multilocus structure (also see Liberman and Karlin 1984). He also concludes that the mapping functions of Haldane, Sturt and Felsenstein (for $1 \le K \le 2$) are valid multilocus mapping functions if a global relationship between recombination fraction and map distance exists.

The Sturt mapping function was derived based on the assumption that there is one obligatory crossover and an additional crossover following the Poisson process, which is

$$F^{-1}(m) = r$$

$$= \begin{cases} \frac{1}{2}\left[1-\left(1-\frac{m}{L}\right)e^{\frac{m}{L}(1-2L)}\right], & m < L \\ 0.5, & m \geq L \end{cases} \qquad (10.31)$$

where $L \geq 0.5$ is genetic length of the chromosome arm in Morgans. This mapping function can be applied to a genome segment representing a single chromosome arm. Sturt also derived a mapping function for a segment occupying two chromosome arms. She assumed one obligatory crossover on each arm of a chromosome. If a segment has a length of $m = m_R + m_L$, where m_R and m_L are map distances from the centromere to the right end and the left end of the segment, respectively, the Sturt mapping function for two obligatory crossovers is

$$r = F^{-1}(m_R, m_L)$$

$$= \begin{cases} \frac{1}{2}\left[1-\left(1-\frac{m_R}{L_R}\right)\left(1-\frac{m_L}{L_L}\right)e^{\frac{m_R}{L_R}(1-2L_R)+\frac{m_L}{L_L}(1-2L_L)}\right] & \text{for } m_R \leq L_R \text{ and } m_L \leq L_L \\ 0.5 & \text{for } m_R > L_R \text{ or } m_L > L_L \end{cases} \quad (10.32)$$

where $L_R \geq 0.5$ and $L_L \geq 0.5$ are total lengths in Morgans for the right and left chromosome arms, respectively.

Table 10.8 Some of the Karlin's mapping functions based on the crossover count location (C-L) process. c_k is the probability that k crossover events occur in an interval.

c_k	Expected m	$F^{-1}(m) = r$
Zero or one $(c_0 + c_1 \equiv 1)$	c_1	m
Poisson (λ)	λ	$\frac{1}{2}(1-e^{-2\|m\|})$
Binomial (N, p)	Np	$\frac{1}{2}\left[1-\left(1-\frac{2m}{N}\right)^N\right]$
Negative binomial (α, p)	$\frac{\alpha p}{1-p}$	$\frac{1}{2}\left[1-\left(1+\frac{2m}{\alpha}\right)^{-\alpha}\right]$

Karlin (1984) summarized the commonly used mapping functions and described some new mapping functions based on the count location (CL) process (Karlin and Liberman 1978 and 1979; Risch and Lange 1979). Table 10.8 shows how some of the mapping functions can be interpreted using the CL process. If zero or only one crossover event can be observed in an interval, then the estimated recombination fraction is the expected number of crossovers. So, the corresponding mapping function is Morgan's mapping function. If c_k is distributed as an independent Poisson variable, then the corresponding mapping function is Haldane's mapping function. Karlin described a mapping function based on a binomial distribution of the crossover events as

$$m = \frac{1}{2}N[1 - (1 - 2r)^{1/N}] \qquad (10.33)$$

with a inverse of

$$r = \frac{1}{2}\left[1 - \left(1 - \frac{2m}{N}\right)^{N}\right] \qquad (10.34)$$

where N is the maximum number of crossovers in an interval.

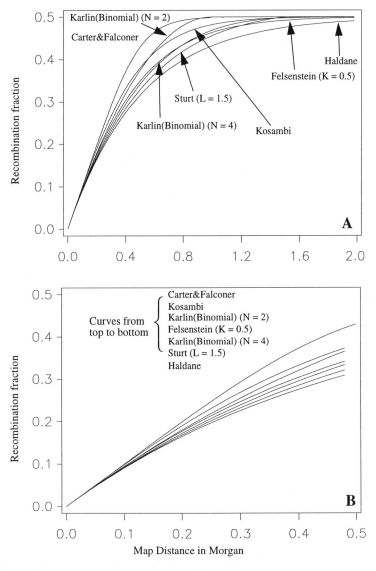

Figure 10.10 The relationship between recombination fraction and some commonly used mapping functions.

10.3.3 COMPARISON OF COMMONLY USED MAPPING FUNCTIONS

Commonly used mapping functions and the concept of crossover interference are largely derived from a three-locus model and certain assumptions regarding the distribution of crossover events over the genome. The mapping functions have often been used without consideration of experimental conditions and genetic configurations. The central issue regarding the use of a particular mapping function in practice is the fitness of the mapping function to the observed data. There are also important biological arguments for or against a particular mapping function. Comparisons among the commonly used mapping functions under some limited conditions are given in Figure 10.10 and Table 10.9.

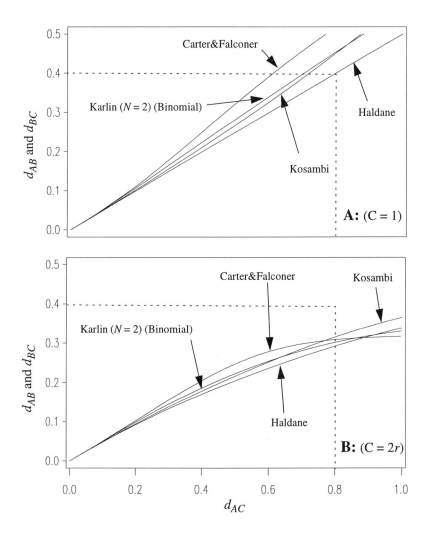

Figure 10.11 Relationship between map distance of segment AC and map distances of segments AB and BC when the three recombination fractions are converted to map distance using the commonly used mapping functions under the assumption of absence of crossover interference (A) and the coefficient of coincidence of two times the recombination fraction (B). The order of the three loci is ABC.

Figure 10.10 shows the relationships between recombination fraction and commonly used mapping functions. When recombination fraction is below 0.15, the relationship is similar to that of the mapping functions. Map distances using different mapping functions do not differ significantly from the value of the recombination fraction. In the range of genomic significance, say $r \leq 0.4$, given an estimated recombination fraction, the mapping functions may give different values of map distance. The value increases using mapping functions of Carter and Falconer, Kosambi, Karlin (Binomial) with N = 2, Felsenstein with K = 0.5, Karlin (Binomial) with N = 4, Sturt with L =1.4, to Haldane (Figure 10.10B). For example, the map distances corresponding to a recombination fraction of 0.2 are 0.20, 0.21, 0.225, 0.231, 0.24 and 0.26 Morgans using the mapping functions in the order mentioned before (Table 10.9). These comparisons mean that given the estimated recombination fractions between adjacent segments, the length of the genome in Morgans will be the longest using Haldane's mapping function and the shortest using the Carter and Falconer mapping function.

The differences among the commonly used mapping functions are due to the assumptions about distribution of crossovers on the genome, crossover interference and length of the chromosome segment considered. If complete crossover interference is assumed, then the map distance is equal to the observed recombination fraction because multiple-crossovers are ignored. If absence of crossover interference is assumed, then map distance (expected crossover) is much greater than the observed recombination fraction.

Mapping functions work only for specific conditions. There is no universal mapping function. Mapping functions do not estimate physical distance for most cases and have a limited advantage over recombination fraction. Figure 10.11 shows how the mapping functions work for only three-locus situations under two different assumptions for crossover interference. For absence of crossover interference, Haldane's mapping function converts the non-additive recombination fraction to additive map distance precisely (Figure 10.11A). All other mapping functions fail to make map distance additive. For a situation in which the coefficient of coincidence is two times the recombination fraction, none of the mapping functions can make map distance additive, even though Kosambi's mapping function was derived based on the assumption regarding crossover interference (Figure 10.11B).

Table 10.9 Values of map distances converted using different mapping functions and recombination fraction.

Map Function	r					
	0.01	0.1	0.2	0.3	0.4	0.45
Haldane	0.01	0.11	0.26	0.46	0.81	1.96
Kosambi	0.01	0.10	0.21	0.35	0.55	1.15
Carter & Falconer	0.01	0.10	0.20	0.31	0.44	0.77
Sturt (L = 1.5)	0.01	0.11	0.25	0.43	0.72	0.96
Felsenstein (K = 0.5)	0.01	0.11	0.23	0.39	0.65	1.44
Karlin (binomial) N=2	0.01	0.11	0.23	0.37	0.55	0.86
Karlin (binomial) N=4	0.01	0.11	0.24	0.41	0.66	1.25

10.4 ESTIMATION OF MULTI-LOCUS MAP DISTANCE

It is well known that the multi-locus model provides more information than does the two-locus model in linkage analysis. However, there are three major obstacles to implementation of true multi-locus models in linkage analysis and linkage map construction.

(1) The multi-locus likelihood function is complex and the number of unknown parameters is large when high levels of crossover interference occur. For example, there are 10 possible pairwise two-point recombination fractions and 45 possible two-way crossover interference parameters for a five-locus model. If three-way interference is considered, then the model is extremely complex.

(2) Data needed to build a multi-locus model are often missing. A multi-locus model is based on possible crossover combinations among the loci (Figure 10.12). For three loci, ABC, there are 4 possible crossover combinations: 11, 10, 01 and 00, if "1" is used to denote that crossover happens in the segment and "0" to denote that crossover does not happen. For seven loci, there are 81 possible crossover combinations (Figure 10.12). There are 512 possible crossover combinations for 10 loci. In general, there are 2^{n-1} combinations for n loci. It is impossible to obtain all the possible crossover combinations in practice when n is large, due to limited population size, uninformative markers and unknown linkage phases.

(3) It is computationally intensive to build a multi-locus model. As discussed previously in Chapter 9, there are $n!/2$ possible locus orders for n loci. It is computationally greedy to obtain a most likely locus order among a large number of possibilities. Even when a single locus order is considered, computation is still intensive to solve the multi-locus likelihood for maximum likelihood estimates for a large number of unknown parameters.

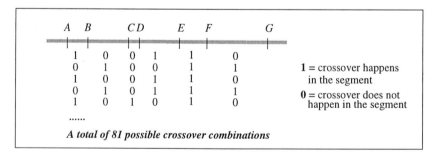

Figure 10.12 Possible crossover combinations for 7 loci.

Mapping functions are an important step toward a useful multi-locus model. However, the adequacy of mapping functions depends on whether the assumptions underlying the mapping functions are true or false. During the last several decades, many approaches have been developed for building a multi-locus model. The least square method (Jensen and Jorgensen 1995; Stam 1993), the EM Algorithm (Lander and Green 1987), the joint maximum likelihood method (Zhao *et al.* 1990) and a simulation approach (Weeks *et al.* 1993) for multi-locus analysis will be briefly introduced.

10.4.1 LEAST SQUARES

To estimate multi-locus map distance, Stam (1993) proposed a least squares method, which is a modified version of the algorithm originally introduced by Jensen and Jorgensen (1975a and b) to estimate multi-locus map distances in barley. The least squares method was implemented in a computer package for genomic mapping, JoinMap (Stam 1993). Similar methods were also described by Lalouel (1977) and Weeks and Lange (1987).

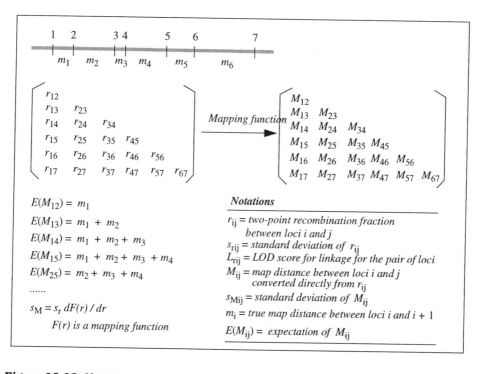

Figure 10.13 Notations and their relationships used to derive the least-square algorithm.

Notation

Figure 10.13 lists notations and their relationships for deriving the least square algorithm. For n loci with a certain locus order, $n - 1$ multi-locus recombination fractions between adjacent loci are the parameters to be estimated. Jensen and Jorgensen (1975) assume that the quantity

$$\frac{M_{ij} - \sum_{k=i}^{j-1} m_k}{s_{Mij}}$$

is distributed as standard normal with mean of zero and variance equal to one, where $\sum m_k$ is the expected value of M_{ij} (see Figure 10.13) and s_{Mij} is the standard deviation of M_{ij}. s_{Mij} can be derived approximately from the standard deviation of the corresponding two-point recombination fraction (s_r) by

$$s_M = s_r \left| \frac{dF(r)}{dr} \right|$$

where $F(r)$ is the function converting recombination fraction to map distance. For example, the equations for Haldane's and Kosambi's mapping functions are

$$s_M = s_r \left| \frac{d\left[-0.5\log(1-2r)\right]}{dr} \right| \qquad (Haldane)$$

$$= \frac{s_r}{1-2r}$$

$$s_M = s_r \left| \frac{d\left(0.5\tanh^{-1}2r\right)}{dr} \right| \qquad (Kosambi)$$

$$= \frac{s_r}{1-4r^2}$$

Likelihood

Jensen and Jorgensen (1975) define a likelihood function

$$L = constant \times \exp\left\{ -\frac{1}{2}\sum_{j=2}^{n-1}\sum_{i=1}^{j}\left[\frac{M_{ij} - \sum_{k=i}^{j-1} m_k}{s_{Mij}} \right]^2 \right\} \qquad (10.35)$$

The multi-locus map distances can be obtained by solving $n-1$ linear equations generated by differentiating the logarithm of Equation (10.35) with respect to m_k. These equations have the form of

$$\frac{dLog(L)}{dm_k} = \sum_{j=2}^{n-1}\sum_{i=1}^{j}\left[\frac{M_{ij} - \sum_{k=i}^{j-1} m_k}{s_{Mij}} \right] = 0 \qquad (10.36)$$

Table 10.10 An example for estimating multi-point map distance using the least square method. The values in parentheses are variances for the recombination fractions or the map distances.

Marker Combination	r	M Haldane	Kosambi	Expectation	LOD
12	0.10 (.03)	0.11 (.038)	0.1 (.031)	m_1	15.99
23	0.15 (.04)	0.18 (.057)	0.16 (.044)	m_2	16.44
13	0.3 (.13)	0.46 (.325)	0.35 (.203)	$m_1 + m_2$	4.29

Example

Table 10.10 lists the values needed to estimate multi-locus map distances for three loci. The logarithms of the likelihood functions are

$$Log\,(L) \;=\; \left(\frac{0.11-m_1}{0.038}\right)^2 + \left(\frac{0.18-m_2}{0.057}\right)^2 + \left(\frac{0.46-m_1-m_2}{0.325}\right)^2 \qquad Haldane$$

$$Log\,(L) \;=\; \left(\frac{0.11-m_1}{0.031}\right)^2 + \left(\frac{0.16-m_2}{0.044}\right)^2 + \left(\frac{0.35-m_1-m_2}{0.203}\right)^2 \qquad Kosambi$$

Two linear equations are

$$\frac{0.11-m_1}{0.038^2} + \frac{0.46-m_1-m_2}{0.325^2} = 0$$

$$\frac{0.18-m_2}{0.057^2} + \frac{0.46-m_1-m_2}{0.325^2} = 0$$

for using Haldane's mapping function and

$$\frac{0.11-m_1}{0.031^2} + \frac{0.35-m_1-m_2}{0.203^2} = 0$$

$$\frac{0.16-m_2}{0.044^2} + \frac{0.35-m_1-m_2}{0.203^2} = 0$$

for using Kosambi's mapping function. By solving the linear equations, we obtain the estimates of the multi-locus map distances

$$\left.\begin{array}{l} \hat{m}_1 = 0.115 \\ \hat{m}_2 = 0.188 \end{array}\right\} \qquad Haldane$$

$$\left.\begin{array}{l} \hat{m}_1 = 0.120 \\ \hat{m}_2 = 0.164 \end{array}\right\} \qquad Kosambi$$

Variance of Estimated Map Distance

Jensen and Jorgensen (1975) also gave methods to estimate the variance of the estimated multi-locus map distance. The information matrix has the form of

$$I = -\frac{d^2 Log\,(L)}{dm_i dm_j}$$

$$= \sum_{j=2}^{n-1}\sum_{i=1}^{j}\left[\frac{1}{s^2_{Mij}}\right] \tag{10.37}$$

The inverse of the information matrix is the variance-covariance matrix for the estimated multi-locus map distances. For the data in Table 10.10, the information matrix is

$$I = \begin{bmatrix} 0.038^{-2}+0.325^{-2} & 0.325^{-2} \\ 0.325^{-2} & 0.057^{-2}+0.325^{-2} \end{bmatrix}$$

$$= \begin{bmatrix} 701.99 & 9.47 \\ 9.47 & 317.25 \end{bmatrix}$$

for Haldane's mapping function and is

$$I = \begin{bmatrix} 0.031^{-2} + 0.203^{-2} & 0.203^{-2} \\ 0.203^{-2} & 0.044^{-2} + 0.203^{-2} \end{bmatrix}$$

$$= \begin{bmatrix} 1064.85 & 24.27 \\ 24.27 & 540.80 \end{bmatrix}$$

for Kosambi's mapping function with inverses

$$I^{-1} = \begin{bmatrix} 0.0014 & -0.000042 \\ -0.000042 & 0.0032 \end{bmatrix} \qquad Haldane$$

$$I^{-1} = \begin{bmatrix} 0.00094 & -0.000042 \\ -0.000042 & 0.00185 \end{bmatrix} \qquad Kosambi$$

The standard deviations are

$$\left. \begin{aligned} sd(m_1) &= 0.037 \\ sd(m_2) &= 0.057 \end{aligned} \right\} \qquad Haldane$$

$$\left. \begin{aligned} sd(m_1) &= 0.031 \\ sd(m_2) &= 0.043 \end{aligned} \right\} \qquad Kosambi$$

Least Square Approach Using Lod Score

The Jensen-Jorgensen algorithm can be considered a least squares approach with the inverse of the estimated standard deviation as weight. Stam (1993) proposed using a LOD score for linkage as weight for the least squares approach. Equation (10.36) becomes

$$\frac{dLog(L)}{dm_k} = \frac{d\left[\sum\limits_{j=2i=1}^{n-1} \sum\limits_{i=1}^{j} L_{rij}\left(M_{ij} - \sum\limits_{k=i}^{j-1} m_k \right)^2 \right]}{dm_k} = 0 \qquad (10.38)$$

For the data in Table 10.10, the equations are

$$\left. \begin{aligned} 15.99\,(0.11 - m_1) + 4.29\,(0.46 - m_1 - m_2) &= 0 \\ 16.44\,(0.18 - m_2) + 4.29\,(0.46 - m_1 - m_2) &= 0 \end{aligned} \right\} \qquad Haldane$$

$$\left. \begin{aligned} 15.99\,(0.11 - m_1) + 4.29\,(0.35 - m_1 - m_2) &= 0 \\ 16.44\,(0.16 - m_2) + 4.29\,(0.35 - m_1 - m_2) &= 0 \end{aligned} \right\} \qquad Kosambi$$

The multi-locus map distance estimates are

$$\left. \begin{aligned} \hat{m}_1 &= 0.139 \\ \hat{m}_2 &= 0.209 \end{aligned} \right\} \qquad Haldane$$

$$\left. \begin{aligned} \hat{m}_1 &= 0.124 \\ \hat{m}_2 &= 0.174 \end{aligned} \right\} \qquad Kosambi$$

10.4.2 EM ALGORITHM

The EM algorithm is a set of procedures for obtaining a maximum likelihood estimate when experimental data are incomplete (Dempster *et al.* 1977). In genomics, the EM algorithm has been used for estimating recombination fraction and allelic frequency (Ott 1977; Weir 1996; Chapters 4 to 7 this book). Estimation of multipoint recombination fractions via the EM algorithm (Lander-Green algorithm, Lander and Green 1987 and 1991) includes four steps:

(1) Make an initial guess, $r^{old} = (r_1, r_2, ..., r_{l-1})$

(2) E step: Using r^{old} as if it were the true recombination fraction, compute the expected number of recombinant for each interval

(3) M step: Using the expected value as if it were the true value, compute the maximum likelihood estimate r^{new} for the recombination fraction

(4) Iterate the E and M steps until the likelihood converges to a maximum ($r^{new} \approx r^{old}$)

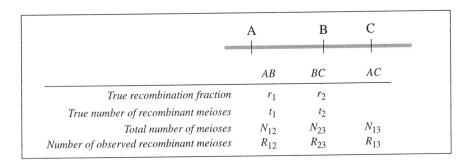

	AB	BC	AC
True recombination fraction	r_1	r_2	
True number of recombinant meioses	t_1	t_2	
Total number of meioses	N_{12}	N_{23}	N_{13}
Number of observed recombinant meioses	R_{12}	R_{23}	R_{13}

Figure 10.14 Notation used for deriving Lander-Green algorithm for three loci.

We can use an example with three loci to illustrate how the Lander-Green algorithm works. Figure 10.14 shows notations needed for deriving the Lander-Green algorithm for three loci. The locus order for the three loci is ABC. Numbers of recombinant meiosis R_{12}, R_{23} and R_{13} were observed among total numbers of meiosis N_{12}, N_{23} and N_{13} for the three possible genome segments AB, BC and AC, respectively. We wish to estimate the true recombination fractions r_1 and r_2 for segments AB and BC, respectively. The true numbers of recombinant meiosis (t_1, t_2) are sufficient statistics for the recombination fractions. Assuming absence of crossover interference, the Lander-Green algorithm can be implemented as follows:

1) $r^{old} = (r_1, r_2)$

2) E step: Contribution to t_1 from AB is R_{12} and from AC is

$$a_1 R_{13} + a_2 (N_{13} - R_{13})$$

where

$$a_1 = \frac{r_1 (1 - r_2)}{r_1 (1 - r_2) + r_2 (1 - r_1)}$$

is the probability of a recombination event in interval AB conditional on a recombination event in interval AC and

$$a_2 = \frac{r_1 r_2}{r_1 r_2 + (1 - r_1)(1 - r_2)}$$

is the probability of a recombination event in interval AB conditional on a recombination event not occurring in interval AC. The estimated number of recombinant meioses for this iteration are

$$t_1 = R_{12} + a_1 R_{13} + a_2 (N_{13} - R_{13})$$

$$t_2 = R_{23} + (1 - a_1) R_{13} + a_2 (N_{13} - R_{13})$$

3) Maximization: The new recombination fractions are

$$r_1^{new} = \frac{t_1}{N_{12} + N_{13}}$$

$$r_2^{new} = \frac{t_2}{N_{23} + N_{13}}$$

4) Iterate the E and M steps until $r^{new} \approx r^{old}$.

Table 10.11 Lander-Green algorithm is used to obtain multi-locus recombination fractions for 14 intervals flanked by 15 loci. The initial values (iteration 0) are the two-point recombination fractions. The estimates converge in 14 iterations.

Iteration	Recombination Fractions between Adjacent Loci	L
0	.086 .079 .112 .037 .003 .026 .121 .143 .072 .045 .154 .133 .048 .012	
1	.089 .088 .124 .038 .005 .023 .106 .119 .067 .043 .150 .142 .050 .011	-619.15
2	.086 .084 .122 .036 .005 .024 .111 .124 .073 .045 .148 .143 .049 .011	-617.96
3	.087 .086 .123 .036 .005 .024 .110 .121 .071 .044 .147 .144 .049 .011	-616.04
...	...	
14	.087 .085 .123 .036 .005 .024 .110 .121 .072 .044 .147 .144 .049 .011	-615.80

For more than 3 loci, the number of recombinant meioses can be estimated using

$$t_1 = R_{12} + a_1 R_{13} + a_2 (N_{13} - R_{13}) + a_3 R_{14} + a_4 (N_{14} - R_{14}) + \ldots$$

where a_3 is the probability of a recombination event in interval AB conditional on a recombination event in interval AD and a_4 is the probability of a recombination event in interval AB conditional on a recombination event not occurring in interval AD. Table 10.11 shows an estimation of multi-locus recombination fractions for 14 intervals flanked by 15 loci. The corresponding two-point recombination fractions are used as the initial estimates. The estimates converge in 14 iterations. The multi-point log likelihood increases from -619.15 in iteration 1 to -615.8 in the final iteration. Under the assumption of absence of crossover interference, the log likelihood is defined as

$$L(\hat{r}_1, \hat{r}_2, ..., \hat{r}_{l-1}) = \sum_{i=1}^{l-1} n_i [\hat{r}_i \log \hat{r}_i + (1 - \hat{r}_i) \log (1 - \hat{r}_i)] \qquad (10.39)$$

where $\hat{r}_i$ is the maximum likelihood estimate of recombination fraction between loci i and $i+1$, l is the total number of loci in this linkage group and n_i is the number of informative observations between loci i and $i+1$, which is the total sample size if the data are fully informative and no observation is missing.

The E-step is both the key and the difficulty of the Lander-Green algorithm. Lander and Green (1987) called the difficulty a "genetic reconstruction problem." The main task of the genetic reconstruction is to estimate the expected number of recombinant meiosis given the true recombination fractions. Lander and Green explained methods for genetic reconstruction for situations of known genotype. We have discussed the known genotype situation here. Most genomic experiments using mapping populations of controlled crosses can be considered to be this type. For general pedigree situations, Lander and Green explain approaches using the classical Elston-Stewart algorithm (Elston and Steward 1971; Ott 1972) and using hidden Markov chain algorithm (Lander and Green 1987). The Lander-Green algorithm greatly reduces the computational complexity of obtaining multi-locus recombination fractions from traditional approaches. It has been further implemented in computer packages such as Mapmaker, Gmendel and PGRI.

10.4.3 JOINT ESTIMATION OF RECOMBINATION AND INTERFERENCE

Zhao et al. (1990) proposed a method to estimate recombination fractions and crossover interference simultaneously using a multiplicative model. They defined the model for n loci with $n-1$ segments as

$$Pr(c) = \Delta^{-1} \exp\left(\sum_{i=1} \theta_i c_i + \sum_{i>j} \omega_{ij} c_i c_j + \sum_{ijk} \omega_{ijk} c_i c_j c_k + ... \right) \qquad (10.40)$$

where $c_i = 1$ or $c_i = 0$ denote crossover occurring or not occurring in the ith genome segment, θ_i, ω_{ij} and ω_{ijk} are canonical parameters which are functions of recombination fraction in the ith genome segment, two-way crossover interference between the ith and the jth genome segments and three-way crossover interference among the corresponding segments and

$$\Delta = \sum \exp\left(\sum_{i=1} \theta_i c_i \times \sum_{i>j} \omega_{ij} c_i c_j \times \sum_{ijk} \omega_{ijk} c_i c_j c_k \times ... \right) \qquad (10.41)$$

is a normalizing constant with summation over all possible values of c. Equation (10.40) provides a framework to explore potential high order crossover interference. The likelihood function can be written as

$$L = \sum \left(\sum_{i=1} \theta_i c_i + \sum_{i>j} \omega_{ij} c_i c_j + \sum_{ijk} \omega_{ijk} c_i c_j c_k + ... - \log \Delta \right)$$

where summation is overall observations. The canonical parameters can be estimated using a computer package for generalized linear regression, GLIM (McCullagh and Nelder 1989). GLIM allows user-specified distribution and link functions. For solving the canonical parameters in Equation (10.40), the Poisson distribution with an exponential link function can be used. Commonly, the canonical parameters θ_i, ω_{ij} and ω_{ijk} are interpreted as main effects, two-way interactions and three-way interactions. Zhao et al. (1990) also gave alternative interpretations of the canonical parameters in terms of conditional odds, odds ratio or ratio of odds ratio, such as

$$\theta_i = \log \frac{Pr[c_i = 1]}{Pr[c_i = 0]}$$

$$= \log \frac{r_i}{1 - r_i} \tag{10.42}$$

$$\omega_{ij} = \log \frac{Pr[c_i = 1, c_j = 1] \, Pr[c_i = 0, c_j = 0]}{Pr[c_i = 1, c_j = 0] \, Pr[c_i = 0, c_j = 1]}$$

The canonical parameters can be converted to recombination fractions and crossover interference using Equation (10.42).

10.4.4 SIMULATION APPROACH

Another way to approach multi-locus map distance is to find a mapping function which fits the data well. This approach involves comparing multi-locus likelihoods of the data using different mapping functions (Weeks *et al.* 1993). Because the distribution of the likelihood difference is unknown, simulation is needed to determine significance of the likelihood difference. I call this the simulation approach.

Crossover Distribution

The expected multi-locus gametic frequencies produced by parents in terms of multi-locus recombination fractions or map distances are key elements to build a multi-locus likelihood. Under the assumption of absence of crossover interference, the expected multi-locus gametic frequency is just the product of appropriate two-locus gametic frequencies. For a four-locus example, the expected frequency of ABCd produced by a parent with genotype ABCD/abcd is

$$0.125 \, (1 - r_1) \, (1 - r_2) \, r_3$$

where r_1, r_2 and r_3 are recombination fractions corresponding to the three intervals. The probability can also be interpreted as the product of gametic frequencies $0.5 \, (1 - r_1), 0.5 \, (1 - r_2)$ and $0.5 r_3$ for AB, BC and Cd, respectively. The multi-locus log likelihood of the experimental data can be defined as

$$L = \sum_{i=1}^{2^n} g_i \log p_i \tag{10.43}$$

where 2^n is the number of possible haplotypes, g_i is the count for the *ith* haplotype and p_i is the corresponding expected frequency.

If the assumption of absence of crossover interference does not hold, the expected gametic frequency cannot be obtained using the simple product. It can be obtained through a crossover distribution. The crossover distribution contains the expected frequencies for all the possible crossover combinations. For n loci in a linkage group, there are 2^{n-1} possible crossover combinations (Figure 10.12). The expected gametic frequencies can be obtained from the crossover distribution. Use matrix notation to denote the crossover combinations (**C**), the crossover distribution (**P**) and map distances (**M**) among the n loci; they are

$$
C = \begin{bmatrix} 0 & 0 & \dots & 0 \\ 0 & 0 & \dots & 1 \\ \dots & \dots & \dots & \dots \\ 1 & 1 & \dots & 0 \\ 1 & 1 & \dots & 1 \end{bmatrix}_{(n-1) \times 2^{n-1}}
$$

$$
P = \begin{bmatrix} Pr\,(00\dots 0) \\ Pr\,(00\dots 0) \\ \dots \\ Pr\,(11\dots 0) \\ Pr\,(11\dots 1) \end{bmatrix}_{1 \times 2^{n-1}} \tag{10.44}
$$

$$
M = \begin{bmatrix} m_1 \\ m_2 \\ \dots \\ m_{n-1} \end{bmatrix}
$$

where "0" and "1" in the crossover combination matrix denote "no-crossover" and "1" in an genome interval, respectively. An equation, which was developed originally by Schnell (1961) (see also Liberman and Karlin 1984; Weeks *et al.* 1993), derives a crossover distribution which can be written as

$$
P = \frac{1}{2^{n-1}} (-1)^{CC'} [1 - 2f(CM)] \tag{10.45}
$$

using the matrix notation, where $f(°)$ is a mapping function used to translate map distance into recombination fraction. For a 3-locus situation, the information needed for Equation (10.45) is

$$
CC' = \begin{bmatrix} 0 & 0 \\ 0 & 1 \\ 1 & 0 \\ 1 & 1 \end{bmatrix} \begin{bmatrix} 0 & 0 & 1 & 1 \\ 0 & 1 & 0 & 1 \end{bmatrix} = \begin{bmatrix} 0 & 0 & 0 & 0 \\ 0 & 1 & 0 & 1 \\ 0 & 0 & 1 & 1 \\ 0 & 1 & 1 & 2 \end{bmatrix}
$$

$$
(-1)^{cc} = \begin{bmatrix} 1 & 1 & 1 & 1 \\ 1 & -1 & 1 & -1 \\ 1 & 1 & -1 & -1 \\ 1 & -1 & -1 & 1 \end{bmatrix}
$$

$$
1 - 2f(CM) = 1 - 2f \left\{ \begin{bmatrix} 0 & 0 \\ 0 & 1 \\ 1 & 0 \\ 1 & 1 \end{bmatrix} \begin{bmatrix} m_1 \\ m_2 \end{bmatrix} \right\} = \begin{bmatrix} 1 \\ 1 - 2f(m_2) \\ 1 - 2f(m_1) \\ 1 - 2f(m_1 + m_2) \end{bmatrix}
$$

If the map distances are 0.1 and 0.2 for the two intervals, the crossover distribution is

$$
\begin{bmatrix} p\,(00) \\ p\,(01) \\ p\,(10) \\ p\,(11) \end{bmatrix} = \frac{1}{4} \begin{bmatrix} 1 & 1 & 1 & 1 \\ 1 & -1 & 1 & -1 \\ 1 & 1 & -1 & -1 \\ 1 & -1 & -1 & 1 \end{bmatrix} \begin{bmatrix} 1 \\ 1-2f(0.2) \\ 1-2f(0.1) \\ 1-2f(0.3) \end{bmatrix} = \frac{1}{4} \begin{bmatrix} 1 & 1 & 1 & 1 \\ 1 & -1 & 1 & -1 \\ 1 & 1 & -1 & -1 \\ 1 & -1 & -1 & 1 \end{bmatrix} \begin{bmatrix} 1 \\ 0.6703 \\ 0.8187 \\ 0.5488 \end{bmatrix} = \begin{bmatrix} 0.7595 \\ 0.1499 \\ 0.0757 \\ 0.01595 \end{bmatrix}
$$

using Haldane's mapping function and is

$$
\begin{bmatrix} p\,(00) \\ p\,(01) \\ p\,(10) \\ p\,(11) \end{bmatrix} = \frac{1}{4} \begin{bmatrix} 1 & 1 & 1 & 1 \\ 1 & -1 & 1 & -1 \\ 1 & 1 & -1 & -1 \\ 1 & -1 & -1 & 1 \end{bmatrix} \begin{bmatrix} 1 \\ 0.8100 \\ 0.9013 \\ 0.7315 \end{bmatrix} = \begin{bmatrix} 0.86065 \\ 0.0900 \\ 0.0443 \\ 0.00505 \end{bmatrix}
$$

using Kosambi's mapping function. Expected gametic frequencies can be obtained from the crossover distributions. For example, a parent with genotype AbC/aBc expects to produce

$$
Pr \begin{bmatrix} ABC \\ ABc \\ AbC \\ Abc \\ aBC \\ aBc \\ abC \\ abc \end{bmatrix} = \begin{bmatrix} 0.5p\,(11) \\ 0.5p\,(10) \\ 0.5p\,(00) \\ 0.5p\,(01) \\ 0.5p\,(01) \\ 0.5p\,(00) \\ 0.5p\,(10) \\ 0.5p\,(11) \end{bmatrix} = \begin{bmatrix} 0.0080 \\ 0.0379 \\ 0.3798 \\ 0.0750 \\ 0.0750 \\ 0.3798 \\ 0.0379 \\ 0.0080 \end{bmatrix} (Haldane) \quad \begin{bmatrix} 0.0025 \\ 0.0222 \\ 0.4303 \\ 0.0450 \\ 0.0450 \\ 0.4303 \\ 0.0222 \\ 0.0025 \end{bmatrix} (Kosambi)
$$

Likelihood functions for the observed data can be constructed using different mapping functions. A comparison among the likelihood functions may determine which mapping function fits the data the best. However, as mentioned by Weeks *et al.* (1993), the comparison among the likelihoods may not be able to be formulated as a well-defined log likelihood ratio test because the models involved may not be nested. The distribution of the likelihood difference in this situation is largely unknown. It may be possible to determine the distribution empirically by simulation.

The objective of the comparison among the likelihoods is to test if models with crossover interference fit the observed data better than does Haldane's mapping function. In practice, simulation under no interference can be implemented easily. This approach will be described including a simulation using the following example data.

Example

In a mapping experiment, the genotypic frequencies shown in Table 10.10 are observed in progeny of a classical backcross between a parent with genotype AbC/aBc and another homozygous parent (abc/abc). Here are the estimated recombination fractions and map distances for the two segments

$$(\hat{r}_1 = 0.1),\ (\hat{r}_2 = 0.16)$$

$$(\hat{m}_1 = 0.112),\ (\hat{m}_2 = 0.193) \qquad Haldane$$

$$(\hat{m}_1 = 0.101),\ (\hat{m}_2 = 0.166) \qquad Kosambi$$

The crossover distribution using Haldane's mapping function is

$$
\begin{bmatrix} p(00) \\ p(01) \\ p(10) \\ p(11) \end{bmatrix} = \frac{1}{4}\begin{bmatrix} 1 & 1 & 1 & 1 \\ 1 & -1 & 1 & -1 \\ 1 & 1 & -1 & -1 \\ 1 & -1 & -1 & 1 \end{bmatrix}\begin{bmatrix} 1 \\ 1-2f(0.\dot{1}93) \\ 1-2f(0.112) \\ 1-2f(0.305) \end{bmatrix} = \frac{1}{4}\begin{bmatrix} 1 & 1 & 1 & 1 \\ 1 & -1 & 1 & -1 \\ 1 & 1 & -1 & -1 \\ 1 & -1 & -1 & 1 \end{bmatrix}\begin{bmatrix} 1 \\ 0.68 \\ 0.80 \\ 0.544 \end{bmatrix} = \begin{bmatrix} 0.756 \\ 0.144 \\ 0.084 \\ 0.016 \end{bmatrix}
$$

and using Kosambi's mapping function is

$$
\begin{bmatrix} p(00) \\ p(01) \\ p(10) \\ p(11) \end{bmatrix} = \frac{1}{4}\begin{bmatrix} 1 & 1 & 1 & 1 \\ 1 & -1 & 1 & -1 \\ 1 & 1 & -1 & -1 \\ 1 & -1 & -1 & 1 \end{bmatrix}\begin{bmatrix} 1 \\ 1-2f(0.\dot{1}66) \\ 1-2f(0.101) \\ 1-2f(0.267) \end{bmatrix} = \frac{1}{4}\begin{bmatrix} 1 & 1 & 1 & 1 \\ 1 & -1 & 1 & -1 \\ 1 & 1 & -1 & -1 \\ 1 & -1 & -1 & 1 \end{bmatrix}\begin{bmatrix} 1 \\ 0.68 \\ 0.80 \\ 0.5113 \end{bmatrix} = \begin{bmatrix} 0.7478 \\ 0.1522 \\ 0.0922 \\ 0.0078 \end{bmatrix}
$$

Expected gametic frequencies can be obtained from the crossover distributions as

$$
Pr\begin{bmatrix} ABC \\ ABc \\ AbC \\ Abc \\ aBC \\ aBc \\ abC \\ abc \end{bmatrix} = \begin{bmatrix} 0.5p(11) \\ 0.5p(10) \\ 0.5p(00) \\ 0.5p(01) \\ 0.5p(01) \\ 0.5p(00) \\ 0.5p(10) \\ 0.5p(11) \end{bmatrix} = \begin{bmatrix} 0.008 \\ 0.042 \\ 0.378 \\ 0.072 \\ 0.072 \\ 0.378 \\ 0.042 \\ 0.008 \end{bmatrix}(Haldane) \quad \begin{bmatrix} 0.0039 \\ 0.0461 \\ 0.3739 \\ 0.0761 \\ 0.0761 \\ 0.3739 \\ 0.0461 \\ 0.0039 \end{bmatrix}(Kosambi)
$$

In this example, the expected genotypic frequencies in the progeny are the same as the expected gametic frequencies. The log likelihoods for using the two mapping functions are

$$
\begin{aligned}
148\log 0.756 + 32\log 0.144 + 20\log 0.084 &= -152.951 \quad Haldane \\
148\log 0.7478 + 32\log 0.1522 + 20\log 0.0922 &= -150.934 \quad Kosambi
\end{aligned} \tag{10.46}
$$

Simulation

Traditionally, we use

$$
-2\left(\log L_{Haldane} - \log L_{Kosambi}\right)
$$

which is distributed as a chi-square with certain degrees of freedom to test the difference between the two models. However, we cannot determine the degrees of freedom here, because there is no parameter difference between using Haldane's and Kosambi's mapping functions (or the models are not nested). Kosambi's mapping function may fit the data in Table 10.12 better than does Haldane's, because the likelihood for using Kosambi's mapping function (-150.934) is larger than for using Haldane's (-152.951) in Equation (10.46). However, we cannot say that the likelihood difference is statistically significant because the statistical distribution of the difference is unknown. Specifically, we do not know how frequently we can observe a likelihood difference which is greater than or equal to

$$-2\,(-152.951 + 150.934) \;=\; 4.033$$

For this reason, simulation can be used to obtain an empirical distribution of the test statistic. For the example in Table 10.12, we can simulate data with 200 observations under assumptions of absence of crossover interference and that the true recombination fractions for the two segments are 0.1 and 0.16. For each simulated data set, the likelihood difference can be obtained. If the simulation is replicated a large number of times, then an empirical distribution of the statistic can be obtained.

Table 10.12 Observed counts and expected frequencies in progeny of a triple backcross between two parents with genotypes of AbC/aBc and abc/abc.

	Count	Expected Frequency	
		Haldane	Kosambi
AaBbCc	0	0.0079	0.00395
AaBbcc	15	0.0421	0.04605
AabbCc	75	0.3779	0.37395
Aabbcc	14	0.0721	0.07605
aaBbCc	18	0.0721	0.07605
aaBbcc	73	0.3779	0.37395
aabbCc	5	0.0421	0.04605
aabbcc	0	0.0079	0.00395

Figure 10.15 shows an empirical distribution obtained through 5,000 replications of the simulation for the example in Table 10.12. The observed likelihood difference 4.033 is ranked at 4,969 in the 5,000 simulated values. So, the probability that an observed likelihood difference is greater than or equal to 4.033 is approximately (5000 - 4969)/5000 = 0.0062. If we use 0.01 as a significance level, then Kosambi's mapping function fits the example data significantly better than does Haldane's. We also can interpret the crossover interference result in the data as significant.

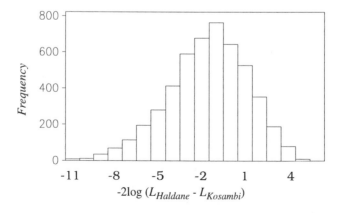

Figure 10.15 Empirical distribution of -2log ($L_{Haldane}$ - $L_{Kosambi}$) obtained through 5,000 replications of simulation.

Here, only the comparison between Haldane's and Kosambi's mapping functions has been described. This approach can be applied to any existing mapping function. For example, besides Haldane's and Kosambi's mapping functions, comparisons between the mapping functions of Sturt (1976), Felsenstein (1979), Rao *et al.* (1977), Karlin and Liberman (1983) and Goldgar and Fain (1989) were implemented in the computer program LINKAGE (Weeks *et al.* 1993).

Table 10.13 Crossover distribution for a six-locus model with map distances 0.1, 0.2, 0.15, 0.3 and 0.25 for the five segments.

Crossover Combination	Haldane	Kosambi	Sturt	Felsenstein (K = 0.5)	Karlin (Binomial, N = 2)
00000	0.4112057	0.3057180	0.3864752	0.3550201	0.30625
00001	0.1007120	0.1150025	0.1075643	0.1099266	0.125
00010	0.1197894	0.1370078	0.1281400	0.1309739	0.15
00011	0.0293387	0.0437131	0.0297946	0.0357099	0.0375
00100	0.0612224	0.0694839	0.0652320	0.0665572	0.075
00101	0.0149945	0.0209570	0.0151321	0.0176542	0.01875
00110	0.0178349	0.0255834	0.0180439	0.0212803	0.0225
00111	0.0043681	-0.0039409	0.0038403	0.0037927	0.0
01000	0.0811619	0.0924041	0.0865691	0.0884077	0.1
01001	0.0198781	0.0282894	0.0200940	0.0236250	0.025
01010	0.0236435	0.0345050	0.0239605	0.0284675	0.03
01011	0.0057907	0.0056589	0.0051012	0.0051885	0.0
01100	0.0120838	0.0164979	0.0121689	0.0140582	0.015
01101	0.0029595	0.0020318	0.0025874	0.0023399	0.0
01110	0.0035202	0.0028348	0.0030870	0.0029324	0.0
01111	0.0008622	-0.0023160	0.0006238	-0.0004980	0.0
10000	0.0409841	0.0463942	0.0436344	0.0444841	0.05
10001	0.0100378	0.0138486	0.0101176	0.0117362	0.0125
10010	0.0119392	0.0169131	0.0120645	0.0141500	0.015
10011	0.0029241	0.0024954	0.0025671	0.0024820	0.0
10100	0.0061019	0.0080721	0.0061272	0.0069792	0.0075
10101	0.0014945	0.0008756	0.0013021	0.0011147	0.0
10110	0.0017776	0.0012366	0.0015534	0.0013996	0.0
10111	0.0004354	-0.0012170	0.0003138	-0.0002800	0.0
11000	0.0080892	0.0109006	0.0081364	0.0093442	0.01
11001	0.0019812	0.0012776	0.0017296	0.0015293	0.0
11010	0.0023565	0.0017905	0.0020635	0.0019180	0.0
11011	0.0005772	-0.0015750	0.0004169	-0.0003490	0.0
11100	0.0012044	0.0006128	0.0010466	0.0008583	0.0
11101	0.0002950	-0.0008910	0.0002113	-0.0002230	0.0
11110	0.0003508	-0.0010130	0.0002522	-0.0002430	0.0
11111	0.0000859	-0.0010330	0.0000492	-0.0003380	0.0

Multilocus Feasible Map Function

From Equation (10.45), a sufficient condition that a mapping function is multi-locus feasible can be derived from the inequalities

$$P = \frac{1}{2^{n-1}} (-1)^{cc} [1 - 2f(CM)] \geq 0$$

for all 2^{n-1} possible crossover combinations, which is

$$(-1)^k \frac{d^k}{dm^k} f^{-1}(m) \leq 0 \qquad k = 1, 2, \ldots \qquad (10.47)$$

for all $m \geq 0$, where $f^{-1}(m)$ is a function converting map distance to recombination fraction (Liberman and Karlin 1984). The sufficient condition of Equation (10.47) for multi-locus feasibility is not necessary in general. Liberman and Karlin (1984) gave a necessary condition for multi-locus feasibility as

$$(-1)^k \frac{d^k}{dm^k} f^{-1}(0) \leq 0 \qquad k = 1, 2, \ldots \qquad (10.48)$$

Some of the commonly used mapping functions, particularly those of Haldane, Sturt and Karlin and Liberman, are multi-locus feasible. The mapping functions of Kosambi, Rao *et al.* and Felsenstein are not multi-locus feasible.

Table 10.13 shows the crossover distributions for a six-locus model with map distances 0.1, 0.2, 0.15, 0.3 and 0.25 Morgan for the five segments. It is obvious that the mapping functions of Kosambi and Felsenstein are not multi-locus feasible because of the negative probabilities.

Practical Implementation

In practice, the difficult part of the simulation approach is obtaining the crossover distribution. Here are the SAS source codes for computing the crossover distribution listed in Table 10.13.

```
data a;
  do i1=0,1;
    do i2=0,1;
      do i3=0,1;
        do i4=0,1;
          do i5=0,1;
          output;
    end; end; end; end; end;
proc iml;
use a;
read all into c;
y=(-1)##(c*c`);
d={0.1,0.2,0.15,0.3,0.25};
m=c*d;
h=1-exp(-2#m);
ks=tanh(2#m);
st=1-(1-m/1.5)#exp(-2#m/1.5);
fs=(1-exp(-3#m))/(1+0.5#exp(-3#m));
bi=1-(1-m)##2;
ph=0.03125#(y*(1-h));
pks=0.03125#(y*(1-ks));
pst=0.03125#(y*(1-st));
pfs=0.03125#(y*(1-fs));
pbi=0.03125#(y*(1-bi));
print c ph pks pst pfs pbi;
run;
```

Readers may need to modify the source codes for their own specific problems. The parts which may need modification are the data procedure to generate the crossover combinations and the matrix containing the map distances for the segments (matrix d in the source codes).

10.5 MARKER COVERAGE AND MAP DENSITY

10.5.1 DEFINITIONS

The quality of a genomic map can be quantified using the confidence of estimated locus order and locus distribution on the map. Ideally, we want a genomic map with a high level of confidence in the estimated locus order (*e.g.*, 99%), with markers evenly distributed on the map and with sufficient density (*e.g.*, at least one marker in a 5cM segment). The above criteria are the properties of a genomic map itself. Marker coverage has two important aspects:

- proportion of genome covered by markers located on the genomic map
- maximum genome segment length between two adjacent markers

The second aspect can be interpreted as a function of map density in terms of the maximum gap on the map. If markers are randomly located on the genome, then the two aspects have a positive correlation. The higher the map density (or the smaller the maximum gap is), the higher the marker coverage. The lower the map density (or the larger the maximum gap is) the lower the marker coverage. However, this relationship may be not true when markers are not randomly distributed. For example, markers may be screened and selected using some criteria related to marker distribution.

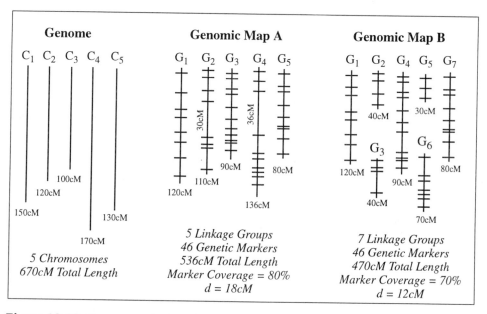

Figure 10.16 Genome and genomic map.

There are several aspects to low marker coverage. For example, let us consider a set of 46 markers on linkage groups G_1, G_2, G_3, G_4 and G_5 of the genomic map A in Figure 10.16 that cover 4/5, 11/12, 9/10, 13/17 and 8/13 of chromosomes C_1, C_2, C_3, C_4 and C_5, respectively. A locus linked within an 18cM distance is considered to be covered by markers. On average, 80% of the genome is covered by the 46 markers. In this case, the uncovered region of the genome corresponds to the two ends of the chromosomes. Even the large gap on linkage group G_4 is considered covered by markers, because the loci in the gap are linked with one of the markers flanking the gap within 18cM distance maximum. To increase the coverage, markers located at the ends of the linkage groups can be used to screen additional markers. If only loci linked with a marker within a 12cM distance are considered to be covered by markers, then a portion of both chromosomes C_2 and C_4 is not covered. Each of the linkage groups G_2 and G_4 of the genomic map A is broken into two separate linkage groups (genomic map B in Figure 10.16). The average marker coverage is reduced to 70% because of the coverage criterion. In practice, a low marker coverage can be a result of many factors, such as low linkage information content, crossover hot spots or tight linkage grouping criteria. In these cases, markers which can bring the small linkage groups together increase the coverage.

10.5.2 FACTORS INFLUENCING MARKER COVERAGE AND MAP DENSITY

There are many factors influencing marker coverage and genomic map density, such as genome length, number of markers, distribution of marker polymorphism, distribution of markers on genome, crossover distribution on genome, mapping population size and type and mapping strategy. There are also several ways to estimate marker coverage and map density according to the objectives of the estimation and available information. Estimating the marker coverage and map density is not only a way to quantify the quality of a genomic map, but also a way to design a more efficient genomic experiment.

Number of Markers

Figure 10.17 shows a simulated distribution of 50 markers on a genome based on the assumption that the markers are randomly distributed. The 50 markers cover an average of 88.5% of the genome. The maximum distance between adjacent markers is 22.7cM, so, the largest genome segment flanked by two markers is, at most, 22.7cM. Under the same assumptions, 25 markers cover 63.6% of the genome and the maximum map distance between adjacent markers is 26.5cM (Figure 10.18). The five linkage groups for the 50 markers correspond to the five chromosomes. When the number of markers is reduced to 25, some genome segments fail to join together to form complete linkage groups.

Marker and Crossover Distribution

How are markers distributed on a genome? This question is important for experimental design and data analysis. Low marker coverage or large gaps between adjacent markers may be the result of nonrandom marker distribution. It is common to assume that markers are randomly distributed on a genome and that polymorphism is independent of genome location. Figures 10.17 and 10.18 show marker distribution under the assumption of randomness. When this assumption is violated, the marker coverage and map density may change. For example, as shown by Figure 10.19, markers can be more frequent in some segments of the genome than others. This can be a result of more polymorphism in the segments or a change in the ratio between physical distance and crossover frequency. In general, marker coverage and map density decrease if markers are distributed on the genome nonrandomly. The 50 simulated markers in Figure 10.19 cover 75.6% of the genome instead of the 88.5% in Figure 10.17. The nonrandom distribution of markers on a genome also limits the estimation of marker coverage and map density.

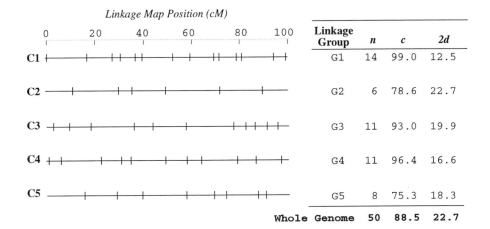

Figure 10.17 Simulated marker distribution based on the assumption that markers are randomly distributed on the genome. Total genome length (L) = 500cM and total number of markers (n) = 50. The result is a genomic map with an average of 88.5% marker coverage (c), maximum map distance between adjacent markers (2d) of 22.7cM at a whole genome level and an average map distance between adjacent markers of 10cM.

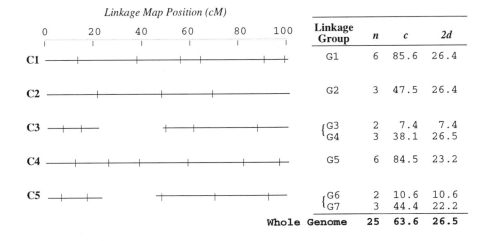

Figure 10.18 Simulated marker distribution based on the assumption that markers are randomly distributed on the genome. Total genome length (L) = 500cM and total number of markers (n) = 25. The result is a genomic map with an average of 63.6% marker coverage (c), maximum map distance between adjacent markers (2d) of 26.5cM at a whole genome level and an average map distance between adjacent markers of 17.7cM. Seven linkage groups correspond to five chromosomes, based on linkage grouping criterion that two markers are linked if the recombination fractions between markers is less than 30%.

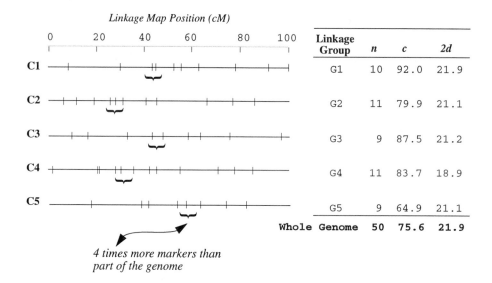

Figure 10.19 Simulated marker distribution based on the assumption that there are more markers in the indicated segments than in other parts of the genome. Total genome length (L) = 500cM and total number of markers (n) = 50. The result is a genomic map with an average of 75.6% marker coverage (c), maximum map distance between adjacent markers (2d) of 21.9cM at a whole genome level and an average map distance between adjacent markers of 8.5cM.

Mapping Population

Population size and type have an effect on marker coverage and map density through linkage information content. Small population size and low linkage information content can cause undetected linkage when a small number of markers are used and unresolved linkage when a large number of markers are used. In general, linkage information content decreases as recombination fraction increases. Linkage among a small number of markers is usually loose. Unattached genome segments (because of undetected linkage) are the major source of low marker coverage. Increasing population size or use of a mapping population with high linkage information can increase the probability of joining the genome segments and increasing marker coverage. When a large number of markers are used in a genomic mapping experiment, map density is not limited by the number of markers, but rather by number of recombination events that occurred in meiosis and are represented in the progeny of the mapping population. There is a limitation on the linkage that can be resolved for a relatively small population size. Even if a large number of markers are available, a high density genomic map may not be obtainable. In such a case, a large portion of markers are clustered and large gaps are left between the clusters.

Data Analysis

Data analysis also affects marker coverage and map density. The wrong gene order may inflate the size of a genomic map, resulting in overestimated map coverage and under-estimated map density. Different mapping functions provide different estimates of genomic map length and may over or under estimate the true marker coverage and map density. Different linkage grouping criteria can result in different linkage group patterns and different estimates of marker coverage and map density. In general, more stringent criteria (*e.g.*, high lod score and low recombination fraction) will result in a large number of linkage groups, low marker coverage and high map density.

10.5.3 PREDICTION OF MARKER COVERAGE AND MAP DENSITY

With known total genome length and a genomic map, marker coverage and map density can be easily estimated. Marker coverage is the simple ratio between genomic map length and the total genome length. Map density is the average or the maximum map distance between adjacent markers (gap).

Prediction of Map Density and Marker Coverage

Based on the assumption that markers are randomly distributed, methods for predicting marker coverage and map density have been developed (Elston and Lange 1975; Botstein *et al.* 1980; Lange and Boehnke 1982). There are basically two approaches:

- marker coverage approach
- confidence probability approach

Marker coverage is the proportion of the genome flanked by two markers with a certain minimum map distance (say, less than $2d$ M) between them over the whole genome. The confidence probability is that at least one marker is located within a $2d$ M genome segment. These approaches are approximately the same when the ratio between the minimum map distance and genome length is small. They differ when the ratio is large.

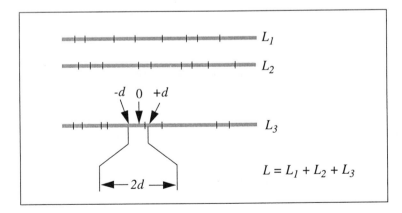

Figure 10.20 Probability that a random marker is located within a $2d$ cM genome segment is $2d/L$.

Let us first assume that a genome is a total of L cM long with three chromosomes (Figure 10.20). If we want any point on the genome to be within a map distance d cM from a marker, then the markers should be located at most $2d$ cM apart. Assuming that markers are evenly spaced on a genome, $L/(2d)$ markers are needed to have markers $2d$ cM apart. For example, 100 markers will cover a whole genome 2M long with one marker every 10 cM distance. Following this logic, 165 markers are sufficient to cover the whole human genome, on the average 3.3M long with one marker every 10 cM. If markers are randomly distributed, tightly linked marker clusters and relatively large gaps between adjacent markers usually exist simultaneously in typical genomic maps. We often assume that markers are randomly distributed. If we randomly sample a marker from the genome, then the probability that the marker is located at a specific site is equal for all possible genome sites.

Lange and Boehnke (1982) estimate the marker coverage (c) given a maximum map distance (d M) (at least one marker is located on the d M genome segment). First, assume a genome with a unit length (*e.g.*, 1 Morgan). The probability that a random marker will be located beyond a minimum distance d M is $1 - 2d$. The probability that n random markers will be located beyond the minimum distance, or that the genome is not covered by the markers, is

$$1 - c = (1 - 2d)^n$$

If $n = 1/(2d)$ (there is a marker within a $2d$ genome segment), then the probability becomes

$$1 - c = (1 - 2d)^{1/(2d)} \approx e^{-1} = 0.368$$

This is the probability that the genome is not covered by n markers. The marker coverage can be estimated using

$$c = 1 - e^{-2dn/L} \tag{10.49}$$

Given the marker coverage, genome length and the minimum distance, the number of markers needed can be solved from Equation (10.49) and estimated using

$$n = \frac{-L \log(1-c)}{2d} \tag{10.50}$$

If we randomly sample a marker, the probability that the marker is located on a $2d$ M genome segment (or the probability that the marker is located within d M) is approximately $2d/L$ (Figure 10.20). The probability that the marker is not located on a $2d$ M genome segment is $1 - 2d/L$. If n markers are sampled, the probability that none of the markers is located on a $2d$ M genome segment is $(1 - 2d/L)^n$. So, the probability that at least one marker is located within a $2d$ M genome segment is

$$P = 1 - \left(1 - \frac{2d}{L}\right)^n \tag{10.51}$$

The number of markers needed to have at least one marker within a $2d$ M genome segment, given genome length and a confidence probability, can be solved from Equation (10.51) and is

$$n = \frac{\log(1-P)}{\log(1 - 2d/L)} \tag{10.52}$$

The map density or the distance between adjacent markers can also be solved from Equation (10.51) and is

$$2d = L[1 - (1 - P)^{1/m}]$$

When the distance is small relative to genome size (say $2d/L \leq 0.1$), Equation (10.52) can be approximated as

$$n = \frac{\log(1 - P)}{\log(1 - 2d/L)} \approx -\frac{L\log(1 - P)}{2d} \qquad (10.53)$$

which has the same form as Equation (10.50). Under this condition, the confidence probability is approximately equal to marker coverage.

Figure 10.21 shows the predicted number of markers needed to have a certain map density and the confidence probabilities for a genome of one Morgan. The number of markers needed increases rapidly as the confidence probability increases. To increase the confidence probability from 90% to 99%, the number of markers needed for a certain map density increases approximately onefold (Figure 10.21 and Table 10.14). For example, to reach a density of 20cM for a 1M long genome, 57 and 113 markers are needed for confidence probabilities of 90% and 99%, respectively (Table 10.14). The number of markers needed changes as genome length and map density change in an approximately simple manner for constant confidence probability. For example, with a confidence probability of 80%, to increase the density 10 times, from 10cM to 1cM, the number of markers needed increases approximately 10 times from 16 to 161 for a 1M genome. Similarly, for a 5M genome, the number of markers needed increases approximately 10 times, from 81 to 804. This ratio can be used as an approximate guideline for designing a genomic experiment. If one wants to increase map density twofold, the number of markers needed should increase twofold.

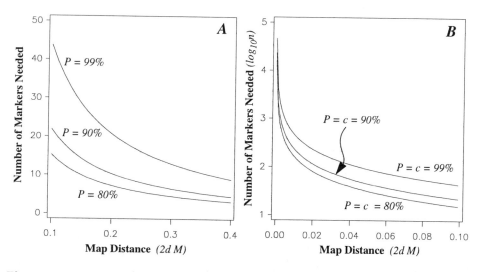

Figure 10.21 Predicted number of markers needed is plotted against the map density (map distance between adjacent markers) for different confidence probabilities (P). A: for confidence probabilities of 80%, 90% and 99% and map distance greater than 10cM. B: For map distance less than 10cM. In B, the number of markers needed is approximately equal for equal marker coverage (c) and the confidence probability (P).

Table 10.14 Number of markers needed for some combinations of map distance 2d (cM) and the confidence probability (P) or the marker coverage (c), assuming genome length is one Morgan.

2d (cM)	P			c		
	0.8	0.9	0.99	0.8	0.9	0.99
1	161	230	459	161	231	461
5	32	45	90	33	47	93
10	16	22	44	17	24	47
20	8	11	21	9	12	24
30	5	7	13	6	8	16
40	4	5	10	5	6	12

For a small map distance between adjacent markers, the ratio between the map distance and genome length $(2d/L)$ is small. In this case, the marker coverage and the confidence probability are approximately the same and they have the same effect on the number of markers needed (Figure 10.21B and Table 10.14) because $\log(1 - 2d/L) \approx -2d/L$. Figure 10.21B is applied to both the confidence probability and the marker coverage approaches. If the ratio is large, Figure 10.22 plots the number of markers needed for certain marker coverages under different minimum map distance criteria. Figure 10.22 has a similar shape to Figure 10.21A. However, the number of markers needed for certain marker coverage is consistently larger than the number needed for equal confidence probability.

Both the marker coverage and the confidence probability approaches are applied under the assumption that the markers are randomly distributed on the genome. This assumption implies:

- markers are randomly obtained without screening
- marker polymorphism is randomly distributed on the genome

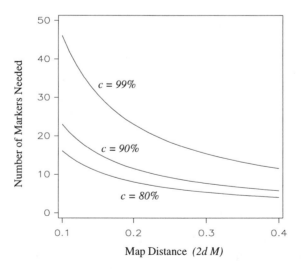

Figure 10.22 Predicted number of markers needed is plotted against the map density (map distance between adjacent markers) for different confidence probabilities (P) or marker coverage (c).

- crossover occurs randomly over the genome

as discussed before, these implications are usually not true in practice. In practical experiments, the number of markers needed may be much less than the predicted number using Equation (10.50) and Equation (10.52), if previous information and screening procedures are used. The predictions are most useful for genomes with little information.

Simulation Approach

In most cases, the total genome length of an organism is not estimated accurately or precisely, but sometimes even approximate genome lengths are not available. In any case, an approach for estimating marker coverage and map density not based on a pre-determined genome length is ideal. What is needed to design future experiments may not be the marker coverage, but rather how much the coverage and density increase if a certain number of markers or offspring are added to the data. The above arguments lead to a simulation approach to estimate changes of marker coverage and map density. The simulation approach uses the observed data to estimate changes of marker coverage and map density by simulating different experimental conditions. For example, we can sample from the data for different numbers of markers and see how the marker coverage changes.

In a mapping experiment, 500 random markers are obtained. We first analyze the 500 markers to construct a genomic map. Then we set the estimated genome length using all the 500 markers as the estimated whole genome length ($\hat{L}$). Finally, we randomly draw markers from the data set with replacement. For example, we sample 100 markers from the 500 markers. A genomic map can be constructed using the 100 markers and map length can be obtained ($\hat{L}_{100}$). If this sampling process is repeated k times for different numbers of markers, an estimated mean marker coverage in proportion to the

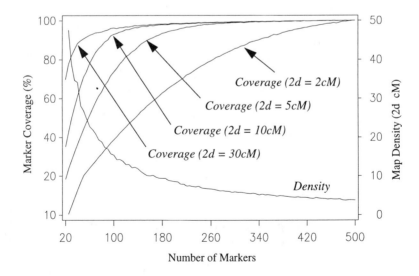

Figure 10.23 Marker coverage and map density are plotted against number of markers. There are 500 random markers in the original experiment. The means of 50 replications for each number of markers are used to produce the curves.

whole data set can be obtained using

$$\hat{c}_n = \frac{L_n}{\hat{L}} = \frac{1}{k\hat{L}}\sum_{i=1}^{k}\hat{L}_{ni} \tag{10.54}$$

where n is number of markers in the sample and L_n is average map length for n markers.

For example, Figure 10.23 shows the simulation results for a data set with 500 markers. In Figure 10.23, estimated marker coverages using samples from the original data set are plotted against the number of markers in a sample for three different criteria of minimum map distances. When the number of markers is greater than 150, marker coverage increases slowly as the number of markers increases for $2d = 10\text{cM}$ and $2d = 30\text{cM}$. For $2d = 5\text{cM}$, marker coverage increases slowly as the number of markers exceeds 300 (Table 10.14). For $2d = 2\text{cM}$, marker coverage increases quickly even when number of markers is close to 500.

From the data in Figure 10.23 and Table 10.14, we can conclude that 300 markers are sufficient if the goal of the experiment is to have an average of one marker in a 5cM genome segment. The estimated coverage increases 1.1% (from 98.9% to 100%) as the number of markers increases from 300 to 500. The estimated genome length using the 500 markers ($\hat{L}$) may not be the true length of the genome. However, we can conclude that the true genome length is close to the estimate (Table 10.15).

Table 10.15 Marker coverage under different criteria of minimum map distance between adjacent markers and maximum gap between adjacent markers for sampling different numbers of markers in the original data set. There are 500 random markers in the original experiment..

Number of	2d (cM)				Maximum Gap
Markers	2	5	10	30	(cM)
20	8.20	19.19	35.22	69.82	50.79
40	16.40	36.55	61.06	88.42	34.72
60	24.15	51.51	78.23	92.62	22.76
80	31.56	63.55	87.28	94.28	18.85
100	38.79	73.67	92.99	96.29	15.39
150	54.50	88.97	97.17	97.76	10.57
200	67.35	95.05	98.31	98.59	8.59
300	85.65	98.91	99.24	99.36	5.97
400	95.40	99.69	99.73	99.77	4.52
500	100.00	100.00	100.00	100.00	3.77

The simulated marker coverage plot if there are only 200 markers in the original data for the same genome and experimental conditions is shown in Figure 10.24. It is clear that 200 markers are not sufficient to have a high marker coverage using 5cM as the criterion.

From the simulation, the changing trend of the maximum gap between adjacent markers as the number of markers changes can also be plotted. There is no limit for the change. The maximum gap decreases as number of markers increases.

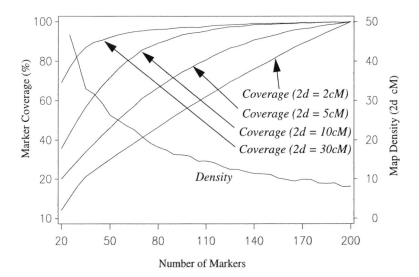

Figure 10.24 Marker coverage and map density are plotted against the number of markers. There are 200 random markers in the original experiment. Means of 50 replications for each number of markers are used to produce the curves.

In practice, the resampling approach can also be applied to experimental factors other than the number of markers, *e.g.*, to population size. The simulation approach can be used to determine the effect of any experimental factor on marker coverage and map density if resampling can be applied to the factor. The advantages of the simulation approach are:

• no assumption on the marker distribution is needed and no previous information is needed
• the rationale behind the approach is intuitive
• the approach can be applied for many different types of problems

The major disadvantage is the intensive computation needed for the approach.

EXERCISES

Exercise 10.1

Explain why and when mapping functions are needed or not needed. Pay attention to:

(1) Relationship between mapping function and recombination fraction
(2) Relationship between mapping function and physical genome maps
(3) Applications of genome information in molecular biology
(4) Applications of genome information in plant and animal breeding

Exercise 10.2

Crossover heterogeneity between two parents of a mapping population creates problems for conventional linkage analysis.

(1) List possible biological reasons for crossover heterogeneity between two parents.

(2) Is there any evidence of genetic control of crossover in a species you are familiar with?

(3) List some possible biological and statistical solutions for the heterogeneity.

Table 10.16 Pairwise two-point recombination fractions.

Locus	WG622	CDO669	MWG635A	TubA1	Dhn6
CDO669	0.17				
MWG635A	0.34	0.19			
TubA1	0.41	0.25	0.09		
Dhn6	0.45	0.27	0.12	0.05	
MWG058	0.48	0.28	0.17	0.10	0.06

Exercise 10.3

Table 10.16 contains pairwise two-point recombination fractions among six loci. The variance of each of the two-point recombination fractions is

$$\hat{\sigma}_r^2 = \frac{\hat{r}(1-\hat{r})}{100}$$

where $\hat{r}$ is the recombination fraction estimate in the Table 10.16. The most likely locus order is

WG622 - CDO669 - MWG635A - Tub1 - Dhn6 - MWG058

Do the following based on the data:

(1) Convert all 15 pairwise recombination fractions to map distances using the mapping functions of Haldane and Kosambi.

(2) Check additivity of the map distances for all six loci and for the four possible three-locus models on the diagonal of the table.

(3) Use the above results to explain the possible sources of non-additivity for some of the models, paying attention to the mapping functions themselves and to sampling error.

(4) Estimate multi-locus map distances using the least square approach of Jensen and Jorgensen.

(5) Test which mapping function, Haldane or Kosambi, fits the data better. Hints: Use the information in Table 10.13. The empirical distribution of the test statistic is not required.

Exercise 10.4

Why are multi-locus models needed? Besides some statistical reasons, are there any biological reasons?

Exercise 10.5

Tables 10.17 and 10.18 contain pairwise two-point recombination fractions among 14 loci and an estimated linkage map. The variance of each of the two-point recombination fractions is

$$\hat{\sigma}_r^2 = \frac{\hat{r}(1-\hat{r})}{100}$$

where $\hat{r}$ is the recombination fraction estimate (Table 10.17). The pairwise recombination fractions are tabulated according to the most likely locus order. Table 10.18 is a linkage map, from the computer program PGRI, corresponding to the pairwise recombination fractions. The maximum likelihood multi-locus recombination fractions were obtained using the Lander-Green algorithm. The multi-locus recombination fractions were converted to map distances using the mapping functions of Haldane and Kosambi.

Table 10.17 Two-point recombination fraction.

Locus	MWG934	Nir	Amy1	MWG820	ABC163	ksuD17
Nir	0.09					
Amy1	0.16	0.08				
MWG820	0.25	0.20	0.11			
ABC163	0.28	0.24	0.15	0.04		
ksuD17	0.31	0.25	0.19	0.05	0.02	
ABG458	0.33	0.30	0.25	0.17	0.13	0.12
MWG916	0.40	0.38	0.30	0.24	0.22	0.21
ABG387B	0.37	0.31	0.31	0.25	0.22	0.19
cMWG652A	0.40	0.41	0.38	0.35	0.33	0.32
ABG378	0.49	0.49	0.49	0.45	0.42	0.43
Nar1	0.49	0.48	0.48	0.46	0.45	0.46
MWG663B	0.51	0.48	0.50	0.47	0.47	0.47
ABG466	0.49	0.48	0.49	0.47	0.45	0.47
PSR167	0.50	0.50	0.52	0.51	0.51	0.53
Locus	ABG458	MWG916	ABG387B	cMWG652A	ABG378	MWG663B
MWG916	0.14					
ABG387B	0.16	0.07				
cMWG652A	0.30	0.20	0.16			
ABG378	0.40	0.34	0.29	0.13		
MWG663B	0.45	0.41	0.34	0.17	0.05	
ABG466	0.46	0.37	0.33	0.18	0.06	0.01
PSR167	0.51	0.47	0.40	0.26	0.12	0.07
Locus	ABG466					
PSR167	0.06					

(1) Obtain multi-locus map distances using the least square approach of Jensen and Jorgenson and both Haldane's and Kosambi's mapping functions. (Hint: Commonly used statistical computer software, such as SAS, can be used to implement the analysis. The least square approach can be implemented using the SAS IML procedure.)

(2) Compare the results of the least square approach with the output from PGRI using the Lander-Green EM algorithm. Explain why they are the same or different.

(3) Explain why multi-locus recombination fractions are sometimes different from two-point recombination fractions.

Table 10.18 A linkage map for a linkage group with 14 loci. Output from PGRI computer software.

Locus (code)	2-R	ML-R		H-D		K-D	
		Point	Sum	Point	Sum	Point	Sum
WG934 (17)	0.000	0.000	0.000	0.000	0.000	0.000	0.000
Nir (15)	0.086	0.087	0.087	0.095	0.095	0.088	0.088
Amy1 (14)	0.079	0.085	0.172	0.094	0.189	0.086	0.174
MWG820 (13)	0.112	0.122	0.294	0.140	0.329	0.125	0.299
ABC163 (11)	0.040	0.040	0.334	0.042	0.370	0.040	0.338
ksuD17 (9)	0.024	0.026	0.361	0.027	0.397	0.026	0.365
ABG458 (8)	0.121	0.116	0.477	0.132	0.530	0.118	0.483
MWG916 (19)	0.143	0.124	0.600	0.142	0.672	0.126	0.609
ABG387B (7)	0.072	0.062	0.662	0.066	0.738	0.063	0.672
cMWG652A (5)	0.163	0.169	0.831	0.206	0.944	0.176	0.848
ABG378 (4)	0.133	0.145	0.976	0.171	1.115	0.149	0.996
MWG663B (18)	0.052	0.053	1.029	0.056	1.170	0.053	1.049
ABG466 (2)	0.012	0.011	1.040	0.011	1.181	0.011	1.060
PSR167 (1)	0.057	0.057	1.097	0.061	1.242	0.057	1.118

"2-R" = two-point recombination fraction
"ML-R" = maximum likelihood multipoint recombination fraction
"H-D" = Haldane map distance
"K-D" = Kosambi map distance

Exercise 10.6

For the species listed in Table 3.1 (in Chapter 3), do the following, assuming the markers are randomly distributed on the genomes.

(1) Estimate the numbers of markers needed for the marker coverages 70%, 90%, 95% and 99%.

(2) Estimate the marker coverage for the numbers of markers 20, 100, 200 and 500.

Exercise 10.7

Using the results in Exercise 10.6, explain the relationships among the genetic size of the genomes in cM, physical size of the genomes in Kb, marker coverage and the numbers of markers.

Exercise 10.8

Explain the biology and statistics behind genetic marker distribution.

CHAPTER 11

LINKAGE MAP MERGING

11.1 INTRODUCTION

11.1.1 LINKAGE MAPPING

Linkage maps have been made for many scientifically or economically important species. Linkage maps for each of these species have been built from different mapping populations, in different laboratories using different marker systems and adopting different mapping strategies and even by different statistical procedures and computer packages.

There is a great need to make comparisons of the linkage maps within a species (Hauge *et al.* 1993; Stam 1993; Kleinhofs 1993; Helentjaris *et al.* 1986; Weber and Helentjaris 1989; Burr *et al.* 1988; Hoisington and Coe 1989; Coe *et al.* 1990; Murray *et al.* 1988; Beavis and Grant 1991). Furthermore, there is significant interest in the relationship of maps of different species (Ahn and Tanksley 1993; Tanksley *et al.* 1992; Whitkus *et al.* 1992; Hulbert *et al.* 1990; Bonierbale *et al.* 1988). A summary of linkage information in the form of a genomic map is clearly a desirable goal. Comparisons of linkage maps are important for validation and development of linkage map methodology, general applications of linkage map information to plant and animal improvement, understanding of genome evolution of different species and genome database construction. These comparisons can be done by several methods, two of which are pooling and bridging. Pooling is the process of obtaining a whole linkage map by jointly analyzing data from multiple mapping populations. Bridging is the process of obtaining linkage maps for some segments of the genome by jointly analyzing data from multiple populations and analyzing the other parts of the genome independently for each mapping population. Before these comparisons can be made, we must know:

- Is there any genetic or statistical basis to combine map information from different models of inheritance, from different mapping populations and even from different species?

- Is there any real need to make the pooling and bridging?

The rudimentary concept of a single "species genomic map" or even "plant genomic map" requires more statistical rigor than a map of one organism to furnish the level of genomic information required for breeding and for genome research.

Pooled linkage maps for *Arabidopsis thaliana* and *Zea mays* have been reported (Hauge *et al.* 1993; Helentjaris *et al.* 1986; Weber and Helentjaris 1989; Burr *et al.* 1988; Hoisington and Coe 1989; Coe *et al.* 1990; Beavis and Grant 1991). Three basic approaches were used to generated the pooled linkage maps:

- The first is the visual approach. Given linkage maps with some common markers, locus order and map distances between adjacent loci are examined visually. Linkage maps sharing similar locus order are pooled together based on homologous loci that serve as the anchor markers and pooled map distances are estimated by the weighted mean of the individual linkage maps. This is an effective method when original data for the linkage maps are not available. This approach is simple and may be valid when there are no genetic or statistical complications. However, independently derived linkage maps for the same genotype are rarely identical.

- The second approach was explained by Beavis and Grant (1991). First, heterogeneity of segregation ratios was tested using Pearson's Chi-square, on pairwise two-point recombination fractions using a method originally developed by Fisher (1949) and on multipoint recombination fractions using a method modified from Morton (1956). To generate the pooled map, it was assumed that markers that mapped to similar chromosomal regions in different populations were identifying the same chromosomal locations. Since the populations were all F2 type, the marker scores were pooled across all populations and were analyzed using MAPMAKER (Lander *et al.* 1987). The missing scores were filled with monomorphic bands (Ott 1985).

- The third approach and the computer package JoinMap are described by Stam (1993). The assumption of the program is that the true recombination fractions are the same in all experiments. Even if there is a violation of the assumption, it is not likely to influence the locus order on a combined linkage map. The pooled pairwise recombination fractions were estimated by maximizing the combined likelihood function for all mapping populations. An approach that is similar to seriation was used to achieve locus order (Buetow *et al.* 1987ab; Stam 1993; see Chapter 9). The core of JoinMap is the estimation of map distance for a given order of the markers. A slightly modified version of the least squares procedure first described by Jensen and Jorgensen (1975) and later by Lalouel (1977) and Weeks and Lange (1987) is used to estimate map distances (see Chapter 10).

There have to be some common genetic bases or common statistics among the mapping populations for linkage map pooling and bridging to be effective. For some situations, the pedigrees of mapping populations are known, for example, one mapping population being the daughter of a parent used for the other mapping population, or the evolutionary relationship is known among the populations, such as rye and common wheat. If this is so, then there is a genetic basis for pooling or bridging the two linkage maps. A genetic relationship leads to the assumption of colinear structure of the genomes. When there is no pedigree or evolutionary information among the mapping populations, or when the comparative mapping populations are in different plant species, statistical hypotheses on the linkage maps will play an important role in inferring genetic relatedness.

The focus of this chapter is the development of theory and methodology for linkage map pooling and bridging. Methods to evaluate differences among the mapping populations will be presented. Linkage map pooling is appropriate when there are no significant differences among the populations. When differences are detected among the populations, the problem of combining linkage information from different individuals and populations is similar to the problem of constructing phylogenetic trees. Based on the proposed hypothesis tests and a string-matching approach, the common segments among the linkage maps can be identified and the possible genetic changes, such as chromosome translocation and inversion, can be inferred. In this chapter, likelihood methods and nonparametric methods will be presented for statistical hypotheses on gene orders.

11.1.2 HYPOTHESIS TESTS ARE NEEDED

Hypothesis tests among the mapping populations at different levels of linkage map construction are needed to make decisions on whether the linkage maps can be pooled or bridged and if so, on how to do it. Hypothesis tests are based on the statistical properties of the estimates. The statistical properties are not only important for linkage map pooling and bridging, but also for single population mapping. Hypothesis tests will help to achieve a better understanding of linkage map construction and the nature of linkage maps. Among the hypothesis tests are those for single locus segregation, two-point recombination fractions, locus orders and multipoint map distances. Except for tests of locus orders and multipoint map distances, the principles of the tests were developed a half century ago.

Locus ordering is a centerpiece of linkage map construction. Several methods for locus ordering have been studied and implemented in several mapping packages (Olson and Boehnke 1990; Lander *et al.* 1987; Stam 1993; Liu and Knapp 1992; see Chapter 9). The main focus of the studies has been the computational hurdles and comparisons among the methods for small linkage groups. The statistical properties of locus order have not been adequately studied and hypothesis tests on locus orders from different mapping populations have not been extensively employed.

Linkage maps can be pooled when differences among the mapping populations are not statistically significant and there is no obvious genetic change among the genetic materials from which the mapping population is made. Linkage maps can be bridged when some of the hypothesis tests are not statistically significant and the genetic changes among the genetic materials can be inferred from the statistics. The concept of the "species map," that there should be one linkage map for a plant species, may be not adequate. For some applications of genome linkage maps, it may not be necessary or efficient for the applications to pool the linkage maps.

11.1.3 WHY LINKAGE MAP POOLING AND BRIDGING?

Cross Validation of Mapping Strategies

Linkage map pooling and bridging provide not only ways to make the connections among the linkage maps, but also to validate the efficiency of the mapping strategies. In general, screening procedures, statistical treatments, computer mapping packages used, types of mapping populations, marker systems, sample sizes and other factors may influence the validity and efficiency of the mapping process. For some particular species, there may be specific strategies designed for their problems. For example, in forest tree species, half-sib megagametophyte methods, full-sib pseudo-testcross methods, three generation pedigree methods and half-sib open-pollinated diploid tissue methods have been used to construct linkage maps. For barley, doubled-haploid lines and for maize, F2 progenies and recombinant inbred lines have been used to generate linkage maps.

Applications of Genome Information to Applied Plant and Animal Breeding

Great attention has been paid to genes controlling economically important traits. These genes have been referred to as Quantitative Trait Loci (QTL) (see

Chapters 12-17). The majority of economically important traits are quantitatively inherited. One of the debates on the usefulness of QTLs in practical breeding regards QTL expression or detection in different genetic backgrounds. Does a QTL found in one breeding population exist in others? This issue is important for a breeding program and it is even more important when germplasm utilization is under consideration. Linkage map pooling and bridging is an efficient way to deal with these problems.

Comparative Mapping

Linkage mapping is a relatively expensive and slow process and probably will be for a long time. Comparative mapping has been used for human and mouse genome mapping in order to increase the efficiency and reduce the cost. Linkage map pooling and bridging will create greater opportunities for comparative mapping in plants and provide ways to find homologous maps among the plant species.

Structures of Genome Database

Much effort has been put into building genome databases for human and several other plant and animal species. A comprehensive USDA plant genome database is at the National Agriculture Library (NAL) and satellite genome data bases for several plant species are in different institutes (Bigwood 1992). Sybase, a commercial database software program and ACEDB, software originally designed for *C. elegans*, are used for genome data base constructions. Plant genome databases will be more complex and larger in volume than the human and *C. elegans* databases, because of the large number of plant species and the large number of varieties within each of these. Linkage map pooling and bridging may make the design of the genome databases simpler and more logical.

11.2 FACTORS RELATED TO LINKAGE MAP MERGING

11.2.1 BIOLOGY AND LINKAGE MAP MERGING

Mating and Genetic Marker Systems

Linkage information is obtained from genetic markers that follow different models of inheritance in common mating systems. When mapping in inbred plants, F2 populations, where codominant markers segregate in a 1:2:1 ratio and dominant markers in a 3:1 ratio, are often used. Sometimes, F2 individuals are inbred for additional generations to produce recombinant inbred lines, each of which is homozygous for a particular set of recombinant chromosomes. The advantage of recombinant inbred lines is that the mapping population is "immortalized" as a collection of true-breeding lines. Other inbred mapping populations are derived from backcrossing, which yields a 1:1 ratio for segregating markers. In outbred plants, 3-generation pedigrees may be needed to establish linkage phase, but the markers can segregate in 1:1, 1:2:1, 3:1 and other ratios because the parental genotypes are heterozygous. Combining linkage information from different models of inheritance in different kinds of families is straightforward in principle, but difficult in practice.

There are several kinds of genetic marker systems commonly used in genomic research, but none is ideally suited to all purposes. RFLPs are ideally codominant and multiallelic, with gene-specific probes. In practice, however, RFLP probes commonly recognize small gene families and one probe can provide several segregating markers. The research problem then becomes one of recognizing specific loci and allelic forms in different mapping families, or even in the same outbred family. Similarly, the PCR-based marker system called RAPD generates several bands for each primer and it can be difficult to determine which segregating bands in different families correspond to the same locus. Microsatellites are another PCR-based marker system which provides hypervariable markers with high heterozygosity in populations. These markers could be useful for "bridging" the information obtained from other marker systems.

Cytogenetics

Inversions, translocations and other chromosome rearrangements have been observed in many species, including *Drosophila* (Sturtevant 1931; Bridges and Brehme 1944), maize (Russel and Burnham 1950; Morris 1955), barley (Das 1951; Kasha 1961; Caldecott and Smith 1952) and oats (Koo 1958). These phenomena change the linear structure of chromosomes, which is the cytological basis for linkage map construction and certainly for linkage map pooling and bridging. Changes in the linear structure can change the gene orders on the genome and also alter the map distances involving genes located at or near breakpoints.

Until recently, research on these phenomena has been limited to microscope observation and linkage genetics of limited morphological markers on a handful of organisms. The developments of genomic mapping and molecular marker technology have generated a large amount of information on the genome structure of many plant and animal species. Chromosome rearrangements can now be studied in a great many organisms. For some plant species such as maize and wheat, cytogenetic variants have been well documented (Carlson 1977). Maize B-A translocations have been successfully used for isozymic and RFLP mapping in this species (Newton and Schwartz 1980; Burr *et al.* 1988; Weber and Helentjaris 1989). For plant species without rich cytogenetics, molecular marker technology has created the opportunities for advanced cytogenetic research.

Recent developments in biochemistry and molecular biology have had a great impact on understanding the effects of transposons, insertions and deletions on the linear structure of the genome. For example, transpositions may not affect gene order, but may alter the physical map distance between some genes. Banding patterns of commonly used molecular markers, such as RFLP and PCR based markers, can often be affected by transposable elements.

11.2.2 STATISTICS AND LINKAGE MAP MERGING

Sampling Variation

The statistical aspects of linkage map construction involve testing for segregation ratio distortion, two-point linkage analysis, recombination fraction estimation, linkage grouping, locus ordering and multipoint map distance estimation. Sampling variation is associated with every step of linkage map construction. Linkage map pooling and bridging will include hypothesis tests between linkage maps on every aspect of the linkage mapping. For some particular linkage map pooling problems, the linkage maps were often constructed in different laboratories. For some situations, data generated from different laboratories can be treated as independent samples. However, in some cases, procedures used to generate the mapping data were different, adding additional variation that is difficult to estimate.

Different Screening Strategies

In practice, screening procedures, such as screening probes and primers, are important steps for generating sufficient data within limited resources. In some situations, different screening strategies may result in the same set of markers for the final linkage map. In other situations, they may result in different sets of markers and different linkage maps. Attention should be paid to the screening procedures when different linkage maps are pooled into a single map.

Missing Data and Missing Linkage Information

Linkage map construction can be considered as a missing data problem even for a single population (Lander and Green 1992). For several mapping populations, there is usually a portion of loci in common across the populations. Whenever there are loci not in common between populations, there is a good chance that some of the pairwise recombination fractions for the pooled data cannot be estimated. The missing recombination fractions have serious impacts on locus ordering and multipoint map distance estimation. For crosses between heterozygous parents, even when there is no missing data, a portion of pairwise recombination fractions cannot be estimated because many combinations of markers are not informative for linkage.

Sample Size and Data Quality

Sample size and data quality influence the amount of information on linkage. Different mapping populations may have different sizes and be of different quality in aspects such as segregation ratio distortion and scoring error. Sample size can easily be taken into consideration in linkage map pooling and bridging. Data quality, which can have very large impact on quality of a linkage mapping (Lincoln and Lander 1992), is more difficult to evaluate.

11.3 HYPOTHESES ABOUT GENE ORDERS

Statistical hypothesis tests about gene orders can be performed only on the common genetic markers among mapping populations. In some situations, different markers or fragments may be mapped to essentially the same site, such as the conversion of a RAPD band to a RFLP or microsatellite band. Even when genetic information about the markers is clear and they can be treated as common markers, they may have different likelihood functions for linkage.

Hypothesis tests between two mapping populations are analogous to pairwise two-population problems. Statistical hypotheses between two mapping populations can be tested at different levels of linkage map construction, such as single-locus models, two-locus models and multipoint models. The hypothesis tests for single-locus models and two-locus models have been used since early this century (Mather 1938; Weir 1996), but the tests have not been carried out at a whole genome level. The methods will be reviewed and possible genetic inferences from the statistics at a whole genome level will be discussed. Using step-by-step hypothesis testing between the two mapping populations, a genetic relation between the two genomes can be inferred. The central issue of the hypothesis is locus order. Several methods were proposed in the literature (Edwards 1987; Smith 1990) to assign a probability to a locus order, or to do a hypothesis test on a locus order. However, the proposed methods are difficult to apply to more than about 6 loci at a time (Smith 1990).

11.3.1 HETEROGENEITY TEST BETWEEN TWO-POINT RECOMBINATION FRACTIONS

For a single locus, heterogeneity for segregation ratios among multiple populations can be tested using a log likelihood ratio test statistic

$$G = 2 \sum_{i=1}^{n_c} \sum_{k=1}^{n_p} f_{oik} \log \frac{f_{oik}}{f_{eik}} - 2 \sum_{i=1}^{n_c} f_{oi\circ} \log \frac{f_{oi\circ}}{f_{ei\circ}} \tag{11.1}$$

where subscripts $k = 1, 2, ..., n_p$ and $i = 1, 2, ..., n_c$ indicate mapping population k and genotype category i, respectively, n_p is the number of mapping populations, n_c is the number of categories for the genotypes, subscript '$\circ$' indicates that the frequency is marginal over the term occupied by the dot, subscript e indicates f_{eik} is the expected count under a standard segregation ratio and f_{oik} is the observed count for population k and genotype i. The G statistic follows a chi-square distribution with $n_c(n_p - 1)$ degrees of freedom. In the formula, the first term is the total G statistic and the second is the pooled G statistic, assuming homogeneity of segregation ratios. The heterogeneity test for segregation ratios at a whole genome level among the populations may show some genetic differences due to mutation, lethal genes, genetic drift or other mechanisms.

For a two-locus model with the same allelic configuration and two populations with the same mating system, the total G-statistics can be partitioned into pooled and heterogeneity G-statistics for segregation and linkage. The total G-statistics for the model can be computed by

$$G_t = 2 \sum_{i=1}^{n_c} \sum_{j=1}^{n_c} \sum_{k=1}^{n_p} f_{oijk} \log \frac{f_{oijk}}{f_{eijk}} \tag{11.2}$$

where subscripts i and j are for locus A and locus B, respectively, k is for the population, f_{oijk} is an observed count and f_{eijk} is an expected count under the assumption of no segregation ratio distortion for the two loci and no linkage between A and B. The pooled G-statistic can be estimated by using the marginal frequencies or by substitution, such as pooled total G-statistic

$$G_{pt} = 2 \sum_{i=1}^{n_c} \sum_{j=1}^{n_c} f_{oij\circ} \log \frac{f_{oij\circ}}{f_{eij\circ}} \tag{11.3}$$

The G-statistic for pooled linkage is $G_{pl} = G_{pt} - G_{pa} - G_{pa}$, where G_{pa} and G_{pb} are the G-statistics for testing segregation distortion for locus A and locus B, respectively. The G-statistic for linkage heterogeneity is $G_{hl} = G_{1l} + G_{2l} - G_{pl}$, where G_{1l} and G_{2l} are G-statistics for testing linkage between locus A and locus B for mating population 1 and population 2, respectively.

When two populations have different mating systems or allelic configurations, the total G-statistic cannot be partitioned clearly. The generalized likelihood ratio approach can be used for the hypothesis tests. The likelihood functions for the two populations having different recombination fractions between loci A and B are defined as

$$L(r_1, r_2) = \sum_{i=1}^{n_{c1}} \sum_{j=1}^{n_{c1}} f_{oij1} \log p_{ij1} + \sum_{i=1}^{n_{c2}} \sum_{j=1}^{n_{c2}} f_{oij2} \log p_{ij2} \tag{11.4}$$

where p_{ij1} is the expected frequency evaluated using $\hat{r}_1$, which is the maximum likelihood estimate of recombination fraction for population 1 and p_{ij2} evaluated using $\hat{r}_2$ is for population 2. The likelihood function for the two populations having common recombination fraction is

$$L(r) = \sum_{k=1}^{n_p} \sum_{i=1}^{n_{ck}} \sum_{j=1}^{n_{ck}} f_{oijk} \log p_{ijk} \qquad (11.5)$$

where p_{ijk} is the expected frequency evaluated under $\hat{r}$, which is the estimate for the pooled likelihood function. The likelihood ratio statistic

$$G = 2[L(\hat{r}_1, \hat{r}_2) - L(\hat{r})]$$

will test the hypothesis that the two populations share a common recombination fraction between locus A and B.

11.3.2 LIKELIHOOD RATIO TESTS AMONG LOCUS ORDERS AND MULTIPOINT MAP DISTANCES

Consider a linear sequence of l loci, $a_1, a_2, ..., a_l$, with multipoint recombination fractions $\theta = (r_1, r_2, ..., r_{l-1})$ between the adjacent loci. The likelihood function for two linkage maps sharing same locus order and equal multipoint map distances is

$$L(\hat{\theta}) = \sum_{i=1}^{l-1} (n_{1i} + n_{2i}) [\hat{r}_i \log \hat{r}_i + (1 - \hat{r}_i) \log (1 - \hat{r}_i)] \qquad (11.6)$$

where n_{1i} and n_{2i} are the number of informative sample sizes for locus combination i and $i+1$ and for the two mapping populations, respectively. The number of informative sample sizes may be different from locus to locus, due to missing data or missing linkage information for the two adjacent loci.

The likelihood function for the two linkage maps having different locus orders and unequal multipoint map distances is

$$L(\hat{r}_1, \hat{r}_2) = \sum_{i=1}^{l-1} n_{1i} [\hat{r}_{1i} \log \hat{r}_{1i} + (1 - \hat{r}_{1i}) \log (1 - \hat{r}_{1i})] \qquad (11.7)$$
$$+ \sum_{i=1}^{l-1} n_{2i} [\hat{r}_{2i} \log \hat{r}_{2i} + (1 - \hat{r}_{2i}) \log (1 - \hat{r}_{2i})]$$

where $\hat{r}_{1i}$ and $\hat{r}_{2i}$ indicate the multipoint map distance between the adjacent locus i and $i+1$ for populations 1 and 2, respectively. The likelihood ratio statistic

$$G = 2[L(\hat{\theta}_1, \hat{\theta}_2) - L(\hat{\theta})]$$

will test the differences between the two linkage maps, including locus order difference and multipoint map distance differences. Given the complex nature of the likelihood functions, this likelihood ratio test may be not sufficient. The profile of the likelihood function for locus ordering and multipoint map distance estimation has multiple peaks. The likelihood function cannot be defined for a common multipoint map distance with different locus orders because different orders give different map distances. Tests for locus orders and multipoint map distances are confounded.

The likelihood functions for two linkage maps sharing same locus order but unequal multipoint map distances is

$$L'(\hat{\theta}_1, \hat{\theta}_2) = \sum_{k=1}^{2} \sum_{i=1}^{l-1} n_{ki} [\hat{r}_{ki} \log \hat{r}_{ki} + (1 - \hat{r}_{ki}) \log (1 - \hat{r}_{ki})] \qquad (11.8)$$

where $\hat{r}_{ki}$ indicates the multipoint map distance between the adjacent loci i and $i+1$ for populations k. This equation has the same form as Equation (11.7). However, the components are different between the two equations. In Equation (11.7), locus orders are different for the two populations and the map distances are estimated based on the orders. In Equation (11.8), the map distances for the two populations are estimated based on the same locus order. The likelihood statistic

$$G = 2[L(\hat{\theta}) - L'(\hat{\theta}_1, \hat{\theta}_2)]$$

will test differences in the multipoint map distances between the two linkage maps. Restrictions can be applied to the test; such tests can be done on a subset of the multipoint map distances.

Distribution of the likelihood ratio test statistic for testing locus orders and multiple-point map distances is complex. For each test statistic, three distributions can be used. The first is based on the test statistic being distributed as a chi-square with $l - 1$ degrees of freedom. The second is based on a chi-square distribution with an estimated degrees of freedom which is the mean of a bootstrapped likelihood ratio test statistic based on bootstrapped samples from homogenous populations. The last uses a 95 percentile of bootstrapped likelihood ratio test statistics as a critical value to declare a statistical significance at 0.05 level.

11.3.3 NONPARAMETRIC HETEROGENEITY TESTS FOR LOCUS ORDERS

If a locus order for a linkage group with l loci is generated at random, then each of the loci has a probability of $1/l$ to occupy a position on the genome (see Chapter 9). For t bootstrap gene orders, the probability of locus i being located at genome position j is P_{ij}. Bootstrapping (Efron 1992) was applied to the observations by randomly drawing n times from the original data set with replacement, where n is sample size of the mapping population. The probability of bootstrap gene locations agreeing with the true position is

$$P_o = \frac{1}{l} \sum P_{ii}$$

and is defined as the percentage of correct gene order (see Chapter 9). If the gene order is random or there is no inference about gene order from the data, then the probability of bootstrap gene locations agreeing with the true genome positions is $1/l$. A significant difference between P_o and $1/l$ will measure the confidence of the estimated gene order. The statistical significance can be tested using a kappa statistic (Cohen 1960)

$$\kappa = \frac{P_o - 1/l}{1 - 1/l} \qquad (11.9)$$

When $P_o = 1$, the κ corresponds to perfect agreement. When $P_o = 1/l$, the agreement equals that expected by chance. For multinomial sampling, the sample measure has a large-sample normal distribution. Its asymptotic variance can be estimated by the formula developed by Fleiss *et al.* (1969)

$$\hat{\sigma}^2(\kappa) = \frac{P_o(1-P_o)}{l(1-1/l)^2}\left[\frac{4P_o(1-P_o)}{(1-1/l)^2} - 3\right] \tag{11.10}$$

Statistic $\hat{\kappa}/\sqrt{\hat{\sigma}^2(\kappa)}$ approximately follows a standard normal distribution and tests the difference between the two orders.

For testing gene order difference between two populations, the same procedure can be applied by treating the gene order achieved from one population as genome position. When comparing the two populations, the κ statistic is the same no matter which population's gene order was used as the genome position. A significant κ statistic is evidence of a difference between the two gene orders.

11.4 LINKAGE MAP POOLING

If the hypothesis tests do not show significant differences among the mapping populations, the linkage maps can be pooled into a common linkage map. It may not be necessary to pool the linkage maps even if there is no significant difference among them. In some situations, statistically significant results may not make any biological sense. It is important to establish the genetics before the pooling. Two basic approaches to pool linkage maps will be described.

11.4.1 ANCHOR MAP APPROACH

The markers have to be common across all the mapping populations to carry out the hypothesis tests. If the hypothesis tests are not significant for heterogeneity, then the common markers can be pooled across the populations. The pooled map can be considered an anchored map and the common markers are defined as anchored markers. The pairwise two-point recombination fractions among the anchor markers can be estimated by maximizing the pooled likelihood functions. The populations involving different crosses and different allelic configurations can be pooled to estimate recombination fraction because of the additive nature of the log likelihood function. Methods such as the sum of adjacent recombination fraction and the maximum likelihood can be used to determine the pooled locus order. The pooled order is not expected to be different from the individual locus orders. The maximum likelihood estimate of the multipoint map distance for the anchored map can be estimated using an EM algorithm (Dempster *et al.* 1977; Lander and Green 1987) or the least squares method (Jensen 1975; see Chapter 10).

If the markers are common across all of the populations, then the anchored map is the final pooled linkage map. In practice, there is usually a portion of markers not common across all populations. In most cases, these markers will create missing two-point pairwise recombination fractions for the whole data set. When the portion of the missing two-point recombination fractions is large, commonly used locus ordering algorithms, such as sum of adjacent recombination fraction and maximum likelihood, do not yield adequate locus orders.

In this section, an iterative approach based on the anchored map and the framework map concept (Keats *et al.* 1991) to locate markers that are not in common is described. The EM approach can also be used to simultaneously estimate the missing recombination fractions and locus order.

An assumption for the anchor map approach is that the anchored map has the correct locus order. The anchored locus order will be treated as a basis for the placement of the rest of the loci. There will be $l - l_c$ loci to be placed on the anchored map, where l is the total number of loci and l_c is the number of common loci. The placements will start at each of the $l - l_c$ loci. Different start points may result in different final orders. The nature of the placements is similar to the seriation approach for locus ordering with different start points.

For each locus among the $l - l_c$ loci, there are $l_c + 1$ possible placements. For each of the placements, the method explained in the next section can be used to estimate the multipoint map distances and there will be a corresponding log likelihood. Of the log likelihoods corresponding to the placements, that with the highest value will be chosen. If the highest value is 3 more than the next highest value, then the locus it corresponds to will be placed at the position with the highest value. If the difference is less than 3.0, then the conclusion will be that the locus cannot be placed on the map with confidence. This is the concept of a framework map. By repeating the above procedures with different start points, different orders may be achieved. A weighted log likelihood, which is the log likelihood divided by the number of loci, will be used to determine which of the orders will be the best order for the pooled linkage map. It will be described in the next section how the multipoint map distances are estimated simultaneously with the locus order determination.

11.4.2 ESTIMATION OF MISSING RECOMBINATION FRACTIONS USING EM ALGORITHM

For a two mapping population example with a total of l loci, $l(l-1)/2$ total pairwise recombination fractions can be estimated if all loci are common for the two populations. If there are l_c common loci among l_a loci for one population and l_b loci for the other, then there are

$$\frac{1}{2}[l_a(l_a - 1) + l_b(l_b - 1) - l_c(l_c - 1)]$$

estimable pairwise recombination fractions and

$$(l_a - l_c)(l_b - l_c)$$

non-estimable recombination fractions. Those missing pairwise recombination fractions can be estimated along with locus order determination using a modified EM algorithm. Locus ordering algorithms, such as sum of adjacent recombination fractions (SAR) and maximum likelihood, can also be used. The algorithm of SAR will be used here to explain the approach.

Given l loci in a linkage group, for any locus order of the $l!/2$ possibilities, SAR is

$$\sum_{i=1}^{l} r_{i,i+1}$$

where $r_{i,i+1}$ is the estimated recombination fraction between loci i and $i+1$. The order with minimum SAR is the best order for the data. For map pooling with missing recombination fractions, SAR can be evaluated only for a subset of the possible orders. The best order among the subset may not be the global best.

For this example, there is a linkage group with 6 loci, A, B, C, D, E and F, where recombination fractions between AD, BE and ED cannot be estimated because of missing data or because the combinations are not informative about linkage. SAR cannot be estimated for the orders with AD, BE or ED adjacent to

each other, such as ADBCEF and ADEBCF. One possible data structure will be
the following:

	A	B	C	D	E	F
Population#1	x	+	+	+	x	+
Population#2	+	x	+	x	+	+
Population#3	+	+	+	x	x	+

where C and F are anchor loci with data across all three populations. For an
order ABCDEF, recombination fractions of AD, BE and DE can be estimated by
the following EM steps:

(1) Make an initial guess: The missing recombination fractions are

$$R^0 = (r^0_{ad}, r^0_{be}, r^0_{de})$$

and the true map distances between adjacent loci for the order are

$$\theta^0 = (\theta^0_1, \theta^0_2, \theta^0_3, \theta^0_4, \theta^0_5)$$

(2) Expectation step (E): Using R^0 as if it were the true recombination
fractions and θ^0 as if it were the true map distance, compute the
expected number of recombinant for each related interval.

(3) Maximization step (M): Using the expected value as if it were the true
value, compute the maximum likelihood estimate R^1 for the
recombination fractions and θ^1 for the map distances.

(4) Iterate the E and M steps until the likelihood converges to a maximum.

Step (2) will follow the methods explained by Lander and Green (1987).
The missing recombination fractions and the SAR for the order of ABCDEF are
estimated simultaneously. The same procedures will be applied for all possible
orders or combined with the simulated annealing approach to determine the
best pooled order. The multipoint map distances are also estimated in the
procedures for the pooled map. There is a tolerance limit on how much missing
data a combined data set can have for a meaningful pooled linkage map.

11.4.3 LINKAGE MAP BRIDGING

If no genetic difference can be inferred from the hypothesis tests on locus
orders and multipoint map distances among the mapping populations, then
the pooled map can be achieved by the methods described in the previous
section. When genetic difference is detected among the mapping populations,
the linkage maps cannot be pooled at a whole genome level, but this does not
mean that the linkage maps cannot be pooled partially. Once the segments to
be pooled are identified, the question becomes the same as the pooling problem
explained in the previous section. So, the bridging problem becomes the
problem of identification of common segments among the linkage maps and the
possible genetic relation among the segments not in common.

P_{ij} has been defined as the probability of a locus ordered at position i in
one population and at position j in the other. The difference between the
probabilities of agreement between the two orders and the agreement between
the two orders purely by chance is $P_{ii} - 1/l$. If a difference is found for the
orders between the two linkage maps, permutations based on genetic changes
between the two maps can be performed on the orders to get maximum
agreement. The problem becomes a classical string matching problem (Lovasz
and Plummer 1986). Given a non-negative matrix P, the problem is to get a

maximum of $P_{ii} - 1/l$ by permuting the orders of the matrix. The maximum corresponds to the maximum agreement between the two locus orders under the genetic assumptions made in the permutations.

EXERCISES

Exercise 11.1

Explain the needs of linkage map merging using a specific species with which you are familiar. Pay attention to the following points in your explanation.

(1) Linkage map merging is needed to advance biological research in the species.

(2) Linkage map merging is needed for practical applications of genome information.

(3) Linkage map merging is needed for efficient management of massive genome information.

Exercise 11.2

Explain the biological rationale behind linkage map merging. Pay attention to the following points.

(1) Genome evolution

(2) Genome structure variation (see Chapter 3)

(3) Genome variation and marker polymorphism

Exercise 11.3

Explain the biological reasons that linkage maps can or cannot be merged among species. How about within a species? For species having extensive genome evolutionary history, arguments supporting or against the idea of linkage map merging may be readily available. How about a species with a little known genome evolution? Can linkage map merging be a tool for the study of genome evolution?

Exercise 11.4

Seven polymorphic markers were obtained in two mapping populations generated using two different crosses within a species. Among the seven loci, three of them, L2, L3 and L5, are polymorphic in both populations. All three loci are fully informative in both populations. The other four loci, L1, L4, L6 and L7, are informative only in one of the populations. Three linkage maps are obtained. One linkage map is obtained using 5 informative loci, using each of the two populations. One linkage map is obtained by treating all the markers coming from a single population (or pooled data). The above illustration shows the three linkage maps and the estimated multiple-locus map distances in cM.

(1) What is the evidence for and against merging the two linkage maps into one pooled linkage map from the three estimated linkage maps?

(2) A hypothesis test on two-locus recombination fractions between the two populations shows a highly significant difference for locus combination L2-L3. Does this difference mean that the two linkage maps cannot be merged?

(3) Bootstrap samples were drawn from the original data for 100 replications. The following shows the locus ordering information. Is the bootstrap evidence strong enough to merge the two linkage maps?

Genome Position

Population	Marker	1	2	3	4	5	6	7
Population 1	1	90	5	3	2	0		
	2	5	89	3	2	1		
	3	4	5	89	2	0		
	4	1	1	3	92	3		
	5	0	0	2	2	96		
Population 2	2		98	2		0	0	0
	3		2	97		1	0	0
	5		0	1		96	2	1
	6		0	0		3	95	2
	7		0	0		0	3	97

Genome Position

Pooled	Marker	1	2	3	4	5	6	7
	1	88	6	4	2	0	0	0
	2	5	86	5	3	1	0	0
	3	3	4	87	4	2	0	0
	4	2	3	3	90	2	0	0
	5	2	1	1	1	93	2	0
	6	0	0	0	0	2	95	3
	7	0	0	0	0	0	3	97

Ex. 11.4

Exercise 11.5

The table on the following page shows gene ordering information using the bootstrap approach. There are three populations of barley in this data set. The number of bootstrap replications is 100. The evidence is strong that the locus orders estimated using populations #1 and #2 is homogenous. However, two segments of linkage maps obtained using population #3 data corresponding to genome positions from 5 to 7 and from 14 to 16 show departure from the pooled genome position. Please answer the following questions related to population #3.

(1) Rearrange the frequency matrix of population #3 corresponding to genome positions from 5 to 7 and from 14 to 16 to see if you can get a frequency distribution which agrees more between the rows and columns than does the original one. (Hint: Ignore the genome positions when you change the order of the markers or ignore the marker order when you change the genome positions.)

(2) Can you make any biological interpretation from what you have done in (1)? (Hint: Pay attention to possible genome rearrangement.)

(3) If what you concluded in (2) is true, can you design another experiment to validate your conclusion?

(4) We know that the three doubled-haploid (DH) populations were generated in the U.S., Germany and Canada, respectively. Population #1 was produced in the U.S. using a cross between two varieties, Steptoe and Morex, with a population size of 150 DH lines. The second population was produced in Canada (Harington x Tr-306) and has a population size of 150 DH lines. The last population was generated in Germany (Franka x Igri) and includes 73 DH lines. Can you think of any evolutionary reason for the results?

(5) If the genome rearrangement among the three barley populations is not likely, can you think of any other reason for the results? (Hint: Pay attention to sampling error, experimental error, data manipulation error, genome variation identified by different marker systems, *etc.*)

(6) If (5) is true, how can the problems be solved? Biologically or statistically?

Marker   **Genome Position**

Population #1

	1	2	3	4	5	6	7	8	9	11	12	13	14	15	16
17	100	0	0	0	0	0	0	0	0	0	0	0	0	0	0
15	0	100	0	0	0	0	0	0	0	0	0	0	0	0	0
14	0	0	100	0	0	0	0	0	0	0	0	0	0	0	0
13	0	0	0	100	0	0	0	0	0	0	0	0	0	0	0
12	0	0	0	0	94	5	1	0	0	0	0	0	0	0	0
11	0	0	0	0	5	95	0	0	0	0	0	0	0	0	0
9	0	0	0	0	1	0	99	0	0	0	0	0	0	0	0
8	0	0	0	0	0	0	0	100	0	0	0	0	0	0	0
7	0	0	0	0	0	0	0	0	100	0	0	0	0	0	0
5	0	0	0	0	0	0	0	0	0	100	0	0	0	0	0
4	0	0	0	0	0	0	0	0	0	0	91	3	6	0	0
3	0	0	0	0	0	0	0	0	0	0	3	91	0	6	0
18	0	0	0	0	0	0	0	0	0	0	0	0	94	0	6
2	0	0	0	0	0	0	0	0	0	0	0	6	0	93	1
1	0	0	0	0	0	0	0	0	0	0	6	0	0	1	93

Population #2

	1	3	4	5	6	8	10	11	14	15
17	100	0	0	0	0	0	0	0	0	0
14	0	100	0	0	0	0	0	0	0	0
13	0	0	100	0	0	0	0	0	0	0
12	0	0	0	100	0	0	0	0	0	0
11	0	0	0	0	100	0	0	0	0	0
8	0	0	0	0	0	100	0	0	0	0
19	0	0	0	0	0	0	100	0	0	0
7	0	0	0	0	0	0	0	100	0	0
5	0	0	0	0	0	0	0	0	100	0
2	0	0	0	0	0	0	0	0	0	100

Population #3

	1	2	3	4	5	6	7	9	12	13	14	15	16
17	99	0	0	0	0	0	0	1	0	0	0	0	0
15	0	99	0	0	1	0	0	0	0	0	0	0	0
14	0	0	99	0	0	1	0	0	0	0	0	0	0
13	0	0	0	99	0	0	1	0	0	0	0	0	0
12	0	0	0	1	3	38	58	0	0	0	0	0	0
11	0	0	1	0	36	61	2	0	0	0	0	0	0
9	1	0	0	0	60	0	39	0	0	0	0	0	0
19	0	1	0	0	0	0	0	99	0	0	0	0	0
4	0	0	0	0	0	0	0	0	99	0	1	0	0
3	0	0	0	0	0	0	0	0	0	99	0	0	1
18	0	0	0	0	0	0	0	0	0	0	1	73	26
2	0	0	0	0	0	0	0	0	0	1	73	26	0
1	0	0	0	0	0	0	0	0	1	0	25	1	73

CHAPTER 12

QTL MAPPING: INTRODUCTION

12.1 HISTORY

A quantitative trait is traditionally defined as a trait with a continuous distribution, in contrast to a discrete distribution. The trait values are usually obtained by measuring instead of counting. The trait is often considered to be controlled by many genes and each of the genes has a small effect on the trait. However, recent findings, using the combination of genomic mapping and the traditional quantitative genetics, show that a small number of genes can produce a trait with continuous distribution.

Searching for genes controlling complex or quantitative traits plays an important role in applying genomic information to clinical diagnosis, agriculture and forestry because a large portion of the traits related to human diseases and agronomic importance are quantitative. The loci controlling quantitative traits have commonly been referred to as QTLs (quantitative trait loci). The procedures for finding and locating the QTLs are called QTL mapping.

QTL mapping involves construction of genomic maps and searching for a relationship between traits and polymorphic markers. A significant association between the traits and the markers may be evidence of a QTL near the markers. The genetics of quantitative traits are more complex than those of the single factor Mendelian traits. Quantitative traits are usually controlled by more than one gene and influenced by the environment, although traits controlled by a single gene with incomplete penetrance can be treated as quantitative traits for mapping and isolating the gene.

The literature on QTL mapping from the past decade is rich. A list of the key references and three sources for general review on QTL mapping are provided (Table 12.1). Much of the remaining literature related to QTL mapping can be found in the references provided. Papers by Tanksley (1993) and Lander and Schork (1994) and *Trends in Genetics* (December 1995 Vol.11 No.12 pp. 463-524) are good sources for information on QTL mapping.

The simple t-test, simple linear regression, multiple linear regression, nonlinear regression and interval test approach using partial regression have been proposed and used to map QTLs (Stuber *et al.* 1992; Weller 1988; Lander and Botstein 1989; Knapp *et al.* 1992; Lande and Thompson 1991; Zeng 1993; Zeng 1994). For QTL mapping using human populations, the sib-pair approach has been used (Haseman and Elston 1972; Lange 1986; Weeks and Lange 1988; Fulker and Cardon 1994; Cardon and Fulker 1994). To apply models of inheritance, least square, maximum likelihood and EM algorithms have been used. To carry out the data analysis, the following software packages are available: MAPMAKER/QTL (Lander *et al.* 1987), QTLSTAT (Liu & Knapp 1992), QTL Cartographer (Basten, Weir and Zeng 1995), PGRI (Liu and Lu 1995), MAPQTL (Van Ooijen and Maliepaard 1996), Map Manager QT (Manly and Cudmore 1996) and QGENE (Tanksley and Nelson 1996) (Chapter 18). Commonly used approaches for QTL mapping, such as the single-marker t-test, are for single-QTL models. The number of markers in the single-QTL

Table 12.1 A list of the key references for QTL mapping.

Subject	Author
Methodology	
Historical	Sturtevant 1913; Sax 1923; Penrose 1938
Single-marker: Linear model	Soller *et al.* 1976; Edwards *et al.* 1987; Stuber *et al.* 1987
Single-marker: Likelihood	Weller 1986
Interval mapping: Regression	Knapp *et al.* 1990; Knott & Haley 1992; Martinez & Curnow 1992; Jansen 1992 & 1993
Interval mapping: Likelihood	Lander & Botstein 1989; Jensen 1993; Luo & Kearsey 1989; Knott & Haley 1992
Interval mapping: Composite	Jansen 1993; Zeng 1993 & 1994
Experimental design	Knapp & Bridges 1990; Knapp 1994
Multi-QTL	Moreno-Gonzalez 1992; Jansen 1993; Rodolphe & Lefort 1993; Zeng 1993 & 1994
Sib-pair: Single marker	Haseman & Elston 1972; Cockerham & Weir 1983; Lange 1986; Weeks & Lange 1988
Sib-pair: Interval mapping	Fulker & Cardon 1994; Cardon & Fulker 1994
Sib-pair: Multi-locus	Weeks & Lange 1992; Fulker *et al.* 1995
Resampling	Churchill & Doerge 1994
QTL-environment interactions	Hayes *et al.* 1993; Knapp 1994; Jiang & Zeng 1995
Statistical power and resolution	Soller *et al.* 1976; Rebai *et al.* 1994 & 1995; Lander & Botstein 1989; Zeng 1993 & 1994; Boehnke 1994; Jansen & Stam 1994; Kruglyak & Lander 1995
Computer Software	
MAPMAKER/QTL	Lander *et al.* 1987; Lander & Botstein 1989
QTLSTAT	Knapp *et al.* 1992
LINKAGE	Terwilliger & Ott 1994
PGRI	Liu 1995
QTL Cartographer	Basten, Weir and Zeng 1996
MAPQTL	Van Ooijen & Maliepaard 1996
Map Manager QT	Manly & Cudmore 1996
QGENE	Tanksley & Nelson 1996
Experiments	
Drosophila	Mackay 1995
Mice	Frankel 1995; Schork *et al.* 1995
Cattle	Haley 1995
Human	Lander & Schork 1994
Maize	Stuber 1995
Tomato	Nienhuis *et al.* 1987; Paterson *et al.* 1988 & 1991
Rice	McCouch & Doerge 1995
Barley	Hayes *et al.* 1993
Trees	Groover *et al.* 1994; Bradshaw & Stettler 1995; Grattapaglia *et al.* 1995

models can vary from one to a large number. However, only one or two markers are directly related to the putative QTL and the others are used in the models to control genetic background effects and sampling error.

QTL mapping has been recognized as a multiple test problem. The test statistic of QTL mapping does not follow standard probability distributions. The tests are not independent among marker loci because of the linkage relationships and possible gene interactions. Traditional adjustment on the test statistic cannot be applied to QTL mapping. A shuffling approach can be used to determine the empirical distributions of the statistics (Churchhill and Doerge 1994).

In this chapter, the quantitative genetics concepts of quantitative trait inheritance will be introduced and an example data set that will be used in the following chapters for illustrating the QTL mapping methodology will be provided. QTL mapping is a combination of qualitative linkage analysis and quantitative genetics analysis. An understanding of classical definitions of genetic effects in quantitative genetics terms is essential for understanding QTL mapping.

In the next five chapters, the essential statistics behind QTL mapping will be introduced. In Chapter 13, single-marker analysis using linear models and the likelihood approaches will be discussed. In Chapter 14, interval mapping and composite interval mapping approaches will be described. Alternative approaches for QTL mapping, such as the sib-pair method and the nonparametric methods, will be dealt with in Chapter 15. Statistical power, which is also related to the QTL resolution, will be discussed in Chapter 16. The future of QTL analysis, including the multiple QTL model and combination of metabolic pathways, cDNA mapping and QTL mapping, will be discussed in Chapter 17. Computer resources for QTL mapping, along with the general linkage analysis, will be presented in Chapter 18.

12.2 QUANTITATIVE GENETICS MODELS

QTL mapping is a combination of linkage mapping and traditional quantitative genetics analysis. Traits can be modeled using a modified quantitative genetics approach, using population parameters such as variances among and within genetically related but different populations, because individual gene identity is usually not inferred. QTL mapping creates the possibility for modeling quantitative traits at the individual gene level. This is the foundation of modeling at the population level, even though individual gene effects cannot be inferred (Falconer and Mackay 1996). In this section, genotypes of QTLs are assumed known. A practical single-locus or single-QTL model and a multiple-locus model for quantitative traits will be described.

12.2.1 SINGLE-QTL MODEL

One of the QTL mapping strategies is to search the whole genome by hypothesis test for a single marker or a single genome position and then to build a multiple-QTL model based on the results from single QTL analysis. Certainly, searching the whole genome simultaneously is better than scanning individual points, if information content is adequate to do so. However, the information content of typical experiments rarely is adequate.

Focus on a single-QTL model first. The definitions of the gene effects for single-QTL models are the same as the traditional quantitative genetics definitions (Cockerham 1954; Falconer and Mackay 1996). The genotypic values for the three genotypes (QQ, Qq and qq) in an F2 population, which is

Table 12.2 Notations for single-QTL models in backcross and F2 populations. n_{QQ}, n_{Qq} and n_{qq} are counts of genotypes QQ, Qq and qq, respectively.

Model	Genotype	Value	Variance
Backcross (Qq X QQ)	QQ	μ_1	σ^2
	Qq	μ_2	σ^2
	Genetic Effect	$g = 0.5\,(\mu_1 - \mu_2)$	$\sigma^2\,(1/n_{QQ} + 1/n_{Qq})\,/4$
F2 (Qq X Qq)	QQ	μ_1	σ^2
	Qq	μ_2	σ^2
	qq	μ_3	σ^2
	Additive	$a = 0.5\,(\mu_1 - \mu_3)$	$\sigma^2\,(1/n_{QQ} + 1/n_{qq})\,/4$
	Dominance	$d = 0.5\,(2\mu_2 - \mu_1 - \mu_3)$	$\sigma^2\,(1/n_{QQ} + 4/n_{Qq} + 1/n_{qq})\,/4$

the selfed progeny of a heterozygous parent Qq, are μ_1, μ_2 and μ_3, respectively, and are shown in Table 12.2. The additive (a) and dominance (d) effects are defined as

$$a = 0.5\,(\mu_1 - \mu_3)$$
$$d = 0.5\,(2\mu_2 - \mu_1 - \mu_3) \tag{12.1}$$

The additive effect is the same as the average effect of the gene-substitution because the expected allelic frequencies for the two alleles are the same in the F2 population ($p = q = 0.5$, see Chapter 2).

For the backcross progeny produced by a cross between a heterozygous parent Qq and a homozygous parent QQ, the additive and dominance effects are confounded. The genetic effect (mixed with additive and dominance effects) is defined as

$$g = 0.5\,(\mu_1 - \mu_2) \tag{12.2}$$

From Equation (12.1), we have $\mu_2 = 0.5\,(\mu_1 + \mu_3 + 2d)$ and

$$g = 0.5\,(\mu_1 - \mu_2)$$
$$= 0.5\,[\mu_1 - 0.5\,(\mu_1 + \mu_3 + 2d)\,] \tag{12.3}$$
$$= 0.5\,(a + d)$$

So the genetic effect defined in backcross progeny is a combination of additive and dominance effects.

In the next few chapters, the backcross and the classical F2 populations will be used to illustrate the rationale and methodology for QTL mapping using experimental populations. Commonly used mapping populations obtained by controlled matings can usually be classified as one of these two population types at whole genome or individual genome segment model levels. For example, doubled-haploid lines and recombinant inbred lines can be treated as a backcross model in terms of the data analysis because the expected genotypic frequencies are same as a backcross. However, the interpretations of the QTL mapping results may be different. QTL effects in a backcross population are a mixture of additive and dominance effects. QTL effects in the doubled haploid and recombinant inbred lines are purely additive. When mapping using hybrids of two heterozygous populations, the progeny set is a mixture of the backcross and F2. When mapping using open-pollinated populations, the progeny represents a mixture of F2, backcross and random mating progeny.

Table 12.3 Number of genetic effects for a l-locus model.

Effect	Number
Main effect	$2\binom{l}{1}$
2-way interaction	$4\binom{l}{2}$
3-way interaction	$8\binom{l}{3}$
4-way interaction	$16\binom{l}{4}$
i-way interaction	$2^i\binom{l}{i}$
Total	$\sum_i 2^i\binom{l}{i} = 3^l - 1$

12.2.2 MULTIPLE-LOCUS MODEL

A genetic model for a quantitative trait is usually defined in terms of number of genes, gene effects and frequencies, relationship among the genes and relationship between environment and gene action. For gene actions and relationship among the genes, there are, by classical quantitative genetic definitions, additive, dominance and epistatic genetic effects. Classical quantitative genetics focuses on additive and dominant genetic variation. Epistatic interactions have been very difficult to estimate and detect.

Assume that l genes control a quantitative trait. For a conventional F2 population, the possible genetic effects are listed in Table 12.3. For a two-locus model, there are 4 main effects, one additive effect and one dominant effect for each locus and there are 4 epistatic interactions, one additive by additive, one dominance by dominance and two additive by dominance interactions. In general, there are

$$2^i\binom{l}{i} = \frac{2^i l!}{i!\,(l-i)!} \tag{12.4}$$

possible i-way effects for an l-locus model. The total number of genetic effects is

$$\sum_{i=1}^{l} 2^i\binom{l}{i} = 3^l - 1 \tag{12.5}$$

Using matrix notation, the multiple-locus model for a quantitative trait can be written as

$$Y = A + D + I + E \tag{12.6}$$

where Y is the trait value, A, D, I and E are additive genetic, dominant genetic, epistatic genetic and error effects for the trait, respectively. E is the result of experimental error or environmental effects. Using notations of Cockerham (1954), components of the matrices are

$$A = \sum_i f_i c_{1i} a_i$$

$$D = \sum_i f_i c_{2i} d_i \tag{12.7}$$

$$I = \sum_i \sum_{j=1}^{i-1} f_j c_{1i} c_{1j} a_{ij} + \sum_i \sum_{j=1}^{i-1} c_{2i} c_{2j} d_{ij} + \sum_{i=1, i \neq j}^{l} \sum_{j=1, j \neq i}^{l} c_{1i} c_{2j} e_{ij}$$

where a_i and d_i are additive and dominant main effects for locus i and a_{ij}, d_{ij} and e_{ij} are additive by additive, dominance by dominance and additive by dominance interactions between loci i and j, respectively. Definitions of the coefficients (dummy variables) are listed in Table 12.4.

Table 12.4 Definitions of the coefficients (dummy variables) and gene effects.

QTL Genotype	Frequency	c_1	c_2	Effects
AA	f_1	1	1/2	$a + d/2$
Aa	f_2	0	-1/2	$-d/2$
aa	f_3	-1	1/2	$-a + d/2$

Equation (12.7) is a complete model for a quantitative trait. However, this model is difficult to implement using traditional quantitative genetic approaches. Even for the recent QTL mapping approaches, the model is difficult to obtain when the number of genes is more than two.

12.3 DATA FOR QTL MAPPING

12.3.1 DATA STRUCTURE

In a quantitative genetic model, it is not always necessary to know the genotypes of the individuals in a population. However, the genotypes are needed to estimate parameters of a genetic model. In previous sections, genotypes of QTL are assumed to be known, but such knowledge does not exist for most situations in practice. The purpose of QTL mapping is to infer the QTL genotypes in order to estimate the QTL effects and locations from association with known genetic markers. What we can observe are the genetic markers located at certain positions on the genome for each of the individuals in the mapping populations and the trait values corresponding to the individuals.

For example, Figure 12.1 shows a segment of a hypothetical genome. Five genetic markers and a QTL are located on this genome segment. The QTL genotypes are not observed. The markers are polymorphic in the mapping population and their genotypes can be determined a laboratory analysis. Let us say that a sample with size N is obtained. The locus order CABDE and the linkage relationships of these loci are obtained based on genotypes of the N

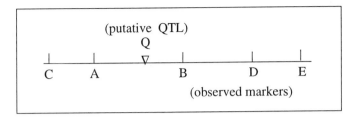

Figure 12.1 A hypothetical segment of a genome contains five genetic markers and a QTL.

individuals for the five loci. It is also required to obtain phenotypic values for the quantitative trait to map the QTL. Most importantly, the phenotypic values have to correspond to all or some of the N individuals. QTL mapping is based on the association between the trait values and the marker genotypes.

Depending on the mating systems of the experiment and the methodology of QTL mapping, statistical models for analyzing the trait and the marker association can be different. In general, single-marker types of analysis do not require locus order and do not yield precise QTL locations. The interval mapping approaches require locus order and provide estimates of QTL locations.

12.3.2 THE BARLEY DATA

This data set will be used in the next 5 chapters to illustrate methodologies of QTL mapping.

Marker Data

Table 12.5 contains 26 molecular markers on barley chromosome IV. Marker genotypes were obtained for 150 doubled haploid lines. The doubled haploid lines were generated from a selfed F1 plant between two barley inbred lines Steptoe X Morex (Hayes *et al.* 1994; Kleinhofs *et al.* 1994). The data were generated by the North American Barley Genome Mapping Project managed by Washington State University and Oregon State University. The markers include several types, *e.g.*, isozymes, RFLPs, PCR based and some morphological markers.

Phenotype Data

Table 12.6 contains phenotypic data corresponding to the marker data in Table 12.5. The phenotypic data for malt extract (a malting quality trait) was obtained from four experimental locations in Idaho, Oregon, Washington and Montana. Figure 12.2 shows the distributions of the trait values. Table 12.7 shows the linkage map obtained using computer software PGRI (Liu 1995). QTL methodology for the doubled haploid progeny set is same as the methodology used for a backcross. However, the interpretations of the genetic effects are different for the two progeny types. For a backcross, the genetic effect is the mixture of additive and dominance effects. For doubled haploid, the genetic effect is pure additive effect.

Table 12.5 Genotypes for 26 molecular markers on barley chromosome IV using 150 doubled haploid lines generated by using selfed F1 between Steptoe X Morex. Codes: "1" is for Steptoe genotype, "2" is for Morex genotype and "0" is missing data. Data was generated by the North American Barley Genome Mapping Project (Washington State University and Oregon State University).

```
WG622     1112211122120121111111122212221211112111212212221221112211221212112112221210
          1220212211122012121111121011022112000022111220211211211111121111212121211221
ABG313B   1012011022122122211101112212121221121112122122212221111121122121121211221211
          1222212111122212122111212101221122111221112212102112111212111112121211110
CDO669    1112111100102222200011110002121221121112122122220021111121222121112112211221210
          1222212111121112122111212111221122111221111221212112112111121212112121211110
BCD402B   1010110022020002222221102121121221022122122212222221211121222121122211221212
          1200110110121112122112121211222112201022211021121121121121111222122121211110
BCD351D   1112111122112022221112111212111122112212212222212222222121112122212112221021212
          1222112111121112122112121211222112211122211121121121121112222221212121211111
BCD265B   1012111100102222200021112121111221222222212222122212212121212211221222112112
          1212010111121112122110021211212112211122210121111021122111222222210122221111
TubA1     0002110000102222020021112121111221222222212222122212121212121211221222112112
          1212111111121112122112121211212112111221112121111121122111222222212122221111
Dhn6      1112111221101222221212111211111221222222112222122210012121212211121222112112
          1212111111121112122112121211212112111221112121111121122111222222212122221112
ABG003    1112111221121022221010111210112202222221122212201221212121212211121222112112
          1112111111112011212200121212112101121111222101210110112112210222222121202210010
WG1026B   1111111122112121222221212111211111221222221122212221221212121221111212220021112
          1112111111121112122112122112121111221112111112112211122222222212122221112
Adh4      1011112100102222020212121102111122122222112222122212121211221112122111121112
          1112111111121112122112122112121211112221012111112112210222222212122222112
ABA003    1111112122112122212121211112111022122221122012221212121211221112122111121112
          1112111111121012102112122112121111221110111111211122222212120222121
ABG484    1111112122112122212121211112111122122221122012221212101221112112211012211121112
          1110111111121102122112122112121211211112221112121112112211122222212122022112
Pgk2A     1111111202201012221012121112111122122221122012221212121212211121122112111121112
          1112010111122010212211212221121121112111221112121121121122211122222212122222110
ABR315    0101111221121222121212111211112212220011222222222212121212211121212211121112
          1122111111220010212211210221121210211102211121121121122111222222212122222100
WG464     1111212122002122212121211021111221222211122222222211212122221112122111121110
          1122111111221112122112020211212112011122211210121211220111121212022222112
BCD453B   1111211122122112221211121111211112212222111222210221112221222211121212211221112
          1120111111221102122112122121211211111121122111212111211111121212122222111
ABG472    1111211100102112010112212121111220222211122221122012222122221212122110211112
          1121111111221122122112122121211201111111221112121211121111121121212122222111
iAco2     1111212112122112221112212221211121222211222221122112222211222121112211221112
          1111211111121211121122122112112111111111121121212121111111221212122212111
ABG500B   1112121111221122111221221211021222211222221122112222112221211122111211121110
          1111211111210111211221221121121111111111121212121211111111101212122212111
ABG498    1211212111122112211112212211211121222211222221022112222112221211120011021110
          1111211111121111212122122112121111111111121212121111121111111212122212111
WG114     1211212111120112211112212011211121222211222221122112222112221211120011021110
          1110121111112101121122120211211211111111111211012121111121111112121212122212111
ABG054    1211212111122112211112212210201121222211222221122112222112221211122112211221110
          1111211111121112112201121211211111111112121212111112100121202122212111
ABG394    1211212111122112211112212211211121222211222221122112222112221211122112211221110
          1111211111121212121122122112121111111101211212121111121111121212122212111
ABG366    1211212111122112211112212211211212122221222221122112222112221211122112211221112
          111112011111212121211021220112111111101101121102201111120111121211121000000
ABG397    1211212111022112222111221221112111212222211222221122112222112222011102112211112
          1111211111102121212112221221121111121111112122221211112121121212122212111
```

Table 12.6 Phenotypes for malting quality corresponding to the marker genotypes in Table 12.5 in four environments. ID = Idaho, OR = Oregon, WT = Washington and MT = Montana. Data was generated by the North American Barley Genome Mapping Project.

Line	ID	OR	WT	MT	Line	ID	OR	WT	MT
1	74.40	73.70	74.40	72.90	76	73.10	73.50	73.90	71.60
2	72.30	71.10	75.50	70.70	77	74.20	73.40	74.10	74.40
3	72.90	71.20	74.80	72.40	78	74.10	72.60	74.30	73.50
4	73.20	71.80	74.20	71.50	79	76.20	75.20	77.40	74.30
5	73.50	74.70	75.70	71.60	80	74.30	75.70	75.00	72.60
6	71.80	72.80	74.90	69.50	81	73.70	71.10	72.00	70.30
7	75.00	71.50	73.80	70.00	82	74.70	74.20	75.10	71.20
8	71.30	70.40	73.00	69.60	83	73.40	75.20	75.60	71.60
9	77.30	76.50	79.90	74.30	84	71.90	74.50	73.90	69.70
10	73.90	74.30	74.90	72.90	85	72.30	70.80	73.20	72.10
11	76.60	71.20	77.40	72.40	87	77.00	77.70	79.40	76.70
12	73.50	74.70	75.90	73.10	88	73.10	70.10	73.50	70.30
13	74.60	74.50	75.90	73.30	89	74.50	75.40	77.50	74.50
14	75.30	73.30	75.50	71.90	90	76.10	72.70	76.90	74.00
15	73.40	71.50	74.60	71.40	91	74.40	73.80	75.80	73.00
16	72.70	73.70	75.70	72.90	92	73.00	72.00	74.30	72.30
17	72.10	73.80	74.40	74.20	93	75.80	74.70	77.20	75.80
18	74.60	73.50	74.90	72.10	94	75.10	75.10	75.60	74.20
19	75.90	75.50	77.80	73.80	95	74.80	73.80	75.20	72.20
20	75.60	73.70	74.70	71.90	96	72.10	71.00	74.60	70.20
21	73.30	73.40	74.90	71.00	97	74.30	73.30	75.50	72.10
22	74.60	75.60	75.50	73.20	98	75.90	74.00	77.30	73.40
23	76.00	75.50	75.70	73.80	99	76.60	76.80	77.80	75.00
24	75.40	74.10	75.50	71.40	100	77.50	76.20	77.30	74.90
25	74.40	75.60	74.30	71.10	101	74.90	75.40	76.10	74.10
26	74.30	71.40	75.20	71.60	102	75.20	73.70	75.20	73.80
27	74.00	74.10	73.80	72.60	103	74.00	73.60	75.00	72.10
28	74.20	73.20	74.80	72.50	104	74.20	74.80	75.20	74.40
29	73.30	73.40	74.20	71.80	105	75.50	74.60	75.10	74.20
30	74.80	74.50	77.10	72.30	106	74.80	73.70	76.30	75.00
31	72.80	69.60	73.20	69.00	107	72.50	73.10	73.00	71.60
32	76.90	74.90	78.40	75.10	108	74.20	74.50	75.80	73.10
33	76.80	74.90	77.80	76.30	109	74.10	74.30	74.40	72.70
34	72.40	72.90	75.50	73.30	110	75.10	74.70	74.10	72.90
35	75.60	75.00	77.00	75.20	111	73.00	74.10	76.30	73.40
36	73.40	75.90	74.70	73.80	112	76.00	74.10	76.70	74.50
37	73.60	73.80	74.90	74.10	113	74.80	75.00	76.00	74.70
38	76.70	74.50	75.10	75.40	114	74.40	72.60	74.90	71.80
39	78.50	74.60	77.80	76.60	115	74.30	72.70	74.20	72.80
40	75.70	74.60	74.80	71.80	116	73.80	73.70	74.50	72.60
41	75.50	73.80	76.80	74.70	117	73.60	73.50	73.60	72.20
42	75.10	75.40	76.60	76.50	118	73.10	71.80	74.10	72.00
43	74.40	73.60	76.10	74.50	119	74.50	72.70	75.40	72.80
44	74.00	71.90	73.70	73.50	120	73.10	71.00	73.80	71.80
45	73.40	73.20	74.90	72.70	121	73.90	76.90	76.70	73.20
46	74.60	71.10	77.10	74.40	122	72.70	72.10	74.40	71.80
47	75.20	73.50	76.20	74.20	123	75.10	73.00	75.50	74.50
48	75.90	72.80	74.60	73.80	124	73.60	72.40	71.80	71.10
49	74.60	74.40	76.60	74.40	125	71.10	71.00	76.30	70.50
50	75.20	73.10	76.20	74.90	126	74.90	71.70	78.10	74.30
51	72.80	72.90	74.20	73.90	127	74.10	74.00	75.70	72.60
52	72.80	73.00	73.90	74.00	128	74.30	74.40	76.10	72.60
53	75.30	72.80	73.60	75.30	129	73.60	74.90	75.00	73.70
54	76.30	74.20	76.30	75.50	130	75.30	71.20	77.00	73.40
55	76.20	75.10	76.60	76.50	131	72.90	71.20	71.90	70.70
56	73.80	73.30	72.70	72.80	132	73.60	73.80	74.50	72.20
57	75.10	74.10	75.40	75.50	133	74.80	72.50	74.60	73.80
58	75.40	76.00	77.10	73.60	134	74.50	75.10	77.00	72.70
59	76.00	73.70	76.10	74.90	135	75.00	73.90	75.90	73.50
60	75.80	74.00	75.20	74.00	136	74.50	74.00	75.10	73.60
61	73.50	73.30	74.70	73.80	137	73.60	70.20	74.30	72.10
62	74.70	74.30	75.40	74.10	138	72.40	73.00	73.60	71.10
63	74.30	73.90	76.30	74.60	139	71.70	69.40	70.60	69.10
64	74.30	73.20	75.20	74.30	140	74.30	73.90	74.20	71.80
65	74.80	75.10	75.50	74.30	141	75.30	75.90	76.70	73.20
66	73.20	73.80	72.80	74.50	142	73.70	74.30	73.70	70.50
67	74.80	74.90	76.20	75.70	143	76.70	76.60	77.50	75.50
68	72.90	72.10	75.50	73.70	144	72.10	74.40	74.20	73.00
69	74.70	74.00	74.20	75.10	145	74.80	72.80	74.40	73.40
70	76.50	74.80	75.70	76.00	146	73.50	72.90	72.10	71.40
71	74.50	75.90	74.80	75.00	147	75.10	74.70	75.00	73.10
72	71.60	71.90	72.70	71.70	148	73.00	74.90	75.40	70.70
73	75.00	73.00	78.70	74.20	149	75.10	75.50	73.40	70.10
74	73.20	73.80	74.10	73.90	150	73.20	71.20	72.40	70.30
75	75.60	74.60	75.50	74.90					

Table 12.7 Linkage map for data in Table 12.5. Two-Point = two-point recombination fractions between the marker and the marker above. Multi-Point = recombination fractions between the marker and the marker above estimated by multi-point analysis. The locus order and the multiple-point recombination fractions were obtained using PGRI (Liu 1995).

Marker	Two-Point	Multi-Point
WG622	-----	-----
ABG313B	0.105	0.108
CDO669	0.046	0.055
BCD402B	0.140	0.139
BCD351D	0.038	0.030
BCD265B	0.097	0.098
TubA1	0.000	0.000
Dhn6	0.051	0.049
ABG3	0.008	0.007
WG1026B	0.015	0.015
Adh4	0.029	0.030
ABA3	0.022	0.015
ABG484	0.022	0.015
Pgk1	0.007	0.007
ABR315	0.030	0.030
WG464	0.078	0.074
BCD453B	0.102	0.103
ABG472	0.051	0.058
iAco2	0.170	0.166
ABG500B	0.027	0.026
ABG498	0.021	0.022
WG114	0.000	0.000
ABG54	0.007	0.007
ABG394	0.000	0.000
ABG397	0.036	0.031
ABG366	0.045	0.034

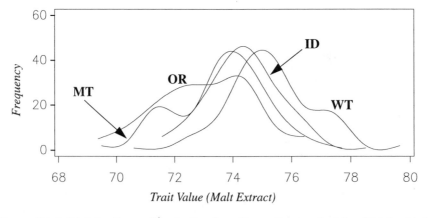

Figure 12.2 Distributions of the trait values for malt extract of the 150 doubled haploid lines in four environments: Idaho, Oregon, Washington and Montana.

In the phenotypic data for the malt extract, line # 86 was missing (Table 12.6). The marker data (Table 12.5) contains all 150 doubled haploid lines but with scattered missing marker information for some of the markers. It is important that the missing values are treated appropriately in the analysis. These types of missing values which are seen in the data sheets can be classified as apparent missings. As mentioned in the genetic mapping chapters, missing values are a serious problem for genetic mapping. This is especially true for QTL mapping. When several loci are considered in an analysis, missing genotypic combinations cannot be avoided and it may not apparent that there is missing data even when the data sheets are complete. This is hidden missing data. In practice, the problem of missing information for genomic mapping cannot be fixed through experimentation because an unrealistic number of samples may be needed.

In the next chapters, this example will be used to illustrate methodologies in QTL mapping. Readers can reproduce the results using the data given in Tables 12.5 to 12.7. Reproducing of analyses is a good way to understand the methodology.

EXERCISES

Exercise 12.1

QTL mapping is a set of procedures to identify and locate potential genes (or QTLs) controlling a quantitative trait using the expected association between the putative genes and known genetic markers. The distribution of the trait values for the known marker genotypes is usually a mixture of several distributions for the true QTL genotypes. If we can observe the true QTL genotypes, the distribution will be unique and the QTL mapping will locate the QTL only. Can you find real-life situations where the true QTL genotypes can be observed? If you can observe the true QTL genotypes, are the methodologies for linkage analysis explained in Chapters 6 to 10 sufficient to analyze the QTL data?

Exercise 12.2

If you are working on quantitative traits, find a data set which contains trait values segregating in a population from a defined mating. Draw some distribution plots of values of the traits using real data.

(1) If you assume that the trait is controlled by a single gene and a significant environmental component, can you fit a quantitative genetics model to the data?

(2) If the genetic effect is 10 times bigger than the environment effect, how do you expect the distribution plot to look? How about in situations in which the genetic effect and the environment effect are the same magnitude, or alternatively, where the genetic effect is one-tenth of the environment effect?

(3) If two genes control the trait, what do you expect regarding the distribution of the trait? How about 10 genes?

Exercise 12.3

Review the history of QTL mapping in the species on which you are working. What are the alternatives for finding and locating genes controlling quantitative traits in this species, besides using genetic markers?

Exercise 12.4

Has any DNA clone encoding a QTL been obtained using the QTL mapping approach in the species you are working on? If yes, review the procedures used to obtain the clone of the gene. If no, do you expect that this will happen soon?

CHAPTER 13

QTL MAPPING: SINGLE-MARKER ANALYSIS

13.1 RATIONALE

The relationship between quantitative trait variation and qualitative traits can be quantified using statistical models. Qualitative traits are usually observed more easily and more accurately than quantitative traits because quantitative traits are usually controlled by a number of genes and the genetic effects usually interact with environment. Genetics of qualitative traits usually can be inferred at the individual genotype level. However, the genetics for quantitative traits can only be studied at the population level. Scientists have tried to gain more understanding of the inheritance of complex traits by comparing the relationships between the complex traits and the simple traits with known genetics, for example, relating complex human diseases to blood types and relating economically important traits of field crops to simple morphological traits. The rationale behind these simple relationships is the fundamental basis of QTL mapping. Genetic markers can be considered traits with simple inheritance.

Using biological knowledge and linkage analysis, we can determine the mode of the marker inheritance and their genome locations. Now, the question becomes whether or not we use that large amount of linkage information to infer the genetics of quantitative traits. The underlying genetic assumptions for finding the relationships between quantitative trait inheritance and the genetic markers are:

(1) Genes controlling the quantitative traits can be mapped on the genome like simple genetic markers.

(2) If the markers cover a large portion of the genome, then there is a good chance that some of the genes controlling the quantitative traits will be linked to some of the genetic markers.

(3) If the genes and the markers are segregating in a genetically defined population, then the linkage relationships among them may be discoverable by looking at the association between the trait variation and the marker segregation pattern.

Certainly, if the genotypes of the genes controlling the traits can be scored, then the problem becomes one of simple linkage analysis. In practice, the genotypes of the genes cannot be observed; instead, what we can usually observe are the continuous trait values.

The single-marker analysis is a good start not only for learning QTL mapping, but also for practical data analysis. The important information readers should get from this chapter is how to derive genetic inference for a putative QTL from the analysis of markers. The same rationale will be applied in the next chapter to more advanced analyses, such as interval mapping and composite interval mapping. Comparisons with the other methods will be described in the next few chapters.

Early work on the association between the trait value and marker segregation patterns has been based on linear models, such as

$$y_j = \mu + f(\text{marker}_j) + \varepsilon_j \tag{13.1}$$

where y_j is the trait value for the jth individual in the population, μ is population mean, $f(\text{marker}_j)$ is a function of marker genotype and ε_j is the residual associated with the jth individual. For example, a gene Q is located near a marker A (Figure 13.1) and the trait controlled by Q can be modelled by the marker A

$$y_j = \mu + f(A) + \varepsilon_j$$

As we will see, the expectation of $f(A)$ contains genetic values of genotypes of Q and the linkage relationship between A and Q.

The marker genotypes can be treated as classification variables for a t-test or an analysis of variance. The marker genotypes also can be coded as dummy variables for regression analysis. Equation (13.1) can also be solved using the likelihood approach, by finding the joint distribution of marker genotypes and the putative QTL genotypes.

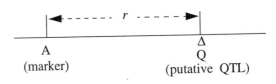

Figure 13.1 Single-marker genetic model.

In this chapter, some of the early work on QTL mapping using single-marker analysis will be described. Single-marker analysis for QTL mapping is a set of procedures to solve Equation (13.1). The single-marker analysis can be implemented as

- a simple t-test
- an analysis of variance
- a linear regression
- a likelihood ratio test and maximum likelihood estimation

Single-marker analysis is simple in terms of data analysis and implementation. It can be performed using common statistical software such as SAS. Gene orders and a complete linkage map are not required. However, linkage map will help in the presentation of the results. The disadvantages of the single-marker analysis are:

(1) The putative QTL genotypic means and QTL positions are confounded. This confounding causes the estimator of QTL effects to be biased and

the statistical power to be low, particularly when linkage map density is low.

(2) QTL positions cannot be precisely determined, due to the nonindependence among the hypothesis tests for linked markers that confound QTL effect and position.

The two sections of this chapter are arranged according to the backcross and F2 types of mapping populations. Experimental populations used for QTL mapping in most plant and animal species can be classified as a backcross or an F2 type of population. If the populations cannot be classified into these types of populations at a whole genome level, they usually can be classified as a mixture of backcross and F2 at a whole genome level and as either backcross or F2 at a single-locus level.

Simple sib-pair analysis and analysis using populations of mixed self and random matings are also related to single-marker analysis. These topics will be covered in Chapter 15 on mapping using a natural population. The methods in this chapter are for analyses using experimental populations obtained by controlled matings.

13.2 SINGLE-MARKER ANALYSIS IN BACKCROSS PROGENY

Single-marker analysis is based on comparisons between marker genotypic means through a t-test, an analysis of variance, a likelihood ratio test, or a simple regression for trait on coded marker genotype. These approaches are equivalent or similar under some assumptions. The single-marker analysis is done by analyzing one marker at a time. The QTL is determined to be located near a marker if phenotypic values for the trait are significantly different among the marker genotypes.

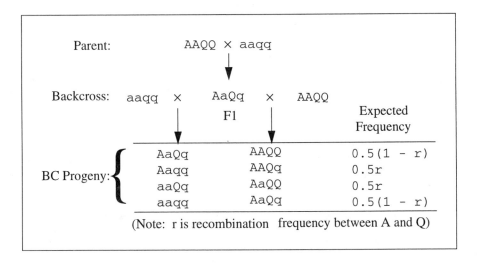

Figure 13.2 Conventionally defined backcross progeny.

13.2.1 JOINT SEGREGATION OF QTL AND MARKER GENOTYPES

For a classical backcross design, in which the population is generated by a heterozygous F_1 backcrossed to a homozygous parent (for example, a cross of AaQq X AAQQ) (Figure 13.2), the rationale behind the single marker analysis can be explained using the co-segregation patterns listed in Table 13.1. Marker A and the QTL are assumed to be linked with r recombination units apart. The expected frequencies for the four marker-QTL genotypes (AAQQ, AAQq, AaQQ and AAQq) are listed in Table 13.1. The conditional frequencies of the QTL genotypes (QQ and Qq) on the marker genotypes (AA and Aa) can be obtained by dividing the joint marker-QTL genotypical frequencies by the marginal marker genotypic frequencies (they are 0.5 for this case). The expected phenotypic values for the observable marker genotypes can be obtained by modified conditional frequencies by the expected trait values (see definitions in Table 12.2). For example, the expected trait value for marker genotype AA is

$$\mu_{AA} = p\,(QQ|AA)\,\mu_1 + p\,(Qq|AA)\,\mu_2$$
$$= (1-r)\,\mu_1 + r\mu_2$$

where $p\,(QQ|AA)$ and $p\,(Qq|AA)$ are probabilities that an individual with marker genotype AA has genotype QQ or Qq, respectively, and μ_1 and μ_2 are the expected genotypic values for the two QTL genotypes QQ and Qq, respectively.

Table 13.1 Expected QTL genotypic frequencies conditional on genotypes of a nearby marker in backcross populations with no double crossovers. r is recombination fraction between the marker and the QTL. See Table 12.2 for other notation.

Marker Genotype	Observed Count	Marginal Frequency	QTL Genotype		Expected Trait Value
			QQ	Qq	
			Joint Frequency		
AA	n_1	0.5	$0.5\,(1-r)$	$0.5r$	
Aa	n_2	0.5	$0.5r$	$0.5\,(1-r)$	
			Conditional Frequency		
AA	n_1	0.5	$1-r$	r	$(1-r)\,\mu_1 + r\mu_2$
Aa	n_2	0.5	r	$1-r$	$r\mu_1 + (1-r)\,\mu_2$

Table 13.2 Analysis of variance for a single QTL analysis using a backcross population. N is the progeny size, b is number of replications and c is a coefficient estimated using $c = N - [n_1^2 + n_2^2]/N$.

Source	df	MS (Mean Square)	Expected MS	F-test
Total Genetics	$N-1$	MSG		
QTL	1	MSQ	$\sigma_e^2 + b\sigma_{G(QTL)}^2 + bcg^2$	MSQ/MSG(Q)
G(QTL)	$N-2$	MSG(Q)	$\sigma_e^2 + b\sigma_{G(QTL)}^2$	
Residual	$N(b-1)$	MSE	σ_e^2	

13.2.2 SIMPLE T-TEST USING BACKCROSS PROGENY

An analysis of variance table for a single-QTL using backcross progeny is given in Table 13.2. A linear model for this analysis can be written as

$$Y_{i(i)k} = \mu + Q_i + g(Q)_{j(i)} + e_{i(j)k}$$

where $Y_{i(i)k}$ is trait value for an individual j with QTL genotype i in the replication k, μ is population mean, Q_i is the effect of QTL genotype i, $g(Q)_{j(i)}$ is the genotypic effect which cannot be explained by the QTL (or genotypic effect within QTL genotype), and $e_{i(j)k}$ is the error term. If QTL genotypes can be observed, a t-test statistic

$$t_Q = \frac{\hat{\mu}_1 - \hat{\mu}_2}{\sqrt{\hat{s}^2\left(\frac{1}{n_1} + \frac{1}{n_2}\right)}} \tag{13.2}$$

can be used to test the QTL effect, where a hat is used to denote an estimate, the difference between the two classes of QTL genotypes is $\mu_1 - \mu_2$, the sample sizes of the two classes of QTL genotype QQ and Qq are n_1 and n_2, $n_1 + n_2 = N$ is the total sample size and the pooled estimate of the variance within the two classes is $\hat{s}^2$. The expectation of the pooled variance is

$$E[s^2] = \frac{\sigma_e^2}{b} + \sigma_{G(QTL)}^2 \tag{13.3}$$

where b is the number of replications for a genotype, σ_e^2 is the error variance and $\sigma_{G(QTL)}^2$ is part of the genetic variance which cannot be explained by the QTL.

Table 13.3 Analysis of variance table for a typical single marker analysis using a backcross population. N is the progeny size, b is number of replications and c is a coefficient estimated using $c = N - [n_1^2 + n_2^2]/N$.

Source	df	MS (Mean Square)	E(MS)
Genetics	$N-1$	MSG	$\sigma_e^2 + b\sigma_G^2$
Marker	1	MSM	$\sigma_e^2 + b[\sigma_{G(QTL)}^2 + 4r(1-r)a^2] + bc(1-2r)^2a^2$
G(Marker)	$N-2$	MSG(M)	$\sigma_e^2 + b[\sigma_{G(QTL)}^2 + 4r(1-r)a^2]$
Residual	$N(b-1)$	MSE	σ_e^2

Let us assume the recombination fraction between a marker and the QTL is r. The linear model becomes

$$Y_{i(i)k} = \mu + M_i + g(M)_{j(i)} + e_{i(j)k}$$

where $Y_{i(i)k}$ is trait value for an individual j with marker genotype i in the replication k, μ is population mean, M_i is the effect of marker genotype i, $g(M)_{j(i)}$ is the genotypic effect which cannot be explained by the marker (or genotypic effect within marker genotype), and $e_{i(j)k}$ is the error term. The t-test statistic in Equation (13.2) becomes

$$t_M = \frac{\hat{\mu}_{AA} - \hat{\mu}_{Aa}}{\sqrt{\hat{s}_M^2 \left(\frac{1}{n_1} + \frac{1}{n_2} \right)}} \qquad (13.4)$$

where $\hat{s}_M^2$ is the pooled variance within the two classes of marker genotype. t_M is distributed as t-distribution with $N-2$ degrees of freedom. If t_M is significant, then a QTL is declared to be near the marker. The expectation of the pooled variance is

$$E[s_M^2] = \frac{\sigma_e^2}{b} + \sigma_{G(QTL)}^2 + 4r(1-r)g_{QTL}^2 \qquad (13.5)$$

where g_{QTL} is the fixed QTL effect (Table 13.3).

The expectation of difference between the two marker classes is

$$\begin{aligned} E[\mu_{AA} - \mu_{Aa}] &= [(1-r)\mu_1 + r\mu_2] - [r\mu_1 + (1-r)\mu_2] \\ &= (1-2r)(\mu_1 - \mu_2) \qquad (13.6) \\ &= 2g(1-2r) = (a+d)(1-2r) \end{aligned}$$

Table 13.4 Results for the malt extract data from Table 12.6 using a simple t-test. The t-statistic and the significant p-values are given for 26 markers at 4 locations (Table 12.5).

Marker	Idaho		Oregon		Washington		Montana	
	t	p	t	p	t	p	t	p
WG622	-2.20	0.0298	-3.33	0.0011	-2.56	0.0115	-2.30	0.0227
ABG313B	-1.75	0.0824	-2.86	0.0049	-3.56	0.0005	-3.35	0.0010
CDO669	-2.83	0.0053	-3.24	0.0015	-3.18	0.0018	-4.09	0.0001
BCD402B	-2.78	0.0062	-1.81	0.0731	-2.87	0.0048	-4.18	0.0001
BCD351D	-2.68	0.0082	-1.74	0.0842	-2.24	0.0266	-4.72	0.0000
BCD265B	-2.47	0.0146	-1.49	0.1374	-1.96	0.0516	-4.52	0.0000
TubA1	-2.53	0.0125	-1.34	0.1814	-1.82	0.0710	-4.44	0.0000
Dhn6	-2.64	0.0091	-1.57	0.1188	-2.01	0.0458	-4.87	0.0000
ABG003	-2.43	0.0164	-1.43	0.1545	-2.23	0.0276	-4.95	0.0000
WG1026B	-3.34	0.0011	-2.12	0.0355	-2.50	0.0137	-5.68	0.0000
Adh4	-3.35	0.0010	-1.53	0.1282	-2.15	0.0334	-4.52	0.0000
ABA003	-3.98	0.0001	-2.82	0.0055	-2.66	0.0088	-4.84	0.0000
ABG484	-3.53	0.0006	-2.02	0.0458	-2.34	0.0209	-4.94	0.0000
Pgk2A	-3.74	0.0003	-1.71	0.0895	-2.44	0.0158	-5.01	0.0000
ABR315	-3.14	0.0021	-1.97	0.0509	-2.46	0.0150	-5.55	0.0000
WG464	-3.09	0.0024	-2.07	0.0405	-1.92	0.0567	-4.05	0.0001
BCD453B	-3.79	0.0002	-3.78	0.0002	-2.27	0.0245	-6.11	0.0000
ABG472	-3.66	0.0004	-3.11	0.0023	-1.79	0.0759	-5.48	0.0000
iAco2	-2.17	0.0320	-1.75	0.0823	-0.17	0.8648	-2.96	0.0036
ABG500B	-1.95	0.0528	-1.86	0.0645	-0.14	0.8878	-2.70	0.0077
ABG498	-1.76	0.0806	-1.44	0.1532	-0.26	0.7915	-2.43	0.0165
WG114	-1.77	0.0793	-1.78	0.0770	-0.14	0.8865	-2.49	0.0139
ABG054	-1.60	0.1129	-1.10	0.2738	-0.21	0.8317	-2.31	0.0221
ABG394	-1.58	0.1159	-0.94	0.3468	-0.04	0.9689	-2.44	0.0158
ABG397	-2.43	0.0162	-1.09	0.2763	-0.88	0.3819	-2.84	0.0051
ABG366	-1.78	0.0778	-0.54	0.5928	0.17	0.8655	-2.38	0.0186

For the null hypothesis

$$H_0: [\mu_{AA} - \mu_{Aa}] = 0$$

there are two possible interpretations

$$(a + d) = 0$$

$$r = 0.5$$

The biological meaning for the first one is that there is no genetic effect, for the other, that the QTL and the marker are independent (no linkage). The single-marker analysis is valid for the backcross progeny. However, the power of the test is low when the marker is loosely linked with the QTL and the genetic effect cannot be estimated without bias.

Table 13.5 Estimated differences of trait values between 26 Steptoe and Morex markers at 4 locations.

Marker	Mean Difference $(\hat{\mu}_{Steptoe} - \hat{\mu}_{Morex})$			
	ID	OR	WT	MT
WG622	-0.48	-0.81	-0.67	-0.66
ABG313B	-0.38	-0.68	-0.89	-0.87
CDO669	-0.64	-0.77	-0.80	-1.10
BCD402B	-0.64	-0.48	-0.74	-1.14
BCD351D	-0.58	-0.44	-0.56	-1.21
BCD265B	-0.56	-0.39	-0.51	-1.24
TubA1	-0.56	-0.35	-0.46	-1.18
Dhn6	-0.59	-0.39	-0.52	-1.27
ABG003	-0.56	-0.37	-0.61	-1.33
WG1026B	-0.73	-0.53	-0.63	-1.42
Adh4	-0.74	-0.39	-0.55	-1.21
ABA003	-0.87	-0.67	-0.68	-1.26
ABG484	-0.78	-0.50	-0.60	-1.31
Pgk2A	-0.82	-0.42	-0.62	-1.30
ABR315	-0.68	-0.50	-0.63	-1.40
WG464	-0.68	-0.53	-0.50	-1.10
BCD453B	-0.82	-0.87	-0.59	-1.53
ABG472	-0.79	-0.76	-0.46	-1.44
iAco2	-0.47	-0.41	-0.04	-0.80
ABG500B	-0.44	-0.43	-0.04	-0.76
ABG498	-0.40	-0.34	-0.07	-0.68
WG114	-0.42	-0.42	-0.04	-0.73
ABG054	-0.36	-0.27	-0.06	-0.67
ABG394	-0.36	-0.23	-0.01	-0.68
ABG397	-0.53	-0.26	-0.22	-0.78
ABG366	-0.42	-0.13	0.05	-0.70

Example: Analysis of the Barley Malt Extract Data Using t-Test

The results of a t-test using the data in Tables 12.5 and 12.6 for the malt extract data in barley are given in Table 13.4 and 13.5. A large number of markers are significantly associated with malting quality. The peak of the t-statistic is located near marker BCD453B for three of the four environments. The results have notable differences among the environments. The significant p-values were obtained using standard t-distribution. There are possible QTLs affecting the amount of malt extract on Barley chromosome IV and possible QTL by environment interactions. However, conclusions about how many QTLs, their precise genome locations and their genetic effects cannot be obtained from these simple t-tests. A single QTL could result in several significant t-tests for nearby markers. The large number of significant markers on chromosome IV could result from a single QTL. The significant p-values computed using a standard t-distribution might not correspond to an actual significance level because the t-tests for linked markers are not independent tests.

13.2.3 ANALYSIS OF VARIANCE USING BACKCROSS PROGENY

Analysis of variance also can be used for testing the model. An F-statistic is

$$F = \frac{MSM}{MSG\,(M)} \tag{13.7}$$

following F-distribution with degrees of freedom 1 and $N-2$ (Table 13.3). The expectation of the F-statistic is

$$E[F] = \frac{\sigma_e^2 + b\,[\sigma_{G\,(QTL)}^2 + 4r\,(1-r)\,a^2] + bc\,(1-2r)^2 a^2}{\sigma_e^2 + b\,[\sigma_{G\,(QTL)}^2 + 4r\,(1-r)\,a^2]}$$

$$= 1 + \frac{bc\,(1-2r)^2 a^2}{\sigma_e^2 + b\,[\sigma_{G\,(QTL)}^2 + 4r\,(1-r)\,a^2]} \tag{13.8}$$

For the barley data in Table 12.6, the results from using the F-test are the same as those from the t-test in terms of the significance levels and the QTL effect estimations. When the degrees of freedom for the numerator of the F-statistic is one, the F-statistic is the square of the t-statistic ($F = t^2$).

13.2.4 LINEAR REGRESSION USING BACKCROSS PROGENY

The model also can be tested using simple linear regression by regressing the trait values on a dummy variable for the marker genotypes. The regression model is

$$y_j = \beta_0 + \beta_1 x_j + \varepsilon_j \tag{13.9}$$

where y_j is the trait value for the jth individual in the population, x_j is the dummy variable taking 1 if the individual is AA and -1 for Aa, β_0 is the intercept for the regression which is the overall mean for the trait, β_1 is the slope for the regression line and ε_j is the random error for the jth individual. The expected means, variances and covariance needed for estimating the regression coefficient for the two variables are (see Table 13.1 for notation)

$$E\,(\bar{x}) = 0.5 \times 1 + 0.5 \times (-1) = 0$$

$$E\,(\hat{s}_x^2) = 0.5 \times 1^2 + 0.5 \times (-1)^2 = 1$$

$$E\,(\bar{y}) = 0.5\,[\,(1-r)\,\mu_1 + r\mu_2 + r\mu_1 + (1-r)\,\mu_2] = 0.5\,(\mu_1 + \mu_2) \tag{13.10}$$

$$E\,(\hat{s}_{xy}) = 0.5 \times 1 \times [\,(1-r)\,\mu_1 + r\mu_2] + 0.5 \times (-1) \times [r\mu_1 + (1-r)\,\mu_2]$$

$$= 0.5\,(1-2r)\,(\mu_1 - \mu_2)$$

The expectations of the estimated intercept and slope for the model are

$$E(\beta_0) = E(\bar{y}) = 0.5(\mu_1 + \mu_2)$$

$$E(\beta_1) = \frac{E(\hat{s}_{xy})}{E(\hat{s}_x^2)} = 0.5(1 - 2r)(\mu_1 - \mu_2) \qquad (13.11)$$

It is not difficult to see that the expectation of the slope is the expectation for the difference between the two marker classes

$$E(\beta_1) = (1 - 2r)g$$
$$= 0.5(a + d)(1 - 2r) \qquad (13.12)$$

The hypothesis test $H_0: \beta_1 = 0$ is equivalent to testing if the putative QTL and the marker A are unlinked or that the genetic effects are equal to zero, which is

$$r = 0.5$$

or

$$g = 0.5(a + d) = 0$$

Table 13.6 Estimated regression coefficients. Steptoe marker genotypes are coded as "1" and Morex marker genotypes are coded as "-1" for 26 markers at 4 locations.

Marker	Regression Coefficient			
	ID	OR	WT	MT
WG622	-0.24	-0.40	-0.33	-0.33
ABG313B	-0.19	-0.34	-0.44	-0.44
CDO669	-0.32	-0.39	-0.40	-0.55
BCD402B	-0.32	-0.24	-0.37	-0.57
BCD351D	-0.29	-0.22	-0.28	-0.60
BCD265B	-0.28	-0.20	-0.25	-0.62
TubA1	-0.28	-0.17	-0.23	-0.59
Dhn6	-0.29	-0.20	-0.26	-0.64
ABG003	-0.28	-0.18	-0.30	-0.66
WG1026B	-0.36	-0.26	-0.32	-0.71
Adh4	-0.37	-0.19	-0.27	-0.61
ABA003	-0.43	-0.34	-0.34	-0.63
ABG484	-0.39	-0.25	-0.30	-0.65
Pgk2A	-0.41	-0.21	-0.31	-0.65
ABR315	-0.34	-0.25	-0.31	-0.70
WG464	-0.34	-0.26	-0.25	-0.55
BCD453B	-0.41	-0.44	-0.30	-0.77
ABG472	-0.39	-0.38	-0.23	-0.72
iAco2	-0.24	-0.21	-0.02	-0.40
ABG500B	-0.22	-0.21	-0.02	-0.38
ABG498	-0.20	-0.17	-0.03	-0.34
WG114	-0.21	-0.21	-0.02	-0.36
ABG054	-0.18	-0.13	-0.03	-0.33
ABG394	-0.18	-0.11	-0.00	-0.34
ABG397	-0.27	-0.13	-0.11	-0.39
ABG366	-0.21	-0.07	0.02	-0.35

Depending on how the dummy variables are coded, the expectations of the slopes are different. For example, AA and Aa can be coded as 1 and 0 or as 2 and 1. Biological interpretations for the different slopes are the same. Here, AA and Aa are coded as 1 and -1.

Example: Analysis of the Barley Data Using Linear Regression

Table 13.6 shows the results from regression analysis using data in Tables 12.5 and 12.6 for malt extract phenotypes and markers on chromosome IV of barley. For the independent variable, Steptoe marker genotypes are coded as "1" and Morex genotypes are coded as "-1". The regression coefficients are approximately half of the estimates of the differences between the two marker classes.

13.2.5 A LIKELIHOOD APPROACH USING BACKCROSS PROGENY

The likelihood approach is also used for the single-marker analysis (Weller 1986). The likelihood is constructed based on the distribution of trait values. Figure 13.3 shows theoretical distributions for the trait value in a backcross population. A marker A is assumed to be linked with a QTL with a recombination fraction $r = 0.2$. The QTL has a genetic effect of $g = 0.5(\mu_1 - \mu_2) = 0.5\sigma$. Trait variances for the two QTL genotypes (QQ and Qq) are assumed equal (σ^2). For each of the two marker classes (AA and Aa), the distribution of the trait value is a mixture of two normal distributions with different means and expected proportions. A distribution with mean μ_1 is for QTL genotype QQ and the other, with mean μ_2, is for Qq. For marker genotype AA, 80% $(1-r)$ of the individuals have QQ genotypes and 20% have Qq. For marker genotype Aa, 20% have QQ and 80% have Qq. The mean difference between the two marker classes is $0.6(\mu_1 - \mu_2)$.

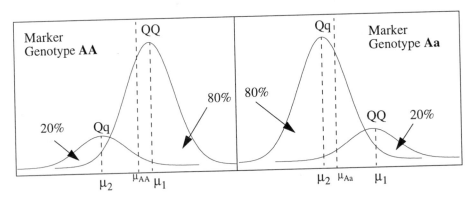

Figure 13.3 Theoretical distributions of trait value for QTL genotypes within each of the marker genotypes AA and Aa in a backcross population. Marker A is linked with a QTL with $r = 0.2$.

For the backcross model, the likelihood is

$$L = \frac{1}{\{\sqrt{2\pi}\sigma\}^N} \prod_{i=1}^{N} \sum_{j=1}^{2} p(Q_j|M_i) \exp\left[-\frac{(y_i-\mu_j)^2}{2\sigma^2}\right] \tag{13.13}$$

if we assume that each of the four marker-QTL classes has equal variance σ^2 and the trait values are normally distributed, where y_i is an observed trait phenotypic value for the ith individual, $p(Q_j|M_i)$ is the conditional probability listed in Table 13.1 and μ_j is the trait value for the jth QTL genotype. Equation (13.13) is a joint distribution function for two normal variables with unequal proportions.

The logarithm of Equation (13.13) is

$$Log[L(\mu_1, \mu_2, \sigma^2, r)] = \sum_{i=1}^{N} Log\left\{ \sum_{j=1}^{2} p(Q_j|M_i) \exp\left[-\frac{(y_i-\mu_j)^2}{2\sigma^2}\right]\right\} \\ -\frac{N}{2}Log(2\pi\sigma^2) \tag{13.14}$$

Under the null hypothesis $H_0: \mu_1 - \mu_2 = 0$ or $\mu_1 = \mu_2 = \mu$, the log likelihood is

$$Log[L(\mu_1 = \mu_2 = \mu)] = -\frac{1}{2\sigma^2}\sum_{i=1}^{N}(y_i-\mu)^2 - \frac{N}{2}Log(2\pi\sigma^2)$$

and for the null hypothesis $H_0: r = 0.5$, the log likelihood is

$$Log[L(r=0.5)] = \sum_{i=1}^{N} \log\left\{ \exp\left[-\frac{(y_i-\mu_1)^2}{2\sigma^2}\right] + \exp\left[-\frac{(y_i-\mu_2)^2}{2\sigma^2}\right]\right\} \\ -\frac{N}{2}Log(2\pi\sigma^2) \tag{13.15}$$

It is easy to see that the log likelihood under $H_0: \mu_1 - \mu_2 = 0$ also implies $r = 0.5$. It is common to use the log likelihood ratio

$$G = 2\{Log[L(\hat{\mu}_1, \hat{\mu}_2, \hat{\sigma}^2, \hat{r})] - Log[L(r=0.5)]\} \tag{13.16}$$

as a test statistic for testing the hypothesis $H_0: r = 0.5$ or if the marker and the QTL are independent, where the log likelihoods are evaluated using the maximum likelihood estimates of the genotypic values for the two QTL genotypes ($\hat{\mu}_1$ for QQ and $\hat{\mu}_2$ for Qq), the variance ($\hat{\sigma}^2$) and the recombination fraction ($\hat{r}$). The G is distributed asymptotically as a chi-square variable with one degree of freedom. The likelihood ratio test using Equation (13.16) has lower statistical power than does the t-test. The t-test is approximately equivalent to the likelihood ratio test using

$$G = 2\{Log[L(\hat{\mu}_1, \hat{\mu}_2, \hat{\sigma}^2, \hat{r})] - Log[L(\mu_1 = \mu_2 = \mu)]\} \tag{13.17}$$

It is common to use lod scores for QTL detection. The differences between the G-statistic and the lod score are the bases for the logarithm and the interpretations. G-statistic is computed using the natural logarithm. Lod score is computed using the base 10 logarithm. For example, the lod score corresponding to the G of Equation (13.16) is

$$lod = Log_{(10)}[L(\hat{\mu}_1, \hat{\mu}_2, \hat{\sigma}^2, \hat{r})] - Log_{(10)}[L(r=0.5)] \tag{13.18}$$

The G-statistic is interpreted as a probability of occurrence of the data under the null hypothesis, which is computed using a theoretical chi-square distribution. However, the lod score is interpreted using the concept of an odds ratio. No theoretical distribution is needed for interpreting a lod score. For example, a lod score of 2 means that the alternative hypothesis is $10^2 = 100$ times more likely than the null hypothesis. As mentioned in Chapter 4, the lod score and the G-statistic have a one to one relationship. The interpretations for these two statistics should be the same (see Chapter 4).

In theory, the maximum likelihood estimates of the two QTL genotypic values, the variance and the recombination fraction can be obtained by setting the partial differentials of the log likelihood Equation (13.14) with respect to the four unknown parameters equal to zero. However, the analytical solutions for the parameters are not readily obtained. The iterative approaches are needed to obtain the maximum likelihood estimates. In practice, the variance can be estimated using the pooled within marker class variance, which is

$$\hat{\sigma}^2 = \frac{\sum (y_{1j} - \bar{y}_1)^2}{n_1} + \frac{\sum (y_{2j} - \bar{y}_2)^2}{n_2}$$

Table 13.7 Estimated QTL genotypic differences using the likelihood approach assuming the QTL are located at three fixed genome positions based on the data in Table 12.6. Results are for Montana only. m is recombination fraction between the marker and the next marker. See beginning of example for notation on the QTL location.

	QTL Location		
Marker	0	0.25m	0.5m
WG622	-0.6	-0.7	-0.8
ABG313B	-0.9	-0.9	-0.9
CDO669	-1.1	-1.2	-1.2
BCD402B	-1.2	-1.2	-1.2
BCD351D	-1.2	-1.2	-1.3
BCD265B	-1.2	-1.2	-1.2
TubA1	-1.2	-1.2	-1.2
Dhn6	-1.3	-1.3	-1.3
ABG003	-1.3	-1.3	-1.3
WG1026B	-1.4	-1.4	-1.4
Adh4	-1.2	-1.3	-1.3
ABA003	-1.3	-1.3	-1.3
ABG484	-1.3	-1.3	-1.3
Pgk2A	-1.3	-1.3	-1.4
ABR315	-1.4	-1.4	-1.4
WG464	-1.1	-1.1	-1.2
BCD453B	-1.6	-1.6	-1.6
ABG472	-1.5	-1.5	-1.6
iAco2	-0.8	-0.8	-0.8
ABG500B	-0.7	-0.7	-0.8
ABG498	-0.7	-0.7	-0.7
WG114	-0.7	-0.7	-0.7
ABG054	-0.7	-0.7	-0.7
ABG394	-0.7	-0.7	-0.7

By setting the variance as a constant, the computation for solving Equation (13.14) for the remaining parameters is greatly reduced. It is also logical to fix the recombination at a meaningful number of points, so that the only unknown parameters in Equation (13.14) are the two means. For example, the QTL can be tested at positions right on the marker ($r = 0$), one fourth of the distance from the marker relative to the next marker, if for example, the interval flanked by the adjacent markers is divided into four sections, and in the middle of the two adjacent markers (Tables 13.7).

Table 13.8 The likelihood ratio test statistic for the 26 markers using the likelihood approach assuming the QTL locates at three fixed genome positions for data in Tables 11.5 to 11.7. Results are for Montana only. See beginning of example for notation on the QTL location.

Marker	0		0.25m		0.5m	
	G	p	G	p	G	p
ABG313B	9.90	0.0017	9.88	0.0017	9.84	0.0017
CDO669	13.78	0.0002	13.54	0.0002	13.22	0.0003
BCD402B	15.58	0.0001	15.56	0.0001	15.52	0.0001
BCD351D	19.28	0.0000	19.02	0.0000	18.72	0.0000
BCD265B	17.80	0.0000	17.80	0.0000	17.80	0.0000
TubA1	17.34	0.0000	17.24	0.0000	17.10	0.0000
Dhn6	20.58	0.0000	20.56	0.0000	20.54	0.0000
ABG003	20.66	0.0000	20.60	0.0000	20.54	0.0000
WG1026B	26.66	0.0000	26.52	0.0000	26.38	0.0000
Adh4	17.90	0.0000	17.88	0.0000	17.86	0.0000
ABA003	20.52	0.0000	20.48	0.0000	20.46	0.0000
ABG484	21.22	0.0000	21.22	0.0000	21.20	0.0000
Pgk2A	21.36	0.0000	21.28	0.0000	21.22	0.0000
ABR315	25.18	0.0000	24.86	0.0000	24.46	0.0000
WG464	14.70	0.0001	14.68	0.0001	14.56	0.0001
BCD453B	29.18	0.0000	28.98	0.0000	28.72	0.0000
ABG472	24.64	0.0000	24.04	0.0000	23.08	0.0000
iAco2	8.16	0.0043	8.14	0.0043	8.12	0.0044
ABG500B	6.92	0.0085	6.90	0.0086	6.88	0.0087
ABG498	5.66	0.0174	5.66	0.0174	5.66	0.0174
WG114	6.10	0.0135	6.08	0.0137	6.08	0.0137
ABG054	5.20	0.0226	5.20	0.0226	5.20	0.0226
ABG394	5.78	0.0162	5.78	0.0162	5.78	0.0162
ABG397	7.74	0.0054	7.72	0.0055	7.70	0.0055
ABG366	5.68	0.0172				

Example: Analysis of the Barley Data Using a Likelihood Approach

Tables 13.7 to 13.9 and Figure 13.4 show results for the barley data in Tables 12.5 and 12.6 using the likelihood approaches. The results are for one environment (Montana). The likelihood ratio tests and the maximum likelihood estimations are based on three fixed QTL locations relative to each of the 26 markers. The three relative positions are

- right on the marker (0)
- one fourth of the distance from the marker relative to the next marker (0.25m)
- in the middle of the two adjacent markers (0.5m)

Table 13.7 shows the maximum likelihood estimates of hypothetical QTL genotypic differences between Steptoe and Morex genotypes, assuming three possible QTL locations relative to each of the 26 genetic markers. The estimates for the QTL right on the marker are the same as the estimates using simple means. The differences among different QTL locations are not significant. When marker density is high, little additional information can be gained by assuming different QTL positions. However, significant information can be gained by assuming different QTL relative positions when linkage density is low.

Table 13.8 shows the log likelihood ratio tests for the 26 markers. For this data set, there is no difference between the likelihood ratio test statistics using Equation (13.16) and (13.17). Only small differences are found among the different QTL relative positions. Small differences are also found between the results of a simple t-test (Table 13.4) and the likelihood approach. In practical QTL mapping analysis, a simple t-test may provide all of the essential information.

Table 13.9 Lod score for the 26 markers using the likelihood approach, assuming that QTLs are located at three fixed genome positions. Results are for Montana only. See beginning of example for notation on the QTL location.

Marker	0	0.25m	0.5m
WG622	1.09	1.07	1.07
ABG313B	2.15	2.15	2.14
CDO669	2.99	2.94	2.87
BCD402B	3.38	3.38	3.37
BCD351D	4.19	4.13	4.06
BCD265B	3.87	3.87	3.87
TubA1	3.77	3.74	3.71
Dhn6	4.47	4.46	4.46
ABG003	4.49	4.47	4.46
WG1026B	5.79	5.76	5.73
Adh4	3.89	3.88	3.88
ABA003	4.46	4.45	4.44
ABG484	4.61	4.61	4.60
Pgk2A	4.64	4.62	4.61
ABR315	5.47	5.40	5.31
WG464	3.19	3.16	3.12
BCD453B	6.34	6.29	6.24
ABG472	5.35	5.22	5.01
iAco2	1.77	1.77	1.76
ABG500B	1.50	1.50	1.49
ABG498	1.23	1.23	1.23
WG114	1.32	1.32	1.32
ABG054	1.13	1.13	1.13
ABG394	1.26	1.26	1.26
ABG397	1.68	1.68	1.67

Table 13.9 shows lod scores corresponding to the log likelihood ratio test statistics in Table 13.8 computed using Equation (13.18). The general interpretations for the two types of test statistic have similar trends. However, different interpretations for each of the 26 loci can be made by using the two types of statistics. For example, a segment of chromosome IV of barley with 15

genetic markers on it is statistically significantly associated with trait values of malt extract if 0.0001 is used as the critical p-value. Ten loci on 4 segments will be declared to be QTL candidates if lod score 4 is the critical value. Caution is needed in interpreting the two types of statistics for same data set.

It has been common to present results of the likelihood analysis using a plot of the test statistic against genome position. Results in Table 13.8 and 13.9 are shown in Figure 13.4. The horizontal axis is the genome position for chromosome IV of barley. The total length of this chromosome is 114 recombination units. The 26 markers are located on the chromosome and ordered as shown in Table 12.7. Some of the 26 markers, such as the markers at the peaks of the lod score plot, are shown.

A common mistake is to treat each of the lod score peaks as genome location corresponding to a different QTL. The peaks of the lod curve may not be the results of QTLs under the peaks. When there is more than one QTL located near the group of linked markers, the lod score peak may be very wide or the peak may not correspond to the QTL location. The lod score peak may correspond to the QTL location when a single QTL contributes to the peak. The lod score plot of Figure 13.4 has little information on precise QTL locations. Even though the plot shows the whole chromosome, the results are still based on single-marker analysis.

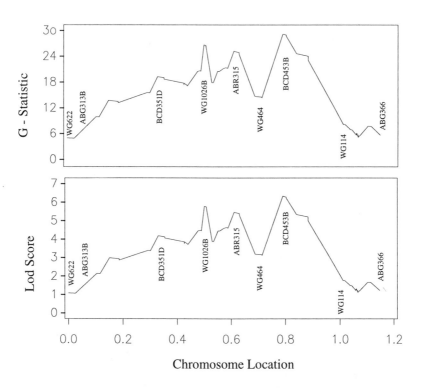

Figure 13.4 Lod score and the log likelihood ratio test statistic (G) are plotted against genome locations using the likelihood approach.

13.3 SINGLE-MARKER ANALYSIS USING F2 PROGENY

13.3.1 JOINT SEGREGATION OF QTL AND MARKER GENOTYPES

Figure 13.2 shows a conventional F2 mating for a two-locus model. A QTL Q and a marker A are linked by r recombination units. The difference between the backcross progeny and the F2 progeny is that in the F2, each of the markers and the QTLs has three possible genotypes and in the backcross, each of the markers and QTLs has two possible genotypes.

Table 13.10 Expected QTL genotypic frequency conditional on genotypes of a nearby marker in F2 populations. See Table 12.2 for notation.

	n_i	$p_{i\circ}$	$p(Q_j\|M_i)$		
			QQ	Qq	qq
AA	n_1	0.25	$(1-r)^2$	$2r(1-r)$	r^2
Aa	n_2	0.5	$r(1-r)$	$(1-r)^2+r^2$	$r(1-r)$
aa	n_3	0.25	r^2	$2r(1-r)$	$(1-r)^2$

Table 13.10 shows the expected frequencies of the three possible QTL genotypes conditional on the marker genotypes in the F2 progeny. The conditional frequencies are the joint frequencies divided by the marginal frequencies corresponding to the marker genotypes

$$p(Q_j|M_i) = \frac{p_{ij}}{p_{i\circ}}$$

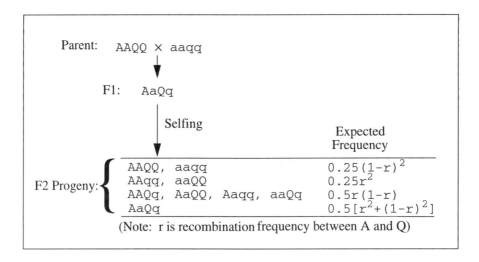

Figure 13.5 Conventionally defined F2 progeny.

where $p(Q_j|M_i)$ is the frequency of the jth putative QTL genotypic class conditional on the ith genotypic class of marker A, p_{ij} is the joint frequency of the jth putative QTL genotypic class and the ith genotypic class of marker A (Figure 13.5) and $p_{i\circ}$ is the marginal frequency of the ith genotypic class of marker A. For the F2 progeny, the marginal frequencies for the marker genotypes AA, Aa and aa are 0.25, 0.5 and 0.25, respectively.

Table 13.11 shows the expected trait values for the three marker genotypic classes (AA, Aa and aa). The trait values for the putative QTL genotypes were defined in Table 12.2 in Chapter 12. For example, the expected trait values for the putative QTL genotypes QQ, Qq and qq are μ_1, μ_2 and μ_3, respectively and

$$a = 0.5 (\mu_1 - \mu_3) \text{ and}$$

$$d = 0.5 (\mu_1 + \mu_3 - 2\mu_2)$$

were defined as the additive and dominant genetic effects of the QTL. The expected trait values for the marker genotypic class are the summation of the product of the putative QTL genotypic value and the frequency conditional on the marker genotypes, *e.g.*, the expected trait value for marker class AA is

$$\mu_{AA} = \mu_1 (1-r)^2 + 2\mu_2 r (1-r) + \mu_3 r^2 \tag{13.19}$$

If we use μ to denote the population mean, then we have

$$\mu_1 = \mu + a$$
$$\mu_2 = \mu + d$$
$$\mu_3 = \mu - a$$

Equation (13.19) becomes

$$\mu_{AA} = \mu + (1 - 2r) a + 2r (1 - r) d$$

In the same way, the expected trait values in terms of the putative QTL genetic effects can be derived (Table 13.11).

Table 13.11 Expected genotypic values for marker classes in an F2 population when codominant marker is used.

	n_i	$p_{i\circ}$	Expected Trait Value
AA	n_1	0.25	$\mu_{AA} = (1-r)^2 \mu_1 + 2r(1-r)\mu_2 + r^2\mu_3$ $= \mu + (1-2r) a + 2r(1-r) d$
Aa	n_2	0.5	$\mu_{Aa} = r(1-r)\mu_1 + [(1-r)^2 + r^2]\mu_2 + r(1-r)\mu_3$ $= \mu + [(1-r)^2 + r^2] d$
aa	n_3	0.25	$\mu_{aa} = r^2\mu_1 + 2r(1-r)\mu_2 + (1-r)^2\mu_3$ $= \mu - (1-2r) a + 2r(1-r) d$

Table 13.12 gives the corresponding putative QTL genotypic frequencies conditional on the two marker phenotypic classes (A_ and aa) when a dominant marker is used. The expected trait values for the marker phenotypes are given in Table 13.13.

Table 13.12 Expected QTL genotypic frequency conditional on genotypes of a nearby marker in F2 populations when dominant marker A is used.

	n_i	$p_{i°}$	$p(Q_j\|M_i)$		
			QQ	Qq	qq
A_	n_1	0.75	$(1-r^2)/3$	$2((1-r+r^2)/3)$	$r(2-r)/3$
aa	n_2	0.25	r^2	$2r(1-r)$	$(1-r)^2$

Table 13.13 Expected genotypic values for marker classes in F2 populations when dominant marker A is used.

	n_i	$p_{i°}$	Expected Trait Value
A_	n_1	0.75	$\mu_{A_-} = (1-r^2)\mu_1/3 + 2(1-r+r^2)\mu_2/3 + r(2-r)\mu_3/3$ $\quad = \mu + (1-2r)a/3 + 2(1-r+r^2)d/3$
aa	n_2	0.25	$\mu_{aa} = r^2\mu_1 + 2r(1-r)\mu_2 + (1-r)^2\mu_3$ $\quad = \mu - (1-2r)a + 2r(1-r)d$

13.3.2 ANALYSIS OF VARIANCE USING F2 PROGENY

Codominant Marker Model

An analysis of variance of the trait values, using marker genotypes as a classification variable, can be used to detect the association between the trait values and the marker, as shown in Table 13.14. The among marker genotype F-statistic is

$$F_M = \frac{MSM}{MSG(M)} \tag{13.20}$$

with degrees of freedom 2 and $N-2$. The number of replications for each individual has less effect on the statistical power than does the number of individuals (Knapp and Bridges 1992). The mean square among the three marker genotypes (AA, Aa and aa) can be partitioned into the hypothetical marker additive and dominance effects using orthogonal contrasts as shown in Table 13.15 and genetic effects within marker genotypes.

Table 13.14 Analysis of variance table for a typical single marker analysis using F2 population.

Source	df	MS	F - statistic
Genetics	$N-1$	MSG	
Marker	2	MSM	MSM / MSG(M)
A	1	MSA	MSA / MSG(M)
D	1	MSD	MSA / MSG(M)
G(Marker)	$N-3$	MSG(M)	
Residual	$N(b-1)$	MSE	

Table 13.15 Two linear contrasts for marker genotypes using F2 progeny.

| | Marker Genotype | | |
Contrast	AA	Aa	aa
Additive	1	0	-1
Dominance	1	-1	1
Expected Frequency	0.25	0.5	0.25

The expectations for the two contrasts are

$$E\,[\text{additive}] = [\mu + (1-2r)\,a + 2r\,(1-r)\,d]$$
$$- [\mu - (1-2r)\,a + 2r\,(1-r)\,d]$$
$$= 2\,(1-2r)\,a$$
$$E\,[\text{dominance}] = [\mu + (1-2r)\,a + 2r\,(1-r)\,d]$$
$$-2\,\{\mu + [\,(1-r)^2 + r^2]\,d\}$$
$$+ [\mu - (1-2r)\,a + 2r\,(1-r)\,d]$$
$$= 2\,(1-2r)^2 d$$

The hypothesis test based on the contrasts for the marker genotypes corresponds to the marker and the QTL being independent or no genetic effects can be detected for the putative QTL, which is

$$r = 0.5$$

or

$$\begin{cases} a = 0 \\ \text{and} \\ d = 0 \end{cases}$$

Dominant Marker Model

When a dominant marker is used, only one contrast can be tested, which is $H_0 : \mu_{A\_} - \mu_{aa} = 0$. The expectation of this contrast is

$$E\,(\mu_{A\_} - \mu_{aa}) = \left[\mu + \frac{1}{3}\,(1-2r)\,a + \frac{2}{3}\,(1 - r + r^2)\,d\right]$$
$$- [\mu - (1-2r)\,a + 2r\,(1-r)\,d] \qquad (13.21)$$
$$= \frac{1}{3}\,[4\,(1-2r)\,a + 2\,(1-2r)^2 d]$$

The contrast tests a confounded additive, the dominant effect and the linkage between the marker and the putative QTL, which are

$$r = 0.5$$

or

$$\begin{cases} a = 0 \\ \text{and} \\ d = 0 \end{cases}$$

If $r = 0.5$ or $a = 0$ and $d = 0$, the expectation goes to zero. The test on the contrast is valid. However, the power of the test could be significantly lower than the test using a codominant marker. If an additive model is assumed ($d = 0$), the contrast tests for an additive effect and Equation (13.21) becomes

$$\frac{4}{3}(1 - 2r)\,a$$

which is $2/3$ less than the $2(1 - 2r)\,a$ for using codominant marker.

13.3.3 LINEAR REGRESSION USING F2 PROGENY

Codominant Marker Model

A linear regression model also can be used for single marker analysis using F2 progeny. The model is

$$y_j = \beta_0 + \beta_1 x_{1j} + \beta_2 x_{2j} + \varepsilon_j \qquad (13.22)$$

where y_j is a trait value for the jth individual in the population, x_{1j} is the dummy variable for the marker additive effect taking 1, 0, and -1 for marker genotypes AA, Aa and aa, respectively, x_{2j} is a dummy variable for the marker dominant effect taking 1, -1 and 1 for marker genotypes AA, Aa and aa, β_0 is the intercept for the regression which is the overall mean for the trait, β_1 is the slope for the additive regression line, β_2 is the slope for the dominant regression line and ε_j is random error for the jth individual. The expected means, variances and covariances needed for estimating regression coefficients for the two variables are

$$E(\bar{x}_1) = 0$$
$$E(\bar{x}_2) = -0.5$$
$$E(\bar{y}) = 0.25\,(\mu_1 + 2\mu_2 + \mu_3)$$
$$E(\hat{s}_{x1}^2) = 0.5$$
$$E(\hat{s}_{x2}^2) = 1$$
$$E(\hat{s}_{x1x2}) = 0$$
$$E(\hat{s}_{x1y}) = 0.5\,(1 - 2r)\,(\mu_1 - \mu_3)$$
$$E(\hat{s}_{x2y}) = 0.5\,(1 - 2r)^2\,(\mu_1 + \mu_3 - 2\mu_2)$$

In matrix notation, we have

$$X'X\beta = XY \qquad (13.23)$$

where X is a matrix form of coded marker genotypes corresponding to the intercept and the genetic effects, β is a vector containing the intercept and the regression coefficients for the model and Y is a vector for the observed trait values. So we have

$$\hat{\beta} = (X'X)^{-1}XY \qquad (13.24)$$

For readers who are not familiar with linear regression, I suggest consulting a reference book on regression, for example, the one by Draper and Smith (1981).

For the regression model using F2 progeny, the values of Equation (13.23) are

$$\begin{bmatrix} 1 & 0 & 0 \\ 0 & 0.5 & 0 \\ 0 & 0 & 1 \end{bmatrix} \begin{bmatrix} \beta_0 \\ \beta_1 \\ \beta_2 \end{bmatrix} = \begin{bmatrix} 0.25\,(\mu_1 + 2\mu_2 + \mu_3) \\ 0.5\,(1 - 2r)\,(\mu_1 - \mu_3) \\ 0.5\,(1 - 2r)^2\,(\mu_1 + \mu_3 - 2\mu_2) \end{bmatrix}$$

The expectation of the intercept and the slopes are

$$E\begin{bmatrix} \hat{\beta}_0 \\ \hat{\beta}_1 \\ \hat{\beta}_2 \end{bmatrix} = \begin{bmatrix} 1 & 0 & 0 \\ 0 & 2 & 0 \\ 0 & 0 & 1 \end{bmatrix} \begin{bmatrix} 0.25\,(\mu_1 + 2\mu_2 + \mu_3) \\ 0.5\,(1 - 2r)\,(\mu_1 - \mu_3) \\ 0.5\,(1 - 2r)^2\,(\mu_1 + \mu_3 - 2\mu_2) \end{bmatrix}$$

$$= \begin{bmatrix} 0.25\,(\mu_1 + 2\mu_2 + \mu_3) \\ (1 - 2r)\,a \\ (1 - 2r)^2\,d \end{bmatrix}$$

Hypothesis tests can be performed using an F-statistic, which is the ratio between the residual mean squares for the reduced-model and the full-model. The residual mean square for the full model of Equation (13.22) is

$$S^2_{full} = (Y - X\beta)'\,(Y - X\beta)$$

where β is a vector containing the intercept and the two regression coefficients for the model. Three meaningful reduced-models are

$$y_j = \beta_0 + \beta_1 x_{1j} + \varepsilon_j \qquad for \qquad \beta_2 = 0$$
$$y_j = \beta_0 + \beta_2 x_{2j} + \varepsilon_j \qquad for \qquad \beta_1 = 0$$
$$y_j = \beta_0 + \varepsilon_j \qquad for \qquad \beta_1 = 0, \beta_2 = 0$$

If the corresponding residual mean squares are denoted by

$$S^2_{\beta_2 = 0} \qquad for \qquad \beta_2 = 0$$
$$S^2_{\beta_1 = 0} \qquad for \qquad \beta_1 = 0$$
$$S^2_{\beta_1 = 0, \beta_2 = 0} \qquad for \qquad \beta_1 = 0, \beta_2 = 0$$

respectively, the F-statistics

$$F_{\beta_2 = 0} = \frac{S^2_{\beta_2 = 0}}{S^2_{full}} \qquad with \qquad df = (1, N - 3)$$

$$F_{\beta_1 = 0} = \frac{S^2_{\beta_1 = 0}}{S^2_{full}} \qquad with \qquad df = (1, N - 3)$$

$$F_{\beta_2 = 0, \beta_1 = 0} = \frac{S^2_{\beta_1 = 0, \beta_2 = 0}}{S^2_{full}} \qquad with \qquad df = (2, N - 3)$$

test the null hypothesis absent a dominant effect, additive effect and both dominant and additive effects, respectively.

Dominant Marker Model

For a dominant marker, a linear regression model similar to Equation (13.22) can be used for single marker analysis. The model is

$$y_j = \beta_0 + \beta_1 x_j + \varepsilon_j \tag{13.25}$$

where y_j is the trait value for the jth individual in the population, x_j is the dummy variable for the marker and taking $+1$ and -1 for marker phenotypes A_ and aa, respectively, β_0 is the intercept for the regression which is the overall mean for the trait, β_1 is the slope for the regression line and ε_j is the random error for the jth individual. The expectation of β_1 is

$$E(\beta_1) = \frac{1}{3}[2(1-2r)a + (1-2r)^2 d]$$

See the analysis of variance approach for interpretation of the regression coefficient. If $r = 0.5$ or $a = 0$ and $d = 0$, the expectation goes to zero. If an additive model is assumed ($d = 0$), the expectation of β_1 is

$$\frac{2}{3}(1-2r)a$$

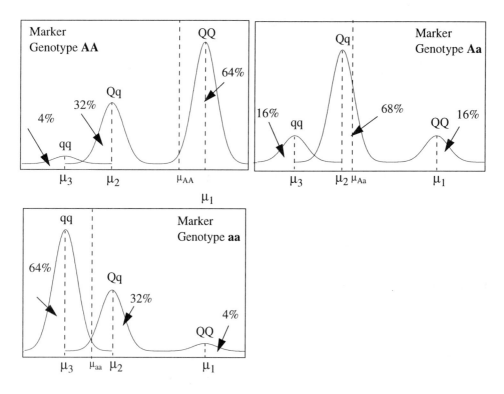

Figure 13.6 Theoretical distributions of trait value for three QTL genotypes within different marker genotypes AA, Aa and aa in an F2 population. Marker A is linked with a QTL with $r = 0.2$. The additive effect is two times that of the dominant effect ($a = 2d$).

13.3.4 LIKELIHOOD APPROACH

Codominant Marker Model

Figure 13.6 shows theoretical distributions of trait value for the three QTL genotypes within marker genotypes when $r = 0.2$. Trait variances for the three QTL genotypes (QQ, Qq and qq) are assumed equal. For each of the three marker classes (AA, Aa and aa), the distribution of the trait value is a mixture of the three normal distributions with different means and expected proportions. Distributions with means μ_1, μ_2 and μ_3 are for QTL genotypes QQ, Qq and qq, respectively. For marker genotype AA, the expected frequencies for QTL genotypes QQ, Qq and qq are

$$\begin{cases} QQ & \begin{aligned} (1-r)^2 &= (1-0.2)^2 \\ &= 0.64 \end{aligned} \\ Qq & \begin{aligned} 2r(1-r) &= 2 \times 0.2 \times (1-0.2) \\ &= 0.32 \end{aligned} \\ qq & r^2 = 0.2^2 = 0.04 \end{cases}$$

For marker genotypes Aa and aa, the corresponding expected QTL genotypic frequencies are

QTL	Aa	aa
QQ	0.16	0.04
Qq	0.68	0.32
qq	0.16	0.64

It is easy to see that the expectations for the two contrasts (additive and dominant) in Table 13.15 are

$$(1-2r)(\mu_1 - \mu_3) = 0.6(\mu_1 - \mu_3)$$
$$(1-2r)^2(\mu_1 + \mu_3 - 2\mu_2) = 0.36(\mu_1 + \mu_3 - 2\mu_2)$$

respectively.

The likelihood for the single-marker analysis using F2 progeny is

$$L = \frac{1}{\{\sqrt{2\pi}\sigma\}^N} \prod_{i=1}^{N} \sum_{j=1}^{3} p(Q_j|M_i) \exp\left[-\frac{(y_i-\mu_j)^2}{2\sigma^2}\right] \tag{13.26}$$

if we assume that each of the nine marker-QTL classes has equal variance σ^2, where y_i is an observed trait value for the ith individual, $p(Q_j|M_i)$ is the conditional probability listed in Table 13.10 and μ_j is the trait value for the jth QTL genotype. The logarithm of Equation (13.26) is

$$Log(L) = \sum_{i=1}^{N} Log\left\{\sum_{j=1}^{3} p(Q_j|M_i) \exp\left[-\frac{(y_i-\mu_j)^2}{2\sigma^2}\right]\right\} - \frac{N}{2}Log(2\pi\sigma^2) \tag{13.27}$$

Under the null hypothesis

$$H_0: \mu_1 - \mu_3 = 0$$

or

$$H_0: a = 0$$

the log likelihood is

$$
Log\,[L\,(a=0)\,] \;=\; \sum_{i=1}^{N} \log\,\{\,[\,p\,(Q_1|M_i) + p\,(Q_3|M_i)\,]\,\exp\!\left[-\frac{(y_i-\mu_1)^2}{2\sigma^2}\right]
$$

$$
+\,p\,(Q_2|M_i)\,\exp\!\left[-\frac{(y_i-\mu_2)^2}{2\sigma^2}\right]\,\} \;-\;\frac{N}{2}Log\,(2\pi\sigma^2) \tag{13.28}
$$

Under the null hypothesis

$$
H_0{:}\mu_1 + \mu_3 - 2\mu_2 = 0
$$

or

$$
H_0{:}d = 0
$$

the log likelihood is

$$
Log\,[L\,(d=0)\,] \;=\; \sum_{i=1}^{N} \log\,\{\,p\,(Q_1|M_i)\,\exp\!\left[-\frac{(y_i-\mu_1)^2}{2\sigma^2}\right]
$$

$$
+\,p\,(Q_2|M_i)\,\exp\!\left[-\frac{(y_i-m)^2}{2\sigma^2}\right]
$$

$$
+\,p\,(Q_3|M_i)\,\exp\!\left[-\frac{(y_i-\mu_3)^2}{2\sigma^2}\right] \tag{13.29}
$$

$$
-\,\frac{N}{2}Log\,(2\pi\sigma^2)
$$

where $m = 0.5\,(\mu_1 + \mu_3)$. Under the null hypothesis

$$
H_0{:}\mu_1 = \mu_2 = \mu_3 = \mu
$$

or

$$
H_0{:}a = 0 \text{ and } d = 0
$$

the likelihood is

$$
Log\,[L\,(\mu_1 = \mu_2 = \mu_3 = \mu)\,] \;=\; -\frac{1}{2\sigma^2}\sum_{i=1}^{N} (y_i-\mu)^2 \tag{13.30}
$$

and for the null hypothesis

$$
H_0{:}r = 0.5
$$

the log likelihood is

$$
Log\,[L\,(r=0.5)\,] \;=\; \sum_{i=1}^{N} \log\,\{\,\exp\!\left[-\frac{(y_i-\mu_1)^2}{2\sigma^2}\right]
$$

$$
+\,2\exp\!\left[-\frac{(y_i-\mu_2)^2}{2\sigma^2}\right] + \exp\!\left[-\frac{(y_i-\mu_3)^2}{2\sigma^2}\right]\,\} \tag{13.31}
$$

It is common to use the log likelihood ratio

$$
G = 2\,[Log\,[L\,(\hat{\mu}_1, \hat{\mu}_2, \hat{\mu}_3, \hat{\sigma}^2, \hat{r})\,] - Log\,[L\,(r=0.5)\,]\,] \tag{13.32}
$$

as a test statistic for testing the hypothesis $H_0: r = 0.5$ that is the independence between the marker and the QTL. The log likelihoods are evaluated using the maximum likelihood estimates of the genotypic values for the three QTL genotypes ($\hat{\mu}_1$, $\hat{\mu}_2$ and $\hat{\mu}_3$ are for QQ, Qq and qq, respectively.), the variance ($\hat{\sigma}^2$) and the recombination fraction ($\hat{r}$). The G is distributed as a chi-square variable with one degree of freedom.

In theory, the maximum likelihood estimates of the three QTL genotypic values, the variance and the recombination fraction can be estimated by setting the partial derivatives of the log likelihood of Equation (13.27) with respect to the five unknown parameters to zero. However, the analytical solutions for the parameters are not readily obtained. Iterative approaches are needed to obtain the maximum likelihood estimates.

Weller (1986) suggests using a combination of moment and ML approach to obtain solutions for the parameters. Six equations for the moment estimators of marker genotypic means and variances can be constructed as follows

$$m_{AA} = (1-r)^2\mu_1 + 2r(1-r)\mu_2 + r^2\mu_3$$

$$m_{Aa} = r(1-r)\mu_1 + [(1-r)^2 + r^2]\mu_2 + r(1-r)\mu_3$$

$$m_{aa} = r^2\mu_1 + 2r(1-r)\mu_2 + (1-r)^2\mu_3$$

$$S_{AA}^2 = \sigma^2 + (1-r)^2(\mu_1 - m_{AA})^2 + 2r(1-r)(\mu_2 - m_{AA})^2 + r^2(\mu_3 - m_{AA})^2$$

$$S_{Aa}^2 = \sigma^2 + r(1-r)[(\mu_1 - m_{Aa})^2 + (\mu_3 - m_{Aa})^2] + [1 - 2r(1-r)](\mu_2 - m_{Aa})^2$$

$$S_{aa}^2 = \sigma^2 + r^2(\mu_1 - m_{aa})^2 + 2r(1-r)(\mu_2 - m_{aa})^2 + (1-r)^2(\mu_3 - m_{aa})^2$$

where m_{AA}, m_{Aa} and m_{aa} are trait means for the three marker genotypes, respectively and S_{AA}^2, S_{Aa}^2 and S_{aa}^2 are variances for the three marker genotypes, respectively. The above equations are different from the equations of Weller (1986) because equal variance for the QTL genotypic classes is assumed here.

Dominant Marker Model

Equations for codominant markers can be used, with minor modifications, for dominant marker models. The expected frequencies of the putative QTL genotype conditional on the marker genotypes and the expected trait values for the observable marker genotypes used for the computation for the dominant marker model are shown in Tables 13.12 and 13.13.

13.3.5 USE OF TRANS DOMINANT LINKED MARKERS IN F2 PROGENY

Dominant markers are gaining more and more attention in genomic research because of the automated analysis of PCR based markers. However, dominant markers provide less information on linkage than do the codominant markers. This is especially true when dominant markers are linked in repulsion phase (or so called trans dominant linked markers, TDLM) (Knapp et al. 1995). However, if two dominant markers are closely linked in repulsion phase, they can be approximately recoded as a codominant marker. Table 13.16 shows the joint genotypic frequency for a pair of TDLMs and a putative QTL near them. The linkage orientation among the three loci (dominant markers A and B and a putative QTL Q) is shown in Figure 13.7. Markers A and B are linked in repulsion phase r_1 recombination units apart. A and B denote the dominant alleles.

If we assume the linkage is tight between the two dominant markers, then the expected frequencies approach 0.5, 0.25, 0.25 and 0.0 for the four genotypic classes (AB, Ab, aB and ab). So, the pair of the TDLMs can be recoded as a single codominant marker. Table 13.17 shows the corresponding coding scheme.

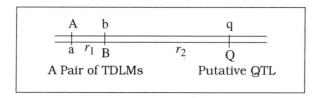

Figure 13.7 Linkage relationship among a pair of TDLMs and a putative QTL. TDLM = trans dominant linked marker.

Table 13.16 Joint frequency for a pair of TDLMs and a putative QTL and the approximate frequency of a putative QTL conditional on the TDLM genotype.

Genotype	p_{ij}	$p\left(Q_j|M_i\right)$
ABQQ	$0.5r_2\left(1-r_1-r_2\right)$	$r_2\left(1-r_2\right)$
ABQq	$0.5\left[1-r_1-2r_2+r_2^2+\left(r_1+r_2\right)^2\right]$	$1-2r_2+2r_2^2$
ABqq	$0.25\left[2r_1+2r_2-r_2^2-\left(r_1+r_2\right)^2\right]$	$r_2\left(1-r_2\right)$
AbQQ	$0.25\left(1-2r_2-r_1^2+r_2^2\right)$	$\left(1-r_2\right)^2$
AbQq	$0.5r_2\left(1-r_2\right)$	$2r_2\left(1-r_2\right)$
Abqq	$0.25r_2^2$	r_2^2
aBQQ	$0.25r_2\left(2r_1+r_2\right)$	r_2^2
aBQq	$0.5\left(1-r_1-r_2\right)\left(r_1+r_2\right)$	$2r_2\left(1-r_2\right)$
aBqq	$0.25\left(1-r_1-r_2\right)^2$	$\left(1-r_2\right)^2$
abQQ	$0.25r_1^2$	NA

The putative QTL genotypic frequency conditional on the TDLM genotype is shown in the last column in Table 13.16. The frequencies corresponding to the frequencies are listed in Table 13.10 for a codominant marker. The expected values for the TDLM genotypes are the same as with the corresponding expectations listed in Table 13.11 (Table 13.18). The expectations for the two contrasts for the TDLM genotypes are

$$E\,[\text{additive}] \approx \mu_{Ab} - \mu_{aB} = 2\,(1 - 2r_2)\,a$$

$$E\,[\text{dominance}] \approx \mu_{Ab} + \mu_{aB} - 2\mu_{AB} = 2\,(1 - 2r_2)^2 d$$

(13.33)

If A and B are sufficiently close, the hypothesis tests on the contrasts have biological meaning for the putative QTL.

Table 13.17 TDLM (A and B) and their corresponding recoded codominant marker genotypes and marginal frequencies.

TDLM Genotype	Expected Frequency	Recoded as C Locus	Approximately Frequency When $r_1 \to 0$
AB	$0.25\,(2 + r_1^2)$	Cc	0.5
Ab	$0.25\,(1 - r_1^2)$	CC	0.25
aB	$0.25\,(1 - r_1^2)$	cc	0.25
ab	$0.25\,r_1^2$		0.0

Table 13.18 Expected genotypic values for different TDLM classes.

TDLM Genotype	n_i	p_{i°	Expected Trait Value (y_{mi})
AB	n_1	0.5	$r_2\,(1 - r_2)\,\mu_1 + [\,(1 - r_2)^2 + r_2^2]\,\mu_2 + r_2\,(1 - r_2)\,\mu_3$ $= \mu + [\,(1 - r_2)^2 + r_2^2]\,d$
Ab	n_2	0.25	$(1 - r_2)^2\mu_1 + 2r_2\,(1 - r_2)\,\mu_2 + r_2^2\mu_3$ $= \mu + (1 - 2r_2)\,a + 2r_2\,(1 - r_2)\,d$
aB	n_3	0.25	$r_2^2\mu_1 + 2r_2\,(1 - r_2)\,\mu_2 + (1 - r_2)^2\mu_3$ $= \mu - (1 - 2r_2)\,a + 2r_2\,(1 - r_2)\,d$

SUMMARY

In this chapter, single marker analysis using backcross and F2 progeny types was described. Approaches including a simple t-test, analysis of variance (ANOVA), simple regression and the likelihood approach can be used for single marker analysis. For the t-test, ANOVA and the regression approaches, results can be presented in tables containing statistics corresponding to each of the genetic markers. For the likelihood approach, plots using lod score or the log likelihood ratio test statistic against the genome positions can be used. For dominant markers, the analysis using the backcross progeny is the same as the use of the codominant markers. However, for F2 progeny, the direct use of the dominant markers provides less information for detecting QTLs. A recoding scheme to convert a pair of TDLMs to a single codominant marker was described. No matter what methods are used to do the single marker analysis and what strategies employed to present the results, precise QTL positions and accurate estimates of QTL effects cannot be obtained unless strong assumptions are made for the trait inheritance.

EXERCISES

Exercise 13.1
Review the biological assumptions of QTL mapping. Are those assumptions valid? Is there any experimental evidence to support the assumptions in the species you are working on? Do you expect that QTL mapping itself will generate experimental evidence for the assumptions?

Exercise 13.2
How does statistics help when searching for genes controlling quantitative traits? (Hint: Pay attention to statistical connections among the biological assumptions of QTL mapping, such as joint, marginal and conditional distributions of marker and QTL genotypes, expected marker genotypic values, etc.)

Exercise 13.3
Single marker analysis for finding genes controlling quantitative traits is based on the expected relationship for linkage between a known marker and a putative gene (or QTL). Derive the following using the illustration below. A gene Q is located r recombination fraction units away from a known marker A.

$$\begin{array}{ccc} | & & | \\ A & r & Q \end{array}$$

(1) Derive the marker genotypic means for AA and Aa using a backcross model AaQq × AAQQ for:

   a. $r = 0.01, \mu_{QQ} = 2, \mu_{Qq} = 1$

   b. $r = 0.01, \mu_{QQ} = 10, \mu_{Qq} = 1$

   c. $r = 0.1, \mu_{QQ} = 2, \mu_{Qq} = 1$

   d. $r = 0.1, \mu_{QQ} = 10, \mu_{Qq} = 1$

(2) Derive the marker genotypic means for AA, Aa and aa using an F2 model AaQq × AaQq for:

   a. $r = 0.01, \mu_{QQ} = 3, \mu_{Qq} = 2, \mu_{qq} = 1$

   b. $r = 0.01, \mu_{QQ} = 3, \mu_{Qq} = 3, \mu_{qq} = 1$

   c. $r = 0.1, \mu_{QQ} = 3, \mu_{Qq} = 2, \mu_{qq} = 1$

   d. $r = 0.1, \mu_{QQ} = 3, \mu_{Qq} = 3, \mu_{qq} = 1$

(3) Compute the expected bias of using marker genotypic means to estimate the corresponding QTL genotypic means, such as AA, Aa and aa for QQ, Qq and qq.

(4) Discuss the bias of using marker genotypic means to estimate QTL genotypic means using the results of (1), (2) and (3). How do linkage map density, mating system and true QTL effects impact on the bias? How will those biases affect QTL mapping using single marker analysis?

Exercise 13.4

It is well known that QTL Q_1, located left of A, will also effect the analysis on marker B, which is located to the right side of A, when loci A and B are linked.

(1) Markers A and B and QTL Q_1 are linked to each other. The two markers are r recombination fraction units away. Q_1 is located on left side of A s recombination fraction units away from A. Derive the expected marker genotypic values as functions of the recombination fractions among the two markers and the QTL and QTL genotypic values for AABB, AABb, AaBB and AaBb in backcross progeny of Q_1q_1AaBb x Q_1Q_1AABB. The genotypic values of Q_1Q_1 and Q_1q_1 are μ_{11} and μ_{12}.

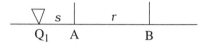

$$\begin{array}{ccc} & s & r \\ Q_1 & A & B \end{array}$$

(2) Derive the expected differences of the genotypic values between AA and Aa and between BB and Bb. Compute the expected differences for:

a. $s = 0.01, r = 0.01, \mu_{11} = 1, \mu_{12} = 0$

b. $s = 0.01, r = 0.3, \mu_{11} = 1, \mu_{12} = 0$

c. $s = 0.3, r = 0.01, \mu_{11} = 1, \mu_{12} = 0$

d. $s = 0.3, r = 0.3, \mu_{11} = 1, \mu_{12} = 0$

(3) Compute the ratio between the expected difference between AA and Aa and the expected difference between BB and Bb for the values in (2). How do the genome positions of the two markers relative to the QTL have an effect on the ratio?

(4) Derive the expected genotypic differences between AABB and AABb and between AaBB and AaBb. Are the genotypic values of the QTL still in the expected difference? Can you think of a way to use this information in QTL mapping?

(5) QTLs Q_1 and Q_2 are both located on the genome segment with the linkage relationship shown in the following illustration. A marker C is located between A and B. Derive the expected marker genotypic values as functions of the recombination fractions among the three markers and the QTLs and QTL genotypic values for AA, Aa, BB, Bb, CC and Cc in backcross progeny of $Q_1q_1AaCcBbQ_2q_2$ x $Q_1Q_1AACCBBQ_2Q_2$. The genotypic values of Q_1Q_1, Q_1q_1, Q_2Q_2 and Q_2q_2 are μ_{11}, μ_{12}, μ_{21} and μ_{22}.

$$\begin{array}{ccccc} & s & r & r & t \\ Q_1 & A & C & B & Q_2 \end{array}$$

(6) Speculate on the QTL analysis results for markers A, B and C using the setting in (5) for the following situations:

a. two QTLs in coupling phase, tight linkage

b. two QTLs in coupling phase, loose linkage

c. two QTLs in repulsion phase, tight linkage

d. two QTLs in repulsion phase, loose linkage

(7) Summarize the results from (1) to (6). Can you see the problem with the single-marker analysis approaches for QTL mapping?

Exercise 13.5

The following graph shows plots of the t-statistic against genome positions in two linkage groups for a trait of interest. A column on the right side of the plot shows the markers corresponding to the "square" of the curves. The reference lines in each of the plots correspond to significant t-values at the 0.05 level using the standard t-distribution with degrees of freedom 98 and the significant t-values at a 95 percentile from empirical distributions using permutation tests (see Chapter 17 and 19 for the permutation test for QTL mapping).

(1) Interpret the plots.

(2) Point to the potential problems in resolving the number of QTLs and their precise genome locations.

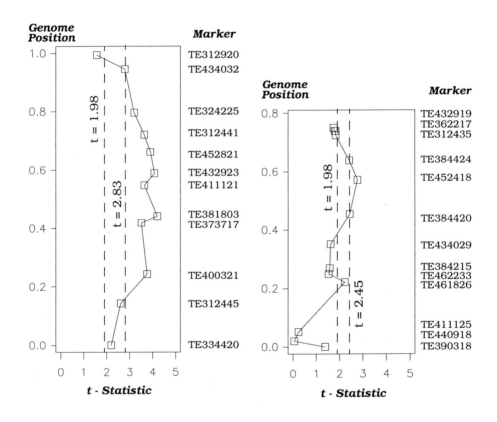

CHAPTER 14

QTL MAPPING: INTERVAL MAPPING

14.1 INTRODUCTION

In Chapter 13, QTL mapping by single-marker analysis using backcross and F2 progeny types was described. The single-marker analysis has the disadvantages of low statistical power and confounding estimates of QTL effects and locations. Lander and Botstein (1986 and 1989) proposed an interval mapping approach to locate QTLs. A large volume of literature has since accumulated on this topic.

The interval mapping approach is based on the joint frequencies of a pair of adjacent markers and a putative QTL flanked by the two markers (Figure 14.1). Figure 14.1 shows a linkage relationship for a pair of adjacent markers A and B and a putative QTL Q. Markers A and B are linked with recombination fraction r and Q is located between the two markers with r_1 recombination fraction from A and r_2 from B. The relationship of the recombination fractions is

$$r = r_1 + r_2 - 2r_1r_2 \tag{14.1}$$

if absence of crossover interference is assumed (see Chapter 10 on crossover interference). When r is small, no double crossover can be assumed and Equation (14.1) is reduced to $r = r_1 + r_2$. For deriving the parameter estimation procedures, the QTL position is usually represented by a position relative to the interval between A and B. That is

$$\rho = r_1/r$$
$$1 - \rho = r_2/r \tag{14.2}$$

This notation will be used through out this chapter.

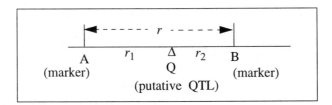

Figure 14.1 Linkage relationship of a QTL and two flanking markers.

According to methods used for parameter estimation, interval mapping methods can be classified into three categories:

- likelihood approach
- regression approach
- a combination of the likelihood and the regression approaches

These methods are equivalent or similar under some assumptions about the markers, the QTLs and the distribution of the trait values. In this chapter, the likelihood approach (Lander and Botstein 1989), the nonlinear regression approach (Knapp *et al.* 1990) and the linear regression approach (Knapp *et al.* 1992) based on the expected cosegregation models among the two flanking markers and the QTL will be described. The subsequent discussion includes the composite interval mapping method, which uses a combination of the likelihood approach and multiple regression (Zeng 1994).

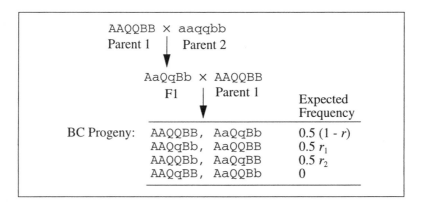

Figure 14.2 Conventionally defined backcross progeny for a QTL and two flanking markers.

14.2 INTERVAL MAPPING OF QTL USING BACKCROSS PROGENY

14.2.1 JOINT SEGREGATION OF QTL AND MARKERS

For a classical backcross design, in which the population is generated by a heterozygous F1 backcrossed to one of the homozygous parents (for example, a cross of AaQqBb × AAQQBB) (Figure 14.2), the rationale behind the interval mapping can be explained using the cosegregation listed in Table 14.1. The linkage relationships for the two flanking markers A and B and the putative QTL is shown in Figure 14.1. If A and B are tightly linked, then the possibility of a double crossover in both segments AQ and QB can be ignored. So the frequency for the marker-QTL genotypes AAQqBB and AaQQBb, which are the results of double crossovers, is relatively small and can be treated as zero for deriving the procedures of interval mapping using the backcross progeny.

The expected frequencies for these possible marker-QTL genotypes are listed in Table 14.1. If the QTL is not located in the interval, the interval

Table 14.1 Expected marker and QTL genotypic frequency in backcross populations with no double crossover.

Marker Genotype	Observed Count	Frequency $p_{i\circ}$	p_{ij}	
			QQ	Qq
AABB	n_1	$0.5(1-r)$	$0.5(1-r)$	0
AABb	n_2	$0.5r$	$0.5r_2$	$0.5r_1$
AaBB	n_3	$0.5r$	$0.5r_1$	$0.5r_2$
AaBb	n_4	$0.5(1-r)$	0	$0.5(1-r)$

mapping analysis is reduced to the single-marker analysis explained in the previous section. If the QTL and markers are in the order QAB, the interval analysis is the same as the single-marker analysis on marker A alone. If the order is ABQ, then the analysis is the same as the single-marker analysis on marker B alone. In this section, the interval mapping analysis is designed for the situations of QTLs being located inside the segment. When the coverage of the genetic map is low and QTLs are located on the genome segments which are not on the linkage map, single-marker analysis is recommended for the markers at two ends of the genetic map and interval mapping analysis for the segments between the two ends.

Table 14.2 Expected QTL genotypic frequency conditional on genotypes of the flanking markers in backcross populations with no double crossover. See Table 12.2 for notation.

Marker Genotype	Frequency $p_{i\circ}$	$p(Q_j|M_i)$		Expected Value (g_i)
		QQ	Qq	
AABB	$0.5(1-r)$	1	0	μ_1
AABb	$0.5r$	$r_2/r = 1-\rho$	$r_1/r = \rho$	$(1-\rho)\mu_1 + \rho\mu_2$
AaBB	$0.5r$	$r_1/r = \rho$	$r_2/r = 1-\rho$	$\rho\mu_1 + (1-\rho)\mu_2$
AaBb	$0.5(1-r)$	0	1	μ_2
Mean	0.25	μ_1	μ_2	$0.5(\mu_1 + \mu_2)$

Table 14.2 shows the conditional frequencies of QTL genotype on the marker genotypes and the expected values for the marker genotypes (see Chapter 13 on single marker analysis using backcross progeny for an approach to obtain the conditional frequencies and the expected values). ρ in the table is defined as the relative position of the putative QTL in the genome segment flanked by the two markers. For example, if $\rho = r_1/r = 0$, the putative QTL is located right on marker A, if $\rho = r_1/r = 0.5$, the QTL is located in the middle of the segment and if $\rho = r_1/r = 1$, the QTL is located on marker B.

14.2.2 A LIKELIHOOD APPROACH FOR QTL MAPPING WITH BACK-CROSS PROGENY

The likelihood function for interval mapping is based on cosegregation among the putative QTL and the two flanking markers (Table 14.2). The likelihood function is

$$L = \frac{1}{\{\sqrt{2\pi}\sigma\}^N} \prod_{i=1}^{N} \sum_{j=1}^{2} p(Q_j|M_i) \exp\left[-\frac{(y_i-\mu_j)^2}{2\sigma^2}\right] \tag{14.3}$$

if we assume that each of the eight marker-QTL classes has equal variance and the trait values are normally distributed, where y_i is an observed trait phenotypic value for the ith individual, $p(Q_j|M_i)$ is the conditional probability listed in Table 14.2 and μ_j is a trait value for the jth QTL genotype.

Figure 14.3 shows theoretical distributions of trait values for four marker genotypes AABB, AABb, AaBB and AaBb in a backcross population. The QTL has a genetic effect of $g = 0.5(\mu_1-\mu_2) = \sigma$. Trait variances for the two QTL genotypes (QQ and Qq) are assumed equal (σ^2). For marker classes AABB and AaBb, the distributions of the trait value are single normal distributions. For marker classes AABb and AaBB, the distributions are mixtures of two normal distributions with different means and expected frequencies. The expected QTL genotypic frequencies conditional on the four marker genotypes are

QTL	AABB	AABb	AaBB	AaBb
QQ	1.0	0.5	0.5	0.0
Qq	0.0	0.5	0.5	1.0

A distribution with mean μ_1 is for QTL genotype QQ and the other with mean μ_2 is for Qq. The logarithm of Equation (14.3) is

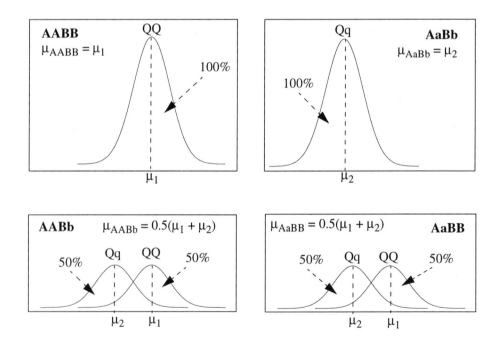

Figure 14.3 Theoretical distributions of trait value for QTL genotypes QQ and Qq for within marker genotypes AABB, AABb, AaBB and AaBb in a backcross population. Markers A and B are linked with $r = 0.3$ and a QTL is in the middle of the two markers.

$$Log\,(L) \;=\; \sum_{i=1}^{N} Log\,\{\,\sum_{j=1}^{2} p\,(Q_j|M_i)\,\exp\left[-\frac{(y_i-\mu_j)^2}{2\sigma^2}\right]\} \;-\frac{N}{2}Log\,(2\pi\sigma^2) \qquad (14.4)$$

Under the null hypothesis $H_0{:}\mu_1-\mu_2 = 0$, the log likelihood is

$$Log\,[L\,(\mu_1 = \mu_2 = \mu)\,] \;=\; -\frac{1}{2\sigma^2}\sum_{i=1}^{4}\sum_{j=1}^{n_i}(y_{ij}-\mu)^2 - \frac{N}{2}Log\,(2\pi\sigma^2) \qquad (14.5)$$

where μ is the population mean. It is common to use the log likelihood ratio

$$G \;=\; 2\,\{\,Log\,[L\,(\hat{\mu}_1, \hat{\mu}_2, \hat{\sigma}^2, \hat{r}_1)\,] - Log\,[L\,(\mu_1 = \mu_2 = \mu)\,]\,\} \qquad (14.6)$$

as a test statistic for the hypothesis that a QTL is not located in the interval, where the log likelihoods are evaluated using the maximum likelihood estimates of the genotypic values for the two QTL genotypes, the variance $(\hat{\sigma}^2)$ and the recombination fraction between marker A and the putative QTL $(\hat{r}_1)$. Under the null hypothesis, the recombination fractions are not involved in the computations for the log likelihood. Under the alternative hypothesis, the recombination fraction between the two flanking markers is assumed for the likelihood computation. The G-statistic is distributed as a chi-square variable with one degree of freedom. It is common to use the lod score for the interval mapping.

$$lod \;=\; Log_{(10)}\,[L\,(\hat{\mu}_1, \hat{\mu}_2, \hat{\sigma}^2, \hat{r}_1)\,] - Log_{(10)}\,[L\,(\mu_1 = \mu_2 = \mu)\,] \qquad (14.7)$$

As mentioned for the likelihood approach for the single-marker analysis, in theory, the maximum likelihood estimates of the two QTL genotypic values, the variance and the recombination fraction between marker A and the putative QTL can be estimated by setting the partial differentials of the log likelihood Equation (14.4) with respect to the four unknown parameters equal to zero. However, the analytical solutions for the parameters are not readily obtained. Iterative approaches are needed to obtain the maximum likelihood estimates. In practice, the variance can be estimated using the pooled within marker class variance, which is

$$\hat{\sigma}^2 \;=\; \frac{\displaystyle\sum_i\sum_j (y_{ij}-\bar{y}_i)^2}{\displaystyle\sum_i n_i} \qquad (14.8)$$

This is not a maximum likelihood estimate of the variance, but an *ad hoc* estimate. By holding the variance constant, the computation needed for solving Equation (14.4) for the remaining parameters is greatly reduced. It is also logical to fix recombination at a number of meaningful points, so that the only unknown parameters in Equation (14.4) are the two means. For example, the QTL can be tested within every recombination unit between two adjacent markers. By doing this for every interval, a log likelihood ratio test statistic plot or a lod score plot can be obtained.

Example: A Likelihood Approach

Figure 14.4 shows the log likelihood ratio test statistic and the lod score plots against genome positions on chromosome IV for malt extract QTL of barley using the interval mapping approach and the data in Tables 12.5 to

12.7. Four curves in each of the plots are for the four environments. The curves are different among the environments. These differences may be the evidence of genotype by environment interactions.

Compared to the likelihood approach for single-marker analysis, the interval mapping approach provides more significance (Figures 13.4 and 14.4). One of the reasons is that the trait value distribution using the two-marker interval mapping model (Figure 14.3) is much clearer than using the single-marker model (Figure 13.2). For the single-marker model, trait distributions for the two classes of marker genotypes are all mixed distributions. For the interval mapping model based on the flanking markers, the trait distributions for two classes of marker genotypes are single-variable normal distributions. When the map density is high, the frequencies of these marker genotype classes are much higher than those of the mixed distribution classes. For example, when recombination fraction between two markers flanking an interval $r = 0.1$, 90% of the marker genotypes for the two markers correspond to unique QTL genotypes and only 10% of the marker genotypes are mixed with different QTL genotypes. If the QTL is located in the middle of the two markers, the expected QTL genotypic frequencies for the four marker genotypes are

QTL	AABB	AABb	AaBB	AaBb
QQ	0.45	0.025	0.025	0.00
Qq	0.00	0.025	0.025	0.45

When linkage map density is high, the marker genotypes with mixed distributions can be ignored in data analysis.

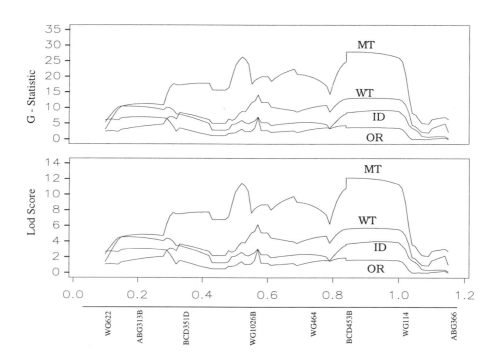

Figure 14.4 Lod score and the log likelihood ratio test statistic are plotted against map location using the likelihood approach for the data in Tables 12.5 to 12.7. High lod scores and likelihood ratio test statistics provide evidence for QTLs in the intervals. MT = Montana, WT = Washington, ID = Idaho, OR = Oregon.

Table 14.3 Dummy variables for interval mapping using regression analysis for backcross.

Maker Genotype	X1	X2	X3	X4
AABB	1	0	0	0
AABb	0	1	0	0
AaBB	0	0	1	0
AaBb	0	0	0	1

14.2.3 A NONLINEAR REGRESSION APPROACH

Nonlinear Regression

Recall what we have in terms of mapping data and what we want to find out. We have

- marker genotypic and trait phenotypic data (for example, Tables 12.5 to 12.7)
- theoretical linkage relationship for a QTL and two flanking markers (Table 14.1)
- expected trait values for four marker genotypic classes (Table 14.2)

We want to know

- if the putative QTL exists in the region flanked by the two markers
- if it exists, what is the magnitude of the QTL effects
- what is the genome position of the QTL

If we code the marker genotypic data as shown in Table 14.3, a nonlinear regression model for the trait value can be written as

$$y_j = \sum_{i=1}^{4} X_{ij} g_i + \varepsilon_j = X_1 \mu_1 + X_2 \frac{1}{r} (r_2 \mu_1 + r_1 \mu_2) + X_3 \frac{1}{r} (r_1 \mu_1 + r_2 \mu_2) + X_4 \mu_2 + \varepsilon_j \quad (14.9)$$

where X_1, X_2, X_3 and X_4 are the dummy variables for the four marker genotypes listed in Table 14.3, g_i is the expected trait value for marker genotypic class i as shown in Table 14.2, r, r_1 and r_2 are the recombination fractions between the two markers, the putative QTL and marker A and the putative QTL and marker B, respectively and ε_j is the residual of the model. If we reparameterize the recombination fraction between marker A and the putative QTL as ρr, then we have

$$y_j = X_1 \mu_1 + X_2 [(1 - \rho) \mu_1 + \rho \mu_2] + X_3 [\rho \mu_1 + (1 - \rho) \mu_2] + X_4 \mu_2 + \varepsilon_j \quad (14.10)$$

where $\rho = r_1 / r$ and $0 \leq \rho \leq 1$. Three unknown parameters are involved in the model: the two QTL genotypic means and the parameter for relative QTL position.

The least-square estimates of the unknown parameters in Equation (14.10) can be solved using an iterative Gauss-Newton algorithm. We want to minimize the sum of squares of deviations

$$SSE = \sum_{j=1}^{N} \varepsilon_j^2 = \sum_{j=1}^{N} \left[y_j - \sum_{i=1}^{4} X_{ij} g_i \right]^2$$

where N is the number of individuals in the mapping population. Using matrix notation, we have

$$SSE = [Y - f(\theta)]'[Y - f(\theta)] \qquad (14.11)$$

where

$$Y = \begin{bmatrix} y_1 \\ y_2 \\ \dots \\ y_N \end{bmatrix}_{N \times 1} \qquad f(\theta) = \begin{bmatrix} \sum X_{i1} g_i \\ \sum X_{i2} g_i \\ \dots \\ \sum X_{iN} g_i \end{bmatrix}_{N \times 1} \qquad and \qquad \theta = \begin{bmatrix} \mu_1 \\ \mu_2 \\ \rho \end{bmatrix}_{3 \times 1}$$

are vectors for observed trait values, the predicted values and the unknown parameters, respectively. If we carry out a Taylor expansion for $f(\theta)$ about the point

$$\theta^0 = \begin{bmatrix} \mu_1^0 & \mu_2^0 & \rho^0 \end{bmatrix}'$$

we have

$$f(\theta) \approx f(\theta^0) + \frac{\partial f(\theta = \theta^0)}{\partial \theta}(\theta - \theta^0) \qquad (14.12)$$

If we set

$$\beta^0 = \theta - \theta^0 \qquad and \qquad Z^0 = \frac{\partial f(\theta)}{\partial \theta} = \begin{bmatrix} Z_{11}^0 & Z_{21}^0 & Z_{31}^0 \\ Z_{12}^0 & Z_{22}^0 & Z_{32}^0 \\ \dots & \dots & \dots \\ Z_{1N}^0 & Z_{2N}^0 & Z_{3N}^0 \end{bmatrix} \qquad (14.13)$$

then we have

$$Y - f(\theta^0) = Z^0 \beta^0 + e \qquad (14.14)$$

The solutions for β^0 are

$$\hat{\beta}^0 = (Z^{0\prime} Z^0)^{-1} Z^{0\prime} (Y - f(\theta^0)) \qquad (14.15)$$

which minimize the sum of square

$$\begin{aligned} SSE &= [Y - f(\theta^0) - Z^0 \beta^0]'[Y - f(\theta^0) - Z^0 \beta^0] \\ &= [Y - f(\theta)]'[Y - f(\theta)] \end{aligned} \qquad (14.16)$$

Then the estimates of the parameters can be updated

$$\theta^1 = \hat{\beta}^0 + \theta^0 \qquad (14.17)$$

By putting the new estimates θ^1 as θ^0, a new iteration can be performed. The iteration continues until the estimates converge, that is

$$\left| \hat{\beta}^0 / \theta^0 \right| < \tau$$

where τ is a tolerance, say 0.00001. For a regression model of interval mapping using backcross progeny, the derivatives needed for the least square estimation are

$$\partial f(\theta) / \partial \mu_1 = X_1 + X_2 (1 - \rho) + X_3 \rho$$
$$\partial f(\theta) / \partial \mu_2 = X_2 \rho + X_3 (1 - \rho) + X_4 \qquad (14.18)$$
$$(\partial f(\theta)) / \partial \rho = X_2 (\mu_2 - \mu_1) + X_3 (\mu_1 - \mu_2)$$

Hypothesis Test

Hypothesis tests can be constructed for linear contrasts of the parameters. For the interval mapping problem, a meaningful contrast is $H = (1, -1)$, which corresponds to $\mu_1 - \mu_2 = 0$. The contrast can be tested using a Wald test statistic

$$W = \frac{h'(\hat{\theta}) \, (\hat{H} \hat{C} \hat{H}')^{-1} h(\hat{\theta})}{q s_e^2} \qquad (14.19)$$

where $h(\hat{\theta})$ is the contrast. For example

$$h(\hat{\theta}) = \mu_1 - \mu_2; \quad \hat{H} = \frac{\partial h(\hat{\theta})}{\partial \theta'}$$

and

$$\text{for } h(\hat{\theta}) = \mu_1 - \mu_2 \quad \hat{H} = (1, -1)$$

q is the degrees of freedom for the numerator. For example, it is one for a single contrast. s_e^2 is the residual mean square. $\hat{C}$ is $(Z'Z)^{-1}$.

The hypothesis test also can be carried out using the log likelihood ratio test statistic. If the null hypothesis is

$$H_0: \mu_1 - \mu_2 = 0 \text{ or } \mu_1 = \mu_2 = \mu$$

the log likelihood ratio test statistic is

$$G = \frac{SSE_{reduced} - SSE_{full}}{SSE_{full} / df_{Efull}} \qquad (14.20)$$

where SSEs are the residual sum of squares for the full and reduced models. The full model is Equation (14.10). The reduced model is the full model under the null hypothesis, which is

$$y_j = (X_1 + X_2 + X_3 + X_4) \mu + \varepsilon_j = \mu + \varepsilon_j \qquad (14.21)$$

For the interval mapping approach, the Wald statistic and the log likelihood ratio test statistics are the same. The F-distribution and the chi-square distribution are the same when the degree of freedom for the numerator is one.

Confidence Interval for the Parameters

Confidence intervals for the parameters can be constructed using the asymptotic property of the estimates. The asymptotic variance for the estimates is

$$\hat{s}_i^2 = \hat{c}_{ii}\hat{s}_e^2 \qquad (14.22)$$

where $\hat{c}_{ii}$ is the diagonal element of the matrix $\hat{C}$ and $\hat{s}_e^2$ is the residual mean square for the whole model. The 95% confidence interval for a parameter θ_i is

$$\hat{\theta}_i \pm 1.96\sqrt{\hat{s}_i^2} \qquad (14.23)$$

where 1.96 is the standard normal deviate corresponding to 95% probability.

Table 14.4 Nonlinear least square iterative phase using Gauss-Newton method.

Iteration	μ_1	μ_2	ρ	SSE
0	75.000	75.000	0.5	936.100
1	72.464	73.829	0.5	347.556
2	72.451	73.812	0	329.655
3	72.418	73.856	0	329.446

Example: Estimation, Hypothesis Tests and Confidence Interval

For the data in Tables 12.5 to 12.7, two adjacent loci, ABG472 and iAco2, are picked for demonstrating the procedures. The matrix $\mathbf{Z}$ is

```
if marker genotype is "11" then Z = (1      0         0        )
if marker genotype is "12" then Z = (1-r    r         μ₂ - μ₁)
if marker genotype is "21" then Z = (r      1 - r     μ₂ - μ₁)
if marker genotype is "22" then Z = (0      0         1        )
```

If $\theta^0 = (75.0, 75.0, 0.5)$ are used as initial values for the parameters, we have

```
if marker genotype is "11" then Z⁰ = (1.0      0.0      0.0)
if marker genotype is "12" then Z⁰ = (0.5      0.5      0.0)
if marker genotype is "21" then Z⁰ = (0.5      0.5      0.0)
if marker genotype is "22" then Z⁰ = (0.0      0.0      1.0)
```

The new estimates can be obtained using Equation (14.17). Table 14.4 lists the iteration history. By three iterations, the estimates converge. The estimated putative QTL mean for the Steptoe genotype is $\hat{\mu}_1 = 72.418$ and for the Morex genotype $\hat{\mu}_2 = 73.856$. The relative position of the putative QTL within the segment is $\hat{\rho} = 0$. So, the recombination fraction between the putative QTL and marker ABG472 is

$$\hat{r}_1 = \hat{\rho}/r = 0/0.17 = 0$$

Therefore, if there is a QTL within the segment, the QTL is most likely located right on marker ABG472. The variance associated with the estimated QTL position is usually large. A large sample size is required to find the precise location of the QTL. Table 14.5 is the analysis of variance table for the nonlinear model.

The residual mean square, s_e^2, is 2.405 for the interval mapping analysis for the two markers. For example, the $\hat{C}$ matrix is

$$\hat{C} = \begin{bmatrix} 0.0135 & -0.0014 & 0.0057 \\ -0.0014 & 0.0188 & -0.0062 \\ 0.0057 & -0.0062 & 0.0179 \end{bmatrix}$$

Table 14.5 Analysis of variance table for the nonlinear model (including all four environments).

Source	DF	Sum of Squares	Mean Square
Regression	3	747036.65432	249012.21811
Residual	557	329.44568	2.40471
Uncorrected Total	560	747366.10000	

Table 14.6 95% confidence intervals for the estimates in Table 14.4.

Parameter	$\hat{\theta}_i$	$\sqrt{\hat{s}_i^2}$	Lower	Upper
μ_1	72.418	0.1802	72.0614	72.7741
μ_2	73.856	0.2126	73.4354	74.2761
ρ	0	0.2476	0	0.4897

For a hypothesis

$$H_0: \mu_1 - \mu_2 = 0$$

the Wald statistic is computed by

$$h(\hat{\theta}) = \hat{\mu}_1 - \hat{\mu}_2 = 72.418 - 73.856 = -1.438$$

$$\hat{H} = \frac{\partial h(\hat{\theta})}{\partial \theta'} = (1, -1)$$

$$\hat{H}\hat{C}\hat{H}' = \begin{bmatrix} 1 & -1 \end{bmatrix} \begin{bmatrix} 0.0135 & -0.0014 \\ -0.0014 & 0.0188 \end{bmatrix} \begin{bmatrix} 1 \\ -1 \end{bmatrix} = 0.0163 \qquad (14.24)$$

$$s_e^2 = 2.405, q = 1$$

$$W = \frac{h'(\hat{\theta})(\hat{H}\hat{C}\hat{H}')^{-1} h(\hat{\theta})}{q s_e^2} = \frac{(-1.438)(0.0163)^{-1}(-1.438)}{2.405} = 52.749$$

The Wald statistic is distributed as F-distribution with 1 degree of freedom for the numerator and $N-4$ degrees of freedom for the denominator. For this example, the significant p-value is

$$P[F_{1,557} > 52.749] < 0.0000001$$

The 95% confidence intervals for the estimates are shown in Table 14.6.

Multiple Environments Model

The nonlinear regression model (Equation 14.10) can be easily extended to multiple environments data. For example, the malt extract QTL was evaluated in four environments for the 150 doubled haploid lines and the data are in Tables 12.5 to 12.7. Equation (14.10) can be rewritten as

$$y_{jk} = \sum_k \{ X_{1k}\mu_{1k} + X_{2k}[(1-\rho)\mu_{1k} + \rho\mu_{2k}] \qquad (14.25)$$
$$+ X_{3k}[\rho\mu_{1k} + (1-\rho)\mu_{2k}] + X_{4k}\mu_{2k} \} + \varepsilon_{jk}$$

where the subscript k is the indicator for the environments; the dummy

Table 14.7 Contrasts for a single-QTL with four environments (for data in Tables 11.5 to 11.7).

	ID		OR		WT		MT	
	μ_{11}	μ_{21}	μ_{12}	μ_{22}	μ_{13}	μ_{23}	μ_{14}	μ_{24}
Standard Contrasts								
Genetic ID	1	-1	0	0	0	0		
Genetic OR	0	0	1	-1	0	0	0	0
Genetic WT	0	0	0	0	1	-1	0	0
Genetic MT	0	0	0	0	0	0	0	0
Genetic Pool	1	-1	1	-1	1	-1	1	-1
							1	-1
Environment	-1	-1	-1	-1	1	1		
	1	1	-1	-1	-1	-1	1	1
	-1	-1	1	1	-1	-1	1	1
							1	1
G x E	-1	1	-1	1	1	-1		
	1	-1	-1	1	-1	1	1	-1
	1	1	1	-1	-1	1	1	-1
							1	-1
Other Contrasts								
ID,OR vs WT,MT	1	1	1	1	-1	-1	-1	-1
G x IO vs WM	1	-1	1	-1	-1	1	-1	1

variables and the parameters are now environmental specific and the QTL position parameter is the same for all four environments. This model has 9 unknown parameters: eight of them are the QTL genotypic means for four environments (two for each environment) and one is the QTL position. The derivatives for solving the model are the simple duplications of Equation (14.18) for the eight means

$$\frac{\partial f(\theta)}{\partial \mu_{1k}} = X_{1k} + X_{2k}(1-\rho) + X_{3k}\rho$$

$$\frac{\partial f(\theta)}{\partial \mu_{2k}} = X_{2k}\rho + X_{3k}(1-\rho) + X_{4k} \qquad (14.26)$$

and for the QTL position, it is the simple summation

$$\frac{\partial f(\theta)}{\partial \rho} = \sum_{k} [X_{2k}(\mu_{2k}-\mu_{1k}) + X_{3k}(\mu_{1k}-\mu_{2k})] \qquad (14.27)$$

For the model based on Equation (14.25), several biologically meaningful contrasts can be tested, as shown in Table 14.7. The standard contrasts which can be tested are the genetic effects in each of the four individual environments, the pooled genetic effect over the four environments, environmental effects and the genetic by environment interactions. The contrasts for the genetic effects are easy to understand. The environmental effects among the four environments can be written as three independent orthogonal contrasts. The contrasts for the genetic by environment interactions are the simple products of the contrast for the pooled genetic effect and the contrasts for environmental effect. Besides those standard contrasts, contrasts can be constructed for many other biologically meaningful effects. For example, contrast can be built for testing a difference between (ID, OR) and (WT, MT) and the possible interaction due to the two areas (Table 14.7). The QTL position parameter is not involved in any meaningful contrasts.

Implementation of the Nonlinear Regression

The regression approach can be carried out using a commonly used statistical package such as SAS, using the software QTLSTAT (Liu and Knapp 1992) or the PGRI software (Liu 1995). When using SAS, the data set requires a certain format. For example, for a single-environment experiment (for Montana in data of Tables 12.5 to 12.7), the data should be arranged as

```
                                             line    trait
  segment  marker1    marker2    g1  g2  number   value
     1      WG622      ABG313B    1   1      1     72.90
     1      WG622      ABG313B    1   0      2     70.70
      ...      ...        ...       ...        ...
```
(Total 25 segments and 150 lines for each segment)

Then, the nonlinear analysis can be carried out using SAS codes similar to

```
data a;
  input seg marker1 $ marker2 $ g1 g2 line y;
  x1=0; x2=0; x3=0; x4=0;
  if g1=1 and g2=1 then x1=1;
  if g1=1 and g2=2 then x2=1;
  if g1=2 and g2=1 then x3=1;
  if g1=2 and g2=2 then x4=1;
  cards;
     1      WG622      ABG313B    1   1      1     72.90
     1      WG622      ABG313B    1   0      2     70.70
      ...      ...        ...       ...        ...;
proc nlin data=a noprint method=gauss
  convergence=0.0000001 outest=output;
  by in;
  parms m1=constantA m2=constantB r=constantC;
  bounds r<=1.0 r>=0;
  model y=x1*m1+x2*((1-r)*m1+r*m2)+x3*(r*m1+(1-
       r)*m2)+x4*m2;
  der.m1=x1+x2*(1-r)+x3*r;
  der.m2=x2*r+x3*(1-r)+x4;
  der.r=x2*(m2-m1)+x3*(m1-m2);
proc print data=output;
run;
```

where the constants are the initial values for the parameters. However, SAS provides neither the hypothesis test for the contrasts nor the matrix $\hat{C}$ needed for the computation (Equation 14.19). To obtain the matrix, a linear regression procedure using the parameters can be done. For example, for segment 18 of the barley data, the following SAS codes were used to generate the matrix

```
data b; set a; if in=18;
  t1=x1+x3;
  t2=x2+x4;
  t3=x2*1.359-x3*1.359;
  y1=y-(x1*72.447+x2*73.806+x3*72.447+x4*73.806);
proc reg all; model y1=t1 t2 t3;
run;
```

where 72.447, 73.806 and 0 are the estimated values for the three parameters, the estimated difference between the two means is -1.359, t1, t2 and t3 are the first derivatives of the model with respect to the three parameters evaluated at the estimated values and y1 is the residual of the predicted value using the estimates from the observed values. The following SAS output is the matrix $\hat{C}$

```
X'X Inverse, Parameter Estimates, and SSE
            T1            T2           T3            Y1
T1    0.013503162  -0.001422537  0.0057376333  0.0274427155
T2   -0.001422537   0.0187884359 -0.006239866  -0.054301096
T3    0.005737633  -0.006239866  0.0251677508  -1.107825897
Y1    0.027442715  -0.054301096  -1.10782589   328.40021666
```

The test statistic can be easily obtained using a hand calculator when the contrasts are simple. When the contrasts are complex with several degrees of freedom, the computation needed to test the contrast may be complex. For example, the contrasts for the environmental effect and the genetic by environment interaction contain three degrees of freedom each. For those cases, specialized software such as QTLSTAT or PGRI is recommended.

Table 14.8 Wald statistics for the data in Tables 12.5 and 12.6 using the nonlinear regression approach. The analysis was performed using PGRI.

Marker1	Marker2	ID	OR	WT	MT	Pooled	G x E
WG622	ABG313B	1.94	8.43	5.89	5.44	1.22	19.75
ABG313B	CDO669	2.85	8.20	12.55	15.14	2.78	34.82
CDO669	BCD402B	7.45	8.19	15.02	30.61	4.41	56.32
BCD402B	BCD351D	7.52	4.54	9.50	24.41	4.49	41.41
BCD351D	BCD265B	3.99	2.75	5.22	28.32	8.79	30.35
BCD265B	TubA1	4.46	1.39	3.40	22.73	7.47	24.50
TubA1	Dhn6	5.35	2.31	3.38	22.49	6.55	26.99
Dhn6	ABG3	5.54	2.22	5.45	28.83	8.74	33.41
ABG3	WG1026B	5.84	2.66	5.86	33.05	10.16	36.95
WG1026B	Adh4	8.18	2.59	6.33	33.67	10.07	41.25
Adh4	ABA3	10.26	3.27	5.87	25.87	6.08	38.60
ABA3	ABG484	13.10	8.01	8.06	26.72	3.65	52.43
ABG484	Pgk1	10.77	3.31	6.19	28.35	6.93	41.68
Pgk1	ABR315	8.65	3.17	4.87	27.51	7.29	36.58
ABR315	WG464	7.72	2.82	6.43	30.13	8.55	37.17
WG464	BCD453B	5.89	5.81	2.74	22.98	5.77	29.94
BCD453B	ABG472	10.93	11.22	4.16	42.07	10.78	57.28
ABG472	iAco2	12.12	8.92	2.69	28.59	8.64	44.26
iAco2	ABG500B	3.64	3.79	0.07	10.38	4.40	13.08
ABG500B	ABG498	3.52	2.59	0.03	8.77	3.93	10.85
ABG498	WG114	2.67	2.55	0.04	8.04	3.52	9.79
WG114	ABG54	2.02	2.35	0.02	6.45	2.87	7.95
ABG54	ABG394	2.33	0.92	0.21	8.77	3.51	8.71
ABG394	ABG366	2.61	1.01	0.11	7.48	3.11	7.95
ABG366	ABG397	2.77	0.08	0.00	7.26	4.86	5.38

Example: The Multiple Environments Problem

Table 14.8 shows the Wald statistic for the nonlinear regression analysis for the 25 genome segments on barley chromosome IV (data in Tables 12.5 to 12.7). The same results are also shown using the Wald statistic plot (Figure 14.5). The results for the individual environments are similar to the results using the likelihood approach (Figure 14.4).

The pooled genetic effect over the four environments shows significance in several segments. The data from the Montana field test shows the highest significance. Genotype by environment interactions are significant for almost every segment. For the data from Montana, there are two Wald statistic peaks at each end of the plot. One peak is located on the segment flanked by markers CDO669 and BCD402B and the other is flanked by BCD453B and ABG472. Another way to show the results is by plotting the distributions of the trait values for each of the marker genotypic classes (Figure 14.6).

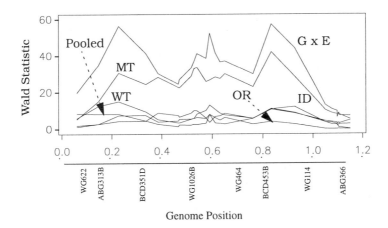

Figure 14.5 Lod score and the log likelihood ratio test statistics are plotted against genetic map locations using the likelihood approach for the data in Tables 12.5 to 12.7. Higher lod scores and likelihood ratio test statistics provide evidence for QTLs in the intervals.

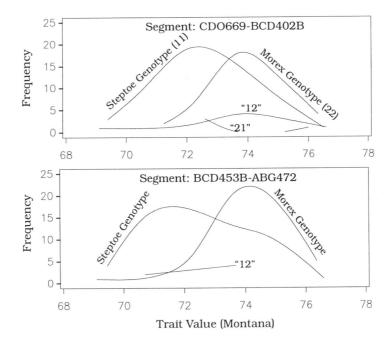

Figure 14.6 Distributions of the trait values for malt extract from 150 doubled haploid lines (in Montana) classified by two genome segments on the barley chromosome IV and four possible marker genotypes for the markers flanking the segments.

Figure 14.6 shows the observed distributions of the malt extract trait values for the 150 doubled haploid lines when grown in Montana. It is clear that the distributions among the marker genotypes are different. The frequency for the recombinant classes are low because the markers are linked tightly with each other. For markers CDO669 and BCD402B, 20 lines among the 150 are recombinant and 7 lines are recombinant for markers BCD453B and ABG472. When linkage map density is high, frequencies of recombinant types for adjacent markers are low. This is the reason that interval mapping approaches have higher statistical powers than do the single-marker approaches.

14.2.4 THE LINEAR REGRESSION APPROACH

Knapp *et al.* (1990) suggest using linear models for interval mapping using regression analysis. If we define

$$\theta_1 = \mu_1$$
$$\theta_2 = (1-\rho)\mu_1 + \rho\mu_2$$
$$\theta_3 = \rho\mu_1 + (1-\rho)\mu_2$$
$$\theta_4 = \mu_2$$

Equation (14.10) can be rearranged as

$$y_j = X_1\theta_1 + X_2\theta_2 + X_3\theta_3 + X_4\theta_4 + \varepsilon_j \tag{14.28}$$

where the θs are trait means for the marker genotypes. If no constraints are imposed, the estimators of $\hat{\theta}_1$ and $\hat{\theta}_4$ are the estimates for the two QTL genotypic means ($\hat{\mu}_1$ and $\hat{\mu}_2$). If constraint

$$\theta_1 + \theta_4 = \theta_2 + \theta_3$$

is imposed, we have

$$\theta_3 = \theta_1 + \theta_4 - \theta_2$$

and the linear model is

$$y_j = \theta_1(X_1 + X_3) + \theta_2(X_2 - X_3) + \theta_4(X_3 + X_4) + \varepsilon_j \tag{14.29}$$

Solving Equation (14.29) minimizing the residual sum of the square

$$SSE = \sum_{j=1}^{N} \{y_j - [\theta_1(X_1 + X_3) + \theta_2(X_2 - X_3) + \theta_4(X_3 + X_4)]\}^2 \tag{14.30}$$

leads us to the estimates $\hat{\theta}_1$, $\hat{\theta}_2$ and $\hat{\theta}_4$. By solving

$$\hat{\theta}_1 = \mu_1$$
$$\hat{\theta}_2 = (1-\rho)\mu_1 + \rho\mu_2 \tag{14.31}$$
$$\hat{\theta}_4 = \mu_2$$

we have

$$\hat{\mu}_1 = \hat{\theta}_1$$
$$\hat{\mu}_2 = \hat{\theta}_4 \qquad\qquad (14.32)$$
$$\hat{\rho} = (\hat{\theta}_1 - \hat{\theta}_2) / (\hat{\theta}_1 - \hat{\theta}_4)$$

As pointed out by Knapp *et al.* (1990), the estimate $\hat{\rho}$ may not satisfy $0 \le \hat{\rho} \le 1$. For the genome segment 18 (flanked by ABG472 and iAco2), the example used for illustrating the nonlinear approach, the estimates $\hat{\theta}_1$, $\hat{\theta}_2$ and $\hat{\theta}_4$ are 72.447, 72.213 and 73.806. The estimates for the putative QTL means and the relative position are

$$\hat{\mu}_1 = \hat{\theta}_1 = 72.447$$
$$\hat{\mu}_2 = \hat{\theta}_4 = 73.806$$
$$\hat{\rho} = \frac{\hat{\theta}_1 - \hat{\theta}_2}{\hat{\theta}_1 - \hat{\theta}_4} = \frac{72.447 - 72.213}{72.447 - 73.806} = -0.172$$

These estimates are different from estimates obtained using the nonlinear approach because the model based on Equation (14.10) has no constraint on ρ.

However, in practice, ρ has been considered a known parameter. Using the relationship

$$\theta_2 = (1 - \rho)\mu_1 + \rho\mu_2$$

Equation (14.29) becomes

$$y_j = \theta_1[X_1 + (1 - \rho)X_2 + \rho X_3] + \theta_4[\rho X_2 + (1 - \rho)X_3 + X_4] + \varepsilon_j \qquad (14.33)$$

The independent variables in the linear regression model based on Equation (14.33) are the probabilities that the individual is a QQ or Qq genotype for the putative QTL, conditional on the flanking marker genotypes (see Table 14.2). In matrix terms, the model is

$$Y = X\beta + \varepsilon \qquad\qquad (14.34)$$

This is

$$
{}_N\!\begin{bmatrix} y_1 \\ y_2 \\ \dots \\ y_N \end{bmatrix}_1 = {}_N\!\begin{bmatrix} 1 & 0_{n_1} \\ 1-\rho & \rho_{n_2} \\ \rho & 1-\rho_{n_3} \\ 0 & 1_{n_4} \end{bmatrix}_2 \begin{bmatrix} \theta_1 \\ \theta_4 \end{bmatrix} + {}_N\!\begin{bmatrix} \varepsilon_1 \\ \varepsilon_2 \\ \dots \\ \varepsilon_N \end{bmatrix}_1 \qquad (14.35)
$$

where the subscripts in the matrix X, such as n_1 for the first row, indicate the number of repeats for the rows. The normal equation for the model based on Equation (14.33) is

$$
\begin{bmatrix} 0.5(1-2c) & c \\ c & 0.5(1-2c) \end{bmatrix} \begin{bmatrix} \theta_1 \\ \theta_4 \end{bmatrix} = \begin{bmatrix} 0.5[(1-2c)\mu_1 + 2c\mu_2] \\ 0.5[2c\mu_1 + (1-2c)\mu_2] \end{bmatrix} \qquad (14.36)
$$

where $c = r\rho(1 - \rho)$. The solutions for the two parameters are

$$\theta_1 = \mu_1$$
$$\theta_4 = \mu_2 \tag{14.37}$$

So, the estimates are unbiased estimators for the QTL means. The expected residual sum of squares for the model is

$$E[SSE_{R\,(full)}] = E[(Y - X\beta)'(Y - X\beta)]$$
$$= N\sigma_e^2 + 0.5rN(\mu_1 - \mu_2)^2[\rho^2 + (1-\rho)^2] \tag{14.38}$$

Under the null hypothesis

$$H_0: \mu_1 = \mu_2 = \mu$$

the model based on Equation (14.33) becomes

$$y_j = \theta + \varepsilon_j \tag{14.39}$$

The hypothesis can be tested at each genome position (a specific ρ value) using the ratio between the residual sum of squares of the models based on Equation (14.33) (full model) and (14.39) (reduced model). The expected residual sum of squares for the reduced model is

$$E[SSE_{R\,(reduced)}] = E[(Y - \mu)'(Y - \mu)]$$
$$= N\sigma_e^2 + N(\mu_1 - \mu_2)^2 + rN(\mu_1 - \mu_2)^2[1 - (0.5 - \rho)^2] \tag{14.40}$$

The linear regression approach has the advantages over the likelihood approach of (1) computational and programming simplicity and (2) feasibility of integrating experimental designs and other factors into QTL mapping (Lander and Botstein 1989; Knapp *et al.* 1990; Haley and Knott 1992; Jansen 1992; Zeng 1993 and 1994; Knapp 1995).

When a linkage map is relatively dense, the power and the parameter estimates using the regression approaches are equal to or closely correlated with those obtained using the likelihood approach (Haley and Knott 1992). However, researchers favoring the likelihood approaches have always preferred them if they can be implemented instead of regression approaches.

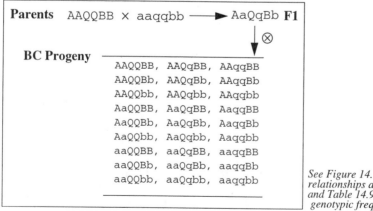

Figure 14.7 F2 progeny for a segregating QTL and two flanking markers.

14.3 INTERVAL MAPPING USING F2 PROGENY

14.3.1 JOINT SEGREGATION OF QTLS AND MARKERS

Let us consider again the model in Figure 14.1. A QTL is located between markers A and B. This time, F2 progeny are used instead of backcross progeny. The classical F2 progeny set is produced by selfing an inbreeding double heterozygous F1 (AaQqBb × AaQqBb) (Figure 14.7). The F1 is typically a hybrid between two inbred lines (AAQQBB × aaqqbb). Linkage phase between a single QTL and the markers is arbitrary and has no effect on the analysis. Linkage phase is a problem only when more than a single QTL is considered and they are linked. For now, a single-QTL model is considered. The approaches for the classical F2 may be applied to other non-classical F2 data, such as the progeny of a cross between two genetically different heterozygous parents. Table 14.9 shows the expected marker-QTL genotypic frequencies in an F2 population assuming absence of double crossover. The recombination fraction between markers A and B is r and between the putative QTL and the marker A is r_1. The frequencies shown in the table are the expected frequencies for a three-locus cosegregation model.

Table 14.9 Expected marker and QTL genotypic frequency in F2 populations assuming no double crossover.

Marker Genotype	Frequency $p_{i\circ}$	p_{ij}		
		QQ	Qq	qq
AABB	$0.25\,(1-r)^2$	$0.25\,(1-r)^2$	0	0
AABb	$0.5\,r\,(1-r)$	$0.5\,r_2\,(1-r)$	$0.5\,r_1\,(1-r)$	0
AAbb	$0.25\,r^2$	$0.25\,r_2^2$	$0.5\,r_1 r_2$	$0.25\,r_1^2$
AaBB	$0.5\,r\,(1-r)$	$0.5\,r_1\,(1-r)$	$0.5\,r_2\,(1-r)$	0
AaBb	$0.5\,[\,(1-r)^2+r^2\,]$	$0.5\,r_1 r_2$	$0.5\,[\,(1-r)^2+r_1^2+r_2^2\,]$	$0.5\,r_1 r_2$
Aabb	$0.5\,r\,(1-r)$	0	$0.5\,r_2\,(1-r)$	$0.5\,r_1\,(1-r)$
aaBB	$0.25\,r^2$	$0.25\,r_1^2$	$0.5\,r_1 r_2$	$0.25\,r_2^2$
aaBb	$0.5\,r\,(1-r)$	0	$0.5\,r_1\,(1-r)$	$0.5\,r_2\,(1-r)$
aabb	$0.25\,(1-r)^2$	0	0	$0.25\,(1-r)^2$

Table 14.10 Expected QTL genotypic frequency conditional on genotypes of the flanking markers in F2 populations with no double crossover.

Marker Genotype	Frequency $p_{i\circ}$	$p\,(Q_j\vert M_i)$		
		QQ	Qq	qq
AABB	$0.25\,(1-r)^2$	1	0	0
AABb	$0.5\,r\,(1-r)$	r_2/r	r_1/r	0
AAbb	$0.25\,r^2$	$(r_2/r)^2$	$2r_1 r_2/r^2$	$(r_1/r)^2$
AaBB	$0.5\,r\,(1-r)$	r_1/r	r_2/r	0
AaBb	$0.5\,[\,(1-r)^2+r^2\,]$	$\dfrac{r_1 r_2}{(1-r)^2+r^2}$	$\dfrac{(1-r)^2+r_1^2+r_2^2}{(1-r)^2+r^2}$	$\dfrac{r_1 r_2}{(1-r)^2+r^2}$

Table 14.10 Expected QTL genotypic frequency conditional on genotypes of the flanking markers in F2 populations with no double crossover.

Marker Genotype	Frequency p_{i°	$p(Q_j\|M_i)$		
		QQ	Qq	qq
Aabb	$0.5r(1-r)$	0	r_2/r	r_1/r
aaBB	$0.25r^2$	$(r_1/r)^2$	$2r_1r_2/r^2$	$(r_2/r)^2$
aaBb	$0.5r(1-r)$	0	r_1/r	r_2/r
aabb	$0.25(1-r)^2$	0	0	1

Table 14.11 Expected phenotypic value for marker classes.

Marker	Count	Expected Value (g_i)
AABB	n_1	μ_1
AABb	n_2	$(r_2/r)\mu_1 + (r_1/r)\mu_2$
AAbb	n_3	$(r_2/r)^2\mu_1 + (2r_1r_2/r^2)\mu_2 + (r_1/r)^2\mu_3$
AaBB	n_4	$(r_1/r)\mu_1 + (r_2/r)\mu_2$
AaBb	n_5	$1/[(1-r)^2+r^2]\{r_1r_2\mu_1 + [(1-r)^2+r_1^2+r_2^2]\mu_2 + r_1r_2\mu_3\}$
Aabb	n_6	$(r_2/r)\mu_2 + (r_1/r)\mu_3$
aaBB	n_7	$(r_1/r)^2\mu_1 + (2r_1r_2/r^2)\mu_2 + (r_2/r)^2\mu_3$
aaBb	n_8	$(r_1/r)\mu_2 + (r_2/r)\mu_3$
aabb	n_9	μ_3

Table 14.12 Expected QTL genotypic frequency conditional on the flanking markers in F2 populations, assuming no double crossover when dominant markers are used.

Marker Genotype	Frequency p_{i°	$p(Q_j\|M_i)$		
		QQ	Qq	qq
A_B_	$\frac{1}{4}(3-2r+r^2)$	$\dfrac{1-r^2+2r_1r_2}{3-2r+r^2}$	$\dfrac{2(1-r+r_1^2+r_2^2)}{3-2r+r^2}$	$\dfrac{2r_1r_2}{3-2r+r^2}$
A_bb	$\frac{1}{4}r(2-r)$	$\dfrac{r_2^2}{r(2-r)}$	$\dfrac{2r_2(1-r+r_1)}{r(2-r)}$	$\dfrac{r_1(2-2r+r_1)}{r(2-r)}$
aaB_	$\frac{1}{4}r(2-r)$	$\dfrac{r_1^2}{r(2-r)}$	$\dfrac{2r_1(1-r+r_2)}{r(2-r)}$	$\dfrac{r_2(2-2r+r_2)}{r(2-r)}$
aabb	$\frac{1}{4}(1-r)^2$	0	0	1

Let us recall the single-QTL model described in Chapter 12. The three QTL genotypes QQ, Qq and qq have means μ_1, μ_2 and μ_3, respectively. The additive and dominance effects are defined as

$$a = 0.5(\mu_1 - \mu_3)$$

$$d = 0.5(2\mu_2 - \mu_1 - \mu_3)$$

In practice, we can observe the 9 marker genotypic classes for the two flanking

Table 14.13 Expected phenotypic value for marker classes for dominant markers.

Marker Genotype	Count	Expected Value (g_i)
A_B_	n_1	$\dfrac{1-r^2+2r_1r_2}{3-2r+r^2}\mu_1 + \dfrac{2(1-r+r_1^2+r_2^2)}{3-2r+r^2}\mu_2 + \dfrac{2r_1r_2}{3-2r+r^2}\mu_3$
A_bb	n_2	$\dfrac{r_2^2}{r(2-r)}\mu_1 + \dfrac{2r_2(1-r+r_1)}{r(2-r)}\mu_2 + \dfrac{r_1(2-2r+r_1)}{r(2-r)}\mu_3$
aaB_	n_3	$\dfrac{r_1^2}{r(2-r)}\mu_1 + \dfrac{2r_1(1-r+r_2)}{r(2-r)}\mu_2 + \dfrac{r_2(2-2r+r_2)}{r(2-r)}\mu_3$
aabb	n_4	μ_3

markers and the trait values for each of the N individuals in the populations. The three-way frequencies in Table 14.9 can be collapsed into the frequencies and by doing so, Table 14.9 is rearranged into Table 14.10. Table 14.10 shows the QTL genotypic frequencies conditional on each of the nine marker genotypic classes. The conditional frequencies are the ratios between the individual cell frequencies and the marginal frequencies for the marker genotypes.

Table 14.10 can be collapsed into Table 14.11. Table 14.11 shows the observed frequencies and the expected trait values for each of the nine observable marker genotypic classes. Shown in Table 14.12 are the expected QTL genotypic frequencies conditional on the flanking marker genotypes when dominant markers are used. The expected trait values for the marker genotypes are shown in Table 14.13.

14.3.2 A LIKELIHOOD APPROACH FOR QTL ANALYSIS USING F2 PROGENY

A likelihood approach for QTL mapping in an F2 progeny set will now be discussed using codominant markers and then using dominant markers.

Codominant Markers

The likelihood function for the data is

$$L = \frac{1}{\{\sqrt{2\pi}\sigma\}^N} \prod_{i=1}^{N} \left\{ \sum_{k=1}^{3} p(Q_j|M_i) \exp\left[-\frac{(y_i-\mu_j)^2}{2\sigma^2}\right] \right\} \qquad (14.41)$$

if we assume that each of the 27 marker-QTL classes has equal variance σ^2 and the trait values are normally distributed, y_i is the observed trait phenotypic value for the ith individual, $p(Q_j|M_i)$ is the conditional probability listed in Table 14.10 and μ_j is trait value for the jth QTL genotype. The logarithm of Equation (14.41) is

$$Log(L) = \sum_{i=1}^{N} \log\left\{ \sum_{j=1}^{3} p(Q_j|M_i) \exp\left[-\frac{(y_i-\mu_j)^2}{2\sigma^2}\right] \right\} - \frac{N}{2}Log(2\pi\sigma^2) \qquad (14.42)$$

Under a null hypothesis $H_0: \mu_1 = \mu_2 = \mu_3 = \mu$ or $a = 0$ and $d = 0$, the log likelihood becomes

$$Log[L(\mu_1 = \mu_2 = \mu_3 = \mu)] = -\frac{1}{2\sigma^2}\sum_{i=1}^{N}(y_i-\mu)^2 - \frac{N}{2}Log(2\pi\sigma^2) \qquad (14.43)$$

Under a null hypothesis of a pure additive model or no dominance effect $d = 0$, the log likelihood is

$$Log \, [L \, (d = 0)] \; = \; \sum_{i=1}^{N} \log \, \{ \sum_{j}^{1,3} p' \, (Q_j | M_i) \exp \left[-\frac{(y_i - \mu_j)^2}{2\sigma^2} \right] \} \; - \frac{N}{2} Log \, (2\pi\sigma^2) \quad (14.44)$$

where $p' \, (Q_j | M_i)$ is the coefficient for the proportional distribution. Their values are the modified conditional frequencies under the constraint of

$$d = 0 \text{ or } \mu_2 = 0.5 \, (\mu_1 + \mu_3)$$

and are shown as the coefficients for the QTL means in the expected value column in Table 14.14. The trait value for Qq genotype is replaced by the function of the trait values for the two homozygotes.

For the dominance model without additive genetic effects

$$\mu_1 = \mu_3 , \; a = 0.5 \, (\mu_1 - \mu_3) = 0$$

and

$$d = 0.5 \, (\mu_1 + \mu_3 - 2\mu_2) = \mu_1 - \mu_2$$

the log likelihood is

$$Log \, [L \, (a = 0)] \; = \; \sum_{i=1}^{N} \log \, \{ \sum_{j}^{1,2} p' \, (Q_j | M_i) \exp \left[-\frac{(y_i - \mu_j)^2}{2\sigma^2} \right] \} \; - \frac{N}{2} Log \, (2\pi\sigma^2) \quad (14.45)$$

where $p' \, (Q_j | M_i)$ are shown in Table 14.15. The trait values for QQ and qq are replaced by a single expectation.

The following log likelihood ratio test statistics can be used

$$G_1 = 2 \, [LogL \, (\hat{\mu}_1, \hat{\mu}_2, \hat{\mu}_3, \hat{\sigma}^2, \hat{r}_1) - LogL \, (\hat{\mu}, \hat{\sigma}^2)]$$

$$G_2 = 2 \, [LogL \, (\hat{\mu}_1, \hat{\mu}_2, \hat{\mu}_3, \hat{\sigma}^2, \hat{r}_1) - LogL \, (\hat{\mu}_1, \hat{\mu}_3, \hat{\sigma}^2, \hat{r}_1)] \quad (14.46)$$

$$G_3 = 2 \, [LogL \, (\hat{\mu}_1, \hat{\mu}_2, \hat{\mu}_3, \hat{\sigma}^2, \hat{r}_1) - LogL \, (\hat{\mu}_1, \hat{\mu}_2, \hat{\sigma}^2, \hat{r}_1)]$$

to test $H_0{:}\mu_1 = \mu_2 = \mu_3 = \mu$, $H_0{:}d = 0$ and $H_0{:}a = 0$, respectively, where $\hat{\mu}_1, \hat{\mu}_2, \hat{\mu}_3, \hat{\sigma}^2, \hat{r}_1$ and $\hat{\mu}$ are the maximum likelihood estimates for the parameters under the different hypotheses: G_1 is distributed as a chi-square variable with two degrees of freedom and G_2 and G_3 are distributed as chi-square variables with one degree of freedom.

If Equation (14.46) is evaluated using base 10 logarithm, then we have the lod scores for genetic effect, dominance effect and additive effect, respectively. They are

$$lod_1 = Log_{10}L \, (\hat{\mu}_1, \hat{\mu}_2, \hat{\mu}_3, \hat{\sigma}^2, \hat{r}_1) - Log_{10}L \, (\hat{\mu}, \hat{\sigma}^2)$$

$$lod_2 = Log_{10}L \, (\hat{\mu}_1, \hat{\mu}_2, \hat{\mu}_3, \hat{\sigma}^2, \hat{r}_1) - Log_{10}L \, (\hat{\mu}_1, \hat{\mu}_3, \hat{\sigma}^2, \hat{r}_1) \quad (14.47)$$

$$lod_3 = Log_{10}L \, (\hat{\mu}_1, \hat{\mu}_2, \hat{\mu}_3, \hat{\sigma}^2, \hat{r}_1) - Log_{10}L \, (\hat{\mu}_1, \hat{\mu}_2, \hat{\sigma}^2, \hat{r}_1)$$

In practice, the variance can be estimated using the pooled within marker class variance. By holding the variance constant, the computation needed for solving Equations (14.42), (14.44) and (14.45) for the remaining parameters is greatly reduced. It is also logical to fix recombination at a number of meaningful points, so the only unknown parameters in these equations are the putative QTL means. For example, the QTL can be tested within every recombination unit between two adjacent markers. By doing this testing for every interval, a log likelihood ratio test statistic plot or a lod score plot can be obtained.

Table 14.14 Expected phenotypic value for additive model or $d = 0$.

Marker Genotype	Count	Expected Value (g_i)
AABB	n_1	μ_1
AABb	n_2	$[\,(2r_2 + r_1)\,/\,(2r)\,]\,\mu_1 + [\,r_1\,/\,(2r)\,]\,\mu_3$
AAbb	n_3	$[\,(r_2^2 + r_1 r_2)\,/\,r^2\,]\,\mu_1 + [\,(r_1^2 + r_1 r_2)\,/\,r^2\,]\,\mu_3$
AaBB	n_4	$[\,(2r_1 + r_2)\,/\,(2r)\,]\,\mu_1 + [\,r_2\,/\,(2r)\,]\,\mu_3$
AaBb	n_5	$0.5\,(\mu_1 + \mu_3)$
Aabb	n_6	$[\,r_2\,/\,(2r)\,]\,\mu_1 + [\,(2r_1 + r_2)\,/\,(2r)\,]\,\mu_3$
aaBB	n_7	$[\,(r_1^2 + r_1 r_2)\,/\,r^2\,]\,\mu_1 + [\,(r_2^2 + r_1 r_2)\,/\,r^2\,]\,\mu_3$
aaBb	n_8	$[\,r_1\,/\,(2r)\,]\,\mu_1 + [\,(2r_2 + r_1)\,/\,(2r)\,]\,\mu_3$
aabb	n_9	μ_3

Table 14.15 Expected phenotypic value for dominance model or $a = 0$.

Marker Genotype	Count	Expected Value (g_i)
AABB	n_1	μ_1
AABb	n_2	$(r_2/r)\,\mu_1 + (r_1/r)\,\mu_2$
AAbb	n_3	$[\,(r_1^2 + r_2^2)\,/\,r^2\,]\,\mu_1 + [\,2r_1 r_2 / r^2\,]\,\mu_2$
AaBB	n_4	$(r_1/r)\,\mu_1 + (r_2/r)\,\mu_2$
AaBb	n_5	$1/[\,(1-r)^2 + r^2\,]\,\{2r_1 r_2 \mu_1 + [\,(1-r)^2 + r_1^2 + r_2^2\,]\,\mu_2\}$
Aabb	n_6	$(r_2/r)\,\mu_2 + (r_1/r)\,\mu_1$
aaBB	n_7	$[\,(r_1^2 + r_2^2)\,/\,r^2\,]\,\mu_1 + [\,2r_1 r_2 / r^2\,]\,\mu_2$
aaBb	n_8	$(r_1/r)\,\mu_2 + (r_2/r)\,\mu_1$
aabb	n_9	μ_1

Dominant Markers

Equations for interval mapping using codominant markers in an F2 can also be used for dominant markers. The corresponding expected trait values for the dominant marker genotypes under the null hypothesis $d = 0$ or $a = 0$ are shown in Tables 14.16 and 14.17. When $d = 0$ and $a = 0$, the expected values for the marker genotypes are the same as the population mean. The log likelihood test statistics (Equations (14.46) and (14.47)) can be implemented for the dominant markers. However, the power of the test using dominant markers is lower than that of the test using codominant markers.

The dominant marker may present a problem due to missing information. Jiang and Zeng (1997) derived a general algorithm to calculate the probability

Table 14.16 Expected phenotypic value for marker classes when dominant markers are used under $d = 0$.

Marker Genotype	Count	Expected Value g_i
A_B_	n_1	$[\,(2-r)\,/\,(3-2r+r^2)\,]\,\mu_1 + [\,(1-r+r^2)\,/\,(3-2r+r^2)\,]\,\mu_3$
A_bb	n_2	$r_2/\,[\,r\,(2-r)\,]\,\mu_1 + [\,r\,(2-r)-r_2\,]\,/\,[\,r\,(2-r)\,]\,\mu_3$
aaB_	n_3	$r_1/\,[\,r\,(2-r)\,]\,\mu_1 + [\,r\,(2-r)-r_1\,]\,/\,[\,r\,(2-r)\,]\,\mu_3$
aabb	n_4	μ_3

Table 14.17 Expected phenotypic value for marker classes when dominant markers are used under $a = 0$.

Marker Genotype	Count	Expected Value (g_i)
A_B_	n_1	$1/\,(3-2r+r^2)\,[\,(1-r^2+4r_1r_2)\,\mu_1 + 2\,(1-r+r_1^2+r_2^2)\,\mu_2\,]$
A_bb	n_2	$1/\,[\,r\,(2-r)\,]\,\{\,[\,r_2^2+r_1\,(2-2r+r_1)\,]\,\mu_1 + 2r_2\,(1-r+r_1)\,\mu_2\,\}$
aaB_	n_3	$1/\,[\,r\,(2-r)\,]\,\{\,[\,r_1^2+r_2\,(2-2r+r_2)\,]\,\mu_1 + 2r_1\,(1-r+r_2)\,\mu_2\,\}$
aabb	n_4	μ_1

distribution of putative QTL genotypes at a given genomic position conditional on observed marker phenotypes. The algorithm takes information from adjacent markers and uses a recursive approach through a Markov chain to obtain the probability distribution of the QTL genotype. Jiang and Zeng concluded that dominant markers can be useful with some codominant markers for QTL mapping.

Table 14.18 Dummy variables for a regression approach to QTL interval mapping using F2 progeny.

Marker Genotype	X_1	X_2	X_3	X_4	X_5	X_6	X_7	X_8	X_9
AABB	1	0	0	0	0	0	0	0	0
AABb	0	1	0	0	0	0	0	0	0
AAbb	0	0	1	0	0	0	0	0	0
AaBB	0	0	0	1	0	0	0	0	0
AaBb	0	0	0	0	1	0	0	0	0
Aabb	0	0	0	0	0	1	0	0	0
aaBB	0	0	0	0	0	0	1	0	0
aaBb	0	0	0	0	0	0	0	1	0
aabb	0	0	0	0	0	0	0	0	1

14.3.3 REGRESSION APPROACH

Nonlinear Regression Approach (Codominant Markers)

To follow the rationale for the regression models using backcross progeny, the models for F2 progeny will be constructed. The marker genotypes for F2 progeny can be coded as shown in Table 14.18. A nonlinear regression model can be written as

$$y_j = \sum_{i=1}^{9} X_{ij} g_i + \varepsilon_j \tag{14.48}$$

where the Xs are the dummy variables for the nine marker genotypes as listed in Table 14.18, g_i is the expected trait value for marker genotypic class i as shown in Table 14.11, r, r_1 and r_2 were defined as the recombination fractions between the two markers, the putative QTL and marker A and the putative QTL and marker B, respectively (Figure 14.1) and ε_j is the experimental error associated with the individual j. As for the backcross progeny, the recombination fractions can be reparameterized. If the recombination fraction between marker A and the putative QTL is defined as ρr, then we have $\rho = r_1/r$ and $0 \le \rho \le 1$. Four unknown parameters are involved in the model: the three QTL genotypic means and the parameter for relative QTL position.

The least-square estimates of the unknown parameters in Equation (14.48) can be solved using an iterative Gauss-Newton algorithm as explained for the backcross progeny. The same is true for the hypothesis tests. The difference between the backcross and the F2 progeny is the number of unknown parameters, the partial derivatives and the contrasts for the hypothesis tests. For the F2 progeny, the vector for the unknown parameters is

$$\theta = \begin{bmatrix} \mu_1 \\ \mu_2 \\ \mu_3 \\ \rho \end{bmatrix}_{4 \times 1}$$

The derivatives needed for the least square estimation using F2 progeny are

$$\partial f(\theta)/\partial\mu_1 = X_1 + (1-\rho) X_2 + (1-\rho)^2 X_3 + \rho X_4$$
$$+ r^2 \rho (1-\rho) / [(1-r)^2 + r^2] X_5 + \rho^2 X_7$$

$$\partial f(\theta)/\partial\mu_2 = \rho (X_2 + X_8) + 2\rho (1-\rho) (X_3 + X_7) + (1-\rho) (X_4 + X_6)$$
$$+ [1 + (1-r)^2 - 2r\rho (1-r\rho)] / [(1-r)^2 + r^2] X_5$$

$$\partial f(\theta)/\partial\mu_3 = \rho X_3 + r^2 \rho (1-\rho) / [(1-r)^2 + r^2] X_5 + \rho X_6 + (1-\rho)^2 X_7 + (1-\rho) X_8$$

$$\partial f(\theta)/\partial\rho = (\mu_1 - \mu_2) (X_4 - X_2) + (\mu_2 - \mu_3) (X_8 - X_6)$$
$$+ [-2 (1-\rho) \mu_1 + 2 (1-2\rho) \mu_2 + 2\rho\mu_3] X_3$$
$$+ [2\rho\mu_1 + 2 (1-2\rho) \mu_2 - 2 (1-\rho) \mu_1] X_7$$
$$+ (r^2 (1-2\rho) (\mu_1 + 2\mu_2 + \mu_3)) / [(1-r)^2 + r^2] X_5$$

For a single-QTL and single-environment model, contrasts can be tested for the additive and dominance effects. They are

$$h_{additive}(\theta) = \mu_1 - \mu_3$$
$$H_{additive} = \partial h_{additive}(\theta)/\partial(\mu_1, \mu_2, \mu_3) = (1, 0, -1)$$
$$h_{dominant}(\theta) = \mu_1 - 2\mu_2 + \mu_3$$
$$H_{dominant} = \partial h_{dominant}(\theta)/\partial(\mu_1, \mu_2, \mu_3) = (1, -2, 1)$$

The contrast can be tested using the Wald statistic of Equation (14.19).

The model based on Equation (14.48) can be easily extended to multiple-environment situations. For a two-environment experiment, the standard contrasts are listed in Table 14.19.

Table 14.19 Contrasts for a single-QTL with the single and two environment(s).

Contrast		Environment 1			Environment 2		
		μ_1	μ_2	μ_3	μ_1	μ_2	μ_3
Single Environment	Additive	1	0	-1			
	Dominance	-1	2	-1			
	Additive				1	0	-1
	Dominance				-1	2	-1
Two Environments	Additive	1	0	-1	1	0	-1
	Dominance	-1	2	-1	-1	2	-1
	Environment	1	1	1	-1	-1	-1
	A x E	1	0	-1	-1	0	1
	D x F	-1	2	-1	-1	2	-1

Linear Regression (Codominant Markers)

As shown by Knapp *et al.* (1990), the model based on Equation (14.48) also can be written as a linear model. It is

$$y_j = \sum_i \theta_i X_i + \varepsilon_j \qquad (14.49)$$

where the θs are the means for each of the nine marker genotypic means and by setting them equal to the expected trait values (Table 14.11), we have

$$\theta_1 = \mu_1$$
$$\theta_2 = (1-\rho)\mu_1 + \rho\mu_2$$
$$\theta_3 = (1-\rho)^2\mu_1 + 2\rho(1-\rho)\mu_2 + \rho^2\mu_3$$
$$\theta_4 = \rho\mu_1 + (1-\rho)\mu_2$$
$$\theta_5 = \frac{r^2\rho(1-\rho)}{(1-r)^2 + r^2}(\mu_1 + \mu_3) + \frac{[(1-r)^2 + r^2(1-2\rho+\rho^2)]}{(1-r)^2 + r^2}\mu_2 \qquad (14.50)$$
$$\theta_6 = (1-\rho)\mu_2 + \rho\mu_3$$
$$\theta_7 = \rho^2\mu_1 + 2\rho(1-\rho)\mu_2 + (1-\rho)^2\mu_3$$
$$\theta_8 = \rho\mu_2 + (1-\rho)\mu_3$$
$$\theta_9 = \mu_3$$

$\hat{\theta}_1$ and $\hat{\theta}_9$ can be used to estimate μ_1 and μ_3, respectively. Knapp *et al.* (1990) gave estimators of μ_2 and ρ as

$$\hat{\mu}_2 = \frac{1}{2}(\hat{\theta}_2 + \hat{\theta}_4 + \hat{\theta}_6 + \hat{\theta}_8 - \hat{\theta}_1 - \hat{\theta}_9)$$

$$\hat{\rho} = \frac{\hat{\theta}_2 + \hat{\theta}_8 - \hat{\theta}_1 - \hat{\theta}_9}{\hat{\theta}_2 + \hat{\theta}_4 + \hat{\theta}_6 + \hat{\theta}_8 - 2\hat{\theta}_1 - 2\hat{\theta}_9} \qquad (14.51)$$

Linear Regression (Dominant Markers)

Four dummy variables can be coded for each of the four possible marker genotypes (A_B_, A_bb, aaB_ and aabb) (see Table 14.18 for codominant markers). The model based on Equation (13.50) can be fitted for the dominant markers as

$$y_j = \sum_{i=1}^{4} \theta_i X_i + \varepsilon_j \tag{14.52}$$

The expectations of the regression coefficients are (see Table 14.13)

$$\theta_1 = \frac{1 - r^2 + 2r_1 r_2}{3 - 2r + r^2} \mu_1 + \frac{2\,(1 - r + r_1^2 + r_2^2)}{3 - 2r + r^2} \mu_2 + \frac{2r_1 r_2}{3 - 2r + r^2} \mu_3$$

$$\theta_2 = \frac{r_2^2}{r\,(2 - r)} \mu_1 + \frac{2r_2\,(1 - r + r_1)}{r\,(2 - r)} \mu_2 + \frac{r_1\,(2 - 2r + r_1)}{r\,(2 - r)} \mu_3 \tag{14.53}$$

$$\theta_3 = \frac{r_1^2}{r\,(2 - r)} \mu_1 + \frac{2r_1\,(1 - r + r_2)}{r\,(2 - r)} \mu_2 + \frac{r_2\,(2 - 2r + r_2)}{r\,(2 - r)} \mu_3$$

$$\theta_4 = \mu_3$$

Equation (14.52) can be re-written as

$$\theta_1\,(3 - 2r + r^2) = (1 - r^2 + 2r_1 r_2)\,\mu_1 + 2\,(1 - r + r_1^2 + r_2^2)\,\mu_2 + (2r_1 r_2)\,\mu_3$$

$$\theta_2 r\,(2 - r) = r_2^2 \mu_1 + 2r_2\,(1 - r + r_1)\,\mu_2 + r_1\,(2 - 2r + r_1)\,\mu_3$$

$$\theta_3 r\,(2 - r) = r_1^2 \mu_1 + 2r_1\,(1 - r + r_2)\,\mu_2 + r_2\,(2 - 2r + r_2)\,\mu_3 \tag{14.54}$$

$$\theta_4 = \mu_3$$

and Equation (14.51) can then be correspondingly changed to

$$y_j = \theta_1' X_1' + \theta_2' X_2' + \theta_3' X_3' + \theta_4' X_4' + \varepsilon_j \tag{14.55}$$

where

$$X_1' = X_1 / (3 - 2r + r^2)$$

$$X_2' = X_2 / [\,r\,(2 - r)\,]$$

$$X_3' = X_3 / [\,r\,(2 - r)\,] \tag{14.56}$$

$$X_4' = X_4$$

If we assume the linkage map density is high (recombination fraction between the two markers is low) and ignore the products involving the recombination fraction between the markers and the putative QTL, Equation (14.53) can be simplified to

$$\theta_1' \approx (1 - r^2)\,\mu_1 + 2\,(1 - r)\,\mu_2$$

$$\theta_2' \approx 2r_2 \mu_2 + 2r_1 \mu_3$$

$$\theta_3' \approx 2r_1 \mu_2 + 2r_2 \mu_3 \tag{14.57}$$

$$\theta_4' = \mu_3$$

Solving this equation, we can obtain the estimates

$$\hat{\mu}_1 = [\hat{\theta}_1' - 2\,(1 - r)]\,[\,(\hat{\theta}_2' + \hat{\theta}_3') / (2r) - \hat{\theta}_4'\,] / (1 - r^2)$$

$$\hat{\mu}_2 = (\hat{\theta}_2' + \hat{\theta}_3') / (2r) - \hat{\theta}_4'$$

$$\hat{\mu}_3 = \hat{\theta}_4' \tag{14.58}$$

$$\hat{r}_1 = (\hat{\theta}_3' - 2\hat{\theta}_4') / [\,(\hat{\theta}_2' + \hat{\theta}_3') / (2r) - \hat{\theta}_4'\,]$$

14.3.4 PROBLEMS WITH THE SIMPLE INTERVAL MAPPING APPROACHES

Interval mapping using the likelihood or the regression approaches is the most commonly used methodology in QTL mapping. However, problems exist, such as

(1) The number of QTLs cannot be resolved

(2) The locations of QTLs are sometimes not well resolved and the exact positions of the QTLs cannot be determined

(3) The statistical power is still relatively low

There are three possible reasons for these problems:

(1) linked QTLs

(2) QTL interactions

(3) limited information in the model

The test statistic plot (log likelihood ratio test statistic, lod score or Wald statistic) may only show one peak even when there are several QTLs in a fairly large genome segment or when the QTLs are located on the segment in coupling (QTL effects are in the same direction). The peak can be low and very wide. This is one of the reasons that the precise location of QTLs cannot be determined.

When QTLs are linked in repulsion (QTL effects are in different directions), the peak of the test statistic plot can be lowered or even eliminated. When there are other QTLs segregating in the mapping population, the genetic effects of these QTLs and their interactions with the QTL in the interval mapping segment are pooled into the experimental error (genetic background effects) if they are not included in the model. All these problems reduce statistical power.

One of the most important reasons for these problems is that interval mapping is not independent for different segments if more than one QTL exists. The assumption for simple interval mapping is that a single segregating QTL influences the trait.

Alternative approaches are needed for QTL mapping. In this section, methodology for controlling genetic background effects and for solving the non-independence problem will be introduced. Multiple-QTL models that include QTL interactions will be discussed along with methods for implementation and possible problems (Chapter 17).

14.4 COMPOSITE INTERVAL MAPPING

14.4.1 MODEL

Composite interval mapping (CIM) is a combination of simple interval mapping and multiple linear regression (Zeng 1993 and 1994). For CIM analysis on a segment between markers i and $i+1$, using backcross progeny, the statistical model is

$$y_j = b_0 + b_i X_{ij} + \sum_{k \neq i, i+1} b_k X_{kj} + \varepsilon_j \qquad (14.59)$$

where y_j is the trait value for individual j, b_0 is the intercept of the model, b_i is the genetic effect of the putative QTL located between markers i and $i+1$, X_{ij} is a dummy variable taking 1 for marker genotype AABB, 0 for AaBb, 1 with a probability of $1 - r_1/r = 1 - \rho$ and 0 with a probability $r_1/r = \rho$ for marker genotype AaBB, 1 with probability of ρ and 0 with a probability of $1 - \rho$ for marker genotype AABb (see Tables 14.10 and 14.11), r is the recombination fraction between the two markers, r_1 is the recombination between the first marker and the putative QTL, b_k is the partial regression coefficient of the trait value on marker k, X_{kj} is dummy variable for marker k and individual j, taking 1 if the marker has genotype AA and 0 for Aa and ε_j is a residual from the model. If ε_j is normally distributed with mean zero and variance σ^2, the likelihood function for the CIM is

$$
L = \frac{1}{\{\sqrt{2\pi}\sigma\}^N} \exp\left\{\frac{1}{-2\sigma^2}\left[\sum_{j=1}^{n_1}(y_{1j}-\mu_1)^2 + \sum_{j=1}^{n_4}(y_{4j}-\mu_2)^2\right]\right\} \tag{14.60}
$$
$$
\times \prod_{j=1}^{n_2}\left\{(1-\rho)\exp\left[-\frac{(y_{2j}-\mu_1)^2}{2\sigma^2}\right] + \rho\exp\left[-\frac{(y_{2j}-\mu_2)^2}{2\sigma^2}\right]\right\}
$$
$$
\times \prod_{j=1}^{n_3}\left\{\rho\exp\left[-\frac{(y_{3j}-\mu_1)^2}{2\sigma^2}\right] + (1-\rho)\exp\left[-\frac{(y_{3j}-\mu_2)^2}{2\sigma^2}\right]\right\}
$$

where y_{ij} is an observed trait phenotypic value for marker class i (AABB, AABb, AaBB or AaBb) and the jth individual in the marker class and $r_1/r = \rho$ is the relative position of the putative QTL on the genome segment between the two markers (see Tables 14.1 and 14.1 for the other notations). Equation (14.60) is the same as Equation (14.3) except for the definitions of μ_1 and μ_2. In Equation (14.3), μ_1 and μ_2 were defined as the means of the two possible genotypes for the putative QTL between the two markers. Here, μ_1 and μ_2 are defined using the regression model of (14.59) as

$$
\begin{aligned}
\mu_1 &= b_0 + b_i + \sum_{k\neq i,i+1} b_k X_{kj} \qquad for \qquad X_{ij} = 1\\
\mu_2 &= b_0 + \sum_{k\neq i,i+1} b_k X_{kj} \qquad\quad for \qquad X_{ij} = 0
\end{aligned} \tag{14.61}
$$

The logarithm of Equation (14.60) is

$$
Log(L) = -\frac{N}{2}Log(2\pi\sigma^2) - \frac{1}{2\sigma^2}\left[\sum_{j=1}^{n_1}(y_{1j}-\mu_1)^2 + \sum_{j=1}^{n_4}(y_{4j}-\mu_2)^2\right] \tag{14.62}
$$
$$
+ \sum_{j=1}^{n_2}\log\left\{(1-\rho)\exp\frac{(y_{2j}-\mu_1)^2}{-2\sigma^2} + \rho\exp\frac{(y_{2j}-\mu_2)^2}{-2\sigma^2}\right\}
$$
$$
+ \sum_{j=1}^{n_3}\log\left\{\rho\exp\frac{(y_{3j}-\mu_1)^2}{-2\sigma^2} + (1-\rho)\exp\frac{(y_{3j}-\mu_2)^2}{-2\sigma^2}\right\}
$$

The derivative of equation (14.62) with respect to b_i is

$$
\frac{\partial Log(L)}{\partial b_i} = \frac{1}{\sigma^2}\sum_{j=1}^{N} P_j\left(y_j - b_0 - b_i - \sum_{k\neq i,i+1} b_k X_{kj}\right) \tag{14.63}
$$

where the P_j is defined as

$$
P_j = \frac{p_{1j}\exp[(y_j-\mu_1')^2/(-2\sigma^{2'})]}{p_{1j}\exp[(y_j-\mu_1')^2/(-2\sigma^{2'})] + p_{0j}\exp[(y_j-\mu_2')^2/(-2\sigma^{2'})]} \tag{14.64}
$$

where p_{1j} and p_{0j} are the conditional probabilities of the dummy variable for the marker interval X_{ij} taking 1 and 0, respectively, conditional on the marker genotype of the individual j and μ_1', μ_1', σ^2 are updated ML estimates of the parameters. For the four possible marker genotypes, the conditional probabilities are

$$
\begin{array}{ccc}
Marker & p_{1j} & p_{0j} \\
AABB & 1 & 0 \\
AABb & 1-\rho & \rho \\
AaBB & \rho & 1-\rho \\
AaBb & 0 & 1
\end{array}
\tag{14.65}
$$

(see Tables 14.10 and 14.11). By setting the derivative of Equation (14.62) to zero, we have

$$
\sum_{j=1}^{N}\left(y_j - b_0 - b_i - \sum_{k \neq i, i+1} b_k X_{kj}\right) P_j = 0
\tag{14.66}
$$

14.4.2 SOLUTIONS

Zeng (1994) gives the solution of b_i in matrix notation as

$$
\hat{b}_i = (Y - X\hat{B})' \hat{P} / \hat{c}
\tag{14.67}
$$

where Y is a vector of $N \times 1$ for the observed trait values, X is a matrix of $N \times (l-1)$ for the dummy variables for the intercept and the l markers except the markers i and $i+1$, $\hat{B}$ is a vector of $(l-1) \times 1$ of the maximum likelihood estimates of the intercept and partial regression coefficients for the l markers except the markers i and $i+1$, $\hat{P}$ is a vector of $N \times 1$ with values defined by Equation (14.64) and

$$
\hat{c} = \sum_{j=1}^{N} P_j
$$

By setting the derivatives of Equation (14.62) with respect to the partial regression coefficients for the rest of the markers and the variance to zero, Zeng (1994) gives the maximum likelihood estimates. The derivative with respect to the partial regression coefficients is

$$
\begin{aligned}
\frac{\partial Log(L)}{\partial b_k} = & \frac{1}{\sigma^2} \sum_{j=1}^{n_1} X_{kj} P_j (y_{1j} - \mu_1) + \frac{1}{\sigma^2} \sum_{j=1}^{n_4} X_{kj} P_j (y_{4j} - \mu_2) \\
& + \frac{1}{\sigma^2} \sum_{j=1}^{n_2} X_{kj} [P_j (y_{2j} - \mu_1) + (1 - P_j)(y_{2j} - \mu_2)] \\
& + \frac{1}{\sigma^2} \sum_{j=1}^{n_3} X_{kj} [P_j (y_{3j} - \mu_1) + (1 - P_j)(y_{3j} - \mu_2)]
\end{aligned}
\tag{14.68}
$$

The derivative with respect to the variance is

$$
\begin{aligned}
\frac{\partial Log(L)}{\partial \sigma^2} = & \frac{1}{2\sigma^4} \sum_{j=1}^{n_1} X_{kj} P_j (y_{1j} - \mu_1)^2 + \frac{1}{2\sigma^4} \sum_{j=1}^{n_4} X_{kj} P_j (y_{4j} - \mu_2)^2 - \frac{N}{2\sigma^2} \\
& + \frac{1}{2\sigma^4} \sum_{j=1}^{n_2} X_{kj} [P_j (y_{2j} - \mu_1)^2 + (1 - P_j)(y_{2j} - \mu_2)^2] \\
& + \frac{1}{2\sigma^4} \sum_{j=1}^{n_3} X_{kj} [P_j (y_{3j} - \mu_1)^2 + (1 - P_j)(y_{3j} - \mu_2)^2]
\end{aligned}
\tag{14.69}
$$

The solutions are

$$\hat{B} = (X'X)^{-1}X'(Y - \hat{P}\hat{b}_i)$$

$$\hat{\sigma}^2 = \frac{1}{N}[(Y - X\hat{B})'(Y - X\hat{B}) - \hat{c}\hat{b}_i^2] \tag{14.70}$$

Zeng also gives the maximum likelihood estimate of the relative position of the putative QTL as

$$\hat{\rho} = \frac{\sum_{i=1}^{n_2}(1 - P_j) + \sum_{j=1}^{n_3}P_j}{n_2 + n_3} \tag{14.71}$$

where the summations are only for the individuals who have marker genotypes AABb and AaBB.

In practice, CIM can be implemented using an iterative ECM (expectation/conditional maximization) algorithm (Meng and Rubin 1993). Zeng (1994) suggests using $\hat{b}_i^0 = 0$ or the least square estimates of b_i and B using $X_j = p_{1j}$. The relative position of the putative QTL can be divided into biologically meaningful points and treated as known constants. For each point, the iteration starts with the E-step: computing the probability of $X_{ij} = 1$ or the genotype of the putative QTL being QQ using Equation (14.65) and then performing the three CM-step to estimate the QTL effect (b_i), the partial regression coefficients (B) and the variance (σ^2) for the next round of iteration using Equations (14.67) and (14.70).

14.4.3 HYPOTHESIS TEST

For each of the relative positions of the putative QTL within the interval flanked by the two markers, the log likelihood ratio test statistic

$$G = 2\{Log[L(\hat{b}_i, \hat{B}, \hat{\sigma}^2)] - Log_{b_i = 0}[L(\hat{B}_0, \hat{\sigma}_0^2)]\} \tag{14.72}$$

where

$$Log[L(\hat{b}_i, \hat{B}, \hat{\sigma}^2)]$$

is the log likelihood of Equation (14.62) for the model based on Equation (14.59) evaluated using the maximum likelihood estimates of the parameters in Equations (14.67) and (14.70) and

$$Log_{b_i = 0}[L(\hat{B}_0, \hat{\sigma}_0^2)]$$

is the log likelihood for the model based on Equation (14.59) under the null hypothesis $b_i = 0$, which is

$$Log_{b_i = 0}(L) = -\frac{1}{2\sigma^2}\sum_j\left(y - b_0 - \sum_{k \neq i, i+1}b_kX_{kj}\right)^2 \tag{14.73}$$

evaluated using the maximum likelihood estimates of the parameters under the null hypothesis ($\hat{B}_0, \hat{\sigma}_0^2$). The estimates are

$$\hat{B}_0 = (X'X)^{-1}X'Y \tag{14.74}$$

$$\hat{\sigma}_0^2 = (Y - X\hat{B}_0)'(Y - X\hat{B}_0)/N$$

Zeng extended Equation (14.72) as

$$G = N(Log\hat{\sigma}^2 - Log\hat{\sigma}_0^2) - \sum_{j=1}^{n_2} \{2\log[(1-\rho)\exp(0.5\hat{d}_j) + \rho] - \hat{P}_j\hat{d}_j\} \tag{14.75}$$

$$- \sum_{j=1}^{n_3} \{2\log[\rho\exp(0.5\hat{d}_j) + (1-\rho)] - \hat{P}_j\hat{d}_j\}$$

where

$$\hat{d}_j = \frac{\hat{b}_i}{\hat{\sigma}^2}\left[2\left(y_j - b_0 - \sum_{k \neq i, i+1} b_k X_{kj}\right) - \hat{b}_i\right] \tag{14.76}$$

If base 10 logarithm is used, the lod score for the hypothesis is

$$lod = Log_{(10)}[L(\hat{b}_i, \hat{B}, \hat{\sigma}^2)] - Log_{(10)b_i=0}[L(\hat{B}_0, \hat{\sigma}_0^2)] \tag{14.77}$$

The markers in the CIM model of Equation (14.59) can control the residual genetic background when they are linked to QTLs. When the size of the mapping population is small and the number of markers is large, it is not practical to fit all the markers in the model. A set of markers which are close to possible QTLs should be fitted in the model. However, the number and locations of QTLs are unknown before the analysis is done. Zeng (1994) suggested a stepwise regression approach to screen markers for use in the model. A small number of markers in each linkage group can be used. In practice, mapping data usually contains missing observations for the marker genotypes. When the number of markers is large, the effective mapping population size is usually small. This situation creates a serious problem for fitting a reasonable model that can control for residual genetic effects.

14.4.4 CIM USING REGRESSION
The CIM can be implemented using the linear regression model for interval mapping and the multiple linear model to control for the residual genetic effects. The model can be written as

$$y_j = X_{i1j}\mu_1 + X_{i2j}[(1-\rho)\mu_1 + \rho\mu_2] + X_{i3j}[\rho\mu_1 + (1-\rho)\mu_2] \tag{14.78}$$

$$+ X_{i4j}\mu_2 + \sum_{k \neq i, i+1} b_k X_{kj} + \varepsilon_j$$

where $X_i s, \mu_1, \mu_2$ and ρ were defined in Tables 14.10, 14.11 and Equation (14.10) and the rest of the notation is the same as in Equation (14.59). Following Knapp et al. (1990)

$$\theta_1 = \mu_1$$

$$\theta_2 = (1-\rho)\mu_1 + \rho\mu_2$$

$$\theta_3 = \rho\mu_1 + (1-\rho)\mu_2$$

$$\theta_4 = \mu_2$$

and $\theta_1 + \theta_4 = \theta_2 + \theta_3$, Equation (14.78) can be rearranged as

$$y_j = \theta_1 (X_{i1j} + X_{i3j}) + \theta_2 (X_{i2j} - X_{i3j}) + \theta_4 (X_{i3j} + X_{i4j}) + \sum_{k \neq i, i+1} b_k X_{kj} + \varepsilon_j \qquad (14.79)$$

If ρ is treated as a known parameter and $\theta_2 = (1-\rho)\mu_1 + \rho\mu_2$, Equation (14.79) becomes

$$y_j = \theta_1 [X_{i1j} + (1-\rho) X_{i2j} + \rho X_{i3j}] \qquad (14.80)$$
$$+ \theta_4 [\rho X_{i2j} + (1-\rho) X_{i3j} + X_{i4j}] + \sum_{k \neq i, i+1} b_k X_{kj} + \varepsilon_j$$

This is a simple linear multiple regression model. The estimates of the two genotypic means for the putative QTL are located at position ρ between the two markers

$$\begin{cases} \hat{\theta}_1 = \hat{\mu}_1 \\ \hat{\theta}_4 = \hat{\mu}_2 \end{cases} \qquad (14.81)$$

Under the null hypothesis $\mu_1 = \mu_2$ or $\theta_1 = \theta_4 = \theta$, Equation (14.80) becomes

$$y_{j(\theta_1 = \theta_4 = \theta)} = \theta [X_{i1j} + X_{i2j} + X_{i3j} + X_{i4j}] + \sum_{k \neq i, i+1} b_k X_{kj} + \varepsilon_j$$
$$= \theta + \sum_{k \neq i, i+1} b_k X_{kj} + \varepsilon_j \qquad (14.82)$$

The hypothesis test can be implemented using the log likelihood ratio test statistic

$$G = \frac{SSE_{reduced} - SSE_{full}}{SSE_{full}/df_{Efull}} \qquad (14.83)$$

where SSEs are the residual sum of squares for the full model of Equation (14.80) and reduced model of Equation (14.82). The test statistic can be estimated for each of the positions on the genome.

Example: CIM Using Regression

CIM using the linear regression models based on Equations (14.80) and (14.82) was performed for the example in Tables 12.5 to 12.7 for the malt extract trait in Montana. The analyses were done for every 1% recombination point by treating the putative QTL position parameters as known constants. The data contain 292 markers for the seven chromosomes of barley. It is impossible to fit all the markers into the models because the mapping population only contains 150 doubled haploid lines. Following the suggestion of Zeng (1994), a subset of markers on different chromosomes (other than chromosome IV) was picked using stepwise regression. Markers Adh7, ABC165, ABR337 and ABR337 (on chromosomes I, II, V and VII, respectively) (Table 14.20) were found to be significantly associated with the malt extract trait and selected as markers to fit the models. On chromosome IV, markers were selected based on their significance of association with the trait and their position relative to each of the 25 segments on the genome (Table 12.7).

Figure 14.8 shows the log likelihood ratio test statistic plot for each genome position on chromosome IV. Compared with single marker analysis (Figure 13.4) and simple interval mapping (Figure 14.4), the CIM approach gives higher resolution of QTL location. All three approaches identify a QTL as

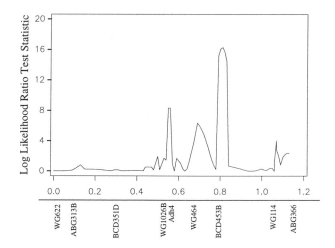

Figure 14.8 The log likelihood ratio test statistics are plotted against genome locations on the chromosome IV of barley using the composite interval mapping approach for the data in Tables 12.5 to 12.7. The results were obtained using Equations (14.80) and (14.82). The markers used to control the residual genetic background and experimental error are listed in Table 14.18.

a peak near marker BCD453B. However, a peak near marker WG1026B and a peak between these markers were identified differently. Both single marker analysis and simple interval mapping identify the peaks located near markers WG1026B and ABR315 (Figures 13.4 and 14.4). However, the CIM approach identifies the peak near markers Adh4 (closely linked with WG1026B by recombination fraction of 3% apart) and WG464 (about 8% recombination units apart from ABR315).

The difference for the peak near WG1026B and Adh4 can be explained by increased QTL resolution with this method. Single marker analysis and simple interval mapping cannot resolve the precise location for the potential QTL in that region of genome. The composite interval mapping approach can use the markers more efficiently as boundaries for locating QTLs. In this case, Adh4 is a marker at a boundary point. The potential QTL is located between Adh4 and ABA3 in a region about 2 cM long.

Regarding the difference for the peak in the middle, the explanation is that the peak of ABR315, using the single marker analysis and the simple interval mapping, is a result of linked markers Adh4 and BCD453B, which are significantly associated with potential QTLs. When the analysis for the segment between ABR315 and WG464 conditional on Adh4 and BCD453B (Adh4 and BCD453B were put in the model to control the genetic background), the peak disappears. For the segment between WG464 and BCD453B, the peak shows up when the analysis is conditional on the adjacent possible QTLs. This is a sign of repulsion linkage phase among the possible QTLs. However, a marker or several markers between WG464 and BCD453B and more individuals in the mapping population are needed (depending on where exactly the QTLs are located in the segment) because the next closest potential major QTL is located in the next segment between markers BCD453B and ABG472. WG464 is about 10 map units from BCD453B of the next peak and there is no marker between WG464 and the marker (BCD453B) at the boundary between two possible QTLs.

Table 14.20 Markers used in the composite interval mapping using the linear regression models for malt extract.

Segment	Markers on Chromosome IV Used in the Models				
WG622 -ABG313B	BCD453B	Adh4	CDO669	BCD402B	BCD265B
ABG313B-CDO669	BCD453B	Adh4	BCD402B	WG622	BCD265B
CDO669 -BCD402B	BCD453B	Adh4	BCD351D	ABG313B	TubA1
BCD402B-BCD351D	BCD453B	Adh4	BCD265B	CDO669	ABG313B
BCD351D-BCD265B	BCD453B	Adh4	Dhn6	BCD402B	CDO669
BCD265B-TubA1	BCD453B	Adh4	Dhn6	BCD351D	CDO669
TubA1 -Dhn6	BCD453B	Adh4	ABG003	BCD265B	BCD402B
Dhn6 -ABG003	BCD453B	ABA003	Adh4	TubA1	BCD265B
ABG003 -WG1026B	BCD453B	ABA003	Adh4	Dhn6	TubA1
WG1026B-Adh4	BCD453B	ABR315	ABG003	ABG484	Dhn6
Adh4 -ABA003	BCD453B	ABR315	ABG003	WG1026B	TubA1
ABA003 -ABG484	BCD453B	Adh4	Pgk2A	ABG003	Dhn6
ABG484 -Pgk2A	BCD453B	Adh4	ABR315	ABA003	ABG003
Pgk2A -ABR315	BCD453B	Adh4	WG464	ABG484	ABG003
ABR315 -WG464	BCD453B	Adh4	ABG472	Pgk2A	ABG500B
WG464 -BCD453B	ABG472	Adh4	iAco2	ABR315	ABG500B
BCD453B-ABG472	ABG500B	Adh4	iAco2	WG464	ABG500B
ABG472 -iAco2	BCD453B	Adh4	ABG498	ABG500B	WG464
iAco2 -ABG500B	BCD453B	Adh4	ABG498	ABG472	WG114
ABG500B-ABG498	BCD453B	Adh4	ABG054	iAco2	ABG394
ABG498 -WG114	BCD453B	Adh4	ABG054	ABG500B	ABG397
WG114 -ABG054	BCD453B	Adh4	ABG397	ABG500B	ABG366
ABG054 -ABG394	BCD453B	Adh4	ABG397	ABG498	ABG366
ABG394 -ABG397	BCD453B	Adh4	ABG366	WG114	ABG498
ABG397 -ABG366	BCD453B	Adh4	ABG054	WG114	ABG498

Table 14.21 Means and frequencies for marker combinations among Adh4, BCD453B, and WG464 for Steptoe and Morex malt extract genotypes.

Adh4 (A)	BCD453B (B)	WG464 (C)		Mean Difference
		Steptoe (CC)	Morex (cc)	
Steptoe (AA)	Steptoe (BB)	72.3673 count = 57	----- 0	----
	Morex (bb)	74.5250 count = 4	72.8167 count = 6	1.7083
Morex (aa)	Steptoe (BB)	73.1143 count = 7	72.5300 count = 10	0.5843
	Morex (bb)	----- 0	74.0044 count = 46	----
	Mean	72.5732 count = 71	73.6631 count = 65	-1.0899

To demonstrate the point, means of combinations of markers Adh4, BCD453B and WG464 for Steptoe and Morex malt extract genotypes are shown in Table 14.21. If the overall means are compared for WG464, the Morex genotype is 1.0899 units higher than the Steptoe genotype. However, if the conditional means on Adh4 and BCD453B are compared for WG464, the Steptoe genotypes are 1.7083 and 0.5843 units higher than the Morex genotype for the two classes which have observations.

When QTLs are linked in the repulsion phase, peaks of the test statistic can be lowered or limited. However, when QTLs are linked in coupling phase, the estimates of the QTL effects could be overestimated. In practice, these cases often happen. For the barley data, Morex is considered a high malt extract variety and is a commonly used malting barley and Steptoe is a low extract variety and a feed barley. Common sense would suggest that Morex has the high malt extract alleles and Steptoe has the low malt extract alleles. Only the conditional analysis can resolve the problem of repulsion linkage phase for linked QTLs. When using QTL mapping for assisting plant and animal breeding, finding bad alleles in good families or varieties and good alleles in bad ones could be more meaningful than finding good alleles in good varieties or bad alleles in bad ones.

14.4.5 IMPLEMENTING CIM

In practice, the original CIM using the ECM algorithm can be performed using the computer software package QTL Cartographer (Basten and Zeng 1995). For composite interval mapping using regression approaches, commercial software such as SAS can be used. For using SAS, the data (called markerdata1) should be arranged as

```
Segment Marker1 Marker2 Positions Lines   Z1    Z2   m1 m2 m3 m4
...
   1     WG622   ABG313B   0.00     147   1.00  0.00  2 2 1 1 1
   1     WG622   ABG313B   0.00     148   0.00  1.00  1 1 1 1 1
   1     WG622   ABG313B   0.00     149   0.00  1.00  1 1 1 1 1
   1     WG622   ABG313B   0.00     150  -1.00 -1.00  1 2 0 0 1
   1     WG622   ABG313B   0.01      1    1.00  0.00  1 1 1 1 1
   1     WG622   ABG313B   0.01      2   -1.00 -1.00  1 0 1 0 0
   1     WG622   ABG313B   0.01      3    1.00  0.00  1 1 1 1 1
   1     WG622   ABG313B   0.01      4    0.00  1.00  1 1 2 0 2
...
```

for the linkage group to be analyzed using CIM, where the genome segment is flanked by the two markers and each of the segments can be divided into a number of genome positions based on percent of recombination fraction. For each genome position, there are N corresponding re-coded variables for the position (Z1 and Z2) and the predetermined marker genotypes for controlling the residual genetic background (m1, m2, m3, ...). The Zs are the coefficients for the two parameters and they are

$$Z1 = X_{i1j} + (1 - \rho) X_{i2j} + \rho X_{i3j} \qquad for \qquad \theta_1$$
$$Z2 = \rho X_{i2j} + (1 - \rho) X_{i3j} + X_{i4j} \qquad for \qquad \theta_4$$

(14.84)

for the model of Equation (14.80). To determine which markers are needed to control the residual genetic background, a stepwise regression can be used to obtain markers linked to potential QTLs. A preconstructed linkage map is needed to find out the relative positions of the target interval and the markers. For stepwise regression using SAS, predetermined variance and covariance matrices for the markers and trait phenotype are recommended if there are missing values for the marker data and the trait data. For this case, if the original data is used for regression analysis, most likely SAS will only use a portion of the data (SAS only uses the observations without any missing values for all the markers and the trait). Two additional data sets are needed to use SAS. They are the data set (called traitdata) containing trait values corresponding to the lines and a data set (called markerdata2) containing the data of markers on the other chromosomes (will be used in the model). The following SAS codes can be used for CIM using the linear regression approach

```
options ps=60 ls=80 nocenter;
/* Read trait data */
data trait;
  infile 'yourtrait.dat';
  input line y;
proc sort; by line;
/* Read marker data for the linkage group */
data markerdata1;
  infile 'yourmarker1.dat';
  input segment name1 $ name2 $ position line z1 z2 m1 m2 m3
...;
proc sort; by line;
/* Read marker data for the rest of the linkage groups  */
data markerdata2;
  infile 'yourmarker2.dat';
  input line l1 l2 l3 ...;
data all; merge trait markerdata1 markerdata2; by line;
proc sort data=all; by segment name1 name2 position;
/* Full model */
proc glm data=all noprint outstat=fullmodel;
by segment name1 name2 position;
model y=z1 z2 m1 m2 m3 ... l1 l2 l3 .../solution noint;
data fullmodel; set fullmodel; if _type_='ERROR';
   keep segment name1 name2  position df ss;
data fullmodel; set fullmodel; rename ss=ssfull;
/* Reduced model */
proc glm data=all noprint outstat=redumodel;
by segment name1 name2 position;
model y=m1 m2 m3 ... l1 l2 l3 .../solution noint;
data redumodel; set redumodel; if _type_='ERROR';
   keep segment name1 name2  position df ss;
data redumodel; set redumodel; rename ss=ssredu;
/* Merge the two data sets and compute the statistic and p-
value */
data model; merge fullmodel redumodel;
by segment name1 name2 position;
gstatistic=df*(ssreduc-ssfull)/ssfull;
if g<0 then g=0.0;
pvalue=1.0-probchi(gstatistic,1.0);
proc print data=cd;
run;
```

Some modifications may be needed for specific computers and versions of the software.

14.4.6 CIM USING F2 PROGENY

For composite interval mapping using F2 progeny, the linear regression model based on Equation (14.80) can be written as

$$y_j = \sum_{l=1}^{9} \theta_l X_{ilj} + \sum_{k \neq i, i+1} b_{ak} X_{akj} + \sum_{k \neq i, i+1} b_{dk} X_{dkj} + \varepsilon_j \qquad (14.85)$$

where the X_{ilj}s are the dummy variables for the interval defined in Table 14.18, X_{akj} and X_{dkj} are dummy variables for the additive and dominance effects for marker k defined in Table 14.19, θ_ls are the partial regression coefficients for the dummy variables for the interval and b_{ak} and b_{dk} are the partial regression coefficients for the additive and dominance effects of marker k used to control the residual genetic background.

Equation (14.85) can be modified using the expectations of the partial regression coefficients for the dummy variables for the interval (see Table 14.11 and Equation (13.51)). If a high linkage map density is assumed, then the probabilities that the individuals with double heterozygous genotypes for the two markers are homozygous for the putative QTL are very small. So, the θ_5 in

Equation (14.50) approximately equals the mean of the heterozygotes for the putative QTL $\theta_5 \approx \mu_2$. For example, if the distances between the two markers are 0.05, 0.1 and 0.15, then the maximum probabilities that an individual with marker genotype AaBb is homozygous (QQ and qq) for the putative QTL in the interval are 0.0007, 0.003 and 0.005 ($0.25r^2 / [(1-r)^2 + r^2]$). So, Equation (14.50) can be written as

$$
\begin{aligned}
\theta_1 &= \mu_1 \\
\theta_2 &= (1-\rho)\mu_1 + \rho\mu_2 \\
\theta_3 &= (1-\rho)^2\mu_1 + 2\rho(1-\rho)\mu_2 + \rho^2\mu_3 \\
\theta_4 &= \rho\mu_1 + (1-\rho)\mu_2 \\
\theta_5 &\approx \mu_2 \\
\theta_6 &= (1-\rho)\mu_2 + \rho\mu_3 \\
\theta_7 &= \rho^2\mu_1 + 2\rho(1-\rho)\mu_2 + (1-\rho)^2\mu_3 \\
\theta_8 &= \rho\mu_2 + (1-\rho)\mu_3 \\
\theta_9 &= \mu_3
\end{aligned}
\tag{14.86}
$$

If ρ is treated as a known parameter and the putative QTL genotypic means (μ_1, μ_2, μ_3) are replaced with their unbiased estimates $(\theta_1, \theta_5, \theta_9)$, Equation (14.84) can be written as

$$
y_j = \theta_1 Z_{1j} + \theta_5 Z_{2j} + \theta_9 Z_{3j} + \sum_{k \neq i, i+1} b_{ak} X_{akj} + \sum_{k \neq i, i+1} b_{dk} X_{dkj} + \varepsilon_j
\tag{14.87}
$$

where the new dummy variables Z are

$$
\begin{aligned}
Z_{1j} &= X_{1j} + (1-\rho)[X_{2j} + (1-\rho)X_{3j}] + \rho(X_{4j} + \rho X_{7j}) \\
Z_{2j} &= X_{5j} + \rho(X_{2j} + \rho X_{8j}) + (1-\rho)[2\rho(X_{3j} + X_{7j}) + X_{4j} + X_{6j}] \\
Z_{3j} &= X_{9j} + (1-\rho)[X_{8j} + (1-\rho)X_{7j}] + \rho(X_{6j} + \rho X_{3j})
\end{aligned}
\tag{14.88}
$$

Under the additive model $d = 0$ or $\theta_5 = 0.5(\theta_1 + \theta_9)$, Equation (14.87) becomes

$$
y_{j(d=0)} = \theta_1\left(Z_{1j} + \frac{1}{2}Z_{2j}\right) + \theta_9\left(Z_{3j} + \frac{1}{2}Z_{2j}\right)
\tag{14.89}
$$
$$
+ \sum_{k \neq i, i+1} b_{ak} X_{akj} + \sum_{k \neq i, i+1} b_{dk} X_{dkj} + \varepsilon_j
$$

Under the dominance model $a = 0$ or $\theta_1 = \theta_9 = \tilde{\theta}$, Equation (14.87) becomes

$$
y_{j(a=0)} = \tilde{\theta}(Z_{1j} + Z_{3j}) + \theta_5 Z_{2j} + \sum_{k \neq i, i+1} b_{ak} X_{akj} + \sum_{k \neq i, i+1} b_{dk} X_{dkj} + \varepsilon_j
\tag{14.90}
$$

Under the model of no genetic effect in the interval $a = 0$ and $d = 0$ or $\theta_1 = \theta_5 = \theta_9 = \theta$, Equation (14.87) becomes

$$
y_{j(a=0, d=0)} = \theta + \sum_{k \neq i, i+1} b_{ak} X_{akj} + \sum_{k \neq i, i+1} b_{dk} X_{dkj} + \varepsilon_j
\tag{14.91}
$$

A set of test statistics can be used to test the biologically meaningful hypothesis using the residual sum of squares for the full and the reduced models. They are

$$F_{a = 0, d = 0} = \frac{SSE_{a = 0, d = 0} - SSE_{full}}{SSE_{full}/df_{Efull}} \qquad H_0: a = 0, d = 0$$

$$F_{a = 0} = \frac{SSE_{a = 0, d = 0} - SSE_{a = 0}}{SSE_{a = 0}/df_{Ea = 0}} \qquad H_0: a = 0 \qquad (14.92)$$

$$F_{d = 0} = \frac{SSE_{a = 0, d = 0} - SSE_{d = 0}}{SSE_{d = 0}/df_{Ed = 0}} \qquad H_0: d = 0$$

where the SSEs are the residual sum of squares for the corresponding models and the dfs are the residual degrees of freedom for the corresponding models. The test statistics are distributed as an F-distribution with degrees of freedom $(2, df_{Efull})$ for $F_{a = 0, d = 0}$, $(1, df_{Ea = 0})$ for $F_{a = 0}$ and $(1, df_{Ed = 0})$ for $F_{d = 0}$. When $df_{Ea = 0}$ and $df_{Ed = 0}$ are large, $F_{a = 0}$ and $F_{d = 0}$ are also distributed as chi-square variables with one degree of freedom.

14.4.7 ADVANTAGES OF THE CIM
The CIM has four advantages over simple interval mapping, single marker analysis and multiple-QTL models using multiple regression (Zeng 1994).

(1) By concentrating on one genome region, the multidimensional search for the multiple-QTL model is reduced to a one-dimensional search for the CIM and the estimates of QTL locations and their effects are asymptotically unbiased.

(2) The resolution of QTL locations obtained using methods conditional on linked markers can be much higher than when simple interval mapping and single marker analysis are used.

(3) There are more variables in the model than in simple interval mapping and single marker analysis. So, the CIM is more informative and efficient.

(4) The log likelihood ratio test statistic plot and the lod score plot for all possible genome positions can still be used to present the results visually.

One of most important advantages of composite interval mapping is that the markers can be used as boundary conditions to narrow down the most likely QTL position. The resolution of QTL locations can be greatly increased. For example, the resolution of the barley malt extract QTLs on chromosome IV (Figure 14.8) can be greatly improved if more markers can be put on the linkage map near the regions where the QTLs are located. Putting more markers in certain regions can be done using the bulk segregating approach (See Chapter 17).

SUMMARY

The regression approach for both simple interval mapping and composite interval mapping is relatively easy to implement and yields similar or the same results as the likelihood approach or the combined approach. The composite interval mapping approach has advantages over simple interval mapping approach, potentially increasing QTL resolution.

EXERCISES

Exercise 14.1

When the linkage map density is high, the interval mapping approach is approximately equivalent to the single marker analysis in Chapter 13. Considering the following illustration below, do the following related to this subject:

(1) A QTL is located in between two markers A and B. Marker A and B are r recombination units apart. Derive the expected differences of marker genotypic means between Aa and AA and between AaBb and AABB using a backcross model of AaQqBb × AAQQBB. How are they different? If you use the mean of AA to estimate the mean of QQ, how big a bias do you expect? If you use the mean of AABB to estimate the mean of QQ, how big a bias do you expect?

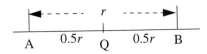

(2) Considering the same situation as in (1), derive the expected differences of marker genotypic means between AA and aa and between AABB and aabb using an F2 model of AaQqBb × AaQqBb. How are they different? If you use the mean of AA to estimate the mean of QQ, how big a bias do you expect? If you use the mean of AABB to estimate the mean of QQ, how big a bias do you expect?

(3) Compute the expected mean differences and the expected means derived in (1) and (2) for

a. $r = 0.02, \mu_{QQ} = 3, \mu_{Qq} = 2, \mu_{qq} = 1$
b. $r = 0.2, \mu_{QQ} = 3, \mu_{Qq} = 2, \mu_{qq} = 1$
c. $r = 0.02, \mu_{QQ} = 3, \mu_{Qq} = 3, \mu_{qq} = 1$
d. $r = 0.2, \mu_{QQ} = 3, \mu_{Qq} = 3, \mu_{qq} = 1$

(4) Discuss the results in (3). Can you draw some conclusions about the relationship between the single marker analysis and the interval mapping approach? How do mating type, map density and genetic effect impact on the relationship?

Exercise 14.2

Let us assume that QTL Q_1 located left of A has no effect on the analysis of marker B conditional on genotypes of marker A. Do the following to prove this point:

(1) Markers A and B and QTL Q_1 are linked with each other. The two markers are r recombination units apart. Q_1 is located on the left side of A and s recombination units apart from A. Derive the expected marker genotypic values as functions of the recombination fractions between the two markers and the QTL and QTL genotypic values for BB and Bb conditional on AA and Aa in backcross progeny of Q_1q_1AaBb ×

Q_1Q_1AABB. The genotypic values of Q_1Q_1 and Q_1q_1 are μ_{11} and μ_{12}.

(2) Assume that there is no QTL to the left of marker A, rather, that a QTL Q_2 is located between A and B at a relative position of $\rho = r_1/r$ from A and $1 - \rho = r_2/r$ from B, and derive the expected marker genotypic values as functions of the recombination fractions among the two markers and the QTL and QTL genotypic values for BB and Bb conditional on AA and Aa in backcross progeny of $Q_2q_2AaBb \times Q_2Q_2AABB$. The genotypic values of Q_2Q_2 and Q_2q_2 are μ_{21} and μ_{22}.

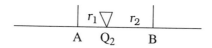

(3) Assume that QTLs Q_1 and Q_2 are both located on the genome segment with the linkage relationship as shown in the following illustration. Derive the expected marker genotypic values as functions of the recombination fractions between the two markers and the QTLs and QTL genotypic values for BB and Bb conditional on AA and Aa in backcross progeny of $Q_1q_1AaQ_2q_2Bb \times Q_1Q_1AAQ_2Q_2BB$. The genotypic values of Q_1Q_1, Q_1q_1, Q_2Q_2 and Q_2q_2 are μ_{11}, μ_{12}, μ_{21} and μ_{22}.

(4) Do (1), (2) and (3) for F2 progeny ($Q_1q_1AaQ_2q_2Bb \times Q_1q_1AaQ_2q_2Bb$) assuming codominant markers and a complete additive model. The marker genotypes for B are BB, Bb and bb. The genotypes for A are AA, Aa and aa. The genotypic values for the QTL genotypes Q_1Q_1, Q_1q_1 and q_1q_1 are a_1, 0 and $-a_1$. The genotypic values for QTL genotypes Q_2Q_2, Q_2q_2 and q_2q_2 are a_2, 0 and $-a_2$.

(5) Evaluate the expected marker genotypic values derived in (1), (3) and (4) using different values of s, for example, 0.01, 0.1, 0.2, 0.3, 0.4 and 0.5.

(6) Compare the conditional expected marker genotypic values with the marginal expected marker genotypic values in Chapter 13. How will the conditional t-test between the marker genotypic values differ from those of the simple single marker t-test (explained in Chapter 13)?

Exercise 14.3

Markers A and B and QTL Q_1 are linked each other. The two markers are r recombination units apart. Q_1 is located on left side of A and s recombination units away from A. The genotypic values of Q_1Q_1 and Q_1q_1 are μ_{11} and μ_{12}. First, derive the expected marker genotypic values as functions of the recombination fractions between the two markers and the QTL and QTL genotypic values for AABB, AABb, AaBB and Bb in backcross progeny of $Q_1q_1AaBb \times Q_1Q_1AABB$, and then do the following problems.

$$\underline{\quad\nabla\, s\, \Big|\quad r\quad \Big|\qquad \Big|\quad}$$
$$\quad Q_1 \quad A \qquad\qquad B$$

(1) Derive the interval mapping model for the segment flanked by A and B using the linear regression approach.

(2) The following model

$$y_j = \theta_1 [X_1 + (1-\rho)X_2 + \rho X_3] + \theta_4 [\rho X_2 + (1-\rho)X_3 + X_4] + \varepsilon_j$$

can be used to model the situation in this problem, where the dummy variables are listed in the following table, and $0 \le \rho \le 1$ is the relative genome position of the point to be tested. Derive the expectations for the parameters in the model θ_1, θ_4 and ρ.

Dummy variables for interval mapping using regression analysis.

Maker Genotype	X1	X2	X3	X4
AABB	1	0	0	0
AABb	0	1	0	0
AaBB	0	0	1	0
AaBb	0	0	0	1

(3) Assume QTLs Q_1 and Q_2 are all located on the genome segment with the linkage relationship as shown in the following illustration. The genotypic values of Q_1Q_1, Q_1q_1, Q_2Q_2 and Q_2q_2 are μ_{11}, μ_{12}, μ_{21} and μ_{22}. Derive the expectations for the parameters using the same model and the dummy variables as in (2).

$$\underline{\quad\nabla\, s\, \Big|\quad r\quad \Big|\, t\, \nabla\quad}$$
$$\quad Q_1 \quad A \qquad\qquad B \quad Q_2$$

(4) Speculate about the QTL analysis results for the interval flanked by A and B using the expectations in (3) for the following situations:

   a. two QTLs in coupling phase, tight linkage
   b. two QTLs in coupling phase, loose linkage
   c. two QTLs in repulsion phase, tight linkage
   d. two QTLs in repulsion phase, loose linkage

(5) More markers are likely to be available. Explain a strategy which can be used to locate the QTLs more precisely and to reduce the chance of a false positive at other genome locations.

CHAPTER 15

QTL MAPPING: NATURAL POPULATIONS

15.1 INTRODUCTION

The methods of QTL mapping described in Chapters 13 and 14 are designed for experimental populations obtained from controlled crosses. However, for many economically important plant and animal species, controlled crosses cannot be produced, or are difficult or time consuming. For example, many tree species need a relatively long time to produce progeny. And obviously, controlled crosses cannot be made for humans. In Chapter 7, the methods for linkage analysis using natural populations were discussed. In this chapter, methods for QTL mapping using natural populations will be discussed.

First, natural population must be defined. "Natural" used here is a term relative to "controlled" cross. Natural, as used here, means that the population is not generated using controlled crosses. So in plants, open-pollinated populations are natural populations. Experimental populations from controlled crosses usually have one or all of the following characteristics:

(1) The progeny can be traced to both parents.

(2) The expected allelic frequency in the population is defined and takes values of 0.0, 0.25, 0.5, 0.75 or 1.0.

(3) Populations have large, genetically homogenous families.

The genetic characteristics of natural populations, compared to those of the populations of controlled crosses, are as follows:

(1) Progeny either cannot be traced to any parent, or can be traced to one of the parents but information for the other parent cannot be obtained or can be obtained at the population, not at the individual level (*e.g.*, open-pollinated populations).

(2) The allelic frequency is a random variable and can be estimated from the population.

(3) For some of the natural populations, families usually have small sizes (*e.g.*, human population).

Because of the characteristics of natural populations, special treatments are needed for QTL mapping. In this chapter, single-marker analysis using open-pollinated populations in plants and the sib-pair approach will be described. The sib-pair method was originally designed for use in humans and has been used extensively for mapping genes controlling complex diseases in humans. The method can also be applied to other animal species and plants. Recently, the sib-pair approach has been extended to interval mapping (Fulker and Cardon 1994) and multiple QTL models (Fulker *et al.* 1995).

The methods of QTL mapping used for natural populations can also be applied to experimental populations. However, the methods for natural populations usually have lower statistical power than do the methods normally used for experimental populations.

15.2 OPEN-POLLINATED POPULATIONS

15.2.1 JOINT SEGREGATION OF QTL AND MARKERS

The QTL mapping methods for experimental populations are based on cosegregation among the putative QTLs and the markers, as are QTL mapping methods using natural populations. The only difference is that the cosegregation patterns for natural populations are more complex than those for experimental populations. Figure 15.1 shows the mating scheme for deriving genotypic frequencies of open-pollinated progeny in plants using a two-locus model (a marker and a QTL). The progeny is produced from both a self and an outcross. It should be mentioned that the maternal parent has to be heterozygous for both the marker and the QTL to find an association between them. Linkage equilibrium for the two loci is assumed in the pollen pool. Marker genotypes can be determined by directly genotyping the maternal plant and by studying the segregation pattern in the progeny (see Chapter 7). Table 15.1 shows the joint frequency of marker-QTL genotypes. The frequencies are classified into selfed and outcrossed genotypes. For the selfs, the expected frequencies are the same as for the classical F2s. For the outcrosses, the frequencies are the products of the corresponding gametic frequencies shown in Figure 15.1. The joint frequency is

$$p_{ij} = tp_{oij} + (1 - t)\, p_{sij}$$

where the subscripts i and j denote the marker and QTL genotypes, o denotes the outcross, s denotes the self and t is the outcrossing rate. The marginal

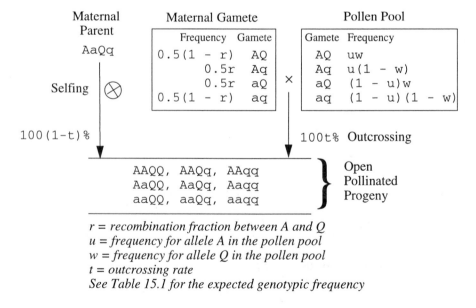

r = recombination fraction between A and Q
u = frequency for allele A in the pollen pool
w = frequency for allele Q in the pollen pool
t = outcrossing rate
See Table 15.1 for the expected genotypic frequency

Figure 15.1 Open-pollinated population with a marker A and a QTL Q.

Table 15.1 Expected and observed genotypic frequencies for codominant markers A and a QTL Q in an open-pollinated population.

Genotype	Frequency	Expected Frequency Outcrossing progeny(p_{oij})	Selfing progeny (p_{sij})
AAQQ	p_{11}	$0.5uw(1-r)$	$0.25(1-r)^2$
AAQq	p_{12}	$0.5u[(1-w)(1-r)+wr]$	$0.5r(1-r)$
AAqq	p_{13}	$0.5u(1-w)r$	$0.25r^2$
AaQQ	p_{21}	$0.5w[(1-u)(1-r)+ur]$	$0.5r(1-r)$
AaQq	p_{22}	$0.5[1-r-(1-2r)(u+w-2uw)]$	$0.5[1-2r+2r^2]$
Aaqq	p_{23}	$0.5(1-w)(u-2ur+r)$	$0.5r(1-r)$
aaQQ	p_{31}	$0.5(1-u)wr$	$0.25r^2$
aaQq	p_{32}	$0.5(1-u)(w-2wr+r)$	$0.5r(1-r)$
aaqq	p_{33}	$0.5(1-u)(1-w)(1-r)$	$0.25(1-r)^2$

r is recombination fraction between A and Q and
u and w are frequencies for allele A and Q in the pollen pool, respectively.

frequencies for the three marker genotypes are

$$p_{1\circ} = 0.5ut + 0.25(1-t) \qquad for \qquad AA$$

$$p_{2\circ} = 0.5 \qquad for \qquad Aa$$

$$p_{3\circ} = 0.5(1-u)t + 0.25(1-t) \qquad for \qquad aa$$

Table 15.2 shows the marginal frequencies for the three marker genotypes under three different outcrossing rates (0.0, 0.5 and 1.0). The frequency of the QTL genotype (j) conditional on the marker genotype (i) is

$$p(Q_j|M_i) = p_{ij}/p_{i\circ}$$

Table 15.2 Expected marginal frequencies for t = 0.0, 0.5 and 1.0.

Marker Genotype	t 0	0.5	1.0
AA	0.25	$0.25(0.5+u)$	$0.5u$
Aa	0.5	0.5	0.5
aa	0.25	$0.25(1.5-u)$	$0.5(1-u)$

15.2.2 MODEL

Consider the situation in Figure 15.1. A QTL locus Q with alleles Q and q is located r recombination units away from a known codominant marker A with alleles A and a. Allelic frequency in the pollen pool for Q is w and for A is u. Therefore, for q it is $1-w$ and for a it is $1-u$. The three QTL genotypes have means μ_1, μ_2 and μ_3, respectively. The additive and dominance effects are defined as

$$a = 0.5\,(\mu_1 - \mu_3)$$
$$d = 0.5\,(\mu_1 + \mu_3 - 2\mu_2)$$

(15.1)

The expected trait value for the three marker genotypes is

$$g_{Mi} = \mu_1 p\,(Q_1|M_i) + \mu_2 p\,(Q_2|M_i) + \mu_3 p\,(Q_3|M_i)$$

(15.2)

For a single marker model, differences among the marker genotypic means need to be tested in order to make inferences about possible QTLs. The tests are H_0: $\mu_{AA} - \mu_{aa} = 0$ for additive effects and H_0: $\mu_{AA} + \mu_{aa} - 2\mu_{Aa} = 0$ for dominance effects, where μ_{AA}, μ_{aa} and μ_{Aa} are trait means for marker genotypes AA, aa and Aa, respectively. These tests can be implemented as an analysis of variance using the model

$$y_{ik} = \mu + \tau_i + \varepsilon_{ik}$$

(15.3)

where y_{ik} is the trait value for the kth individual among the ith marker genotypic class, τ_i is the effect of the ith marker genotypic class and ε_{ik} is the error for the individual. The two null hypotheses can be tested using two contrasts about the marker means (μ_{AA}, μ_{Aa} and μ_{aa}), such as

$$contrast_1 = 1, 0, -1 \quad for \quad \mu_{AA} - \mu_{aa} = 0$$
$$contrast_2 = 1, -2, 1 \quad for \quad \mu_{AA} + \mu_{aa} - 2\mu_{Aa} = 0$$

(15.4)

The question is now whether or not the contrasts on the marker genotypic means correspond to any meaningful contrasts related to the QTL genotypes. This question can be explained by setting the outcrossing rate at some fixed point, because it is difficult to interpret the expected contrasts when the outcrossing rate is treated as a random variable. It is logical that the outcrossing rate be estimated before the QTL analysis (see Chapter 7 for the discussion on outcrossing rate). When $t = 0$ (0% outcrossing), the problem is reduced to the traditional F2 and the expectations of the hypothesis tests reduce to the same situation as a conventional F2. For $\mu_{AA} - \mu_{aa}$, the expectation is

$$E\,(\mu_{AA} - \mu_{aa}) = (1 - 2r)\,(\mu_1 - \mu_3)$$

(15.5)

and for $\mu_{AA} + \mu_{aa} - 2\mu_{Aa}$, the expectation is

$$E\,(\mu_{AA} + \mu_{aa} - 2\mu_{Aa}) = (1 - 2r)^2\,(\mu_1 + \mu_3 - 2\mu_2)$$

(15.6)

15.2.3 COMPLETE OUTCROSSING

When $t = 1$ (100% outcrossing), the expectations of marker genotypic values for the three marker genotypic classes with respect to QTL genotypic means are

$$\mu_{AA} = w\,(1 - r)\,\mu_1 + [\,(1 - w)\,(1 - r) + wr]\,\mu_2 + (1 - w)\,r\mu_3$$
$$\mu_{Aa} = w\,[\,(1 - u)\,(1 - r) + ur]\,\mu_1 + [1 - r - (1 - 2r)\,(u + w - 2uw)\,]\,\mu_2$$
$$\quad + (1 - w)\,(u - 2ur + r)\,\mu_3$$
$$\mu_{aa} = wr\mu_1 + (w - 2wr + r)\,\mu_2 + (1 - w)\,(1 - r)\,\mu_3$$

(15.7)

It is important to know first that the expectations for the three marker genotypes are the same as

$$0.5w\mu_1 + 0.5\mu_2 + 0.5(1-w)\mu_3 \quad for \quad r = 0.5$$
$$\mu \quad for \quad \mu_1 = \mu_2 = \mu_3 = \mu \tag{15.8}$$

under the null hypothesis that the marker and the putative QTL are independent $(r = 0.5)$ and that there are no QTL effects $(\mu_1 = \mu_2 = \mu_3 = \mu$ or $a = 0$ and $d = 0)$. So, the analysis of variance for the trait values using the marker genotypes as classification variables is meaningful to achieve inferences regarding the putative QTL near the marker. Now the question becomes whether or not the variance can be partitioned into meaningful contrasts corresponding to the additive and the dominance effects. The expectation for $\mu_{AA} - \mu_{aa}$ is

$$E[\mu_{AA} - \mu_{aa}] = (1-2r)[w\mu_1 + (1-2w)\mu_2 + (1-w)\mu_3], \tag{15.9}$$

The test is independent of marker allelic frequency in the pollen pool, but is not independent with respect to QTL allelic frequency in the pollen pool.

When $w = 0.5$, Equation (15.9) is reduced to

$$E[\mu_{AA} - \mu_{aa}] = 0.5(1-2r)(\mu_1 - \mu_3) \tag{15.10}$$

which is half the expectation for the single-marker analysis using F2 progeny (see Chapter 13). The test statistic tests the QTL additive effect $\mu_1 = \mu_3$ or $r = 0.5$.

When $w = 1$ or $w = 0$, the hypothesis tests a mixture of additive and dominance effects $(\mu_2 - \mu_3$ or $\mu_1 = \mu_2)$ or $r = 0.5$. They are

$$E[\mu_{AA} - \mu_{aa}] = (1-2r)(\mu_2 - \mu_3) \quad for \quad w = 0$$
$$E[\mu_{AA} - \mu_{aa}] = (1-2r)(\mu_1 - \mu_2) \quad for \quad w = 1 \tag{15.11}$$

They are equivalent to the contrast for the single-marker analysis using the backcross progeny.

The expectation of dominance effect $\mu_{AA} + \mu_{aa} - 2\mu_{Aa}$ is

$$E(\mu_{AA} + \mu_{aa} - 2\mu_{Aa}) = (1-2u)(1-2r)[(1-w)\mu_3 - w\mu_1 - (1-2w)\mu_2]$$
$$= (1-2u)(1-2r)[(\mu_3 - \mu_2) - w(\mu_1 + \mu_3 - 2\mu_2)] \tag{15.12}$$

The test is related to both the marker and QTL allelic frequencies in the pollen pool. When $u \approx 1$ or $u \approx 0$ and $w \approx 0.5$, the hypothesis tests the QTL additive effect $\mu_1 = \mu_3$ or $r = 0.5$. When $w \approx 1$ or $w \approx 0$, the hypothesis tests a mixture of additive and dominance effects $(\mu_1 = \mu_2$ or $\mu_2 = \mu_3)$ or $r = 0.5$.

Table 15.3 Expectations for the additive and dominance contrasts on the marker genotypes as functions of the putative QTL genotypic values, recombination fraction between the marker and the putative QTL and the allelic frequencies of the marker in the pollen pool for complete self (t = 0) and complete outcrossing (t = 1) and different allelic frequencies of the putative QTL in the pollen pool (w).

t	w	$E(\mu_{AA} - \mu_{aa})$	$E(\mu_{AA} + \mu_{aa} - 2\mu_{Aa})$
0	-	$(1-2r)(\mu_1 - \mu_3)$	$(1-2r)^2(\mu_1 + \mu_3 - 2\mu_2)$
1	0	$(1-2r)(\mu_2 - \mu_3)$	$(1-2u)(1-2r)(\mu_3 - \mu_2)$
	0.5	$0.5(1-2r)(\mu_1 - \mu_3)$	$0.5(1-2u)(1-2r)(\mu_3 - \mu_1)$
	1	$(1-2r)(\mu_1 - \mu_2)$	$(1-2u)(1-2r)(\mu_2 - \mu_1)$

Table 15.3 shows these expectations. It is clear that an open-pollinated population with 100% outcrossing is an effective population for QTL mapping. When the frequency of a QTL allele is close to 0.5, the additive contrast on the marker genotypic value tests true QTL additive effect or independence between the marker and the QTL ($r = 0.5$). As the QTL allelic frequency departs from 0.5, the additive contrast on marker genotypic value tests a mixture of QTL additive and dominance effects or independence between the marker and the QTL ($r = 0.5$).

The frequency of a marker allele in the pollen pool has no effect on the expectation of the additive contrast on marker genotypic values. Using the open-pollinated population with 100% outcrossing that is effective for QTL mapping, marker alleles in the pollen pool do not vary the conditional frequency or the putative QTL genotypes on marker genotypes. However, when the marker allelic frequency in the pollen pool is 0.5 ($u = 1 - u = 0.5$), the expectation of the dominance contrast on marker phenotypes (Equation 15.12) is zero. So, the statistical power of detecting QTLs using open-pollinated populations depends on results from screening the markers. Rare markers will have higher statistical power than the common markers. However, if we are willing to ignore the dominance contrast, the marker allelic frequencies have no effect on the additive contrast. Therefore, in terms of QTL mapping, the completely outcrossed population is comparable to a backcross population.

15.2.4 HALF OUTCROSSING

When $t = 0.5$, the expectations of marker genotypic values for the three marker genotypic classes with respect to the QTL genotypic means are

$$\mu_{AA} = \frac{0.5\,(1-r)\,(2uw+1-r)}{0.5+u}\mu_1$$
$$+\frac{u\,[\,(1-w)\,(1-r)+wr\,]+r\,(1-r)}{0.5+u}\mu_2$$
$$+\frac{0.5r\,[\,2u\,(1-w)+r\,]}{0.5+u}\mu_3$$

$$\mu_{Aa} = 0.5\,\{\,w\,[\,(1-u)\,(1-r)+ur\,]+r\,(1-r)\,\}\,\mu_1$$
$$+0.5\,\{\,[\,1-r-(1-2r)\,(u+w-2uw)\,]+[\,1-2r+2r^2\,]\,\}\,\mu_2 \qquad (15.13)$$
$$+0.5\,[\,(1-w)\,(u-2ur+r)+r\,(1-r)\,]\,\mu_3$$

$$\mu_{aa} = \frac{0.5r\,[\,2\,(1-u)\,w+r\,]}{1.5-u}\mu_1$$
$$+\frac{(1-u)\,(w-2wr+r)+r\,(1-r)}{1.5-u}\mu_2$$
$$+\frac{0.5\,(1-r)\,[\,2\,(1-u)\,(1-w)+(1-r)\,]}{1.5-u}\mu_3$$

The expectations become

$$\mu_{AA} = \mu$$
$$\mu_{Aa} = \mu$$
$$\mu_{aa} = \mu$$

under the null hypothesis $\mu_1 = \mu_2 = \mu_3 = \mu$. So, if there is no QTL segregating in the population, both additive and dominance contrasts will be zero. The inference on the putative QTL genotypic means can be drawn from the hypothesis on the marker genotypic means. The expectations become

$$\mu_{AA} = \frac{0.25\,(2uw + 0.5)}{0.5 + u}\mu_1 + 0.5\mu_2 + \frac{0.25\,[2u\,(1 - w) + 0.5]}{0.5 + u}\mu_3$$

$$\mu_{Aa} = 0.125\,(2w + 1)\,\mu_1 + 0.5\mu_2 + 0.125\,[2\,(1 - w) + 1]\,\mu_3$$

$$\mu_{aa} = \frac{0.25\,[2\,(1 - u)\,w + 0.5]}{1.5 - u}\mu_1 + 0.5\mu_2$$

$$+\frac{0.25\,[2\,(1 - u)\,(1 - w) + 0.5]}{1.5 - u}\mu_3$$

(15.14)

under the null hypothesis $r = 0.5$. The interpretation of the hypothesis on the marker genotypic means is not simple in terms of the recombination fraction between the marker and the putative QTL. For some combinations of allelic frequencies of QTLs and markers in the pollen pool, the expectation of additive or dominance contrast may not be zero under $r = 0.5$. The true QTL on another independent genome segment may create bias for the hypothesis test in this segment. If $w = 0.5$, Equation (15.14) becomes

$$\mu_{AA} = \mu_{Aa} = \mu_{aa} = (\mu_1 + 2\mu_2 + \mu_3)\,/4$$

(15.15)

If $w = 1, u = 0.5$, Equation (15.14) becomes

$$\mu_{AA} = \mu_{Aa} = \mu_{aa} = (3\mu_1 + 4\mu_2 + \mu_3)\,/8$$

(15.16)

Table 15.4 shows the expectations under the null hypothesis $r = 0.5$ and $t = 0.5$.

Table 15.4 Expectations of marker genotypic values as functions of the putative QTL genotypic values for mixture of outcrossing and self (t = 0.5) and combinations of allelic frequencies for the marker and the putative QTL in the pollen pool.

w	u	$E\,[\mu_{AA}]$	$E\,[\mu_{Aa}]$	$E\,[\mu_{aa}]$
0	0	$(\mu_1 + 2\mu_2 + \mu_3)\,/4$	$(\mu_1 + 4\mu_2 + 3\mu_3)\,/8$	$(\mu_1 + 6\mu_2 + 5\mu_3)\,/12$
	0.5	$(\mu_1 + 4\mu_2 + 3\mu_3)\,/8$	$(\mu_1 + 4\mu_2 + 3\mu_3)\,/8$	$(\mu_1 + 4\mu_2 + 3\mu_3)\,/8$
	1	$(\mu_1 + 6\mu_2 + 5\mu_3)\,/12$	$(\mu_1 + 4\mu_2 + 3\mu_3)\,/8$	$(\mu_1 + 2\mu_2 + \mu_3)\,/4$
0.5		$(\mu_1 + 2\mu_2 + \mu_3)\,/4$	$(\mu_1 + 2\mu_2 + \mu_3)\,/4$	$(\mu_1 + 2\mu_2 + \mu_3)\,/4$
1	0	$(\mu_1 + 2\mu_2 + \mu_3)\,/4$	$(3\mu_1 + 4\mu_2 + \mu_3)\,/8$	$(5\mu_1 + 6\mu_2 + \mu_3)\,/12$
	0.5	$(3\mu_1 + 4\mu_2 + \mu_3)\,/8$	$(3\mu_1 + 4\mu_2 + \mu_3)\,/8$	$(3\mu_1 + 4\mu_2 + \mu_3)\,/8$
	1	$(5\mu_1 + 6\mu_2 + \mu_3)\,/12$	$(3\mu_1 + 4\mu_2 + \mu_3)\,/8$	$(\mu_1 + 2\mu_2 + \mu_3)\,/4$

Table 15.5 shows the expectations under the null hypothesis $r = 0.5$ and $t = 0.5$. For example, when $w = 0, u = 0$, the expectation of the dominance contrast is

$$\frac{\mu_1 + 2\mu_2 + \mu_3}{4} + \frac{\mu_1 + 6\mu_2 + 5\mu_3}{12} - \frac{2\,(\mu_1 + 4\mu_2 + 3\mu_3)}{8} = \frac{\mu_1 - \mu_3}{12}$$

So, even when the QTL is independent of the marker, a portion of the genetic

Table 15.5 Expectations of null hypothesis ($r = 0.5$) for the additive and dominance contrasts on the marker genotypes as functions of the putative QTL genotypic values for mixtures of outcrossing and self ($t = 0.5$) and combinations of allelic frequencies for the marker and the putative QTL in the pollen pool.

w	u	$E(\mu_{AA} - \mu_{aa})$	$E(\mu_{AA} + \mu_{aa} - 2\mu_{Aa})$
0	0	$(\mu_1 - \mu_3)/6$	$(\mu_1 - \mu_3)/12$
	0.5	0	0
	1	$-(\mu_1 - \mu_3)/6$	$(\mu_1 - \mu_3)/12$
0.5		0	0
1	0	$-(\mu_1 - \mu_3)/6$	$(-(\mu_1 - \mu_3))/12$
	0.5	0	0
	1	$(\mu_1 - \mu_3)/6$	$(-(\mu_1 - \mu_3))/12$

effect could be associated with the marker. When $w = 1, u = 0$, the expectation is

$$\frac{\mu_1 + 2\mu_2 + \mu_3}{4} + \frac{5\mu_1 + 6\mu_2 + \mu_3}{12} - \frac{2(3\mu_1 + 4\mu_2 + \mu_3)}{8} = -\frac{\mu_1 - \mu_3}{12}$$

These undesirable biases are the result of correlation between marker and QTL allelic frequencies in the pollen pool. In theory, these problems can be solved by incorporating the allelic frequencies into the analysis.

15.2.5 EXPECTATION OF THE ADDITIVE CONTRAST

It is clear that the hypothesis test for detecting QTLs using open-pollinated populations can have good statistical power and unbiased results when the outcrossing rate is 100% or 0%. When the progeny is produced by a mixture of selfing and outcrossing, the hypothesis test for $r = 0.5$ is confounded by the correlation between marker and QTL alleles in the pollen pool. How the biases of the additive and dominance contrasts will affect the detection of QTL in an open-pollinated progeny set using a numerical approach will be shown. A completely additive model will be used to illustrate points of interest. Consider

$$\mu_1 = 1$$
$$\mu_2 = 0$$
$$\mu_3 = -1$$

The expected additive contrast is 2 when the gene is located right on a marker. The dominance contrast is zero. When the gene is r recombination units away, the additive contrast is expected to be $2(1 - 2r)$, if classical F2 progeny are used.

Table 15.6 shows the expectations. The additive contrast is expected to be 1.6 if a classical F2 progeny is used. The additive contrast among the marker genotypic values in the progeny of the open-pollinated population has a portion of QTL detection power compared to the classical F2 progeny. The disturbing aspect is the non-zero expectation of the contrast for some combinations of allelic frequencies of markers and QTLs in the pollen pool under the null hypothesis ($r = 0.5$). When $w = 0.5$ or $u = 0.5$, the correlation between the two frequencies is zero and the expectation of the additive contrast is zero. Figure 15.2 shows a more detailed picture of the relationships between the expectation of the additive contrast and the allelic frequencies. The surfaces

Table 15.6 Expectations of the additive and dominance contrasts on the marker genotypes as functions of the putative QTL genotypic values ($\mu_1 = 1, \mu_2 = 0, \mu_3 = -1$) for a mixture of outcrossing and self ($t = 0.5$) and combinations of allelic frequencies for the marker and the putative QTL in the pollen pool when the marker and QTL are linked ($r = 0.1$) or null hypothesis ($r = 0.5$) is true.

t	w	u	Additive		Dominance	
			$r = 0.1$	$r = 0.5$	$r = 0.1$	$r = 0.5$
0			1.6	0	0	0
0.2	0	0	1.6333	0.1667	0.00667	0.0333
		0.5	1.44	0	0	0
		1	1.3	-0.1667	0.06	0.0333
	0.5	0	1.4667	0	-0.02667	0
		0.5	1.44	0	0	0
		1	1.4667	0	0.02667	0
	1	0	1.3	-0.1667	-0.06	-0.0333
		0.5	1.2	0	0	0
		1	1.6333	0.1667	-0.00667	-0.0333
0.5	0	0	1.6667	0.3333	0.03333	0.1667
		0.5	1.44	0	0	0
		1	1.0	-0.3333	0.3	0.1667
	0.5	0	1.3333	0	-0.1333	0
		0.5	1.2	0	0	0
		1	1.3333	0	0.1333	0
	1	0	1.0	-0.3333	-0.3	-0.1667
		0.5	1.2	0	0	0
		1	1.6667	0.3333	-0.0333	-0.1667
0.9	0	0	1.6947	0.4737	0.0853	0.4263
		0.5	0.88	0	0	0
		1	0.7474	-0.4737	0.7674	0.4263
	0.5	0	1.2211	0	-0.3411	0
		0.5	0.88	0	0	0
		1	1.2211	0	0.3411	0
	1	0	0.7474	-0.4737	-0.7674	-0.4263
		0.5	0.88	0	0	0
		1	1.6947	0.4737	-0.0853	-0.4263
1			0.8	0	0	0

are symmetric in two directions. The two directions are the lines corresponding to

$$w = u$$

$$w + u = 1$$

The expectation of the additive contrast reaches a maximum when

$$w = u = 0$$

$$w = u = 1$$

The maximum additive contrast point also corresponds to the maximum false positive point, where the absolute value of the expectation of the contrast is maximum. When either or both the QTL and marker allelic frequencies in the pollen pool are close to 0.5, the false positive frequency is at a minimum. This minimum additive contrast point under the null hypothesis is also the point which has the minimum additive contrast under the alternative hypothesis (low statistical power).

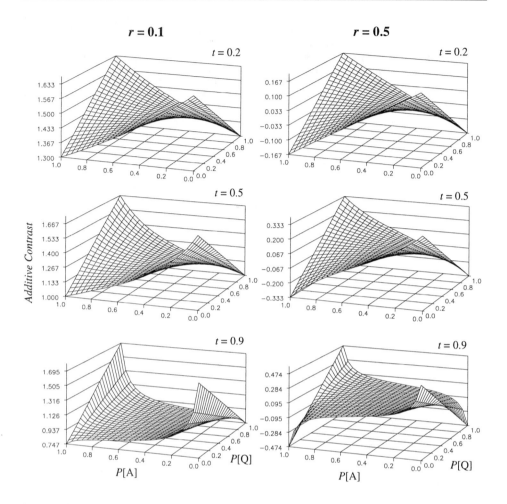

Figure 15.2 The expectation of additive contrast as a function of allelic frequencies of markers and QTLs in the pollen pool and the QTL genotypic values ($\mu_1 = 1, \mu_2 = 0$ and $\mu_3 = -1$) for some combinations of outcrossing rate and recombination fraction between a marker and a QTL.

In this section, the models and the methodology for QTL mapping using the single-marker analysis in open-pollinated populations in plants have been described. Caution should be used in interpreting the results of the single-marker analysis because of the high false positive rate. The false positive is the result of an allelic frequency correlation between the marker and the potential QTL in a pollen pool. A low P-value or a high lod score is recommended for use in declaring QTLs in open-pollinated populations. More research is needed to take the marker and QTL alleles into consideration in the analysis.

A similar approach can be applied to the general half-sib populations in plants and animals. Half-sib populations can be treated as "open-pollinated" populations with an outcrossing rate of 100%. However, this method is not effective for small family sizes and a large number of families in the population. For situations with a small family size and a large number of families, the sib-pair method is more suitable.

15.3 SIB-PAIR METHODS

The basic rationale behind the sib-pair method is that the variation of a trait value between sibs must be related to the probability of identity by descent for the genes or QTLs controlling the trait. Haseman and Elston (1972) used the squared difference between two sibs or a sib-pair to measure the variation of the trait among the sibs. If identity by descent can be measured for a large number of markers, then there is a good chance of finding the association between the squared difference and the identity by descent measure. If the association is significant for a marker locus, then there is a good chance that a QTL is located near the marker. Recently, the sib-pair method has been extended to interval mapping and multiple-QTL models (Goldgar 1990; Fulker and Cardon 1994; Fulker *et al.* 1995).

15.3.1 MODEL FOR QTL LOCATING ON MARKER

Model

Haseman and Elston (1972) first proposed using the sib-pair method for computing linkage between a known allele locus and a hypothetical two-allele locus controlling a quantitative trait. In their model, random mating, linkage equilibrium and no epistatic effects among the potential genes were assumed. The method of the Haseman-Elston sib-pair is based on a regression model, such as

$$E(Y_j|\pi_j) = \alpha + \beta\pi_j \qquad (15.17)$$

where Y_j is the squared sib-pair difference for the *jth* sib-pair conditional on π_j and π_j is the proportion of alleles identical by descent (ibd) for a given marker. The main tasks of the sib-pair approach are to find the estimators of π_j, α and β and to derive their biological inferences. In this section, assume that the QTL is located directly on the marker. In the next section, a model will be developed for QTL located near the markers.

Identity by Descent

First, define identity by descent and find the estimate of π_j. In natural or experimental populations, the two parents may have the same allele from one ancestor. There is a chance that an offspring receives two identical alleles from the two parents. These two alleles are replicates of the same allele from a common ancestor and are defined as alleles ibd, *i.e.*, identity by descent. The inbreeding coefficient is defined as the probability that two alleles at a locus of an individual are identical by descent. If the two alleles have the same function, but are derived from different ancestors, then the two alleles are called identical by state. The inbreeding coefficient of an individual equals the coefficient of coancestry of its two parents.

For cases when the parental and the sib-pair genotypes at a locus can be clearly determined, Haseman and Elston (1972) give the equation to estimate the gene ibd for the model based on Equation (15.17) as

$$\hat{\pi}_j = f_{j2} + 0.5f_{j1} \qquad (15.18)$$

where f_{ji} is the probability that the *jth* sib-pair should have i alleles ibd at a marker locus. They also gave the method for estimating the ibd value for the model based on Equation (15.17) when some of the genotypes are unknown.

The ibd value can be estimated for more than one locus or even for a segment of genome instead of for a marker locus (Thompson 1988; Goldgar 1990; Guo 1994).

Table 15.7 Nine possible sib-pairs for a single-locus model and their expected squared sib-pair difference and frequencies conditional on the probability of identity by descent.

Sib-Pair	$Y_j = (x_{1j} - x_{2j})^2$	$P[SP|\pi_j = 0]$	$P\left[SP\left|\pi_j = \dfrac{1}{2}\right.\right]$	$P[SP|\pi_j = 1]$
GG-GG	$(\varepsilon_{1j} - \varepsilon_{2j})^2 = \varepsilon_j^2$	p^4	p^3	p^2
gg-gg	$(\varepsilon_{1j} - \varepsilon_{2j})^2 = \varepsilon_j^2$	q^4	q^3	q^2
Gg-Gg	$(\varepsilon_{1j} - \varepsilon_{2j})^2 = \varepsilon_j^2$	$4p^2q^2$	pq	$2pq$
GG-Gg	$(a - d + \varepsilon_j)^2$	$2p^3q$	p^2q	0
Gg-GG	$(-a + d + \varepsilon_j)^2$	$2p^3q$	p^2q	0
Gg-gg	$(a + d + \varepsilon_j)^2$	$2pq^3$	pq^2	0
gg-Gg	$(-a - d + \varepsilon_j)^2$	$2pq^3$	pq^2	0
GG-gg	$(2a + \varepsilon_j)^2$	p^2q^2	0	0
gg-GG	$(-2a + \varepsilon_j)^2$	p^2q^2	0	0

Sib-Pair Difference

Trait value of a pair of sibs (x_{1j} for the first sib and x_{2j} for the second sib) can be modeled as

$$\begin{cases} x_{1j} = \mu + g_{1j} + \varepsilon_{1j} \\ x_{2j} = \mu + g_{2j} + \varepsilon_{2j} \end{cases} \tag{15.19}$$

where μ is the population mean and g_{ij} and ε_{ij} are the genetic and the residual effects, respectively. The dependent variable in Equation (15.17) is

$$Y_j = (x_{1j} - x_{2j})^2$$

Following the classical definitions of additive and dominance genetic effects, the genotypic values for a trait controlled by a gene G are

$$\begin{aligned} \mu_1 &= \mu + a \quad & for \quad & GG \\ \mu_2 &= \mu + d \quad & for \quad & Gg \\ \mu_3 &= \mu - a \quad & for \quad & gg \end{aligned} \tag{15.20}$$

(Falconer 1989) (see Chapters 2 and 12), where

$$a = 0.5\,(\mu_1 - \mu_3)$$
$$d = 0.5\,(\mu_1 + \mu_3 - 2\mu_2)$$

were defined as additive and dominance genetic effects of the QTL (G) and μ denotes the population mean. The genetic variance contributed by this locus is

$$\sigma_g^2 = \sigma_a^2 + \sigma_d^2 = 2pq\,[a - d\,(p - q)]^2 + 4p^2q^2d^2 \tag{15.21}$$

where

$$\sigma_a^2 = 2pq\,[a - d\,(p - q)]^2$$
$$\sigma_d^2 = 4p^2q^2d^2$$

are the genetic variances contributed by the additive and the dominance genetic effects, respectively and p and $q = 1 - p$ are the frequencies of alleles G and g in the population, respectively.

For a single locus G, there are nine possible sib-pair combinations (Table 15.7). The squares of the trait value differences between the sib-pairs are given in column two in the table. The means for genotypes GG, Gg and gg are

$$x_{(GG)j} = \mu + a + \varepsilon_j$$
$$x_{(Gg)j} = \mu + d + \varepsilon_j \qquad (15.22)$$
$$x_{(gg)j} = \mu - a + \varepsilon_j$$

The squares of the trait value differences between the possible sib-pairs can be easily obtained. For example, for the sib-pair GG-GG, it is

$$Y_{(GG-GG)j} = (x_{1j} - x_{2j})^2 = [(\mu + a + \varepsilon_{1j}) - (\mu + a + \varepsilon_{2j})]^2$$
$$= (\varepsilon_{1j} - \varepsilon_{2j})^2 = \varepsilon_j^2 \qquad (15.23)$$

and for the sib-pair GG-gg, it is

$$Y_{(GG-gg)j} = (x_{1j} - x_{2j})^2$$
$$= [(\mu + a + \varepsilon_{1j}) - (\mu - a + \varepsilon_{2j})]^2 = (2a + \varepsilon_j)^2 \qquad (15.24)$$

For convenience, set $E(\varepsilon_j^2) = \sigma_\varepsilon^2$.

For diploid systems, the sib-pairs must have 0, 1 and 2 alleles ibd and they must correspond to $\pi_j = 0, 0.5$ and 1, respectively. Frequencies of the sib-pair conditional on the $\pi_j = 0, 0.5$ and 1 are given in the columns 3, 4 and 5.

For $\pi_j = 0$, the sibs are independent at the trait locus and their frequencies are the same as the frequencies in random mating populations. The frequencies of the three genotypes (GG, Gg and gg) are $p^2, 2pq$ and q^2 under the assumption of Hardy-Weinberg equilibrium (Falconer 1989). The frequencies of the sib-pairs are the simple products of the three probabilities. For example, the frequency of sib-pair GG-GG is $p^2 \times p^2 = p^4$ and for sib-pair Gg-gg it is $2pq \times q^2 = 2pq^3$.

For $\pi_j = 1$, the two sibs have the same genotype and their probabilities are the same with the probability of one of the sibs in the population. The probabilities for all other sib-pairs with different genotypes are zero.

For $\pi_j = 0$ or 0.5, Haseman and Elston (1972) gave the following arguments: the sib-pairs GG-gg and gg-GG are impossible. For sib-pair GG-GG, given one allele ibd, the frequency is the probability of three G alleles occurring together (GG-G_, GG-_G, G_-GG, or _G-GG), which is p^3. For the same argument, the frequency of gg-gg is q^3. For GG-Gg, the G allele in the second sib must be ibd with a G allele in the first sib; the frequency is the probability of GG and g occurring which is p^2q. Using the same argument, the frequencies for GG-gG, Gg-GG, and gG-GG can be obtained. For the sib-pair Gg-Gg either allele G is ibd, in which case the frequency is pq^2, or allele g is ibd, in which case the frequency is p^2q. So, the frequency of Gg-Gg is a sum of the two probabilities, which is $pq^2 + p^2q = pq$.

The verification for those conditional frequencies is that the summation for each of the $\pi_j = 0, 0.5$ and 1 is one. For example,

$$\sum_{i=1}^{9} P[SP_i|\pi_j = 0] = p^2 + q^2 + 2pq = 1$$

$$\sum_{i=1}^{9} P[SP_i|\pi_j = 0.5] = p^3 + q^3 + pq + 2p^2q + 2pq^2 = 1 \qquad (15.25)$$

$$\sum_{i=1}^{9} P[SP_i|\pi_j = 1] = 1$$

Expected Square of the Sib-Pair Difference

The expected square of the sib-pair differences conditional on the probability of ibd is the simple summation of the product of the square of the sib-pair difference and its corresponding conditional probability over all possible sib-pairs. It can be written as

$$E[Y_j|\pi_j] = E\left[\sum_{i=1}^{9} (x_{1j} - x_{2j})^2 P[SP_i|\pi_j]\right] \qquad (15.26)$$

where $P[SP_i|\pi_j]$ is the frequency of the *ith* sib-pair conditional on π_j (Table 15.7). The expectations of the square of the sib-pair differences conditional on $\pi_j = 0, 0.5$ and 1 are

$$E[Y_j|\pi_j = 1] = E[\varepsilon_j^2 (p^2 + q^2 + 2pq)] = \sigma_\varepsilon^2$$

$$E[Y_j|\pi_j = 0] = E[\varepsilon_j^2] (p^4 + q^4 + 4pq) + E[(2a + \varepsilon_j)^2 + (-2a + \varepsilon_j)^2] p^2 q^2$$

$$+ E[(a - d + \varepsilon_j)^2 + (-a + d + \varepsilon_j)^2] 2p^3 q$$

$$+ E[(a + d + \varepsilon_j)^2 + (-a - d + \varepsilon_j)^2] 2pq^3 = \sigma_\varepsilon^2 + 2\sigma_a^2 + 2\sigma_d^2 \qquad (15.27)$$

$$E\left[Y_j|\pi_j = \frac{1}{2}\right] = E[\varepsilon_j^2] (p^3 + q^3 + pq) + E[(a - d + \varepsilon_j)^2 + (-a + d + \varepsilon_j)^2] p^2 q$$

$$+ E[(a + d + \varepsilon_j)^2 - (a - d + \varepsilon_j)^2] pq^2 = \sigma_\varepsilon^2 + \sigma_a^2 + 2\sigma_d^2$$

Solutions for the Linear Model

Now, we can derive the expectation for the parameters of the model based on Equation (15.17). In matrix notation, the model based on Equation (15.17) can be written as

$$Y = X\theta \qquad (15.28)$$

where

$$Y = \begin{bmatrix} E(Y_1|\pi_1) \\ E(Y_2|\pi_2) \\ \dots \\ E(Y_N|\pi_N) \end{bmatrix}_{N \times 1} \qquad X = \begin{bmatrix} {}_{n_0}[1 \quad 0] \\ {}_{n_1}[1 \quad 0.5] \\ {}_{n_2}[1 \quad 1] \end{bmatrix}_{N \times 1} \qquad and \qquad \theta = \begin{bmatrix} \alpha \\ \beta \end{bmatrix} \qquad (15.29)$$

Vector Y is the observed square sib-pair difference, vector θ contains the intercept and the slope for the linear model based on Equation (14.15), the first column of matrix X is the coefficient for the intercept and the second column is the coefficient for the slope. If we order the data according to their values of number of alleles ibd, then the first n_0 elements of matrix X are for the individual sib-pairs with $\pi_j = 0$, the middle n_1 elements are for $\pi_j = 0.5$ and the last n_2 elements are for $\pi_j = 1$. The total number of sib-pairs is $N = n_0 + n_1 + n_2$. The parameters can be obtained using the least square method and they are

$$\hat{\theta} = (X'X)^{-1}X'Y \qquad (15.30)$$

where

$$X'X = \left[\begin{bmatrix} 1 \\ 0 \end{bmatrix}_{n_0} \begin{bmatrix} 1 \\ 0.5 \end{bmatrix}_{n_1} \begin{bmatrix} 1 \\ 1 \end{bmatrix}_{n_2} \right] \begin{bmatrix} n_0 \begin{bmatrix} 1 & 0 \end{bmatrix} \\ n_1 \begin{bmatrix} 1 & 0.5 \end{bmatrix} \\ n_2 \begin{bmatrix} 1 & 1 \end{bmatrix} \end{bmatrix} = \begin{bmatrix} N & a \\ a & b \end{bmatrix}$$

and

$$a = n_1/2 + n_2$$
$$b = n_1/4 + n_2$$

so

$$(X'X)^{-1} = \frac{1}{Na - b^2} \begin{bmatrix} b & -a \\ -a & N \end{bmatrix} = \frac{1}{n_1 n_2/4 + n_0 n_1/4 + n_0 n_2} \begin{bmatrix} b & -a \\ -a & N \end{bmatrix}$$

and

$$X'Y = \left[\begin{bmatrix} 1 \\ 0 \end{bmatrix}_{n_0} \begin{bmatrix} 1 \\ \frac{1}{2} \end{bmatrix}_{n_1} \begin{bmatrix} 1 \\ 1 \end{bmatrix}_{n_2} \right] \begin{bmatrix} E(Y_1|\pi_1) \\ E(Y_2|\pi_2) \\ \dots \\ E(Y_N|\pi_N) \end{bmatrix}$$

$$= \begin{bmatrix} N\sigma_\varepsilon^2 + (n_1 + 2n_0)\sigma_a^2 + 2(n_1 + n_0)\sigma_\varepsilon^2 \\ (n_1/2 + n_2)\sigma_\varepsilon^2 + n_1\sigma_a^2/2 + n_1\sigma_\varepsilon^2 \end{bmatrix}$$

The solutions for the parameters are

$$\hat{\theta} = \begin{bmatrix} \hat{\alpha} \\ \hat{\beta} \end{bmatrix} = (X'X)^{-1}(X'Y)$$

$$= \frac{1}{n_1 n_2/4 + n_0 n_1/4 + n_0 n_2} \begin{bmatrix} b & -a \\ -a & N \end{bmatrix} \begin{bmatrix} N\sigma_\varepsilon^2 + (n_1 + 2n_0)\sigma_a^2 + 2(n_1 + n_0)\sigma_\varepsilon^2 \\ (n_1/2 + n_2)\sigma_\varepsilon^2 + n_1\sigma_a^2/2 + n_1\sigma_\varepsilon^2 \end{bmatrix}$$

$$= \begin{bmatrix} \sigma_\varepsilon^2 + 2(\sigma_a^2 + \sigma_d^2) + 2n_1 n_2\sigma_d^2/(n_1 n_2 + n_0 n_1 + 4n_0 n_2) \\ -2(\sigma_a^2 + \sigma_d^2) - 2n_1(n_2 - n_0)\sigma_d^2/(n_1 n_2 + n_0 n_1 + 4n_0 n_2) \end{bmatrix}$$

The solutions lead the model based on Equation (15.17) to

$$E[Y_j|\pi_j] = \sigma_\varepsilon^2 + 2(\sigma_a^2 + \sigma_d^2) + 2n_1n_2\sigma_d^2/(n_1n_2 + n_0n_1 + 4n_0n_2)$$
$$-[2(\sigma_a^2 + \sigma_d^2) - 2n_1(n_2 - n_0)\sigma_d^2/(n_1n_2 + n_0n_1 + 4n_0n_2)]\pi_j$$

$$(15.31)$$

Equation (15.31) asymptotically reduces to

$$E[Y_j|\pi_j] = \sigma_\varepsilon^2 + 2(\sigma_a^2 + \sigma_d^2) - 2(\sigma_a^2 + \sigma_d^2)\pi_j \qquad (15.32)$$

The expectations of the estimated intercept and the regression coefficient are

$$E(\hat{\alpha}) = \sigma_\varepsilon^2 + 2(\sigma_a^2 + \sigma_d^2)$$
$$E(\hat{\beta}) = -2(\sigma_a^2 + \sigma_d^2)$$

$$(15.33)$$

If we know the gene G controls the trait, then the genetic variance which can be explained by the gene is $-0.5\hat{\beta}$.

15.3.2 MARKER MODEL

In the previous sub-sections, the model was developed for the cases in which the genotypes of parents and the sib-pairs could be determined for the genes controlling the trait or the QTL was located right on the marker. However, this is impossible for many experiments. Generally, the genotypes or the gene locations and the gene effects are what we want to find through QTL mapping. To do this, we need to derive the relationship between the QTL model with known genotypes for the genes and the marker model without knowing the genotypes of the genes. In the next several paragraphs, the model will be derived in terms of the marker genotypes. We can write a marker model as

$$E(Y_{Aj}|\hat{\pi}_{Aj}) = \alpha_A + \beta_A\pi_{Aj} \qquad (15.34)$$

where Y_{Aj} is the squared sib-pair difference for the *jth* sib-pair conditional on the proportion of alleles ibd for a given marker π_{jA}. The main tasks of the sib-pair approach are to find the estimators α_A and β_A and to derive their biological inferences.

Table 15.8 Conditional probability $P[\pi_{jG}|\pi_{jA}]$, where $\Psi = r^2 + (1-r)^2$.

π_{jG}	π_{jA}		
	0	0.5	1.0
0	Ψ^2	$2\Psi(1-\Psi)$	$(1-\Psi)^2$
0.5	$\Psi(1-\Psi)$	$1 - 2\Psi + 2\Psi^2$	$\Psi(1-\Psi)$
1	$(1-\Psi)^2$	$2\Psi(1-\Psi)$	Ψ^2

Consider that a marker A is r recombination units away from the putative QTL Q. We can use π_{jG} and π_{jA} to denote the probabilities of the allele ibd at the QTL locus and the marker A locus, respectively. Table 15.8 shows the probability of the allele ibd at the QTL locus π_{jG} conditional on the probability of the allele ibd at the marker locus π_{jA}. See Haseman and Elston's original paper for the detailed steps to obtain the conditional probabilities. The meaning of the conditional probability is simple. For example

$$P\left[\pi_{jG} = 0 \mid \pi_{jA} = 0\right] = \Psi^2$$

means that if $\pi_{jA} = 0$ for the marker locus and the marker locus is r recombination units away from the QTL, then the chance that $\pi_{jG} = 0$ is Ψ^2.

Table 15.9 Conditional probability $P\left[\pi_{jA} \mid \hat{\pi}_{jA}\right]$.

$\hat{\pi}_{jA}$	π_{jA}		
	0	0.5	1.0
0	1.0	0	0
0.25	0.5	0.5	0
0.5	0.25(1-z)	0.5(1+z)	0.25(1-z)
0.75	0	0.5	0.5
1	0	0	1

p is the frequency of allele A in the population, $q = 1 - p$ is frequency of allele a and $z = p^2q^2 / (p^4 + 5p^2q^2 + q^4)$.

There is also a need for distinguishing between the estimated ibd value for the locus, $\hat{\pi}_{jA}$, and the true ibd, π_{jA}. Table 15.9 shows that the probabilities for the true ibd values are 0, 0.5 and 1.0, conditional on the observed ibd values. See Haseman and Elston (1972) for procedures to obtain the probabilities.

If we can replace the ibd value for the QTL in Equation (15.17) with the observed marker ibd value for the marker, we have the expectation

$$E\left[Y_j \mid \hat{\pi}_{jA}\right] = \sum_{\pi_{jG}} E\left[Y_j \mid \pi_{jG}\right] P\left[\pi_{jG} \mid \hat{\pi}_{jA}\right]$$

$$= \sum_{\pi_{jG}} \sum_{\pi_{jA}} E\left[Y_j \mid \pi_{jG}\right] P\left[\pi_{jG} \mid \pi_{jA}\right] P\left[\pi_{jA} \mid \hat{\pi}_{jA}\right] \tag{15.35}$$

In matrix notation it is

$$E_A = P'_{A|\hat{A}} P_{G|A} E_G \tag{15.36}$$

where

$$E_G = \begin{bmatrix} \sigma_\varepsilon^2 + 2\sigma_g^2 \\ \sigma_\varepsilon^2 + \sigma_g^2 \\ \sigma_\varepsilon^2 \end{bmatrix},$$

$$P_{G|A} = \begin{bmatrix} \Psi^2 & 2\Psi(1-\Psi) & (1-\Psi)^2 \\ \Psi(1-\Psi) & 1-2\Psi+2\Psi^2 & \Psi(1-\Psi) \\ (1-\Psi)^2 & 2\Psi(1-\Psi) & \Psi^2 \end{bmatrix},$$

and $P_{A|\hat{A}} = \begin{bmatrix} 1 & 0.5 & 0.25(1-z) & 0 & 0 \\ 0 & 0.5 & 0.5(1+z) & 0.5 & 0 \\ 0 & 0 & 0.25(1-z) & 0.5 & 1 \end{bmatrix}$

The additive model is assumed here, so $\sigma_a^2 = \sigma_g^2$. The expectations are

$$E_A = P'_{A|\hat{A}} P_{G|A} E_G$$

$$= \begin{bmatrix} 1 & 0.5 & 0.25(1-z) & 0 & 0 \\ 0 & 0.5 & 0.5(1+z) & 0.5 & 0 \\ 0 & 0 & 0.25(1-z) & 0.5 & 1 \end{bmatrix}' \begin{bmatrix} \Psi^2 & 2c & (1-\Psi)^2 \\ c & 1-2\Psi+2\Psi^2 & c \\ (1-\Psi)^2 & 2c & \Psi^2 \end{bmatrix} \begin{bmatrix} \sigma_\varepsilon^2 + 2\sigma_g^2 \\ \sigma_\varepsilon^2 + \sigma_g^2 \\ \sigma_\varepsilon^2 \end{bmatrix}$$

$$= \begin{bmatrix} \sigma_\varepsilon^2 + 2\Psi\sigma_g^2 \\ \sigma_\varepsilon^2 + (0.5+\Psi)\sigma_g^2 \\ \sigma_\varepsilon^2 + \sigma_g^2 \\ \sigma_\varepsilon^2 + (1.5-\Psi)\sigma_g^2 \\ \sigma_\varepsilon^2 + 2(1-\Psi)\sigma_g^2 \end{bmatrix} = \begin{bmatrix} E[Y_j|\hat{\pi}_{jA} = 0] \\ E[Y_j|\hat{\pi}_{jA} = 0.25] \\ E[Y_j|\hat{\pi}_{jA} = 0.5] \\ E[Y_j|\hat{\pi}_{jA} = 0.75] \\ E[Y_j|\hat{\pi}_{jA} = 1] \end{bmatrix}$$

where

$$c = \Psi(1-\Psi)$$

Now we can write the marker model in matrix notation as

$$Y_A = X_A \theta_A \tag{15.37}$$

where the subscript A denotes that the model is based on the marker genotype, instead of the QTL genotype and

$$Y_A = \begin{bmatrix} E(Y_1|\hat{\pi}_{A1}) \\ E(Y_2|\hat{\pi}_{A2}) \\ \dots \\ E(Y_N|\hat{\pi}_{AN}) \end{bmatrix}_N \qquad X_A = \begin{bmatrix} n_0 \begin{bmatrix} 1 & 0 \end{bmatrix} \\ n_1 \begin{bmatrix} 1 & 1/4 \end{bmatrix} \\ n_2 \begin{bmatrix} 1 & 1/2 \end{bmatrix} \\ n_3 \begin{bmatrix} 1 & 3/4 \end{bmatrix} \\ n_4 \begin{bmatrix} 1 & 1 \end{bmatrix} \end{bmatrix}_N \qquad and \qquad \theta_A = \begin{bmatrix} \alpha_A \\ \beta_A \end{bmatrix} \tag{15.38}$$

Vector Y_A is the observed square sib-pair difference conditional on the probability of ibd in the marker locus, vector θ_A contains the intercept and the slope for the marker model based on Equation (15.35), the first column of matrix X_A is the coefficient for the intercept and the second column is the coefficient for the slope. If we order the data according to the values of number of alleles ibd, then the first n_0 elements of matrix X are for the individual sib-pairs with $\hat{\pi}_{Aj} = 0$, n_1 elements are for $\hat{\pi}_{Aj} = 1/4$, n_2 elements are for $\hat{\pi}_{Aj} = 1/2$, n_3 elements are for $\hat{\pi}_{Aj} = 3/4$ and the last n_4 elements are for $\hat{\pi}_{Aj} = 1$. The total number of sib-pairs is

$$N = \sum_{k=0}^{5} n_k$$

The parameters can be obtained using the least square method and they are

$$\hat{\theta}_A = (X_A'X_A)^{-1}X_A'Y_A \tag{15.39}$$

where

$$X_A'X_A = \begin{bmatrix} \begin{bmatrix} 1 \\ 0 \end{bmatrix}_{n_0} & \begin{bmatrix} 1 \\ \frac{1}{4} \end{bmatrix}_{n_1} & \begin{bmatrix} 1 \\ \frac{1}{2} \end{bmatrix}_{n_2} & \begin{bmatrix} 1 \\ \frac{3}{4} \end{bmatrix}_{n_3} & \begin{bmatrix} 1 \\ 1 \end{bmatrix}_{n_4} \end{bmatrix} \begin{bmatrix} n_0 \begin{bmatrix} 1 & 0 \end{bmatrix} \\ n_1 \begin{bmatrix} 1 & 1/4 \end{bmatrix} \\ n_2 \begin{bmatrix} 1 & 1/2 \end{bmatrix} \\ n_3 \begin{bmatrix} 1 & 3/4 \end{bmatrix} \\ n_4 \begin{bmatrix} 1 & 1 \end{bmatrix} \end{bmatrix}_1 = \begin{bmatrix} N & a \\ a & b \end{bmatrix}$$

where

$$a = \frac{n_1}{4} + \frac{n_2}{2} + \frac{3n_3}{4} + n_4$$

$$b = \frac{n_1}{16} + \frac{n_2}{4} + \frac{9n_3}{16} + n_4$$

$$(X_A'X_A)^{-1} = \frac{1}{Nb - a^2} \begin{bmatrix} b & -a \\ -a & N \end{bmatrix}$$

$$X_A'Y_A = \begin{bmatrix} N\sigma_\varepsilon^2 + [2a + (2n_0 + n_1 - n_3 - 2n_4)\,\Psi]\,\sigma_g^2 \\ a\sigma_\varepsilon^2 + \left[2b + \left(\dfrac{n_1}{4} - \dfrac{3n_3}{4} - 2n_4\right)\Psi\right]\sigma_g^2 \end{bmatrix}$$

The solutions for the parameters are

$$\hat{\theta}_A = \begin{bmatrix} \hat{\alpha}_A \\ \hat{\beta}_A \end{bmatrix} = (X_A'X_A)^{-1}(X_A'Y_A)$$

$$= \frac{1}{Nb - a^2} \begin{bmatrix} b & -a \\ -a & N \end{bmatrix} \begin{bmatrix} N\sigma_\varepsilon^2 + [2a + (2n_0 + n_1 - n_3 - 2n_4)\,\Psi]\,\sigma_g^2 \\ a\sigma_\varepsilon^2 + \left[2b + \left(\dfrac{n_1}{4} - \dfrac{3n_3}{4} - 2n_4\right)\Psi\right]\sigma_g^2 \end{bmatrix}$$

$$= \begin{bmatrix} \sigma_\varepsilon^2 + 2\Psi\sigma_g^2 \\ 2(1 - 2\Psi)\,\sigma_g^2 \end{bmatrix}$$

The solutions lead the model based on Equation (15.34) to

$$\begin{aligned} E[Y_j | \hat{\pi}_{jA}] &= [\sigma_\varepsilon^2 + 2\Psi\sigma_g^2] + 2(1 - 2\Psi)\,\sigma_g^2\hat{\pi}_{jA} \\ &= [\sigma_\varepsilon^2 + 2(1 - 2r + r^2)\,\sigma_g^2] - 2(1 - 2r)^2\sigma_g^2\hat{\pi}_{jA} \end{aligned} \tag{15.40}$$

if $\sigma_a^2 = \sigma_g^2$ and $\sigma_d^2 = 0$ (additive model) is assumed. The slope of the regression is a function of recombination fraction between the marker and the putative QTL and the genetic variance attributed to the QTL. If the recombination fraction is 0.5 or the genetic variance is zero, then the function is zero.

15.3.3 IMPLEMENTATION OF THE SIB-PAIR METHOD

The sib-pair method has been used extensively for QTL mapping in humans. The theory for the sib-pair method was described earlier. In this section, implementation of sib-pair methods including hypothesis tests and parameter estimation will be discussed. Hypothesis tests can be performed on the slope, using an F-test. The method to estimate the ibd $\hat{\pi}_{ji}$ at marker locus i for each of the individual sib-pair j in the mapping population was described. The slope for the model can be easily estimated using commonly used statistical analysis software such as SAS. As shown by Equation (15.41), the expectation of the slope is $-2(1-2r)^2\sigma_g^2$. The genetic effect and the recombination fraction are confounded using the simple regression analysis.

Table 15.10 Dummy variables for the regression approach for the sib-pair method.

$\hat{\pi}_{ji}$	X_{j1}	X_{j2}	X_{j3}	X_{j4}	X_{j5}
0	1	0	0	0	0
0.25	0	1	0	0	0
0.5	0	0	1	0	0
0.75	0	0	0	1	0
1	0	0	0	0	1

Haseman and Elston (1972) give an approach to obtain maximum likelihood estimates of the recombination fraction and the genetic effects, which required iterative procedures to obtain the solution. Here, a multiple regression approach, which is similar to the approaches used for the experimental populations, will be introduced. For the sib-pair data, five dummy variables can be coded according to their observed ibd value, as shown in Table 15.10. A multiple linear regression model can be written as

$$E[Y_j|\hat{\pi}_{jA}] = \sum_{k=1}^{5} \theta_k X_{jk} + \varepsilon_j \tag{15.41}$$

If we set the estimated partial regression coefficients equal to their expectations, we have

$$\hat{\theta}_1 = \sigma_\varepsilon^2 + 2\Psi\sigma_g^2$$

$$\hat{\theta}_2 = \sigma_\varepsilon^2 + (0.5+\Psi)\sigma_g^2$$

$$\hat{\theta}_3 = \sigma_\varepsilon^2 + \sigma_g^2 \tag{15.42}$$

$$\hat{\theta}_4 = \sigma_\varepsilon^2 + (1.5-\Psi)\sigma_g^2$$

$$\hat{\theta}_5 = \sigma_\varepsilon^2 + 2(1-\Psi)\sigma_g^2$$

Two constraints can be found; they are

$$2\theta_2 - \theta_1 = \theta_3$$

$$2\theta_4 - \theta_5 = \theta_3$$

If we replace θ_3 with $2\theta_2 - \theta_1$ and θ_4 with $0.5(2\theta_2 - \theta_1 + \theta_5)$, Equation (15.41) becomes

$$E[Y_j|\hat{\pi}_{jA}] = \theta_1(X_{j1} - X_{j3} + 0.5X_{j4}) + \theta_2(X_{j2} + 2X_{j3} + X_{j4}) + \theta_5(0.5X_{j4} + X_{j5}) + \varepsilon_j$$

and Equation (15.42) reduces to

$$\hat{\theta}_1 = \sigma_\varepsilon^2 + 2\Psi\sigma_g^2$$

$$\hat{\theta}_2 = \sigma_\varepsilon^2 + (0.5 + \Psi)\,\sigma_g^2 \tag{15.43}$$

$$\hat{\theta}_5 = \sigma_\varepsilon^2 + 2\,(1 - \Psi)\,\sigma_g^2$$

The solutions for the parameters are

$$\hat{\sigma}_g^2 = 2\,(\hat{\theta}_1 - \hat{\theta}_2) + \hat{\theta}_5$$

$$\hat{\sigma}_\varepsilon^2 = 4\hat{\theta}_2 - 3\hat{\theta}_1 - \hat{\theta}_5 \tag{15.44}$$

$$\hat{\Psi} = 0.5 + \frac{\hat{\theta}_1 - \hat{\theta}_2}{\hat{\theta}_1 - \hat{\theta}_2 + 0.5\hat{\theta}_5}$$

Because

$$\Psi = r^2 + (1 - r)^2$$

therefore, the estimate of the recombination fraction between the putative QTL and the marker is

$$\hat{r} = 0.5 \pm 0.5 \sqrt{\frac{2\,(\hat{\theta}_1 - \hat{\theta}_2)}{\hat{\theta}_1 - \hat{\theta}_2 + 0.5\hat{\theta}_5}} \tag{15.45}$$

EXERCISES

Exercise 15.1

In this chapter, QTL mapping in open-pollinated populations using the single-marker analysis has been discussed. The disadvantage of QTL mapping in an open-pollinated population is the high false positive probability caused by the correlation between the allelic frequencies of markers and QTLs in the pollen pool. Please do the following problems related to this subject.

(1) A possible approach to limit this disadvantage statistically is to bring the allelic frequencies of markers and QTLs in the pollen pool into the model of QTL mapping. The marker allelic frequencies in the pollen pool can be estimated if a heterozygous maternal parent is assumed (see Chapter 7). Can QTL allelic frequencies be estimated? (Hint: Answer this question by assuming that the maternal parent is (i) a homozygote for the QTL, (ii) a heterozygote or (iii) an unknown genotype.)

(2) Another possible approach to limit the disadvantage is to control the pollen resources experimentally. Outline some key points of this approach and steps to implement this approach. (Hint: Use maize as an example. Focus on the following questions: (i) Can you control the proportion of outcrossing? (ii) Can you control the pollen composition? (iii) Does it make a difference if you pick different maternal parents?)

(3) It may be argued that QTL mapping should be done in species in which experimental populations from controlled crosses can be easily obtained. Explain some of the ideas behind this argument. Prove or disprove this argument using specific examples.

Exercise 15.2

QTL mapping using open-pollinated populations has its disadvantages. However, it also has many advantages over QTL mapping using controlled crosses. Explain if the following points are true and why.

(1) For some organisms, an experimental population from controlled matings is difficult or impossible to obtain. For example, it will take a long time to generate experimental populations for most of the forest tree species. For such organisms, especially for forest trees, open-pollinated populations are widely available.

(2) Open-pollinated populations are closer to nature than are the populations of controlled crosses. QTL mapping in open-pollinated populations may have more evolutionary significance than in the populations of controlled crosses.

(3) In terms of mating, an open-pollinated population is a mixture of self and random mating (MSR), which is a generalization for different types of populations of controlled crosses. Commonly used mapping populations, such as backcross and F2, are special cases of populations with MSR with different settings of outcrossing rates and allelic frequencies in the pollen pool.

(4) The goal of QTL mapping is to assist selection using genetic markers in the populations. The bias of the hypothesis test and parameter estimation caused by the correlation between allelic frequencies of QTLs and markers in a pollen pool is irrelevant for this application. Actually, the biased estimate is closer to the parameter of interest, the average effect of gene substitution, than is the unbiased estimate, which is the genotypic value difference among QTL genotypes. (Hint: See Chapter 2)

Exercise 15.3

Sib-pair method has been used in humans for detection of genes responsible for complex human diseases. Point out the advantages and disadvantages of this approach.

CHAPTER 16

QTL MAPPING: STATISTICAL POWER

16.1 INTRODUCTION

How many genes are there in the organisms we study as biological systems? To what extent are genes the same in different species? How many genes control a trait of interest? How many genes can be detected by a QTL mapping approach? These are fundamental biological questions of classical genetics that can now be investigated using recent developments in genomic mapping and genomic sequencing.

QTL mapping involves selecting mapping populations, traits of interest, genetic marker assays and trait evaluations, then making inferences about QTLs based on association analysis between the genetic markers and the traits. In terms of statistical treatments of the data, t-test, analysis of variance, regression and generalized likelihood approaches are commonly used methods to detect association between traits and genetic markers. A significant association between the traits and the markers may be an indication of a QTL residing near the markers. However, statistical significance does not always indicate biological significance due to the multiple test problem associated with QTL mapping.

The number of genes which can be found and the precision with which those genes can be located on the genome are two important questions facing QTL mapping. These questions have to be answered in order to design the QTL mapping experiments efficiently. At the very least, the following factors affect power of QTL mapping:

- number of genes controlling the traits and their genome positions
- distribution of genetic effects and existence of gene interactions
- heritability of the trait
- number of genes segregating in mapping populations
- mapping population type and size
- linkage map density and coverage
- statistical methodology and significance level used for QTL mapping

In Chapter 4, the basic theory of statistical power and estimation of a confidence interval for estimated parameters were discussed. The basic methods for QTL mapping were described in Chapters 13, 14 and 15. In this chapter, ways to answer the two fundamental questions of QTL mapping experimental design using the concepts of statistical power will be explored. Detection power on a single-locus level will be discussed, then the methodology will be extended to multi-locus detection power.

16.2 SINGLE QTL DETECTION POWER

16.2.1 SINGLE MARKER ANALYSIS

Single marker analysis (SMA) is based on the comparison between marker genotypic means through a t-test, an analysis of variance, a likelihood ratio test or a simple regression for a trait on coded marker genotypes. Under some assumptions, these approaches are equivalent or similar. If a QTL is located right on a genetic marker, a t-test statistic

$$t_Q = \frac{\hat{\mu}_1 - \hat{\mu}_2}{\sqrt{\hat{s}^2 \left(\frac{1}{n_1} + \frac{1}{n_2} \right)}} \tag{16.1}$$

can be used to test the QTL effect where the hats (^) denote that the values are estimated from samples; $\mu_1 - \mu_2$ denotes the difference between the two classes of QTL genotypes, n_1 and n_2 are the sample sizes of the two classes of QTL genotype and s^2 is the pooled estimated of variance within the two classes. The expectation of the pooled variance is

$$E[s^2] = \frac{\sigma_e^2}{b} + \sigma_{G(QTL)}^2 \tag{16.2}$$

where b is the number of replications for a genotype, σ_e^2 is the error variance and $\sigma_{G(QTL)}^2$ is the part of the genetic variance which cannot be explained by the QTL. Assume that the recombination fraction between a marker and the QTL is r. The t-test statistic becomes

$$t_M = \frac{\hat{\mu}_{MM} - \hat{\mu}_{Mm}}{\sqrt{\hat{s}_M^2 \left(\frac{1}{n_1} + \frac{1}{n_2} \right)}} \tag{16.3}$$

where s_M^2 is the pooled variance within the two classes of marker genotypes and n_1, n_2 are sample sizes for the two classes. The expectation for the pooled variance is

$$E[s_M^2] = \frac{\sigma_e^2}{b} + \sigma_{G(QTL)}^2 + 4r(1-r)\tau_{QTL}^2 \tag{16.4}$$

where τ_{QTL}^2 is the fixed QTL effect. The expectation of the difference between the two marker classes is

$$E[\mu_{MM} - \mu_{Mm}] = (1 - 2r)(\mu_1 - \mu_2)$$
$$= (1 - 2r)a$$

The power of the test for a single marker analysis using the t-test is

$$Pr[t_{n-2, t_{nv}} \geq t_\alpha] + Pr[t_{n-2, t_{nv}} \leq -t_\alpha] \tag{16.5}$$

where α is a significance level, t_α is a critical value for declaring a significance and t_{nc} is a parameter for a noncentral t-distribution. It can be estimated using

$$t_{nc} = \frac{(1 - 2r)a}{\sqrt{[\sigma_e^2/b + \sigma_{G(QTL)}^2 + 4r(1-r)\tau_{QTL}^2]/n}} \tag{16.6}$$

where n is the total sample size and $2a$ is the QTL effect. Relationships among the related components can be formulated as

$$\tau_{QTL}^2 = a^2$$

$$H = \frac{\sigma_G^2}{\sigma_G^2 + \sigma_e^2/b} \tag{16.7}$$

$$\sigma_{G(QTL)}^2 = \sigma_G^2 - \tau_{QTL}^2$$

where H is the heritability of a trait.

Assume that, for any trait, genes with additive effects follow a geometric series (see Chapter 19 for detail). Therefore, for multiple genes, the genetic effects can be considered as a geometric series

$$\sigma_G^2 (1 - \lambda) [\lambda^0, \lambda^1, \lambda^2, \lambda^3, ..., \lambda^{l-1}] \tag{16.8}$$

where σ_G^2 is the total genetic variance, λ is the parameter for the geometric series and $0 < \lambda \leq 1$. The sum of the individual genetic variances is the total genetic variance; it is

$$\sum_{i=1}^{l} \sigma_{Gi}^2 = \sum_{i=1}^{l} [\sigma_G^2 (1 - \lambda) \lambda^i] = \sigma_G^2 \tag{16.9}$$

If σ_G^2 is set to 1, then the summation is unity. It is common to define

$$N_e = \frac{1 + \lambda}{1 - \lambda} \tag{16.10}$$

as the number of effective genes.

Once the genetic variance explained by the locus is obtained, the genetic effects (additive and dominant) can be determined. Table 16.1 lists some examples of the distributions of genetic effects as a geometric series. Table 16.1 contains situations of effective numbers of loci 2, 5 and 10. Figure 16.1 shows the graphic distributions of the genetic effects. There are a small number of genes with large genetic effects and a large number of genes with small genetic effects.

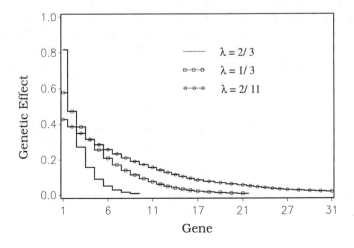

Figure 16.1 Genes and their additive effects will be used to illustrate the statistical power for QTL detection.

Table 16.1 Genetic effect is distributed as a geometric series when the numbers of effective loci are 2, 5 and 10. The total genetic variance is one.

λ	Ne	σ_{gi}
1/3	2	0.82, 0.47, ...
2/3	5	0.58, 0.47, 0.38, 0.31, 0.26, ...
9/11	10	0.43, 0.39, 0.35, 0.32, 0.29, 0.26, 0.23, 0.21, 0.19, 0.17, ...

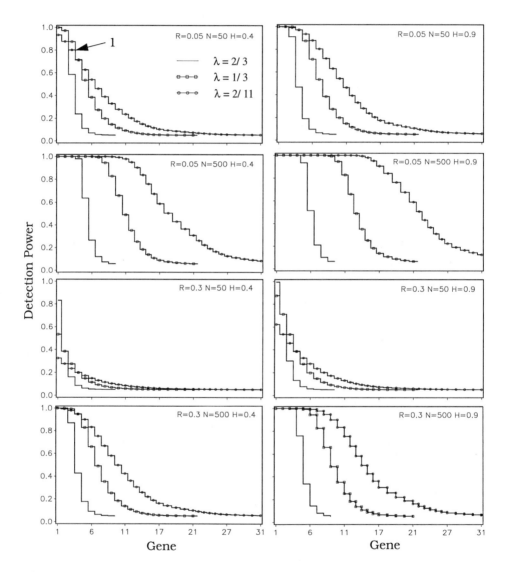

Figure 16.2 Statistical power for detecting the genes in Figure 16.1 using a single marker t-test in backcross progeny for some combinations of the alternative hypothesis. R = recombination fraction between marker and QTL, N = population size and H = heritability of the trait. Probability for declaring significance is 95%.

Figure 16.2 shows the statistical power for detecting the genes in Figure 16.1 using the single marker analysis approach in backcross progeny for some combinations of the alternative hypothesis. For example, for marker density of 0.05 recombination units between markers, a sample size of 50 and the heritability of 0.4, statistical power is 80% for a gene controlling 12.25% of genetic variation (gene number 3 for $\lambda = 2/11$ and $\sigma_{gi} = 0.35$) (see the arrow in Figure 16.2).

It is common to consider that the genetic effects are normally distributed. This means that genes with large and small effects have low frequencies and genes with median effects have relatively high frequencies. It is also common to determine the genetic effects arbitrarily or based on experimental results.

16.2.2 INTERVAL MAPPING

For interval mapping using a linear regression approach with backcross progeny, the null hypothesis $H_0 : \mu_1 - \mu_2 = 0$ (no QTL effect in the interval) can be tested using the log likelihood ratio test statistic

$$G = \frac{SSE_{reduced} - SSE_{full}}{\dfrac{SSE_{full}}{df_{Efull}}} \tag{16.11}$$

where SSEs are the residual sums of the squares for the full and reduced models. The full model is

$$y_j = X_1 \mu_1 + X_2 [(1 - \rho) \mu_1 + \rho \mu_2] + X_3 [\rho \mu_1 + (1 - \rho) \mu_2] + X_4 \mu_2 + \varepsilon_j \tag{16.12}$$

where X_1, X_2, X_3 and X_4 are the dummy variables for the four marker genotypes as listed in Table 14.3 and $\rho = r_1/r$ is the relative position of the putative QTL in the interval flanked by the two markers. Three unknown parameters are involved in the model: the two QTL genotypic means and the parameter for relative QTL position. The reduced model is the full model under the null hypothesis ($H_0 : \mu_1 - \mu_2 = 0$), which is

$$\begin{aligned} y_j &= (X_1 + X_2 + X_3 + X_4) \mu + \varepsilon_j \\ &= (\mu + \varepsilon_j) \end{aligned} \tag{16.13}$$

The expected residual sum of squares for the model is

$$\begin{aligned} E[SSE_{R(full)}] &= E[(Y - X\beta)'(Y - X\beta)] \\ &= n\sigma^2 + 0.5rn (\mu_1 - \mu_2)^2 [\rho^2 + (1 - \rho)^2] \end{aligned} \tag{16.14}$$

If we assume that the QTL is located in the middle of the interval between two markers ($\rho = 0.5$), the expected residual sum of squares is

$$E[SSE_{R(full)}] = n\sigma^2 + 0.25rna^2$$

The expected residual sum of the squares for the reduced model is

$$\begin{aligned} E[SSE_{R(reduced)}] &= E[(Y - \mu)'(Y - \mu)] \\ &= n\sigma^2 + n(\mu_1 - \mu_2)^2 + rn(\mu_1 - \mu_2)^2 [1 - (0.5 - \rho)^2] \\ &= n\sigma^2 + n(1 + r)a^2 \end{aligned} \tag{16.15}$$

So, the expected log likelihood ratio test statistic is

$$E[G] = \frac{[n\sigma^2 + n(1+r)a^2] - [n\sigma^2 + 0.25rna^2]}{[n\sigma^2 + 0.25rna^2] / (n-2)}$$

$$= \frac{(n-2)(1+0.75r)a^2}{\sigma^2 + 0.25ra^2} \tag{16.16}$$

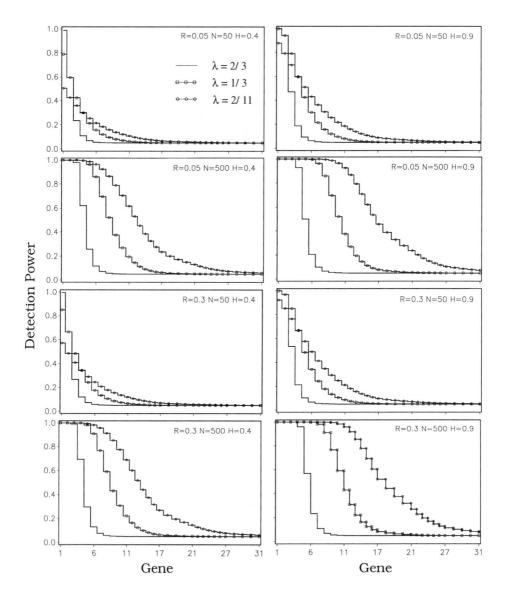

Figure 16.3 Statistical power for detecting the genes in Figure 16.1 using the interval mapping approach in backcross progeny for some combinations of alternative hypothesis. R = recombination fraction between two markers flanking the QTL, N = population size and H = heritability of the trait. Probability of declaring a significance is 95%. An additive model is assumed.

where σ^2 can be determined using the relationships in Equation (16.7) when an alternative hypothesis is specified. Figure 16.3 shows the statistical power for detecting the genes in Figure 16.1 using the interval mapping approach in backcross progeny for some combinations of the alternative hypothesis.

In general, the interval mapping approach has a higher statistical power for detecting QTLs than the single marker t-test when linkage map density is low (recombination fraction is large between markers). However, the single marker t-test is as powerful as or more powerful than the interval mapping approach when the linkage map density is high.

16.3 MULTIPLE QTLS

In the previous section, methods for determining statistical power for QTL detection at a single-QTL level were explained. These methods are useful when the trait is controlled by a single gene. When traits are controlled by a number of genes, the single QTL detection power may be limited. The question of interest changes from "what is the chance that this QTL can be detected?" in the single QTL situation to "what is the chance that x% of the QTL can be detected?." In this section, methods for determining statistical power at a multi-locus level will be explained.

16.3.1 RATIONALE

Assume that there are q_t independent QTLs controlling a trait of interest. There is no genetic linkage among the genes and no genetic interaction among the gene effects. The statistical powers for each of the genes are $\{p_1, p_2, ..., p_{q_t}\}$ at α level. The average statistical power (P_e) for detecting at least q_e QTLs, where $q_e \leq q_t$, can be determined by

$$P_e = \frac{1}{c} \sum_{j=1}^{c} \prod_{i=1}^{q_e} p_{i_{(j)}} \tag{16.17}$$

where

$$c = \sum_{k=q_e}^{q_t} \binom{q_t}{k}$$

$$= \sum_{k=q_e}^{q_t} \frac{q_t!}{k!\,(q_t - k)!} \tag{16.18}$$

is the number of possible combinations and $p_{i_{(j)}}$ is the statistical power for detecting gene i for the combination j. The $p_{i_{(j)}}$ is determined differently than the simple single QTL model. For different combinations, the statistical power for an individual locus could be different depending on the order of the genes detected. For the first gene detected, the residual error used for the test statistic is large. For later genes, the previously detected genes can be used to control the residual error. For genes with the same gene effects, the statistical power required for the later detected genes is generally higher than for the earlier detected genes. In practice, the gene with the largest effect is detected first by screening the whole genome and then using the gene as a control to search for the second largest gene. An iterative approach for detecting-adjusting could increase the power of QTL mapping. The best situation would be that the residual caused by genetics could be totally adjusted. When the number of genes is small, it is relatively easy to achieve the situation. When the number of genes is large, the approach faces a computation problem. The non-centrality parameters for a multiple genes situation can be estimated using Equation (16.6) for a t-test and Equation (16.17) for the interval mapping approach. However, the expectation for the residual mean square is

$$E[s_M^2] = \frac{\sigma^2}{b} + \sigma_{G(QTL)}^2 - \sum_{i=1}^{q_d} (1 - 2r_i)^2 \tau_{QTL_i}^2 + 4r_{i+1}(1 - r)_{i+1} \tau_{QTL_{i+1}}^2 \qquad (16.19)$$

and different from Equation (16.6), where q_d is the number of genes detected, r_i is the recombination between the gene and the reference marker and $\tau_{QTL_i}^2$ is the fixed genetic variance explained by the gene. QTL $i+1$ is the gene under testing. For interval mapping using the likelihood approach, the non-centrality parameter for determining the statistical power for a single QTL under a multiple QTL model can be computed using Equation (16.16). However, the residual mean square should be adjusted by the detected QTLs and can be determined by

$$\sigma^2 = \frac{1 + H(b - 1)}{bH}\sigma_G^2 - \sum_{i=1}^{q_d+1} \tau_{QTL_i}^2 \qquad (16.20)$$

16.3.2 A SIMULATION APPROACH

When the number of genes is large, there are limitations due to the level of computation required. For those cases, P_e can be determined empirically by a Monte Carlo simulation. Let "1" indicate that a QTL is detected and "0" that it is not. For QTL i, if a random uniform $(0, 1)$ is less than or equal to p_i, the gene is detected and set $I_i = 1$. Otherwise, set $I_i = 0$. For each of the simulation replications, p_i is determined independently. For the t-test approach, p_i is determined by the approach explained previously, except that the expectation of the pooled mean square error is determined by

$$E[s_M^2] = \frac{\sigma^2}{b} + \sigma_{G(QTL)}^2 - \sum_{i=1}^{q_s} I_i(1 - 2r_i)^2 \tau_{QTL_i}^2 + 4r_{i+1}(1 - r)_{i+1} \tau_{QTL_{i+1}}^2 \qquad (16.21)$$

where q_s is the number of genes simulated. The expectation depends on the outcomes of the previously simulated loci. By doing the simulation for all loci, the number of QTLs detected is

$$\sum_{i=1}^{q_t} I_i$$

If this process is repeated a large number of times, the statistical power can be determined by counting the frequency of

$$\sum_i I_i \geq q_e \qquad (16.22)$$

The same analog can be applied to the interval mapping approach.

16.3.3 THE PERCENT OF GENETIC VARIATION EXPLAINED BY QTL

For a certain distribution of gene effects, statistical power can be predicted by means of the percentage of genetic variance explained by the detected QTLs.

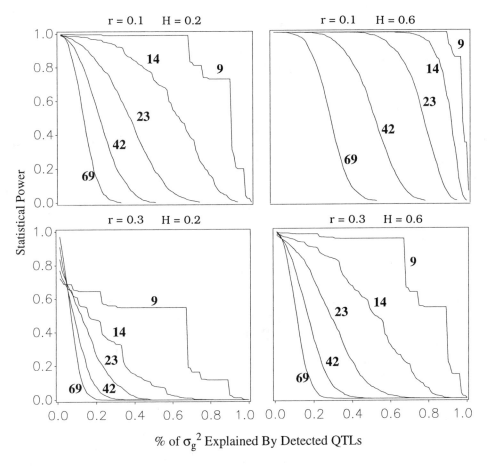

Figure 16.4 Statistical power for the percent of genetic variance that can be explained by detected QTLs using the interval mapping approach. The sample size is 200 and significance level is set to be 95%. The trait is assumed to be controlled by 9, 14, 23, 42 and 69 genes for the five curves, respectively (see Chapter 19). The gene effects are distributed as geometric variables. r is the recombination fraction between adjacent markers and H is the heritability of the trait.

For a large number of QTL problems, the simulation approach explained above can be used. The percentage of genetic variance explained by each of the q_t QTLs is

$$\{\lambda_1, \lambda_2, ..., \lambda_{q_t}\}$$

This yields

$$\sum_{i=1}^{q_t} \lambda_i = 1$$

The percentage of genetic variance explained by the detected QTL is

$$\sum_i \lambda_i I_i$$

The statistical power for detecting QTLs contributing λ percent genetic variance can be determined by counting the frequency of

$$\sum_{i=1}^{q_t} \lambda_i I_i \geq \lambda \qquad (16.23)$$

Figure 16.4 shows simulation results for some of the combinations of numbers of genes controlling the trait, heritability of the trait and linkage map density for a population size of 200 using the interval mapping approach. For example, the chance of having the detected QTL explain about 70% of genetic variation is 0.98 when the population size is 200, the trait heritability is 0.2, the linkage map density is 0.1 recombination units between markers, 9 genes control the trait (see Figure 16.1 and Chapter 19 for the genetic effect of the 9 genes) and the iterative interval mapping approach is used in backcross progeny.

EXERCISES

Exercise 16.1
In Chapter 14, the conditional t-test approach for QTL mapping was discussed. Consider that markers A and B and QTLs Q_1 and Q_2 are linked to each other. The two markers are r recombination units apart. Q_1 is located on the left side of A and s recombination units apart from A. Q_2 is located in between A and B at a relative position of $\rho = r_1/r$ from A and $1 - \rho = r_2/r$ from B.

(1) Derive the expected difference of marker genotypic values as functions of the recombination fractions between the two markers and the QTLs, and QTL genotypic values for marker genotypes BB and Bb conditional on that AA and Aa in backcross progeny of $Q_1q_1AaQ_2q_2Bb$ x $Q_1Q_1AAQ_2Q_2BB$. The genotypic values of Q_1Q_1, Q_1q_1, Q_2Q_2 and Q_2q_2 are μ_{11}, μ_{12}, μ_{21} and μ_{22}.

(2) Derive the statistical power for the conditional t-test.

(3) How does the value of s effect the statistical power?

(4) Use the information from (1) to (3) to design a strategy to increase the resolution of QTL mapping.

Exercise 16.2
Statistical power and the probability of false positives are major concerns in data analysis. Explain the consequences of false positives and false negatives (low power) for QTL mapping.

Exercise 16.3
Multiple-QTL detection power is usually more interesting than single-QTL. However, single-QTL detection power is usually easy to obtain. Answer the following questions.

(1) How is multiple-QTL detection power measured?

(2) What is the relationship between single-QTL and multiple-QTL detection power? Answer this question considering the following four situations. The first situation is that the QTLs are independent in terms of linkage and their genetic effects. The second is that some of the QTLs interact with each other in terms of genetic effects. The third is that the QTLs are independent in terms of their functions, but some of them are linked with each other. The final situation is that some of the QTLs interact with each other in terms of their functions and are also linked to each other.

(3) How can multiple-QTL detection power be derived from single-QTL detection power?

(4) Can multiple-QTL detection power be obtained without single-QTL detection power?

Exercise 16.4

How can the expected single-QTL and multiple-QTL detection powers be used in practical experimental design? Pay attention to population type, population size, genetic marker type and the number of genetic markers when answering this question.

CHAPTER 17

QTL MAPPING: FUTURE CONSIDERATIONS

In this chapter, some problems related to QTL mapping will be reviewed and some future directions for QTL mapping will be discussed, for example, the multiple-test problem, high level QTL interaction and problems with the traditional definitions of genetic effects (additive, dominance and epistatic interactions). QTL resolution problems will also be discussed. Most importantly, issues related to what QTLs are will be discussed.

17.1 PROBLEMS WITH QTL MAPPING

17.1.1 MULTIPLE-QTL MODEL

If a trait is controlled by two QTLs and the QTLs have been detected using QTL mapping, a two-QTL model can be constructed and tested using the contrasts on the QTL genotypic means listed in Table 17.1.

Practical Implementations

In practice, multiple-QTL models are usually obtained by a multi-step scan of the genome or by stepwise regression. For example, for the barley data used as an example in previous chapters (data is partially provided in Chapter 12), multiple QTL models can be constructed by the following steps. In the first pass through the 7 chromosomes, a QTL with the largest genetic effect was found on chromosome 4. By fitting the QTL in the model, the second QTL was found on chromosome 5. By repeating the previous steps, a total of 6 QTLs were found. Table 17.2 shows the analysis of variance table for the final model and Table 17.3 shows the statistics for each of the six loci. The coefficient of determination for the model is 0.536 (53.6% total variation can be explained by the six loci).

Table 17.1 Contrasts for a two-QTL model using F2 progeny.

	QTL Genotype								
QTL1	11	11	11	12	12	12	22	22	22
QTL2	11	12	22	11	12	22	11	12	22
Additive 1	1	1	1	0	0	0	-1	-1	-1
Dominant 1	1	1	1	-2	-2	-2	1	1	1
Additive 2	1	0	-1	1	0	-1	1	0	-1
Dominant 2	1	-2	1	1	-2	1	1	-2	1
Add 1 x Add 2	1	0	-1	0	0	0	-1	0	1
Add 1 x Dom 2	1	-2	1	0	0	0	-1	2	-1
Dom 1 x Add 2	1	0	-1	-2	0	2	1	0	-1
Dom 1 x Dom 2	1	-2	1	-2	4	-2	1	-2	1

Table 17.2 The ANOVA table for a multiple-QTL model for malt extract including 6 possible loci, using linear regression of trait value on markers (output from PGRI).

Source	DF	Sum of Squares	Mean Square	F-Value	Pr > F
Model	6	132.844	22.141	16.19	0.0001
Error	84	114.858	1.367		
Total	90	247.702			

Table 17.3 Markers in the multiple-QTL model for malt extract, their estimates and corresponding P-values, using linear regression of trait value on markers (output from PGRI).

Source	DF	Estimate	Mean Square	F-Value	Pr > F
BCD453B(4)	1	0.928	11.071	8.10	0.0056
WG1026B(4)	1	0.759	6.941	5.08	0.0269
Adh7 (1)	1	0.919	17.417	12.74	0.0006
ABC165 (2)	1	-0.563	6.834	5.00	0.0280
ABR337 (5)	1	0.958	19.420	14.20	0.0003
ABR334 (3)	1	-0.729	11.857	8.67	0.0042

Table 17.4 The ANOVA table for 6 possible loci using a combination of interval mapping and multiple regression on markers (output from PGRI).

Source	DF	Sum of Squares	Mean Square	F-Value	Pr > F
Model	11	151.367	13.761	11.23	0.0001
Error	82	100.492	1.226		
Total	93	251.859			

Table 17.5 QTL mapping using a combination of interval mapping and linear regression of trait value on markers (output from PGRI).

Source	DF	Estimate	Mean Square	F-Value	Pr > F
Dor5 - Bmy1 (4)					
1	1	1.269	8.697	7.10	0.0093
2	1	1.471	11.022	8.99	0.0036
3	1	1.510	10.542	8.60	0.0044
ABG472-BCD453B (4)					
1	1	-1.909	19.442	15.86	0.0001
2	1	-1.633	5.689	4.64	0.0341
3	1	-1.054	5.286	4.31	0.0409
WG1026B(4)	1	0.892	9.998	8.16	0.0054
Adh7 (1)	1	1.041	21.260	17.35	0.0001
ABC165 (2)	1	-0.481	5.090	4.15	0.0448
ABR337 (5)	1	0.931	17.948	14.65	0.0003
ABR334 (3)	1	-0.659	9.744	7.95	0.0060

A better result was obtained by combining the interval mapping and single-marker analysis approach and using the same scanning strategy (Tables 17.4 and 17.5). One more segment on chromosome 4 was identified. The coefficient of determination for the model, including two segments and 5 individual marker loci, is 0.601.

Problems

The total number of genetic effects is $2^l - 1$ for a trait controlled by l QTL, if only additive effects and additive by additive interactions are considered. The multi-locus additive model for a quantitative trait can be written as

$$Y = \sum_{i=1}^{n} f_i c_{1i} a_i + \sum_{i=2}^{n} \sum_{j=1}^{i-1} f_j c_{1i} c_{1j} a_{ij} + E \qquad (17.1)$$

where the notation can be found in Chapter 12. If dominance effects and the epistatic interactions associated with the dominance effects are also considered, then there are a total of $3^l - 1$ possible genetic effects for a trait controlled by l QTLs. In practice, the problem is that the number of QTLs controlling the trait is usually unknown, and furthermore, the QTL genotypes can rarely be observed in experiments. QTL mapping is a set of procedures to find the number of QTLs controlling the trait and to estimate the QTL genotypes for each individual in a mapping population. Polymorphic genetic markers are the tools for solving these problems.

Table 17.6 Numbers of genetic effects and multiple-QTL model screening for different numbers of genes and markers. A = additive, A x A = additive by additive interactions and Complete = additive + dominance + all epistatic interactions.

Genes	A, A x A	Complete	Number of Markers		
			100	200	500
1	1	2	100	200	500
2	3	8	4950	19900	124750
3	7	26	161700	1313400	2.1×10^7
5	31	242	7.4×10^7	2.5×10^9	2.5×10^{11}
10	1023	59048	1.7×10^{13}	2.2×10^{16}	4.9×10^{17}

Since what the QTLs are and how many QTLs control a trait are unknown, the multiple-QTL model with epistatic interactions has to be constructed step-by-step. Each step should include the complete model. For a single-QTL model, the analysis can be performed on each of the markers. However, if we want to build a two-QTL model, for example, when there are 100 markers on the linkage map and single-marker analysis is used for the screening, we have to screen all 4950 possible two-locus combinations (Table 17.6). A one-step model approach including all the markers in the model is not realistic due to the limited size of the mapping population. The number of possible marker combinations increases rapidly as the number of QTLs in the model increases. The success of the model is limited by the computational intensity.

Another important drawback of this step-by-step multiple-QTL model building approach is the inconsistency of the results among different marker combinations. The multiple test problem for the multiple-QTL model building is more problematic than for the single-QTL analysis (see the next section). The unbalanced data problem is also more serious for multiple-QTL model building than for single-QTL analysis.

17.1.2 MULTIPLE-TEST PROBLEM

QTL analysis is a multiple test problem. The number of repeated analyses include the number of markers in the data and the number of small segments (*e.g.*, 1cM) of the genome or some combinations of markers depending on the methods used for QTL analysis. The repeated analyses are not independent, because some of the markers or the segments are linked to each other or epistatic interactions exist. In the two extremes, all markers and segments are linked completely or all markers and segments are independent, the significance level can be simply determined using standard statistical distributions, such as chi-square, t-distribution, F-distribution or a simple lod score test statistic. If they are linked completely, it is a single test. When they are completely independent, the significant criterion should be the product of the independent tests. However, in practical QTL analysis, situations are always between the two extremes. The markers and segments used in QTL analysis are linked to some degree.

Three basic approaches have been proposed to solve the problem. A Monte Carlo simulation approach was suggested to build a distribution of test statistics under the null hypothesis (Lander and Botstein 1989; Knott and Haley 1992; Zeng 1994). It was also suggested that a conservative threshold value be used to reduce false positives (Knapp *et al.* 1990; Jansen 1993). Churchill and Doerge (1994) discussed a threshold value for declaring a significant QTL and proposed a permutation approach to obtain the empirical threshold values. They proposed two levels of threshold values: comparison-wise and experimental-wise. The test statistic is obtained by permuting trait values. N test statistics for each of the analysis points (a marker or a small segment) can be obtained by repeating the permutation N times. A comparison-wise critical value is the 100 (1 - α) percentile in distribution of the test statistic at the analysis point (α is the significance level). For each permutation, a maximum test statistic is obtained over all the analysis points. The 100 (1 - α) percentile in the distribution of the N maximum test statistic is the experimental-wise critical value.

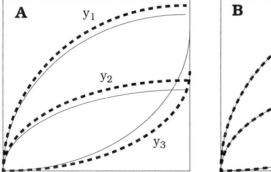
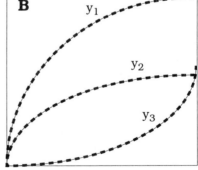

Figure 17.1 Fitted curves for the three dependent variables when they are related to each other. A: The relationships among the three variables are ignored. B: The relationships among the three variables are taken into consideration in the model fitting.

17.1.3 MULTIPLE RELATED TRAITS PROBLEM

In the previous chapters, QTL mapping has been based on a model such as

$$y_j = \mu + f(M, \theta) + \varepsilon_j \tag{17.2}$$

where y_j is the trait value for the *jth* individual in the population, μ is population mean, $f(M, \theta)$ is a function of marker genotypes (M), θ is a vector of parameters related to QTL effects and the QTL locations and ε_j is the residual associated with the *jth* individual. However, in many cases, we are interested in several traits instead of one. Those traits are sometimes related to each other. For example, grain yield and protein content may be negatively related to each other. The relationship may be a result of physiology or genetics. In either situation, those traits should be analyzed simultaneously.

It is well known that the model prediction is biased when the relationship among the dependent variables is ignored in the model fitting process (Chapter 5 of Gallant 1987, Figure 17.1). Intuitively, we can see that, if we analyze the data trait by trait, the QTL found may not be related to the trait, but rather related to other traits. This is caused by the relationship among the traits. The results may have serious consequences in practical applications. For example, if as mentioned above, grain yield and protein content are negatively related, selection may be acting against protein content even when high protein content is desirable. This can occur when selection is based on the results of QTL mapping on grain yield, because a favorable allele for grain yield found by QTL mapping may have nothing to do with high yield, but rather with low protein. The model based on Equation (17.2) should be rewritten as

$$y_{ij} = \mu + f_i(M, \theta) + \varepsilon_{ij} \qquad (17.3)$$

where i is an indicator for the trait. The model has multiple responses. For the previous example, the first trait can be grain yield and the second one can be protein content. When we are interested in growth and development, the dependent variables can be the measurements at different stages of biological development. For example, we could fit growth curves for the dependent variables and take the parameters of the curves into consideration in QTL mapping.

17.1.4 ARE THE QTLS REAL?

The resolution of QTL location is commonly low and the significance level of a test statistic is not precise. Logical questions to ask are: Are the QTLs real? Are the QTLs real in statistical terms? Are the QTLs real in biological terms? Do the estimated gene effects mean anything biologically? Do they mean anything statistically? I will try to answer the statistical questions using the following example.

To demonstrate the point, malt extract means for Steptoe and Morex genotypes of combinations of markers Adh4, BCD453B and WG464 are shown in Table 14.21 in page 451. If the overall means are compared for WG464, the Morex genotype is 1.0899 units higher than the Steptoe genotype. However, if the conditional means of Adh4 and BCD453B are compared for WG464, the Steptoe genotype is 1.7083 and 0.5843 higher than the Morex genotype for the two classes of observations.

When QTLs are linked in repulsion phase, peaks of the test statistic could be lowered. When QTLs are linked in coupling phase, the estimates of the QTL effects could be overestimated. In practice, such cases happen often. For the barley data, Morex is considered a high malt extract variety and so a commonly used malting barley and Steptoe is a low extract variety and so is a feeding barley. Morex is often presumed to have all the high malt extract alleles and Steptoe all the low ones. Only conditional analysis can resolve the problem of repulsion linkage phase for linked QTLs. Finding the bad alleles in the good families or varieties and the good alleles in the bad families or varieties can be more meaningful than finding the good ones in good varieties and the bad ones in bad varieties.

From the barley example, it is obvious that some QTLs may not be real and the estimated QTL effects may not mean anything if the underlying genetic or statistical model is wrong. A genetic model may be wrong due to assumptions regarding the number of genes and the linkage relationships of the genes. A statistical model may be wrong due to the imbalance of the data. In order to have a high confidence in a QTL, extensive analysis of many possible genetic and statistical models is needed. However, most of the computer software for QTL analyses are not suitable for extensive model screening.

17.2 QTL RESOLUTION

In the previous section, some of the problems of QTL mapping have been discussed. The next logical step after QTLs are located on the chromosome is to clone the genes responsible for the quantitative traits. If one's objective is to clone the QTL, an obvious problem is the low resolution of QTL mapping. This is especially true for species with large genomes.

17.2.1 QTL LOCATION

Common mistakes in interpreting QTL analysis are (1) too much confidence in QTL position, (2) inadequate significance level and (3) wrong conclusion due to an over-simplified model.

It is common to present results of QTL analysis by plotting the lod score or another test statistic against genome position. For example, Figure 17.2 shows that F statistics plotted against genome locations on chromosome IV of barley using the non-linear regression approach for simple interval mapping. Genome locations corresponding to peaks of the test statistic have been commonly considered as where a QTL may be located. However, this assumption is true only for some limited situations (Figure 17.3 A and C). The

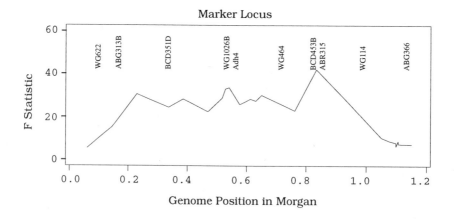

Figure 17.2 F statistics are plotted against genome locations on chromosome IV of barley using the non-linear regression approach for simple interval mapping. The trait is malt extract. For information on experimental materials, see Hayes *et al.* (1994). Data are provided by the North American Barley Genome Mapping Project.

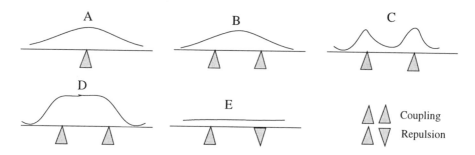

Figure 17.3 Some possible patterns of test statistic plots for one-QTL and two-QTL models in QTL mapping. A: peak corresponds to a single QTL, B: peak corresponds to a genome location between two linked QTLs in coupling phase, C: two peaks correspond to two QTLs, D: a wide plateau corresponds to two linked QTLs in coupling and E: there is a low peak or no peak for two linked QTLs in the repulsion phase.

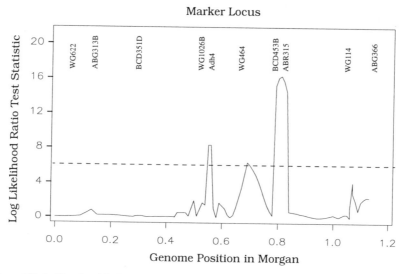

Figure 17.4 The log likelihood ratio test statistics are plotted against genome location on chromosome IV of barley, using composite interval mapping.

peaks of the curve may not be the result of QTLs under the peaks. The peak may correspond to the QTL location when a single QTL contributes to the peak. When more than one QTL are located near the group of linked markers, the peak may be very wide or the peak may not correspond to the QTL location. Figure 17.2 may have a very small amount of information on precise QTL locations.

The Composite Interval Mapping (CIM) approach increases the resolution of QTL location by controlling residual noise in the model using markers other than the markers immediately flanking the segment. Figure 17.4 shows the results for chromosome IV and the same trait as in Figure 17.2 using CIM. The analysis was done for every 1% recombination point by treating the putative

QTL position parameters as known constants. Compared with simple interval mapping (Figure 17.2), the CIM approach gives higher resolution for QTL location.

17.2.2 HIGH RESOLUTION QTL MAPPING

Quantitative Analysis of the Trait

To increase the resolution of a QTL location, a multi-step strategy can be effective (Figure 17.5). It is important to first analyze traits of interest quantitatively. The quantitative analysis has the potential to screen for families with major useful genes segregating and to eliminate some experimental errors. Traditionally, attention has been paid to family mean and variance among families. Within family variation has been considered secondary in plant and animal improvement and in evolutionary biology. This is not because within family variation is not important, but rather because of a lack of valid strategies to study within family variation. QTL mapping mainly explores within family variation. Linkage disequilibrium among genes and markers can be readily studied within a family. For some cases, the existence of segregating major genes can be determined by studying trait value distribution and variation within a family. These segregating genes are targets of QTL mapping. Genes segregating in a family may not show any effect on trait distribution because of small genetic effects, large number of genes, large environmental effects or experimental error. However, it is still a good start to analyze the trait quantitatively, because only a small effort is usually needed relative to future work.

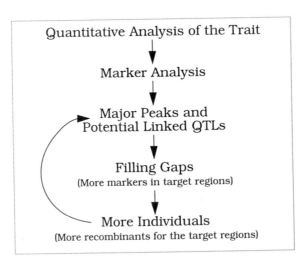

Figure 17.5 Steps to obtain high resolution QTL mapping.

For organisms in which controlled crosses are commonly used for genomic research, quantitative trait analysis can be implemented for parents of the crosses and the progeny. Certainly, large trait differences between the parents may increase the chance of detecting QTLs in their progeny. However, this increase may not always be desirable, depending on the objectives of the experiment. For example, progeny of a cross between a high and a low yield crop variety may increase the chance to find yield QTLs. However, the QTLs found in this cross may be of little use if the objective of the genomic experiment is to assist breeding for yield improvement and all the desirable QTL alleles come from the same parent and are fixed already. Only QTLs with new, desirable alleles from the low yield parent would be useful for further yield improvement. In this case, two high yield parents may be more desirable. Any QTL found in the cross between two high yield parents would be useful for further yield improvement. This is especially true for organisms with a long history of breeding. In most cases, genetic variation within the progeny of a cross between two parents with similar trait values will also be small. The chance of finding QTLs in the progeny will be small. So, it is essential to screen crosses of parents with desirable trait values and within population genetic variation to insure a successful QTL mapping experiment. For example, crosses of A and B may have a better chance for major gene segregation than a cross of either A or B with C (Figure 17.6). Statistical methods for analysis of mixture distribution have been developed (Bohning and Schlattmann 1992). Genetic variation within a population is the most important factor for the quantitative analysis to obtain high resolution QTL mapping. Characteristics of both parents are important for a useful QTL mapping experiment.

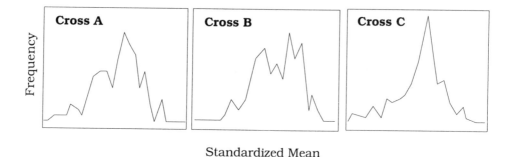

Standardized Mean

Figure 17.6 Trait distributions for three full-sib crosses of loblolly pine (Data provided by Dr. Bailian Li of North Carolina State University).

For organisms in which controlled crosses are difficult to make, quantitative trait analysis must focus on existing populations. For example, for practical applications of QTL mapping in forest tree breeding, it is critical to use the populations that exist and are ready for use in breeding. Theoretically, QTL mapping should be useful in tree breeding. In practice, however, the results may be of little use because of the time needed for making crosses, for evaluating traits and transferring the QTLs to breeding populations. Quantitative screening among available materials will ensure a high resolution and useful QTL analysis.

Besides quantitative analysis of within population variation, another important aspect of quantitative analysis is to control the environmental and experimental error. Contrary to traditional quantitative genetics, QTL mapping depends more on single genotypic value. Precise measurement of the trait at the individual genotype level is a critical aspect of high resolution QTL mapping. Environmental heterogeneity within a population should be minimized before QTL analysis. It is common to obtain least square means for each of the genotypes and then use them as trait values for QTL analysis.

Conditional Marker Analysis

The initial QTL analysis of the data should find the major test statistical peaks and potentially linked QTLs. To find the major peaks, a simple linear model (t-test, linear regression or analysis of variance) or interval mapping should be adequate. To locate the QTLs more precisely, composite interval mapping can be used.

An approach using markers as boundary condition can be an alternative. In this case, QTL effects cannot pass the marker which is used as the conditional boundary. In this way, more precise QTL locations may be determined, because a large portion of genome can be excluded from target QTL locations. It is then needed to search for additional markers in the regions which are not excluded from the conditional analysis, for example, the regions between markers WG464 and BCD453B and between BCD453B and ABR315 for the barley malt extract QTL in Figure 17.4. A commonly used approach is to increase marker density in the region of interest using methods like bulked segregation analysis (see Section 17.3.1).

Mapping Population Extension

More individuals in the mapping population may be needed to observe recombinants when more markers are genotyped in a region. It is also logical to genotype only the recombinant individuals. For example, if AAbb and aaBB are known to be recombinant, only individuals with those genotypes should be used to fill more markers within the region between markers A and B. When the obvious recombinants are rare in the population, individuals such as AaBB and AaBb may also be used for the next step. By repeating the process of QTL analysis by filling gaps with more individuals, a relatively high QTL resolution can be reached.

17.3 MAPPING STRATEGIES

17.3.1 BULK SEGREGANT ANALYSIS

Since Michelmore *et al.* (1991) described the technique, Bulk Segregant Analysis (BSA) has been widely used as a tool to find markers linked to target genes or to other markers. BSA is effective for high resolution mapping to find genes controlling simple inherited traits, such as disease resistance genes. Indirectly, BSA can be used to saturate linkage maps and to screen for markers linked to QTLs.

In BSA, the DNA samples from different individual progeny in a family are pooled into bulked samples by genotypic class or by phenotypic class. The success of bulking by phenotypes is dependent on the correspondence of genotype and phenotype. A specific target allele will occur in one bulked sample, but not the other. This pattern of frequency difference will also be seen for any marker or gene that is tightly linked to the target allele. A polymorphic marker which shows a clear difference between the two bulks is likely to be linked to the target genes or nearby markers. The two bulks show no differences for the rest of the genome.

PCR based markers have been commonly used in BSA. For RAPD markers, the principle behind BSA is that the low frequency allele will not be amplified. The minimum frequency of an allele in a DNA sample that will allow the allele to be amplified in the PCR reaction is defined as sensitivity. For a detailed methodology discussion, see Michelmore *et al.* (1991).

For example, we want to find markers linked to a disease resistance gene (R). A bulked sample is prepared from the diseased (susceptible) progeny and another bulked sample is prepared from the non-diseased (resistant) progeny. A DNA marker that is independent of the resistance gene will be represented as a band in both the resistant and susceptible bulked sample gel lanes. A DNA marker linked to the disease resistance gene should be represented as a band in one of the bulked sample lanes, but not the other. Sometimes, a band from an unlinked marker will be present in the lane for one bulked sample, but not for the other. This outcome is called a "false positive." A "false negative" outcome occurs if a DNA marker linked to the disease resistance gene produces a band in both bulked sample lanes. The control of false positives and false negatives is an important concept in BSA. Probabilities of false positives and false negatives vary, depending on the types of mapping populations (*e.g.*, backcross vs. F2), the bulk size and the sensitivity.

Table 17.7 Expected conditional frequency of marker alleles in different bulked samples in backcross progeny when the target gene R and the marker are independent.

Bulk	Marker	Expected Conditional Frequency (f_i)	Type A	Type B
R	+	0.5	$>p$	$<p$
	-	0.5	$<p$	$>p$
r	+	0.5	$<p$	$>p$
	-	0.5	$>p$	$<p$

Table 17.8 Expected conditional frequency of marker alleles in different bulked samples in backcross progeny when the target gene R and the marker are linked.

Bulk	Marker	Expected Conditional Frequency (f_i)	Type A	Type B
R	+	$1 - r$	$>p$	$<p$
	-	r	$<p$	$>p$
r	+	r	$<p$	$>p$
	-	$1 - r$	$>p$	$<p$

Consider the false positives first in backcross progeny. As mentioned before, if a marker (with two alleles, + and -) is independent of the gene (with alleles R and r), a false positive result occurs when the banding patterns of the two bulks are clearly different. In a backcross progeny produced by Rr+- X rr-- the expected frequencies for the four gene-marker genotypes R+, R-, r+ and r- are all 0.25, since the gene and marker are independent. The expected conditional frequency of the marker genotype in different classes of the disease resistant phenotypes is 0.5 (Table 17.7). The probability of the false positive result will be

$$P[false^{positive}] = 2P[f_i > p] P[f_i < p] \qquad (17.4)$$

where p is the sensitivity. For example, in the Type A column in Table 17.7, the frequency of the + allele of the marker in the R bulk is greater than the sensitivity and consequently that of + allele in the r bulk is less than the sensitivity, so different band patterns will be observed for the two bulks. In the Type B column, the opposite situation occurs.

In backcross progeny, a false negative occurs when the banding patterns for the marker are the same for the two bulks and the marker is linked to the gene. Table 17.8 shows the expected conditional frequencies of the marker genotypes in different classes of the disease resistant phenotypes. The probability of a false negative outcome can be computed using

$$P[false^{negative}] = P[f_i > p]^2 + P[f_i < p]^2 \qquad (17.5)$$

Table 17.9 Frequencies of false positive and false negative BSA tests using dominant or codominant markers in backcross progeny.

n/2	p	False Positive	False Negative for Different r			
			0.05	0.10	0.2	0.3
5	0.30	0.500	0.002	0.017	0.109	0.273
	0.20	0.305	0.023	0.082	0.266	0.474
	0.10	0.305	0.023	0.082	0.266	0.474
	0.05	0.061	0.226	0.410	0.672	0.830
10	0.30	0.285	0.001	0.013	0.122	0.354
	0.20	0.103	0.012	0.070	0.322	0.617
	0.10	0.021	0.086	0.264	0.624	0.851
	0.05	0.021	0.086	0.264	0.624	0.851
20	0.30	0.109	0.000	0.002	0.087	0.392
	0.20	0.012	0.003	0.043	0.370	0.762
	0.10	0.000	0.075	0.323	0.794	0.965
	0.05	0.000	0.264	0.608	0.931	0.992
30	0.30	0.042	0.000	0.000	0.061	0.411
	0.20	0.001	0.001	0.026	0.393	0.840
	0.10	0.000	0.061	0.353	0.877	0.991
	0.05	0.000	0.188	0.589	0.956	0.998
40	0.30	0.016	0.000	0.000	0.043	0.423
	0.20	0.000	0.000	0.015	0.407	0.889
	0.10	0.000	0.048	0.037	0.924	0.997
	0.05	0.000	0.323	0.777	0.992	1.000

Two bulked DNA samples are prepared from n backcross progeny. One bulk contains DNA from $n/2$ resistant plants and the other from $n/2$ susceptible plants. Table 17.9 shows the false positive and false negative frequencies for some combinations of bulk sizes, sensitivities and different true recombination fractions between the gene and the marker in backcross progeny. It is clear that the false positive is a problem when the sensitivity is low (p is large) and the bulk size is small. However, when the sensitivity is high (p is small), the false negative becomes the problem. In practice, the sensitivity may be determined using known bulks. A combination of bulk size and the sensitivity that gives adequate results needs to be considered in mapping strategy (see Michelmore *et al.* 1991).

Table 17.10 Expected conditional frequency of marker alleles in different bulked samples in F2 progeny when the target gene R and the marker are independent.

Bulk	Marker	Expected Conditional Frequency (f_i)	Type A	Type B
R	+	0.75	$>p$	$<p$
	-	0.25	$<p$	$>p$
r	+	0.75	$<p$	$>p$
	-	0.25	$>p$	$<p$

Table 17.11 Expected conditional frequency of marker alleles in different bulked samples in F2 progeny when the target gene R and the marker are linked in coupling.

Bulk	Marker	Expected Conditional Frequency (f_i)	Type A	Type B
R	+	$(3 - 2r + r^2)/3$	$>p$	$<p$
	-	$(2r - r^2)/3$	$<p$	$>p$
r	+	$2r - r^2$	$<p$	$>p$
	-	$(1 - r)^2$	$>p$	$<p$

Table 17.12 Frequencies of false positive and false negative in BSA using dominant markers for F2 progeny.

n/2	p	False Positive	False Negative for Different r			
			0.05	0.10	0.2	0.3
5	0.30	0.186	0.008	0.053	0.258	0.517
	0.20	0.031	0.078	0.242	0.590	0.822
	0.10	0.031	0.078	0.242	0.590	0.822
	0.05	0.002	0.401	0.651	0.893	0.972
10	0.30	0.007	0.012	0.104	0.513	0.844
	0.20	0.001	0.066	0.292	0.759	0.952
	0.10	0.000	0.254	0.593	0.924	0.991
	0.05	0.000	0.254	0.593	0.924	0.991
20	0.30	0.000	0.002	0.069	0.620	0.952
	0.20	0.000	0.039	0.327	0.899	0.995
	0.10	0.000	0.309	0.761	0.990	1.000
	0.05	0.000	0.594	0.916	0.998	1.000
30	0.30	0.000	0.000	0.045	0.684	0.984
	0.20	0.000	0.023	0.340	0.953	1.000
	0.10	0.000	0.335	0.848	0.999	1.000
	0.05	0.000	0.571	0.943	1.000	1.000
40	0.30	0.000	0.000	0.030	0.731	0.994
	0.20	0.000	0.013	0.345	0.978	1.000
	0.10	0.000	0.350	0.900	1.000	1.000
	0.05	0.000	0.762	0.988	1.000	1.000

When using BSA for F2 progeny, Equations (17.4) and (17.5) can be used directly. However, the expected conditional frequencies of marker genotypes in different classes of the disease resistant phenotypes are different from those in the backcross progeny. Tables 17.10 and 17.11 show the expected conditional frequencies needed to compute false positive and false negative frequencies for F2 progeny. Table 17.12 shows the false positive and false negative frequencies for some combinations of bulk sizes, sensitivities and different true recombination fractions between the gene and a marker in F2 progeny. The trends are similar to those for a backcross. In general, BSA using F2 progeny has fewer false positives and more false negatives than using backcross progeny with the same bulk size, sensitivity and true recombination fraction between the marker and the gene.

BSA is effective when the targets are a few simple inherited traits or when the goal is to saturate a region of genome defined by known markers. By replicating the bulks, false negatives and false positives can be reduced.

17.3.2 SELECTIVE GENOTYPING

Lander and Botstein (1989) proposed a method using selective genotyping of the extreme progeny to map QTL. They claimed that the method increases the power of QTL mapping. The rationale behind the method is that in a QTL mapping population, some progeny contribute more linkage information than others. The individuals that provide the most linkage information are those genotypes which can clearly be inferred from their extreme phenotypes.

In general, this approach will bias the hypothesis test and parameter estimation. This is especially true when the data generated from the selected genotypes is analyzed by commonly used computer software which does not take the selection process into consideration. When the mapping population is large and the experiment is designed for a preliminary study, selective genotyping may be effective. It is not recommended for use when experimental objectives include several traits or multiple QTLs. It is also not recommended for use when the objective of the experiment is to obtain precise estimates of genetic effects and gene locations.

17.3.3 INCREASE MARKER COVERAGE

Marker coverage can be increased at the whole genome or a genome segment level. This can be an important step to obtain high resolution QTLs. It is certainly desirable to have a high resolution genomic map. Strategies used to increase marker coverage and map density at the whole genome level do not go beyond

- increasing the number of markers (using markers without number limitation and with high polymorphism) and
- using high informative mapping population and increasing population size.

Strategies for high resolution mapping at the whole genome level are usually straightforward. Only strategies used to increase map resolution at a genome segment level will be discussed here.

Comparative Mapping

Comparative mapping based on genome homology is a strategy to increase marker coverage for species with less genome information or where it is difficult to gain mapping information. This strategy can also be used to obtain information for species which are difficult to study, such as humans, and from species which are much easier to study, such as mice. Information from species with simple genomes can be applied to species with complex genomes, such as from barley to wheat.

Increase Useful Progeny Size

Useful progeny include those who have recombinant genotypes for the target region, identified by screening for recombinants using the known flanking markers of the target region. For example, AAbb or aaBB recombinants in F2 progeny can be screened for if A and B flank the target region and A and B are linked in coupling. For species where random mating can easily be performed, a number of generations of random mating will increase the frequency of recombinants in the population (see Chapter 7).

17.4 WHAT ARE QTLS?

17.4.1 WHAT ARE QTLS?

In recent years, great progress has been made in both molecular and quantitative genetics. In molecular genetics, many specific genes and proteins affecting growth, metabolism, development and behavior of plants and animals have been identified and characterized. New methods of quantitative genetics have similarly identified genetic regions regulating important functions through the genetic dissection of quantitative traits. QTL analysis has fundamentally changed the conventional view of polygenic inheritance and has led to the identification of a new set of loci with important biological roles, but usually lacking molecular identification. The function of many genes identified and characterized at the molecular level is unknown. A lot of genes with presumably known functions have not been subjected to direct tests of their biological roles through loss of function or gain of function experiments. In most cases, the quantitative effects of genes of known functions have not been explored. A few cases exist, such as the study of human hemoglobins, where the underlying molecular events are well understood and the consequences of these changes in metabolism and development have a reasonable molecular basis. In other cases, some individual genes and their effects on complex phenotypic traits are understood, but much less is known about the genetic and biochemical basis of the regulation of coordinately and differentially controlled gene expression and protein regulation of complex biochemical pathways.

Lander and Schork (1994) wrote in their review paper on genetic dissection of complex traits:

> ... one can systematically discover the genes causing inherited diseases without any prior biological clue as to how they function. The method of genetic mapping, by which one compares the inheritance pattern of a trait with the inheritance patterns of chromosomal regions, allows one to find where a gene is without knowing what it is ....

The gene mapping approach has revolutionized the process of finding genes compared to the traditional approaches. However, without knowing what QTLs are, QTL mapping is limited in terms of fundamental biology and knowledge of mechanism behind the phenotypes. A definition of QTL, such as "QTLs are genes located on the genome with significant additive, dominant or epistatic genetic effects," is superficial. Not knowing what QTLs are may lead to wrong genetic models and imprecise definitions. Most importantly, not knowing what QTLs are makes complete genetic models difficult to construct and QTL mapping inefficient in practice. In this section, limitations of QTL models, relationships of the traditional quantitative approach, QTL mapping and molecular biology and the possibility of constructing the metabolic genetic models will be discussed.

Limitations of QTL Mapping

There currently are four possible reasons for an unsuccessful multiple-locus model for a quantitative trait:

(1) QTLs are hypothetical genes based on statistical inference. Genetic effects used in QTL mapping could have very little biological meaning.

(2) The genetic models which QTL mapping is based on are not accurate.

(3) The amount of genetic information contained is not adequate.

(4) The methodology of statistical analysis is not powerful enough. The current statistical tools are not adequate for dealing with high levels of epistatic interactions.

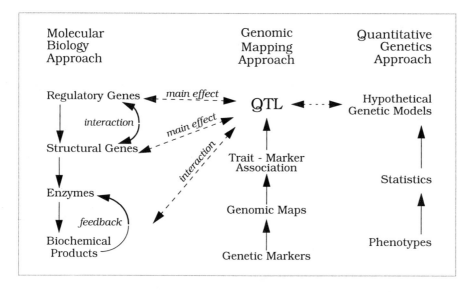

Figure 17.7 A metabolic genetic model that will integrate molecular, QTL mapping and quantitative genetic approaches for a better understanding of quantitative trait inheritance.

17.4.2 QUANTITATIVE GENETICS, GENOMIC MAPPING AND MOLECULAR BIOLOGY

In molecular biology, genes are defined as segments of DNA on genomes and include structural genes encoding enzymes for biosynthetic pathways and regulatory genes regulating gene expressions. Definitions of the genes detected by QTL mapping and in traditional quantitative genetics, such as "the unit of inheritance" or "a functional genome segment", are very imprecise. Terms of genetic effects, such as additive, dominant and epistatic, have been used for

QTL mapping and quantitative genetics. How the definitions of these terms in the different fields are related to each other and how those approaches can be integrated with each other will be important for the future development of biology (Figure 17.7).

Figure 17.7 shows one view of the relationships among different approaches for modeling quantitative traits inheritance. Molecular approaches model quantitative traits such as gene regulation, one-gene-one-enzyme and possible feedback or forward effects between products and enzyme activities. Quantitative genetics models quantitative traits at the population level using the concepts of additive effect, dominant effect and epistatic interaction. The genomic mapping approaches model quantitative traits using hypothetical QTLs obtained by analysis of associations between traits and genetic markers. QTL mapping has shown greater power in finding possible genome positions of genes controlling quantitative traits than classical quantitative approaches alone. The QTL mapping approach is likely to bridge quantitative genetics and molecular biology. To make the connection between "genetic effects" of quantitative genetics and biochemical processes, QTLs should be physically defined instead of genetically defined. It is important to have a precise definition of a genetic effect before an effective quantitative genetic model of a trait or a group of traits is tested.

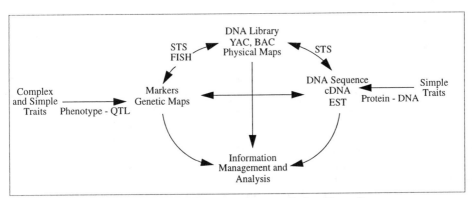

Some Concepts for Making the Connections among Traits - Maps - Sequences:

1. STS, EST and cDNA can be mapped on both genetic and physical maps and sequenced.
2. Genes for simple traits can be cloned using the protein to DNA approach. Genes controlling complex traits can be located coarsely on genetic maps using statistically oriented QTL mapping.
3. FISH brings classical cytogenetics and genetic and physical maps together.
4. Physical mapping using overlapped cloned DNA fragments is an efficient way to get a high resolution genetic map within a short region and is a powerful tool for isolating the target genes.
5. Genomic information goes to information management and analysis systems (IMAS). Ultimately, genomics prevails when the IMAS can resemble a cell, from where the information came.

Figure 17.8 Genomics.

17.5 FUTURE QTL MAPPING

The resolution of QTL mapping can be increased by picking a valid mapping population, using improved statistical procedures and employing a multi-step strategy. However, there is a limit to how high the resolution can be using those approaches, because they are based on the number of recombination events between a marker and the QTL and the genetic variation of the trait. The resolution may still not be sufficiently high for some applications. The number of recombination events between a marker and a gene which can be observed is limited by the number of polymorphic markers which can be obtained in the genome region and by the size of the mapping population. Once the mapping population is chosen, there is limited genetic variation. Alternatives which are not based on recombination are needed for further improvement of QTL resolution. An approach using QTL mapping as an intermediate step to connect the trait and the physical map may be an important alternative (Figure 17.8).

17.5.1 GENETIC AND PHYSICAL MAPS

Both genetic and physical maps have been obtained for several organisms with small genomes, such as *E. coli*, yeast, *Drosophila*, *C. elegans*, *Arabidopsis thaliana* and rice. High density genetic maps have been developed even for some organisms with relatively large genomes, such as human, barley, maize, tomato and loblolly pine. Physical maps have been constructed for most of human chromosomes.

In principle, genetic and physical maps should reflect the same genome constitution, such as the order of genes or markers and distances between the loci. However, the relationship between the distances in units of recombination and in physical length is not straightforward, due to non-randomness of the crossover events. Moreover, the gene orders on the genetic map and the physical map might not agree with each other due to random sampling errors or laboratory errors (including data analysis).

A genetic map is built based on meiotic recombination between homologous chromosomes. Polymorphic markers are needed to identify the genotypes which result from meiotic recombinations. Resolution of a genetic map is limited by the number of crossover events. If no crossover events can be detected within a region of a genome, then genetic mapping cannot resolve the precise genome positions for the genes and markers in the region. To increase the resolution for a genetic map, a relatively large mapping population is usually needed. Physical length of a genome has a relatively small impact on the size of a genetic map. For a small genome, a recombination unit may represent a small physical segment of the genome. For a large genome, a recombination unit may represent a relatively large segment. For example, a recombination unit may represent, on the average, approximately 200 kb in *Arabidopsis thaliana* and 15 to 20 Mb in loblolly pine.

On the other hand, physical mapping is based on overlapping DNA fragments. Polymorphisms and recombinations are not essential for physical mapping. Physical mapping is limited by the size and the number of the cloned DNA fragments. If the size of the DNA fragments is small, then a large number of clones are needed to cover even a small region of the genome. As genome size increases, the number of clones needed to cover the genome increases proportionately.

Genetic maps and physical maps can be related to each other by placing genetic markers on the physical map and the DNA fragments on the genetic map. By doing this, the genetic map and the physical map can cross validate each other. Closely linked genes and markers on a genetic map usually have a low confidence of gene order and can be checked with the physical map at the region. Gaps are usually common for low coverage physical maps and can be filled using the genetic map.

Most importantly, a genetic map is a bridge between phenotypic traits and the physical map. Coordination between a genetic and a physical map is an efficient way to identify and isolate genes of interest.

17.5.2 TRAIT - MAPS - SEQUENCE

It is not logical (at least for now) to relate a complex trait directly to massive amounts of DNA sequence data. It is not possible to determine the molecular basis of complex traits when the QTLs are only mapped to very large segments of DNA. However, DNA sequence is the ultimate information resource for the analysis of complex traits. Reverse-genetics, from protein to DNA, is an elegant way to isolate genes with simple functions. Many genes have been identified and cloned using the reverse-genetics approach. Coordination among genetic and physical maps and DNA sequences may provide more efficient ways to isolate genes with simple functions and to begin to identify and isolate genes controlling complex traits. Genetic and physical maps are bridges between complex traits and DNA sequences (Figure 17.9).

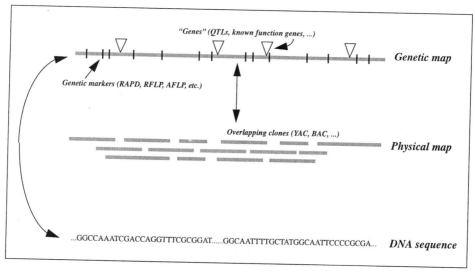

Figure 17.9 The relationships among a genetic map, a physical map and a DNA sequence.

Association between a marker or a group of markers and trait variation is a sign that a gene controlling the trait is located near the marker. To increase the resolution of the gene localization, the increase of the mapping population size, the improvement of statistical methodology and the use of linkage disequilibrium accumulated over generations have been proposed. However, gene mapping obtained with those improvements is still usually not precise enough to identify and isolate the gene. For example, a disease gene may be located within a 1cM region flanked by two markers (Figure 17.10). In humans, 1 cM still represents 1 Mb of DNA, which is not precise enough to clone the disease gene efficiently. If a physical map for the region can be obtained, 12 BAC clones will cover the 1Mb region. A more precise location of the disease gene may be located using the overlapped BAC clones.

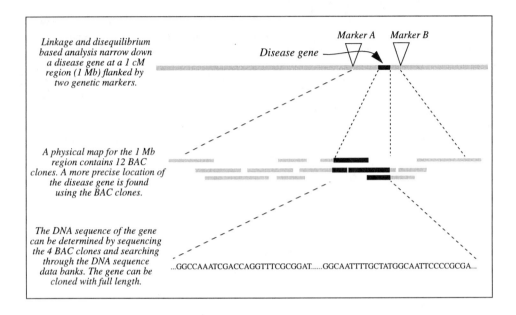

Figure 17.10 Study of a disease gene using the combination of genetic mapping (linkage and disequilibrium), physical mapping and DNA sequencing.

17.5.3 METABOLIC GENETIC MODEL (MGM)

What is an adequate approach for modeling a quantitative trait? A combination of molecular and QTL approaches may be the logical alternative. Let us call the model which integrates a metabolic pathway with QTL mapping a metabolic genetic model (MGM). MGM has at least the following advantages:

(1) cDNAs encoding enzymes or regulatory proteins can be mapped on the genome as regular genetic markers. QTL mapping based on prior information becomes a search for other structural and regulatory genes. MGM can be built based on predetermined trait-gene relation, while the QTL mapping models are based purely on statistical association between traits and molecular markers.

(2) For MGM, complex biological relationships among the components of the trait are determined by biochemical experiments and the information can be integrated into the analysis.

(3) Genetic models and metabolic pathways can be cross-validated with each other. The genetic models will help in searching for additional genes on the pathways and the pathway will help in building more accurate and precise genetic models to predict the basis for quantitative variation in the studied trait.

Let us use the pathway for the biosynthesis of lignin as an example to illustrate the methodology. Enzymes for some key steps on lignin biosynthetic pathway have been purified and their activities can be measured in plant

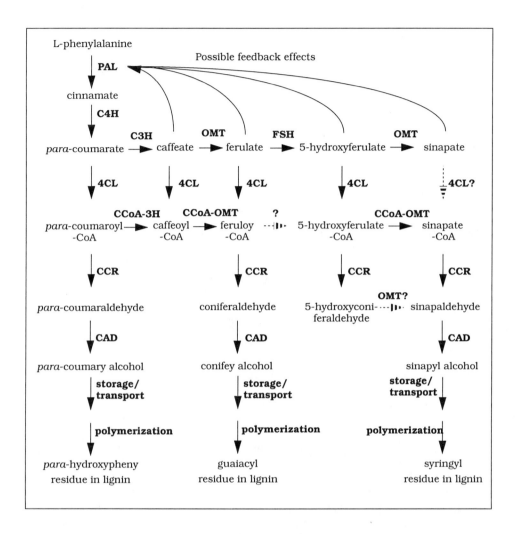

Figure 17.11 The lignin biosynthetic pathway. The enzymes are PAL (phenylalanine ammonia-lyase), C4H (cinnamate 4-hydroxylase), C3H (4-hydroxycinnamate 3-hydroxylase), OMT (S-adenosyl-L-methionine:caffeate O-methyltransferase), F5H (ferulate 5-hydroxylase), 4CL (hydroxycinnamate:CoA ligase), CCoA-3H (4-hydroxycinnamoyl-CoA 3-hydroxylase), CCoA-OMT (SAM:caffeoyl-CoA/5-hydroxyferuloyl-CoA O-methyltransferase), CCR (hydroxycinnamoyl-CoA:NADPH oxidoreductase) and CAD (cinnamyl alcohol dehydrogenase). The pathway is provided by R. Sederoff and M. Campbell at North Carolina State University.

tissues (Figure 17.11). cDNAs that encode most of the the enzymes have been cloned. A similar approach can be used for anthocyanin formation, fatty acid content and composition, photosynthesis, *etc.*, for different species. The MGM for lignin can be formulated as follows: Let

$$y_{ij}\,(i = 1, 2, ..., n;\ j = 1, 2, ..., w)$$

be w values for each of the products on the pathway for n individuals in a population and let

$$x_{ijk} \ (i = 1, 2, ..., n; \ j = 1, 2, ..., w; \ k = 1, 2, ..., m)$$

be w genotypes for genes of the known functions and m marker genotypes (such as RFLP and RAPD) for n individuals in the population. We can use

$$y_{ij} = f_j (x_{ijk}, \theta_l) + \varepsilon_{ij} \tag{17.6}$$

to represent the general MGM, where $\theta_l \ (l = 1, 2, ..., z)$ are z unknown parameters for the model, which include the feedback effects, possible regulatory gene effects and functional gene effects and ε_{ij} denotes random error. For the lignin biosynthesis problem, the y_{ij} can be the measurements of biochemical products in the metabolic pathways, x_{ijk} can be the allelic forms of the known function genes in the pathways and genetic markers in the linkage maps for the populations and θ_l can be the parameters for formulating the flux of the pathways, the relationship between the biochemical products and genotypes of the known functional genes and the association between the molecular markers and the biochemical products. The last association may be a sign of regulatory gene effects or unknown functional genes.

EXERCISES

Exercise 17.1

In quantitative genetics, genetic effects are defined as additive, dominant, and epistatic interactions. A trait is recorded as the content of a simple compound in a plant. Two enzymes are in the biosynthetic pathway and clones of the DNA encoding the enzymes have been obtained. Let us assume that each enzyme has two alternative forms in the population we study and that they are denoted by A and a for Enzyme A and by B and b for Enzyme B. An F2 progeny with 160 individuals was obtained by selfing a cross between two individuals with AABB and aabb genotypes. Contents of the compound were obtained from an experiment including two true replications of the 160 individuals in three different environments. Genotypes, in terms of the two forms of the two genes encoding the enzymes, were also obtained for the 160 individuals. Surprisingly, the segregation ratio of AABB, AABb, AAbb, AaBB, AaBb, Aabb, aaBB, aaBb and aabb in the 160 individuals follows exactly the Mendelian expectation, which is 10:20:10:20:40:20:10:20:10.

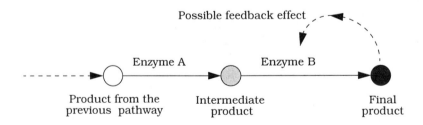

(1) Develop a full quantitative genetics model for the trait, including all possible genetic effects (a total of 8). Fill in some values for the effects and make a perfect model (without error).

(2) How can the definitions of additive, dominance and epistatic interactions in quantitative genetics be related to the biosynthetic pathway? For example, what are additive effects of A and B? (Hint: Use the hypothetical values in (1).)

(3) A null mutation of A is found in nature. It is proved by enzyme assay that A loses its function completely due to the mutation. However, the final product is still found in the individual with the mutation. How should the quantitative genetics model be modified to fit this situation?

(4) If the mutation in (3) is unknown, construct a quantitative genetic model based on the B locus alone. How is this model different from the model in (3)?

(5) Speculate on the reasons for the existence of the final product under the situation in (3).

(6) The intermediate products are toxic. The individuals will die if the intermediate product exceeds certain amounts. Let us assume that genotype bb will somehow partially block the pathway from the intermediate product to the final product. 50% of individuals with genotype bb die before their DNA is sampled and the final product measurement were taken. How should the quantitative genetics model in (1) be modified to fit this situation? If this situation is not known, how serious a mistake will be made in estimating the genetic effects of the genes?

(7) Can genes involved in the previous pathway have effects on the final product? Will the QTL mapping approach find the genes?

Exercise 17.2

Let us consider the situation in Exercise 17.1 again. The two genes A and B are independent in terms of linkage. Develop an approach for finding the genes using the idea of a conditional test. (Hint: test gene A conditional on gene B and test gene B conditional on gene A).

Exercise 17.3

The following graph (top of the next page) shows plots of the t-statistic against genome positions in two linkage groups for a trait of interest. A column on the right side of the plot shows the markers corresponding to the "square" of the curves. The reference lines in each of the plots correspond to significant t-values at 0.05 level using the standard t-distribution with degrees of freedom 98 and the significant t-values at a 95 percentile from empirical distributions using the permutation tests (see this chapter and Chapter 20 for the permutation test for QTL mapping).

(1) How precisely can the potential QTLs be located?

(2) If one QTL can be located on a segment about 25 cM long between markers TE373717 and TE452821 and the other is located on a segment about 20cM long between markers TE384420 and TE38424, give a strategy to improve the resolution of QTL locations.

(3) What are the possible reasons that the test statistic is significant for such a large segment (about 80cM long) on the linkage group on the left?

(4) For the linkage group on the right, two peaks can be observed from the plot. Can you say that two QTLs are located on the genome? Why?

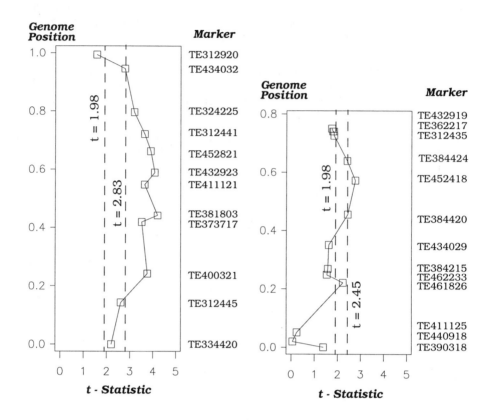

(5) The complete data in this exercise were analyzed using PGRI (see Chapter 18). Two loci were found by the program to build the multiple-QTL model. The above table shows the results of an analysis of variance on these two loci. The coefficient of determination for the model is 0.2221. Explain sources of variation for the term of line within QTL [line(QTL)].

An ANOVA table for a multiple-QTL model (output from PGRI).

Source	DF	Sum of Squares	Mean Square	F-Value	Pr > F
MODEL	2	266.013	2	13.847	0.000005
LINE(QTL)	97	931.746	97		
TOTAL	99	1197.760	99		

(6) If heritability of the trait is 0.5, how significant is the result in (5)? What if the heritability of the trait is 0.95?

(7) The following table shows the analysis for each of the two loci. The estimate in the table is half of the difference between the two marker genotypic values in the multiple-QTL model. The following figure shows the marker genotypic mean difference for the two linkage groups using the

simple single-marker analysis. In theory, how will the estimate in the multiple-QTL model differ from the estimate in the single-QTL model?

Linear regression of trait value on markers (Output from PGRI).

Source	DF	Estimate	Type III Mean Square	F-Value	Pr > F
Intercept		10.690			
TE452418	1	0.921	84.662	8.814	0.0038
TE381803	1	1.328	176.151	18.338	0.00004

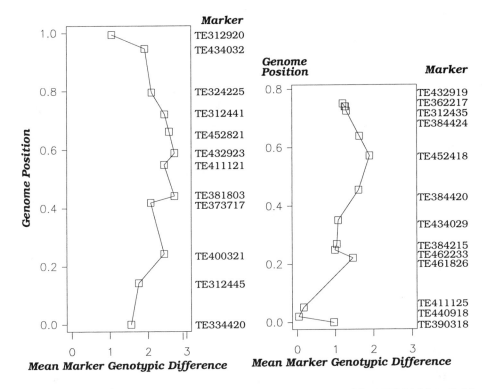

(8) The actual estimate for marker TE452418 is 1.9 and for TE38183 is 2.69, which are close to the expectations. Justify this result using the arguments in (7).

CHAPTER 18

COMPUTER TOOLS

Linkage analysis among a number of loci, linkage grouping, gene ordering and multi-locus model building are some of the key steps for analyzing discrete genetic markers. When the number of loci is large, every step of the analysis is computationally intensive. This is especially true for gene ordering and multi-locus model building (see Chapters 9 and 10). Specialized computer software packages are needed to implement the analyses.

Searching for genes controlling complex or quantitative traits plays an important role in applying genomic information to clinical diagnoses, agriculture and forestry, because a large portion of the traits related to human diseases and agronomic importance are quantitative traits. These traits usually are controlled by more than one gene and are influenced by environmental effects. Traits controlled by a single gene with incomplete penetrance can also be treated as quantitative traits. QTL mapping involves the construction of genomic maps and a search for the relationship between traits and polymorphic markers. A significant association between the traits and the markers may be evidence of a QTL near the region of the markers. Simple t-tests, simple linear regression models, multiple linear regression models, log-linear models, mixture distribution models, nonlinear regression models and interval test approaches using partial regression have been proposed to map QTLs (Stuber *et al.* 1992; Weller 1988; Lander and Botstein 1989; Knapp *et al.* 1992; Lande and Thompson 1991; Zeng 1993). To estimate model parameters, least squares and maximum likelihood methods have been used. For most QTL mapping methods, analyses can be implemented in common statistical software packages, such as SAS. However, when the number of markers is large, specialized software packages are needed.

As information is accumulated, specialized databases are needed to efficiently manage the information. This chapter focuses on computer software for marker analysis, linkage map construction and QTL mapping. Some problems with the available software packages and the future direction of the development of computer software for genomics will also be discussed. In the final section of this chapter, the software package PGRI (plant genome research initiative) will be introduced.

Genome databases are available for *E. coli*, yeast, *C. elegans*, *Drosophila*, mouse, human and some higher plants (Table 18.1). GenBank release 87.0 (15/2/95) contains 269,478 loci and 248,499,214 bases of DNA sequence. EMBL is a European molecular biology laboratory database. DDBJ is a DNA database of Japan. A comprehensive USDA plant genome database is located at the National Agriculture Library (NAL) and satellite genome data bases for several plant species reside in different institutes (Bigwood 1992; Neale 1992). Sybase (relational database), a commercial database software and ACEDB, a public domain database software package originally designed for the *C. elegans* genome project, are used for the plant genome databases. Genome databases have been built in some other institutes, such as The Institute for Genomic Research (TIGR) and Lawrence Livermore National Laboratory.

Table 18.1 Some database tools and specific databases for genome research.

Database Tools:	
ACEDB	Database tool, between hierarchical and object-oriented approach (Example: AATDB, Soybase, GrainGenes, APtDB, DogBase, MycDB, and Flybase)
Sybase	Database tool, relational database (Example: GenBank, GDB, MaizeBase)
Database (Human):	
Genbank (NCBI)	NIH genetic sequence database
EMBL	European molecular biology laboratory database,
DDBJ	DNA database of Japan
ESTDB (TIGR)	Human EST database
Database (Plant):	
AATDB	*Arabidopsis thaliana* database
SoyBase	Soybean database
GrainGenes	Database for wheat related plants
APtDB	A pine tree database
MaizeBase	Maize database

18.1 COMPUTER TOOLS FOR GENOMIC DATA ANALYSIS

18.1.1 LINKAGE ANALYSIS AND MAP CONSTRUCTION

Table 18.2 lists some available computer software packages for linkage analysis and linkage map construction. Methods used for simple linkage detection and recombination fraction estimation are the same or similar among these different packages. For the same data set, all the packages should yield the same pairwise two-point recombination fraction matrix. The differences among these packages concern data format, computer platforms, user interface, graphic output, population types, algorithms for locus ordering and algorithms for multi-locus model building. All packages can handle commonly used mapping populations, such as F2 and backcross. For multi-locus model building, Mapmaker, Gmendel and PGRI use the algorithm of Lander and Green, while JoinMap uses the least squares method (see Chapter 10). The least square method is also being implemented into PGRI. For gene ordering, MapmakerMac and JoinMap use an algorithm similar to the seriation approach. Gmendel uses the simulated annealing algorithm with the minimum sum of adjacent recombination fraction as the target function. PGRI has several choices for gene ordering, such as combinations for both algorithm and criteria (simulated annealing and branch-and bound, minimum sum of adjacent recombination fraction and maximum likelihood). In terms of function, PGRI has some unique functions, such as handling open-pollinated populations, an algorithm for significance tests on two-point recombination fraction and locus orders among multiple populations, a nonparametric approach for estimating confidence on locus order, specific procedures for obtaining gene order with high confidence and subroutines for marker-assisted progeny selection and mating design. All of these packages are constantly changing. Contact the authors for the latest information.

Table 18.2 Some available software packages for QTL mapping.

MapmakerUnix	Function	Linkage analysis and map construction
	Population Type	F2, backcross, RIL, DH, CEPH pedigree
	Computer Platform	Unix
	Graphic Interface	no
	Graphic Output	Postscript
	Contact	Eric Lander (mapmaker@genome.wi.mit.edu)
MapmakerMac	Function	Linkage analysis and map construction
	Population Type	F2, backcross, RIL, DH
	Computer Platform	Mac
	Graphic Interface	yes
	Graphic Output	yes
	Contact	Eric Lander (mapmaker@genome.wi.mit.edu)
Gmendel	Function	Linkage analysis and map construction
	Population Type	F2, backcross, RIL, DH, heterozygous F1
	Computer Platform	Unix
	Graphic Interface	no
	Graphic Output	no
	Contact	Steve Knapp (sknapp@helix.css.orst.edu)
JoinMap	Function	Linkage analysis and map construction
	Population Type	F2, backcross, RIL, DH, heterozygous F1
	Computer Platform	Unix, Mac, PC Windows
	Graphic Interface	no
	Graphic Output	no
	Contact	Piet Stam (Wageningen Agriculture University)
Linkage	Model	Linkage analysis for pedigree data
	Population Type	Pedigree
	Computer Platform	PC
	Graphic Interface	no
	Graphic Output	no
	Contact	Jurg Ott (Columbia University)
PGRI	Function	Linkage analysis, map construction, map merging
	Population Type	F2, backcross, RIL, DH, heterozygous F1, OP
	Computer Platform	Unix
	Graphic Interface	no
	Graphic Output	no
	Contact	Ben Liu (benliu@unity.ncsu.edu)

Information in this table may not be current. Contact the authors of the packages for the
latest information.
DH: doubled haploid
RIL: recombinant inbred line
OP: open-pollinated population

Table 18.3 Some available software packages for QTL mapping.

MAPMAKER/QTL	Function	Interval mapping, multiple QTL modeling
	Population Type	F2, backcross, RIL, DH
	Computer Platform	Unix
	Graphic Interface	no
	Graphic Output	Postscript
	Contact	Eric Lander (mapmaker@genome.wi.mit.edu)
QTLSTAT	Function	Interval mapping using nonlinear regression
	Population Type	F2, backcross, RIL, DH
	Computer Platform	Unix
	Graphic Interface	no
	Graphic Output	no
	Contact	Steve Knapp (sknapp@helix.css.orst.edu)
PGRI	Function	t-test, conditional t-test, linear regression, multiple QTL modeling, permutation test
	Population Type	F2, backcross, RIL, DH, heterozygous F1, OP
	Computer Platform	Unix
	Graphic Interface	no
	Graphic Output	no
	Contact	Ben Liu (benliu@unity.ncsu.edu)
QTL Cartographer	Function	t-test, composite interval mapping, permutation test, bootstrap, jackknife
	Population Type	F2, backcross, RIL, DH
	Computer Platform	Unix, Mac, PC Windows
	Graphic Interface	no
	Graphic Output	GUNPLOT (public domain software)
	Contact	Christopher Basten (basten@statgen.ncsu.edu)
MAPQTL	Function	Interval mapping, MQM, nonparametric mapping
	Population Type	F2, backcross, RIL, DH, heterozygous F1
	Computer Platform	Vax, Unix, Mac, and PC
	Graphic Interface	no
	Graphic Output	no
	Contact	Johan Van Ooijen (J.W.vanOOIJEN@cpro.dlo.nl)
Map Manager QT	Function	Interval mapping using regression, multiple-QTL
	Population Type	F2, backcross
	Computer Platform	MAC OS
	Graphic Interface	yes
	Graphic Output	yes
	Contact	Kenneth Manly (kmanly@mcbio.med.bufflo.edu)
QGENE	Function	Linear regression,
	Population Type	F2, backcross
	Computer Platform	MAC
	Graphic Interface	yes
	Graphic Output	yes
	Contact	James C. Nelson (jcn5@cornell.edu)

DH: doubled haploid
RIL: recombinant inbred line
OP: open-pollinated population

18.1.2 SPECIFIC PACKAGES FOR QTL MAPPING

Compared to the general statistical analyses of biological data, the statistical analyses for the study of genes controlling complex traits have the following characteristics:

(1) many repeated analyses in one task

(2) lack of standard distributions for some test statistics

(3) same level of complexity as models used in QTL mapping

Specialized software packages that make some specific genetic and statistical models have been developed mainly by scientists who are working in the area of statistical genetics, such as MAPMAKER/QTL (Lander *et al.* 1987), QTLSTAT (Liu and Knapp 1992), QTL Cartographer (Basten, Weir and Zeng 1995), PGRI (Liu and Lu 1995), MAPQTL (Van Ooijen and Maliepaard 1996), Map Manager QT (Manly and Cubmore 1996) and QGENE (Tanksley and Nelson 1996). These are all public domain packages except MAPQTL. These packages have some similarities, such as

(1) the interface is not user friendly compared to commercial software

(2) user support is limited due to their non-commercial status

(3) statistical models which can be built using the software are limited

(4) speed of model building is fast for the models which the software can

   build

Because of limitation (3), these software packages usually cannot handle data with complex experimental designs. In practice, means or least square means of the genotypes are used as input data for these packages. These packages can usually perform simple t-tests, linear and nonlinear regressions, interval mapping using the likelihood approach and composite interval mapping (Table 18.3).

For the software packages in Table 18.3, a known linkage map is needed for both running the programs and interpreting results. Companion packages for linkage map construction are available for MAPMAKER/QTL, QTLSTAT, PGRI, Map Manager QT and MAPQTL. They are MAPMAKER/EXP, GMENDEL, PGRI, MapManager and JoinMap, respectively. It is common to analyze marker data and to obtain a linkage map before QTL analysis. For packages QTL Cartographer and QGENE, linkage maps obtained using MAPMAKER/EXP can be incorporated in the analysis.

18.1.3 QTL ANALYSIS USING SAS

Most of the available models for QTL mapping can be implemented using statistical software packages such as SAS (SAS Institute 1990). The advantages of using the general statistical packages are

(1) they are commercially available

(2) user interfaces are friendly

(3) user support is available (with or without charge)

(4) the user can specify models

However, many statistical software packages are not efficient for a large number of repeated analyses and data manipulation. SAS is flexible enough to build any kind of model. Knapp (1994) listed several SAS programs for QTL mapping data with experimental designs. For QTL analysis using SAS, only the regression approaches here are given.

Interval Mapping Using Nonlinear Regression

SAS requires a specific data format. For example, for a single-environment experiment, the data should be arranged as

segment	Marker Name marker1	marker2	Marker Genotype Marker1	Marker2	line number	trait value
1	WG622	ABG313B	1	1	1	72.90
1	WG622	ABG313B	1	2	2	70.70
1	WG622	ABG313B	2	1	3	72.90
1	WG622	ABG313B	2	2	4	74.75
	...	...	...	...		

Then, the nonlinear analysis can be carried out using SAS code similar to

```
data a;
    infile 'mayqtl.dat';
    input seg marker1 $ marker2 $ g1 g2 line y;
    x1=0; x2=0; x3=0; x4=0;
    if g1=1 and g2=1 then x1=1;
    if g1=1 and g2=2 then x2=1;
    if g1=2 and g2=1 then x3=1;
    if g1=2 and g2=2 then x4=1;
proc nlin data=a noprint method=gauss convergence=0.0000001
    outest=output;
    by seg;
    parms m1=constantA m2=constantB r=constantC;
    bounds r<=1.0 r>=0;
    model y=x1*m1+x2*((1-r)*m1+r*m2)+x3*(r*m1+(1-r)*m2)+x4*m2;
    der.m1=x1+x2*(1-r)+x3*r;
    der.m2=x2*r+x3*(1-r)+x4;
    der.r=x2*(m2-m1)+x3*(m1-m2);
proc print data=output;
run;
```

The constants (`parms m1=constantA m2=constantB r=constantC`) are initial values for the parameters. SAS provides neither the hypothesis test for the contrasts nor the matrix $\hat{C}$ needed for the computation. To obtain the matrix, a linear regression procedure using the estimate of the parameters can be used. For example, for segment 18 of the barley data set (see Chapter 12), the following SAS code was used to generate the matrix

```
data b; set a; if seg=18;
    t1=x1+x3;
    t2=x2+x4;
    t3=x2*1.359-x3*1.359;
    y1=y-(x1*72.447+x2*73.806+x3*72.447+x4*73.806);
proc reg all; model y1=t1 t2 t3;
run;
```

In this SAS code, 72.447, 73.806 and 0 are the estimated values for the three parameters, the estimated difference between the two means is -1.359, t1, t2 and t3 are the first derivatives of the model with respect to the three parameters evaluated at the estimated values and y1 is the residual of the predicted value using the estimates from the observed values. The following SAS output is the matrix $\hat{C}$

```
X'X Inverse, Parameter Estimates, and SSE
           T1           T2           T3           Y1
T1     0.013503    -0.001422     0.005737     0.027442
T2    -0.001422     0.018788    -0.006239    -0.054301
T3     0.005737    -0.006239     0.025167    -1.107825
Y1     0.027442    -0.054301    -1.107825   328.400211
```

The test statistic can be easily obtained using a hand calculator when the contrasts are simple. When the contrasts are complex with several degrees of freedom, the computation may be more complex. For example, the contrasts for the environmental effect and the genotype by environment interaction contain three degrees of freedom each. For those cases, specialized software such as QTLSTAT is recommended.

Composite Interval Mapping Using Regression

In practice, the original composite interval mapping using the ECM algorithm can be performed by computer software QTL Cartographer (Basten, Weir and Zeng 1995). To use the linear regression approach for composite interval mapping, both commercial software such as SAS and the specialized software PGRI (Liu 1995) can be used. When using SAS, the data (called markerdata1) should be arranged as

```
Genome                          Genome
Segment   Marker1  Marker2 Positions   Lines  Z1 Z2   m1 m2 m3 m4
...
  1       WG622    ABG313B  0.00         147    1  0    2  2  1  1  1
  1       WG622    ABG313B  0.00         148    0  1    1  1  1  1  1
  1       WG622    ABG313B  0.00         149    0  1    1  1  1  1  1
  1       WG622    ABG313B  0.00         150   -1 -1    1  2  0  0  1
  1       WG622    ABG313B  0.01           1    1  0    1  1  1  1  1
  1       WG622    ABG313B  0.01           2   -1 -1    1  0  1  0  0
  1       WG622    ABG313B  0.01           3    1  0    1  1  1  1  1
  1       WG622    ABG313B  0.01           4    0  1    1  1  2  0  2
...
```

for the linkage group to be analyzed by CIM, where the genome segment is flanked by the two markers and each of the segments can be divided into a number of genome positions, such as each percent of recombination fraction. For each of the genome positions, there are N corresponding re-coded variables for the position (Z1 and Z2) and predetermined marker genotypes for controlling the residual genetic background (m1, m2, m3, ...). The Zs are the coefficients for the two parameters and they are

$$Z1 = X_{i1j} + (1 - \rho) X_{i2j} + \rho X_{i3j} \quad for \quad \theta_1$$
$$Z2 = \rho X_{i2j} + (1 - \rho) X_{i3j} + X_{i4j} \quad for \quad \theta_4$$

To determine which markers are needed to control the residual genetic background, a stepwise regression can be used to obtain markers linked to potential QTLs. A preconstructed linkage map is needed to find out the relative positions of the target interval and the markers. For stepwise regression using SAS, a predetermined variance and covariance matrix for the markers and trait phenotype is recommended if there are missing values in the marker data and the trait data. In that case, if the original data is used for the regression analysis, SAS uses only a portion of the data (the observations without any missing values for all the markers and the trait).

Another two data sets are needed to use SAS. They are the data set (called traitdata) which contains trait values corresponding to the lines and a data set (called markerdata2) containing the data of markers on the other chromosomes which will be used in the model. The following SAS code can be used for CIM using the linear regression approach

```
options ps=60 ls=80 nocenter;
/* Read trait data */
data trait;
  infile 'yourtrait.dat';
  input line y;
proc sort; by line;
```

```
/* Read marker data for the linkage group */
data markerdata1;
  infile 'yourmarker1.dat';
  input segment name1 $ name2 $ position line z1 z2 m1 m2 m3 ...;
proc sort; by line;
/* Read marker data for the rest of the linkage groups  */
data markerdata2;
  infile 'yourmarker2.dat';
  input line l1 l2 l3 ...;
data all; merge trait markerdata1 markerdata2; by line;
proc sort data=all; by segment name1 name2 position;
/* Full model */
proc glm data=all noprint outstat=fullmodel;
by segment name1 name2 position;
model y=z1 z2 m1 m2 m3 ... l1 l2 l3 .../solution noint;
data fullmodel; set fullmodel; if _type_='ERROR';
  keep segment name1 name2  position df ss;
data fullmodel; set fullmodel; rename ss=ssfull;
/* Reduced model */
proc glm data=all noprint outstat=redumodel;
by segment name1 name2 position;
model y=m1 m2 m3 ... l1 l2 l3 .../solution noint;
data redumodel; set redumodel; if _type_='ERROR';
  keep segment name1 name2  position df ss;
data redumodel; set redumodel; rename ss=ssredu;
/* Merge the two data sets and compute the statistic and p-value */
data model; merge fullmodel redumodel;
by segment name1 name2 position;
gstatistic=df*(ssreduc-ssfull)/ssfull;
if g<0 then g=0.0;
pvalue=1.0-probchi(gstatistic,1.0);
proc print data=cd;
run;
```

Some modifications may be needed for the computer used and the specific version of the software.

18.2 FUTURE CONSIDERATIONS

18.2.1 COMMERCIAL QUALITY SOFTWARE IS NEEDED

The available public domain software packages for linkage map construction and for the study of genes controlling complex traits are not adequate in terms of user interface, flexibility and user support for the development of genomic research on complex traits. Software packages of commercial quality are needed to accommodate the growing needs for data analysis and data management in genomic research.

During the last decade, private and public organizations have been working on automated devices for DNA sequencing and marker genotyping. Technologies, such as sequencing using DNA chips, capillary array electrophoresis (CAE), multiplex PCR, whole genome shotgun sequencing and transposon-facilitated direct sequencing, in combination with automated devices, reduce the cost of genome research and increase the speed of DNA sequencing and fragment sizing. Despite rapid advances in technology such as these, applications of genome information are still limited to a small scale and are confined to particular research laboratories. The technology is still expensive and inefficient, relative to its potential. The limitations are mainly due to the speed of processing and integrating the data, which are functions of software design, computer speed and algorithms. The currently available bioinformation analysis and management (BIAM) systems for genome research are not adequate for a great many real world applications because:

(1) there are limitations in the mathematical and statistical algorithms used for analyzing DNA sequence data and genomic mapping

(2) current BIAM systems are designed for gene discovery and not for practical applications

(3) the software packages are not user-friendly or widely available because they are developed mostly by biologists and are not on the commercial market

(4) the packages are not functionally complete and are difficult to incorporate into an automated system

Only a few institutes with in-house informatics teams have been able to semi-automate their genomic analysis process. Data entry, assembly and analysis take more time than does generating the data. The error and inaccuracy of genome research is due largely to human handling of the data. BIAM is becoming a bottleneck for practical applications of genomic research.

Methodologies for DNA sequence polymorphism detection, such as single-strand conformation polymorphism (SSCP), genomic mismatch scanning (GMS), high-density multiplex PCR, cleavage fragment length polymorphism (CFLP) and genetic bit analysis (GBA) are becoming routine biochemical assays for many laboratories. Automated devices for carrying out these processes are being developed. An integrated system which can carry out BIAM from initial biochemical assays to final practical applications is needed. The system can provide a bridge from molecular biologists to practical users, such as clinical physicians, attorneys, plant and animal breeders and environmentalists. A multi-disciplinary collaboration among molecular biologists, physicians, plant and animal biologists, forensic scientists, mathematicians and computer scientists is needed to develop the system (Figure 18.1).

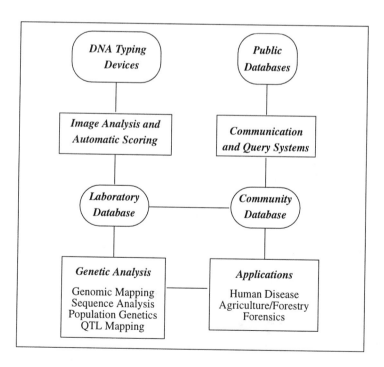

Figure 18.1 Bio-information analysis and management (BIAM) system and its relationship to automated DNA typing devices and public genome databases.

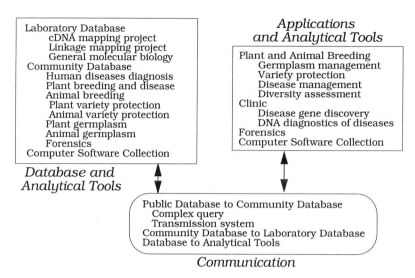

Figure 18.2 BIAMS (Biological Information Analysis and Management System).

18.2.2 STRUCTURE OF BIOINFORMATION ANALYSIS AND MANAGE-MENT SYSTEM (BIAMS)

Information analysis includes analysis of genetics and analysis for imaging data and for practical applications. Intensive computation and advanced statistics are involved in every step of the analysis. Information management includes information maintenance and distribution and adequate query systems for the data analysis. BIAMS includes three major components:

(1) analytical tools

(2) BIAMS database

(3) BIAMS communication and query tools (Figure 18.2)

The analytical tools include gel reading devices, data quality control software, genetic analysis software and computational tools for the application of genomic information. The genetic analysis includes DNA sequence analysis, linkage analysis, population genetic analysis, genomic map construction and searching for genes controlling complex traits. The applications of genomic information include genetic counselling, selection schemes for breeding programs, forensic applications and variety protection analysis. BIAMS databases are not only for bookkeeping purposes, but are also for scientific and practical applications. The communication and query tools play roles as bridges between the analytical tools, BIAMS databases and the public databases, such as Genbank (NCBI) and Plant Genome Database (PGD) (National Agricultural Library), through a complex query system.

18.2.3 DATA QUALITY PROBLEM

Genomic mapping data can have mistakes incorporated at any stage of the mapping process. These stages include DNA preparation, running gels and capturing gel images by automated DNA typing devices, scoring systems and data manipulation. Some of the potential errors in mapping data can be avoided by carefully designing the experiments and the statistical and genetic

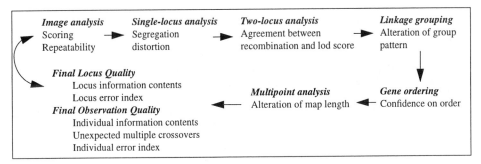

Figure 18.3 Data quality control flow chart.

analyses. There are many different sources of errors. Errors from some sources may be reduced or eliminated, while those from others may not. In some cases, a DNA marker or a single data point may give unexpected results in linkage map construction. These errors may damage the linkage map construction and the applications of the genomic information. An automated data quality control system would be a powerful aid for genomic researchers (Figure 18.3).

The first step is the gel reading (image analysis) step. An image is scored for several initial conditions. Data points with high repeatability over the initial conditions are kept for further genetic analyses. The next step is the testing for segregation distortion. Loci with unexpected segregation ratios and without an adequate biological interpretation will be discarded. A disagreement between recombination fraction and lod score (or significant P-value or chi-square) for linkage may be a sign of data problems. Expected information content on linkage is estimated for different locus combinations. Linkage grouping patterns should not be altered significantly by one or a few loci. When more than two linkage groups are merged together by a few loci, it is a sign of false linkage. If confidence of an estimated gene order increases significantly after a locus is dropped from the analysis, errors may be associated with the locus. Whole map length may be reduced dramatically by dropping a locus in the linkage group (not at either end of the linkage group) and this reduction is a sign of problems with the mapping function or with data quality. An unexpectedly high multiple crossover frequency often indicates errors and these frequencies can be determined when a linkage map is constructed. At the final stage, quality indices can be estimated for each locus and each observation. By iterating the whole process, data quality will be greatly improved.

18.3 PLANT GENOME RESEARCH INITIATIVE (PGRI)

PGRI is a computer package named after the Plant Genome Research Initiative (PGRI). The V1.0 of PGRI includes programs for linkage map construction, QTL mapping and progeny and parent selection for GMAPB (Genome Map Assisted Plant Breeding) for two-parent mating models. Linkage analysis was based on theory for mixed populations with random mating and selfing. Gene ordering was based on an algorithm that combines branch and bound (BB) and simulated annealing (SA). Confidence of the estimated gene order is quantified using the percentage of correct orders obtained by bootstrapping from original data. A combination of jackknife and bootstrap algorithms was incorporated into PGRI to automatically or manually obtain gene order with a specified gene order confidence. A likelihood approach and a nonparametric approach were incorporated into PGRI to test gene order difference between different mapping populations. If the hypothesis test among the gene orders is not significant, the program implements several approaches to merge the linkage maps from different populations. For a single QTL model, a QTL is located based on a nonlinear relationship between a pair of flanking markers and the putative QTL. For multiple-QTL models, the package searches two-way and three-way interactions and builds models using an approach similar to stepwise regression. For marker assisted progeny selection, an index proposed by Lande and Thompson (1990) and an alternative index are used in the package. The program was written in C and Fortran.

18.3.1 DATA TYPE

Six types of populations are suitable for PGRI. Those populations and their codes in PGRI data sets are listed in Table 18.4. Backcross (BC), doubled-haploid (DH) and recombinant inbred lines (RI) are coded as '1' in the PGRI data set, because they share the same expected genotypic frequencies in terms of the function of recombination fractions and likelihood function for recombination fraction estimation. The only difference among them is the estimation of variance for the estimated recombination fraction. The F2 type population is divided into two codes: '2' for codominant markers and '4' for dominant markers. The reason is the different treatments of linkage phases for these two types of markers.

Table 18.4 Types of populations and their codes in PGRI data sets.

Code	Population Type
1	Backcross (BC), Doubled-haploid, Recombined inbred line, etc.
2	F2 (codominant and two-allele)
3	F2 (multiple-allele)
4	Heterozygous F1 and F2 (dominant marker)
5	Open-pollinated (mixed self and random mating)
6	Mixture of above

Table 18.5 Genotype coding convention for PGRI data set.

Population Code	Genotype Code	Comments
1	1	Homozygous
	2	Homozygous (DH, RI), Heterozygous (BC)
2	1	Homozygous
	2	Heterozygous
	3	Homozygous
3	1, 2, 3, 4	Corresponding to each of the four genotypes (diploid)
4	1	Homozygous
	2	Heterozygous (most cases), Homozygous (dominance)
	3	Homozygous
5	1	Homozygous
	2	Heterozygous
	3	Homozygous
6		Same as above

For each of these population types, codes for the genotypes are listed in Table 18.5. '0' is used to code a missing genotype. There is no other rationale for coding the populations and genotypes in the ways explained here than convenience and standardization. Data coded differently from these conventions will either not be recognized by PGRI (data format error), will lead to incorrect results or will be treated as missing values.

18.3.2 DATA FORMAT

Data should be saved in ascii format. Presented below is a backcross example with a single population, 18 markers, 72 individuals in the population and without quantitative trait values. In the first row, the first '1' is the code for population type, the second '1' represents one population, the '18' is number of markers in the data set, the two '0's mean no quantitative trait and zero trait observation, the first '72' is total number of observations and the second '72' shows the number of observations for the first population. Starting the second row, the first 8 columns contain the locus name and after one space are the coded marker genotypes.

```
 1 1 18  0  0 72 72
MWG836   122222121111011111111121221101212111101111122111102212222111121221212121122
MWG8511  111112111222022221221111211212120222211120211112222112121112211122111222221222
RISIC101 111112002220202222011110121202222111201111122221121211122111221112221122222222
RISP103  120222221000012211211121222201211111222222121112112222001212122222121222
Wx       111112122210222222111111212022121111111111122221121211222111221122122122222
iEst5    111112122210222222111111212022121111111111122221121211222111221122122122222
ABC1511  121112122211011222111111121201212111011121121211222111221112212122222
ABC156D  112222121110112111111121221102121112211112211212122212222211122222121122
ABC1671  021222122211011102111121121101212111101112112111222211221112122122121122
ABC3102  122222221222012200121112122220021211112022221211121112212221121222222121122
ABC455   112222221221022112112122200212121122221112221211221222221122222121122
ABG380   121222122211011122111112112110121010111011212112111222211212222222121122
ABG461   112222212220012210211212211011211112022221211222212222211212121212111222
ABG476   112222112210221112112222012212110222112222121112212222211222222121122
Brz      112222111211011111111021221102112110211112211112221222221111222212121122
MWG089   121222122211011122111121121101212111111211211112222112211112212212121222
MWG2018  111112111222022212211112121202220112021110222211212112112211122211222221220
MWG5551  111112112220202222111120212022221112111121222112121111221112211222222222
```

The second example is a doubled-haploid data set example with three populations, a total of 373 individuals (150, 73 and 150 for each of the populations), 13 markers and without quantitative traits. The numbers in the first row show the three populations, the total number of observations, which is 373, and the numbers of observations for each of the populations, which are 150, 73 and 150, respectively. In the body of the data, after each of the locus names, there are 5 rows of data. The first two rows of 150 observations are for the first population, the center row of 73 observations is for the second population and the final two rows are for the third population. The same data layout is recommended to users for reasons of clarity and ease of handling. Extremely long rows and irregular layouts for different loci are not recommended.

```
1 3 13  0  0 373 150  73  150
Plc      222222112211112121212121211211112111122222221111222212211112122122122222222211
         121211221211122112212111121121112122112222112211221122211211221222222211211211122
         000000000000000000000000000000000000000000000000000000000000000000000000
         211112111222212102111221222222222222121111222121212210221111221121122121
         122212121122221211111221111121222111111121222212121122102211111221121122121
ABG704   222222112211112121212121211211112111122222221111222122112122212222221122
         121211221211122112212111121121112122112222112211221122211211221222222211
         000000000000000000000000000000000000000000000000000000000000000000000000
         211112111222212112112212202222222121111222121212211221111221121122121
         122212121122221211111222111112122211111112122211211112221112212212222221122
ABG312   222222010010110020002120020112011121111222202011112222122111121221220222222211
         121211221210022112202111021120122211222110211220002112212222201121121122122
         000000000000000000000000000000000000000000000000000000000000000000000000
         211112111222212112111021222022222221211112221212122112211111221121122121
         122212121122221211111222111112122211111112122211211112221212222221122
dRpg1    222222112211112121212121211211112111112222222111122221221111212212212222222211
```

```
            1212112221211122112212111121121122211222112211222112112212122221121121122122
            0000000000000000000000000000000000000000000000000000000000000000000000000000
            2111121112222121121112212222222222222121111222121212211221111122112112122121
            1222121211222212111112221111121222211111111212211222221121122221212222221122
MWG036B     2222221122111212121212112111211112222222111122212211112122122122222222212
            1212112211211122112112111211122211222112211222112112212222122211211222122
            0000000000000000000000000000000000000000000000000000000000000000000000000000
            2111121112222121121112212222222212111122212112212112111122211121122121
            1222011211222212101112220101121222211111111212112222211211122221212222221122
MWG555A     2222221122111212121212112111211112222122211112222122111121221221222222222212
            1212112211211121121112112112221122112212211211221212222122211211221222122
            1111112112220202222211111202120222211121111121222211212112211122112222220222
            2111121112222121121112212222222212111122212112212112111122211121122121
            1222211211222212111110220111112122221211112222111112222211211221212222221122
BCD129      1222221122111212212121111211121111221212111122221221111121221221222222222202
            2212112221211122112112211021121122211122212221121122120222221221121222122
            0000000000000000000000000000000000000000000000000000000000000000000000000000
            2111121112222121121112212222222122202222201122121212211221112222211211112121
            2222211211222222111122221111121222211111111212211222221121122221112222221120
iEst5       1222221122111212121211121211021111222121210002202121111121221221222222222212
            2122112212111201112111211121122211122211122121201212222222112222122122
            1111112122210222222111111212022212111111111112222112121122211122112122222
            2111121112122121121121122222221222222222211122212121221122111122222112111121
            2222211211222222111122221111121222211121111212222211121121211222222221122
Wx          1222221122110021212121112121112111122212121112222121111212212212222222212
            2121112212111221112111111121111122111222112212222211211221212222222211211222120
            1111121222210222222111111212022121111111111122211212112221112212221212222
            0000000000000000000000000000000000000000000000000000000000000000000000000000
            0000000000000000000000000000000000000000000000000000000000000000000000000000
Prx1A       1022221000010121200021121011121110222120211112222121111121221221222212221222
            2101112212110221112111111101111020101220002212222112112212122222021101222120
            0000000000000000000000000000000000000000000000000000000000000000000000000000
            2111121112122112211211211222222122222222211122212121221121111222221121112121
            2220201201222221112122222101121122021112111222221222221121122221112222221122
His3A       1222221022010021212121121211121112222121211112222121111121221221222221222212
            2101112212111221112111110211110221112221122122221021122121222222211212122120
            0000000000000000000000000000000000000000000000000000000000000000000000000000
            2111121112122121122112102022220202222222011222121212211211112222211211112121
            2222211211222222112122021101212221211121112222212222211201222211210222111222
ABC151A     1222222122111111212111121111211111211122221212011222211211122122121222221222212
            2120112212111221112111111121111122111222112212222112112212122222221121222222
            1211121222110112221111111212012121111011121121211222112221112212212122222
            0000000000000000000000000000000000000000000000000000000000000000000000000000
            0000000000000000000000000000000000000000000000000000000000000000000000000000
MWG089      1222222122111211212111121111121111112222121211112221212111122122212222211222212
            2121112212111221112111111121111122111222112212222112112212122222221121222222
            1212221222110111221111211211011212111111112112111222112221112212212121222
            0000000000000000000000000000000000000000000000000000000000000000000000000000
            0000000000000000000000000000000000000000000000000000000000000000000000000000
```

Another doubled-haploid example from a single population has a total of 150 individuals, 16 markers and three quantitative traits. For the quantitative traits, line 86 is missing. Observations for the quantitative trait are available for a total of 149 individuals. It is important to keep in mind that the line number for the quantitative traits has to match the marker data.

```
1 1 16 3 149 150 150
   1    4669.72    73.850    115.168
   2    5015.76    72.400    105.506
   3    4709.62    72.825    103.168
   4    5761.78    72.675    108.876
......
  80    6192.39    74.400    101.944
  81    5926.48    71.775    103.282
  82    5476.89    73.800    106.044
  83    4968.16    73.950    110.260
  84    4981.14    72.500    122.400
  85    6033.48    72.100    104.114
  87    4866.88    77.700    108.460
  88    5795.08    71.750    110.438
........
 148    4305.77    73.500    121.060
 149    5194.47    73.525    103.868
 150    5699.51    71.775    108.952
Nar1     1222112212212222221211211212212112111121121212211222221112212121212221121212211
         1222211111212112121212111111111111112212121221122212112222212221112222221211112211211
Ale      2122111221221111221112111211121111221221212112222212111100000000121221111221211
         1201122212222111121111111110221111111212222112121112122112121111111221221221212212
Dhn6     1112111122110212222212121112111112222222112221222100121212121221112122211121112
         1212111111112111212211212121112121112111122211121111121122111222222212122221112
Nar7     1122111111111222211111122222212222112222112221122111111112221000000000012111111221212212
```

```
        122212111211122211221221212122211112122112212111212212111222221221221221122
Glx     12222211221100212121211121121112111122212121111222212111112122122122222222212
        21211122121112211121111112111112111112211122211221222211211122121222222111212222120
ABC451  11121211221212121211111111221122211111211222211112121211112121121201212221222212
        2121011121121221121212221211222011121211001112121122122212101211111221221112
ABG452  21111121220212211121112212112121212122222221221111121222221212112222221222221
        2221011211211212122122222111122022111112002212111122112222201122222122221212
ABG453  12121122212212211221212112211212121221212221222121121211112111122121112111122211
        22211112201212122221212121112222112211111112222122212221212212212212121122
ABC454  21121201121212122121111222211122111121222221111112111121222112112121222221111
        11211111221212221212122222222221111212111112121112112222121121211111121221111
ABC455  11112121211221121112222211111211211211222112221121121111221222111212222221011222
        11112111102212111101112121212211111222221111111211212122122121102122112011221
PSR128  21221112212111111221112211211111221121111212221111112111211122202111122211
        12212222121221121211111122211111212122121121111122221211211121112211222122020212
PSR156  12111222122112122121112211212212212122222221212121111121122111122122211
        22211111211121212222121221211222211121111211122221222212122221121211112120021222
ABG458  12222221122122112022220021221211222112211111121111211221121102211112212212
        112211110211122201221122212112112121222122211121112212221111211222222121122
ABG459  21121221221212122121111022211122111111122222211112211112122211211211212222221111
        21211111221212222121221222122222111121211111112111211222112112211111121221111
ABG460  12112221121112211221221222121222112211122221111121112221112212112112211
        212201122212221112212111221122201212121212001122212222212211202221112212121
Prx2    11122110110200121020221121111121221111111111122111221211222121121111111111211
        12112112211211122212112222211111112111121112111122222211222121211112221122111221
```

Three computer screens for the three example data sets are shown in Tables 18.6, 18.7 and 18.8. **pgri** is the command to start the program.

18.3.3 LINKAGE ANALYSIS AND MAP CONSTRUCTION

In Table 18.6, the data set used has one population and is without quantitative trait values. Questions related to linkage map merging and quantitative traits are not asked by the program. The first three questions are for filenames, which include data filename, filename for the single- and two-locus model statistics and one for the results after linkage grouping.

Table 18.6 Computer screen for running simple linkage analysis using PGRI (command line version). The words in bold are commands or values typed by users and the others are generated by the program.

```
pgri  ◄─────── start program

                        PGRI-beta, 1995
                      (Command Line Version)
   Data filename                                          Filenames
Ex1.data                                                  for input
   Filename for storing single and two-locus statistics   and output
Ex1.out1                                                  data sets
   Filename for storing linkage maps
Ex1.out2
   Please wait, two-point analysis
Locus #  10
   Give R and P-value for grouping pleas  Linkage grouping criteria
0.25 0.001
   Group# 1:   1  2  3  4  5  6  7  8  9  10  11  12  13  14  15  16  17  18
   Change R and P to do another grouping.
      1 = change
      2 = no change
1
   Give an R and P-value
0.2 0.0001
   Group# 1:   1  2  3  4  5  6  7  8  9  10  11  12  13  14  15  16  17  18
   Change R and P to do another grouping.
      1 = change
      2 = no change
1
   Give an R and P-value
0.2 0.00001
   Group# 1:   1  2  3  4  5  6  7  8  9  10  11  12  13  14  15  16  17  18
   Change R and P to do another grouping.
      1 = change
```

Table 18.6 (continued)

```
      2 = no change
1
 Give an R and P-value
0.2 0.000001
 Group#  1:    1   2   3   4   5   6   7   8   9  10  11  12  13  14  15  16  17  18


Change R and P to do another grouping.
      1 = change
      2 = no change
1
 Give an R and P-value
0.15 0.0001
 Group#  1:    1   8  15
 Group#  2:    2   3   5   6   7   9  12  16  17  18
 Group#  3:    4  10
 Group#  4:   11  14
 Group#  5:   13
Change R and P to do another grouping.
      1 = change
      2 = no change
1
 Give an R and P-value
0.17 0.0001
 Group#  1:    1   2   3   5   6   7   8   9  12  15  16  17  18
 Group#  2:    4  10
 Group#  3:   11  14
 Group#  4:   13
Change R and P to do another grouping.
      1 = change
      2 = no change
2
 Option for locus ordering
      1 = SA-SAR              Locus ordering options
      2 = SA-ML               (see Chapter 9)
      3 = SA-SAR - BB-SAR
1
 Option for map construction     Locus ordering and
      1 = Simple Linkage Map      linkage map merging
      2 = Manually Interactive    (see Chapters 9 and 11)
      3 = Auto-framework-map
      4 = Map merging and hypothesis test among orders
2

 Jackknifing and Bootstraping

  Bootstraping
 Two-point R of Group  1
---------------------------------------------------------------------------
Locus      ABC156D  Brz     MWG836  ABC1671 ABG380   MWG089   ABC1511
---------------------------------------------------------------------------
Brz         0.09
MWG836      0.18    0.12
ABC1671     0.31    0.26    0.17
ABG380      0.32    0.27    0.18    0.00
MWG089      0.31    0.26    0.18    0.00    0.00
ABC1511     0.38    0.35    0.26    0.12    0.12     0.12
Wx          0.40    0.38    0.35    0.27    0.26     0.26     0.14
iEst5       0.41    0.40    0.37    0.28    0.28     0.27     0.16
RISIC101    0.49    0.48    0.47    0.38    0.38     0.37     0.25
MWG5551     0.51    0.49    0.49    0.39    0.39     0.38     0.26
MWG8511     0.45    0.43    0.46    0.42    0.43     0.42     0.33
MWG2018     0.46    0.44    0.47    0.43    0.43     0.43     0.34
---------------------------------------------------------------------------
Locus      Wx      iEst5   RISIC101 MWG5551 MWG8511
---------------------------------------------------------------------------
iEst5       0.01
RISIC101    0.11    0.09
MWG5551     0.14    0.13    0.02                   Two-point pairwise
MWG8511     0.19    0.17    0.06    0.07           recombination
MWG2018     0.19    0.18    0.06    0.08    0.00   frequency matrix

 Map of Group   1
---------------------------------------------------------------------------
```

Table 18.6 (continued)

Locus		2-R	ML-R		H-D		K-D	
ABC156D	(8)	0.000	0.000	0.000	0.000	0.000	0.000	0.000
Brz	(15)	0.113	0.086	0.086	0.094	0.094	0.086	0.086
MWG836	(1)	0.101	0.118	0.203	0.134	0.228	0.120	0.206
ABC1671	(9)	0.132	0.175	0.378	0.215	0.443	0.182	0.389
ABG380	(12)	0.000	0.000	0.378	0.000	0.443	0.000	0.389
MWG089	(16)	0.000	0.000	0.378	0.000	0.443	0.000	0.389
ABC1511	(7)	0.097	0.121	0.499	0.139	0.582	0.123	0.512
Wx	(5)	0.097	0.153	0.652	0.183	0.764	0.158	0.670
iEst5	(6)	0.056	0.014	0.666	0.014	0.778	0.014	0.684
RISIC101	(3)	0.087	0.103	0.768	0.115	0.893	0.104	0.788
MWG5551	(18)	0.029	0.014	0.783	0.015	0.908	0.014	0.803
MWG8511	(2)	0.083	0.067	0.850	0.072	0.980	0.068	0.870
MWG2018	(17)	0.000	0.000	0.850	0.000	0.980	0.000	0.870

Log(likelihood) (MLE) =   -185.61

Linkage map
2-R = two point R from above locus
ML-R = multi-locus maximum
*   likelihood R*
Next column = cumulative value from
*   the first locus*
H-D = Haldane's map distance
K-D = Kosambi's map distance
(see Chapter 10)

Locus Information (Jackknifing and Bootstrapping)

Locus		Chi	#Missings	PCO(%)
ABC156D	(8)	0.057	2	0.933
Brz	(15)	2.118	4	0.892
MWG836	(1)	4.765	4	0.867
ABC1671	(9)	1.209	5	0.867
ABG380	(12)	1.471	4	0.892
MWG089	(16)	1.429	2	0.867
ABC1511	(7)	1.174	3	0.883
Wx	(5)	0.057	2	0.858
iEst5	(6)	0.229	2	0.800
RISIC101	(3)	0.385	7	0.892
MWG5551	(18)	0.710	3	0.875
MWG8511	(2)	0.014	3	0.767
MWG2018	(17)	0.015	5	0.892

Chi = chi-square for segregation
*   distortion*
#Missing = number of missing
*   observations*
PCO(%) = percentage of correct
*   gene order after the locus is*
*   discarded from the analysis*

Bootstrapping Results

Genome Position	Individual Locus Map Position												
	8	15	1	9	12	16	7	5	6	3	18	2	17
1	94	5	1	0	0	0	0	0	0	0	0	0	0
2	5	95	0	0	0	0	0	0	0	0	0	0	0
3	1	0	99	0	0	0	0	0	0	0	0	0	0
4	0	0	0	67	8	24	0	0	0	0	1	0	0
5	0	0	0	22	63	14	0	0	0	0	0	1	0
6	0	0	0	10	28	61	0	0	0	0	0	0	1
7	0	0	0	0	0	0	99	0	0	1	0	0	0
8	0	0	0	0	0	0	0	75	25	0	0	0	0
9	0	0	0	0	0	0	0	25	75	0	0	0	0
10	0	0	0	0	0	0	1	0	0	93	6	0	0
11	0	0	0	0	0	1	0	0	0	6	87	1	5
12	0	0	0	1	0	0	0	0	0	0	1	92	6
13	0	0	0	0	1	0	0	0	0	0	5	6	88

Frequency of a locus is located
at certain genome positions
among 100 bootstrap
gene orders
(see Chapter 9)

Your Action
  1 = drop
  2 = add
  3 = end
1
Give # of loci to be dropped and ids   *loci 12, 16, and 17*
3 12 16 17   *are dropped from the analysis*
Jackknifing and Bootstrapping

Two-point R of Group 1

Table 18.6 (continued)

```
------------------------------------------------------------------
Locus        MWG8511  MWG5551  RISIC101 iEst5    Wx       ABC1511  ABC1671
------------------------------------------------------------------
MWG5551       0.07
RISIC101      0.06     0.02
iEst5         0.17     0.13     0.09
Wx            0.19     0.14     0.11     0.01
ABC1511       0.33     0.26     0.25     0.16     0.14
ABC1671       0.42     0.39     0.38     0.28     0.27     0.12
MWG836        0.46     0.49     0.47     0.37     0.35     0.26     0.17
Brz           0.43     0.49     0.48     0.40     0.38     0.35     0.26
ABC156D       0.45     0.51     0.49     0.41     0.40     0.38     0.31
------------------------------
Locus        MWG836   Brz
------------------------------
Brz           0.12
ABC156D       0.18     0.09
```

Map of Group   1

```
------------------------------------------------------------------
Locus              2-R      ML-R                H-D                K-D
------------------------------------------------------------------
MWG8511  (   2)   0.000    0.000    0.000    0.000    0.000    0.000    0.000
MWG5551  (  18)   0.077    0.063    0.063    0.067    0.067    0.063    0.063
RISIC101(   3)    0.017    0.015    0.078    0.015    0.082    0.015    0.078
iEst5    (   6)   0.097    0.102    0.180    0.115    0.197    0.104    0.182
Wx       (   5)   0.000    0.014    0.193    0.014    0.210    0.014    0.195
ABC1511  (   7)   0.119    0.154    0.347    0.184    0.394    0.159    0.354
ABC1671  (   9)   0.172    0.132    0.479    0.153    0.547    0.135    0.489
MWG836   (   1)   0.203    0.174    0.652    0.213    0.760    0.181    0.670
Brz      (  15)   0.108    0.118    0.770    0.134    0.894    0.120    0.790
ABC156D  (   8)   0.077    0.086    0.856    0.094    0.988    0.086    0.877
```

Log(likelihood) (MLE) =   -171.00

Locus Information (Jackknifing and Bootstrapping)

```
--------------------------------------------------
Locus              Chi      #Missings   PCO(%)
--------------------------------------------------
MWG8511  (   2)   0.014        3        0.789
MWG5551  (  18)   0.710        3        0.956
RISIC101 (   3)   0.385        7        0.956
iEst5    (   6)   0.229        2        0.933
Wx       (   5)   0.057        2        0.711
ABC1511  (   7)   1.174        3        0.678
ABC1671  (   9)   1.209        5        0.844
MWG836   (   1)   4.765        4        0.911
Brz      (  15)   2.118        4        0.889
ABC156D  (   8)   0.057        2        0.756
```

Bootstrapping Results

```
Genome    Individual Locus Map Position
Position   2 18  3  6  5  7  9  1 15  8

    1     85 14  0  0  0  0  1  0  0  0
    2     11 71 17  0  0  1  0  0  0  0
    3      3 14 82  0  1  0  0  0  0  0
    4      0  0  0 83 17  0  0  0  0  0
    5      0  0  1 17 82  0  0  0  0  0
    6      0  1  0  0  0 99  0  0  0  0
    7      1  0  0  0  0  0 99  0  0  0
    8      0  0  0  0  0  0  0 99  0  1
    9      0  0  0  0  0  0  0  0 94  6
   10      0  0  0  0  0  0  0  1  6 93
```

Your Action
 1 = drop
 2 = add
 3 = end
1
Give # of loci to be dropped and ids   *locus 3 is dropped*
1 3   *from analysis*

Two-point R of Group   1

Table 18.6 (continued)

```
----------------------------------------------------------------------
Locus      MWG8511  MWG5551  iEst5    Wx      ABC1511  ABC1671  MWG836
----------------------------------------------------------------------
MWG5551      0.07
iEst5        0.17     0.13
Wx           0.19     0.14    0.01
ABC1511      0.33     0.26    0.16    0.14
ABC1671      0.42     0.39    0.28    0.27     0.12
MWG836       0.46     0.49    0.37    0.35     0.26     0.17
Brz          0.43     0.49    0.40    0.38     0.35     0.26     0.12
ABC156D      0.45     0.51    0.41    0.40     0.38     0.31     0.18

------------------------
Locus      Brz
------------------------
ABC156D      0.09

Map of Group   1
----------------------------------------------------------------------
Locus           2-R     ML-R           H-D            K-D
----------------------------------------------------------------------
MWG8511 (   2)  0.000   0.000  0.000   0.000  0.000   0.000  0.000
MWG5551 (  18)  0.059   0.071  0.071   0.076  0.076   0.071  0.071
iEst5   (   6)  0.145   0.129  0.200   0.150  0.226   0.133  0.204
Wx      (   5)  0.014   0.014  0.215   0.015  0.241   0.014  0.218
ABC1511 (   7)  0.101   0.153  0.368   0.183  0.424   0.158  0.377
ABC1671 (   9)  0.119   0.132  0.500   0.153  0.577   0.135  0.512
MWG836  (   1)  0.200   0.174  0.673   0.213  0.790   0.181  0.693
Brz     (  15)  0.106   0.118  0.791   0.134  0.925   0.120  0.813
ABC156D (   8)  0.147   0.086  0.877   0.094  1.018   0.086  0.899

Log(likelihood) (MLE) =       -178.74

  Locus Information (Jackknifing and Bootstrapping)
  --------------------------------------------------
  Locus            Chi      #Missings    PCO(%)
  --------------------------------------------------
MWG8511  (   2)    0.014         3        0.975
MWG5551  (  18)    0.710         3        0.975
iEst5    (   6)    0.229         2        0.925
Wx       (   5)    0.057         2        0.950
ABC1511  (   7)    1.174         3        0.775
ABC1671  (   9)    1.209         5        0.950
MWG836   (   1)    4.765         4        1.000
Brz      (  15)    2.118         4        0.925
ABC156D  (   8)    0.057         2        1.000

  Bootstrapping Results

  Genome      Individual Locus Map Position
  Position     2 18   6   5   7   9   1  15   8

     1        88  9   1   1   0   1   0   0   0
     2         9 87   2   1   1   0   0   0   0
     3         0  3  78  19   0   0   0   0   0
     4         2  0  19  79   0   0   0   0   0
     5         0  1   0   0  97   0   0   1   1
     6         1  0   0   0   0  97   0   1   1
     7         0  0   0   0   0   0  98   1   1
     8         0  0   0   0   0   2   1  94   3
     9         0  0   0   0   2   0   1   3  94

  Your Action
    1 = drop
    2 = add        end linkage group 1
    3 = end
  3

  Do you want to use another ordering algorithm?
    1 = yes
    2 = no         end gene ordering and PGRI
  2
```

18.3.4 LINKAGE MAP MERGING

In Table 18.7, the data set has four populations and is without quantitative trait values. PGRI asks for a confidence probability for each linkage group. Results are written into a file.

Table 18.7 Computer screen output analyzing data with multiple mapping populations and merging linkage maps using PGRI (command line version).

```
(See Figure 1)
 Option for locus ordering
  1 = SA-SAR
  2 = SA-ML
  3 = SA-SAR - BB-SAR
1
 Option for map construction
  1 = Simple Linkage Map                  pick option 4 for multiple
  2 = Manually Interactive                  mapping populations
  3 = Auto-framework-map
  4 = Map merging and hypothesis test among orders
4
 Give a confidence for linkage group  1
60.0
Number of bootstrap cycles =  1
Number of bootstrap cycles =  2       the target confidence on
Number of bootstrap cycles =  3       gene order is 60% for
Number of bootstrap cycles =  4       linkage group 1
Number of bootstrap cycles =  5
......
```

18.3.5 QTL ANALYSIS AND BREEDING PLAN

In Table 18.8, the data set has one population and three quantitative traits. PGRI asks for information about the breeding plan.

Table 18.8 Computer screen output for analyzing data with quantitative traits and progeny selection and mating design analysis using PGRI (command line version).

```
 Select Methodology for QTL Analysis
  1 = simple t-test
  2 = conditional t-test
  3 = nonlinear model interval mapping
  4 = interval test
1
 Please wait, QTL analysis
...
 Different QTL Mapping Methodology
  1 = yes
  2 = no
2
 Number of target traits and their relative weight
 Number of target traits
2
Traits
1 2
Relative weight
0.01 1.0
 Do you know heritability for the traits
 0 = no
 1 = yes
1
Input heritability
0.2 0.6
```

Table 18.8 (continued)

```
Methodology of Picking Loci for Selection
  1 = stepwise regression by computer
  2 = manual picking by user
1
 Give a significant level
0.1
                MULTILOCUS MODEL (MAIN EFFECT)
 Trait 1
-----------------------------------------------------------
SOURCE                SS         DF        MS            P
-----------------------------------------------------------
MODEL         31827576.000        2  15913788.000   0.0000
LINE(QTL)     34952072.000      146    239397.750
TOTAL         66779648.000      148    451213.844
COEFFICIENT OF DETERMINATION = 0.4766


-----------------------------------------------------------
MARKER             SS          B        R-SQUARE     P-VALUE
-----------------------------------------------------------
                             5418.724
ABG474        6541382.500    -215.040      0.098       0.000
ABR334       26765066.000     454.805      0.401       0.000

Trait 2
-----------------------------------------------------------
SOURCE                SS         DF        MS            P
-----------------------------------------------------------
MODEL            102.255         6      17.042      0.0000
LINE(QTL)        135.433       142       0.954
TOTAL            237.688       148       1.606
COEFFICIENT OF DETERMINATION = 0.4302


-----------------------------------------------------------
MARKER             SS          B        R-SQUARE     P-VALUE
-----------------------------------------------------------
                               74.226
ABG62            4.258         0.179       0.018       0.036
BCD453B         24.016        -0.416       0.101       0.000
ABC455          24.533        -0.424       0.103       0.000
BCD351F         13.467        -0.315       0.057       0.000
CDO484           8.198        -0.245       0.034       0.004
ABR313           4.576         0.190       0.019       0.030

Here are the loci picked for selection (Single-locus model)
-----------------------------------------------
                               Linkage
      Locus          Trait     Group     Effect
-----------------------------------------------
  1  ABG474 ( 106)     1          1     -188.959
  2  ABR334 ( 229)     1          7      440.074
  1  ABG62  ( 289)     2          1        0.296
  2  BCD453B( 169)     2          3       -0.478
  3  ABC455 (  10)     2          4       -0.494
  4  BCD351F(  55)     2          5       -0.411
  5  CDO484 ( 124)     2          9       -0.328
  6  ABR313 ( 140)     2         11        0.275

 Do you want to repick
  1 = repick
  2 = no
2
 Initial conditions (for mating design):
 Possible progeny size (for the first segregating population)
100
 Expected genetic and economic gains
Genetic gain (% of mean) for trait#  1
0.02
Genetic gain (% of mean) for trait#  2
```

Table 18.8 (continued)

0.02
 Economic gain (% mean value)
0.02
 Acceptable probability of recovering target genotype
0.05
 Here are some initial conditions:
```
------------------------------------------------
Possible progeny size              100
Expected genetic trait# 1          0.020
Expected genetic trait# 2          0.020
Expected economic                  0.020
Acceptable probability             0.050
------------------------------------------------
```

Two Individuals Are Picked as Parents (Phenotypic Selection, Plan# 1)
```
------------------------------------------
Parent#1 = Line#       143
   Marker genotype     11111221211
   Trait Value         7071.24    76.57
Parent#1 = Line#       87
   Marker genotype     22021222222
   Trait Value         4866.88    77.70
------------------------------------------
```

Expected Result from the Mating (Phenotypic Selection, Plan# 1)
```
------------------------------------------
 Chance for achieving expected gain
           Trait# 1              0.7460
           Trait# 2              0.0000
      Economic Value             0.5152
 Expected gain under the probability
           Trait# 1              0.0539
           Trait# 2              0.0131
      Economic Value             0.0299
------------------------------------------
 Marker-assisted-selection, Wait !!!
```

Two Individuals Are Picked as Parents (Marker-assisted Selection, Plan# 1)
```
------------------------------------------
Parent#1 = Line#       57
   Marker genotype     21112222212
   Trait Value         5939.62    75.03
Parent#1 = Line#       9
   Marker genotype     11111222201
   Trait Value         5533.75    77.00
------------------------------------------
```
Expected Result from the Mating(Phenotypic Selection, Plan# 1)
```
------------------------------------------
 Chance for achieving expected gain
           Trait# 1              1.0000
           Trait# 2              0.0000
      Economic Value             0.9982
 Expected gain under the probability
           Trait# 1              0.0558
           Trait# 2              0.0090
      Economic Value             0.0287
------------------------------------------
```

Do you want to change the initial conditions ?
  1 = change
  2 = no
2

18.3.6 OUTPUT SAMPLES

The following tables contain some output samples. These outputs are usually self-explanatory.

Linkage Map Merging

Table 18.9 Output from linkage map merging analysis.

```
Pooled Map of Group #   1 (imcomplete data)
-----------------------------------------------------------------------
Locus              2-R    ML-R          H-D            K-D
-----------------------------------------------------------------------
RISP161A(   39)   0.000  0.000  0.000  0.000  0.000  0.000  0.000
dRpg1     (    4)   0.000  0.000  0.000  0.000  0.000  0.000  0.000
MWG555A   (    6)   0.030  0.038  0.038  0.039  0.039  0.038  0.038
BCD129    (    7)   0.055  0.053  0.091  0.057  0.096  0.054  0.091
Prx1A     (   10)   0.083  0.077  0.168  0.083  0.179  0.077  0.169
MWG089    (   13)   0.063  0.080  0.248  0.088  0.267  0.081  0.250
MWG836    (   16)   0.176  0.217  0.465  0.285  0.552  0.233  0.482
Brz       (   17)   0.074  0.075  0.541  0.082  0.633  0.076  0.558
ABC156D   (   18)   0.068  0.065  0.606  0.070  0.703  0.065  0.624
MWG003    (   19)   0.135  0.129  0.735  0.149  0.852  0.132  0.756
ABC308    (   20)   0.048  0.053  0.788  0.057  0.909  0.054  0.809
MWG511    (   37)   0.007  0.006  0.794  0.006  0.915  0.006  0.816
ABC254    (   30)   0.015  0.016  0.810  0.016  0.931  0.016  0.831
VAtp57A   (   31)   0.051  0.061  0.871  0.065  0.995  0.061  0.892
ABC310B   (   23)   0.323  0.324  1.195  0.523  1.518  0.387  1.279
WG380A    (   28)   0.217  0.224  1.419  0.297  1.815  0.241  1.520
Log(likelihood) (MLE) =        -957.96
```

```
Locus quality information
-----------------------------------------------------
 Locus            Chi      #Missings     PCO(%)
-----------------------------------------------------
RISP161A ( 39)   0.228       158          67.0
dRpg1    (  4)   0.859        75          77.0
MWG555A  (  6)   2.459         7          87.0
BCD129   (  7)   2.488        80          97.0
Prx1A    ( 10)   2.306       102          97.0
MWG089   ( 13)   0.455       153          97.0
MWG836   ( 16)   0.541        11          96.0
Brz      ( 17)   2.382        20          95.0
ABC156D  ( 18)   0.455       153          94.0
MWG003   ( 19)   1.095        77          92.0
ABC308   ( 20)   0.123        81          88.0
MWG511   ( 37)   3.840       154          86.0
ABC254   ( 30)   0.132       101          92.0
VAtp57A  ( 31)   0.514        93          91.0
ABC310B  ( 23)   0.744        29          95.0
WG380A   ( 28)   0.856       110          95.0
```

```
 Information for dropped loci
----------------------------------------------------
 Locus            Chi      #Missing
----------------------------------------------------
Plc      (  1)   1.351        77
ABG704   (  2)   0.973        76
ABG312   (  3)   2.118       101
MWG036B  (  5)   1.646        79
iEst5    (  8)   1.337        11
Wx       (  9)   0.018       155
His3A    ( 11)   2.594        92
ABC151A  ( 12)   0.115       156
ABC167A  ( 14)   0.117       160
ABG380   ( 15)   1.888        15
ABC455   ( 21)   0.005       158
ABG476   ( 22)   0.018       155
RISP103  ( 24)   0.568       160
```

Table 18.9 (continued)

```
ABC305     ( 25)     0.125        85
PSR129     ( 26)     0.126        87
ABG461     ( 27)     1.751        16
MWG635B    ( 29)     0.215        75
MWG851A    ( 32)     1.449         8
iPgd1A     ( 33)     1.805        80
MWG2018    ( 34)     0.667       157
RISIC10A   ( 35)     0.786       158
ABG077     ( 36)     0.973        76
MWG626     ( 38)     5.303       155
RISP144    ( 40)    11.441       196
MWG808     ( 41)     5.558       165
```

Heterogeneity Test on Segregation Ratio

Locus Name   Chi-square   DF   P-Value

None

Heterogeneity Test on Two-point Recombination

Locus Name   Chi-square DF   P-Value

```
MWG089  ( 13)-MWG555A (  6)     9.76    2   0.0076
MWG089  ( 13)-iEst5   (  8)    13.28    2   0.0013
MWG089  ( 13)-Wx      (  9)    13.71    2   0.0011
MWG089  ( 13)-ABC151A ( 12)     7.99    2   0.0184
ABC167A ( 14)-MWG555A (  6)     8.81    2   0.0122
ABC167A ( 14)-iEst5   (  8)    11.21    2   0.0037
ABC167A ( 14)-Wx      (  9)    10.39    2   0.0055
ABG380  ( 15)-iEst5   (  8)    10.05    3   0.0181
ABG380  ( 15)-Wx      (  9)     7.89    2   0.0194
ABG380  ( 15)-His3A   ( 11)     7.41    2   0.0246
MWG836  ( 16)-MWG036B (  5)     9.04    2   0.0109
MWG836  ( 16)-ABG312  (  3)     9.64    2   0.0081
MWG836  ( 16)-Plc     (  1)     8.31    2   0.0157
MWG836  ( 16)-ABG704  (  2)     8.64    2   0.0133
MWG836  ( 16)-dRpg1   (  4)     9.08    2   0.0107
MWG836  ( 16)-MWG555A (  6)     8.36    3   0.0391
MWG836  ( 16)-BCD129  (  7)     7.39    2   0.0248
MWG836  ( 16)-iEst5   (  8)     9.17    3   0.0271
MWG836  ( 16)-Prx1A   ( 10)     7.35    2   0.0254
MWG836  ( 16)-His3A   ( 11)    13.13    2   0.0014
Brz     ( 17)-MWG036B (  5)     8.35    2   0.0154
Brz     ( 17)-ABG312  (  3)     9.33    2   0.0094
Brz     ( 17)-Plc     (  1)     7.82    2   0.0201
Brz     ( 17)-ABG704  (  2)     8.02    2   0.0181
Brz     ( 17)-dRpg1   (  4)     8.59    2   0.0137
Brz     ( 17)-BCD129  (  7)     7.10    2   0.0288
Brz     ( 17)-His3A   ( 11)     7.30    2   0.0260
ABC310B ( 23)-Brz     ( 17)     8.13    3   0.0434
ABC310B ( 23)-MWG003  ( 19)     7.31    2   0.0259
ABC310B ( 23)-ABC308  ( 20)     8.10    2   0.0175
```

Heterogeneity Test among Locus Orders (complete data)
Term   Log-likelihood   DF
Total   -1719.26   87
Pooled   -1283.80   40

LR-test Statistic =   870.92
           DF =   47
        P-value =   0.0000

Table 18.9 (continued)

```
Bootstrapping Results for population#  1
```

Pooled

Position	Individual Population Position											
	4	6	7	10	13	16	17	18	19	20	30	31
4	99	0	1	0	0	0	0	0	0	0	0	0
6	0	100	0	0	0	0	0	0	0	0	0	0
7	1	0	99	0	0	0	0	0	0	0	0	0
10	0	0	0	100	0	0	0	0	0	0	0	0
13	0	0	0	0	100	0	0	0	0	0	0	0
16	0	0	0	0	0	100	0	0	0	0	0	0
17	0	0	0	0	0	0	100	0	0	0	0	0
18	0	0	0	0	0	0	0	100	0	0	0	0
19	0	0	0	0	0	0	0	0	98	2	0	0
20	0	0	0	0	0	0	0	0	2	98	0	0
30	0	0	0	0	0	0	0	0	0	0	100	0
31	0	0	0	0	0	0	0	0	0	0	0	100

```
Bootstrapping Results for population#  2
```

Pooled

Position	Individual Population Position						
	39	6	13	16	17	18	37
39	94	6	0	0	0	0	0
6	6	94	0	0	0	0	0
13	0	0	100	0	0	0	0
16	0	0	0	100	0	0	0
17	0	0	0	0	100	0	0
18	0	0	0	0	0	100	0
37	0	0	0	0	0	0	100

```
Bootstrapping Results for population#  3
```

Pooled

Position	Individual Population Position											
	39	4	6	7	10	16	17	19	20	37	30	31
39	97	0	0	0	3	0	0	0	0	0	0	0
4	0	97	0	3	0	0	0	0	0	0	0	0
6	0	0	100	0	0	0	0	0	0	0	0	0
7	1	2	0	97	0	0	0	0	0	0	0	0
10	2	1	0	0	97	0	0	0	0	0	0	0
16	0	0	0	0	0	94	0	0	1	2	0	3
17	0	0	0	0	0	0	94	0	0	0	4	2
19	0	0	0	0	0	0	0	94	0	3	2	1
20	0	0	0	0	0	0	0	0	94	6	0	0
37	0	0	0	0	0	0	0	6	5	89	0	0
30	0	0	0	0	0	0	6	0	0	0	94	0
31	0	0	0	0	0	6	0	0	0	0	0	94

QTL Analysis

Table 18.10 Output from QTL analysis.

```
Output from QTL Mapping Analysis
----------------------------------------
Significant Level           Critical Value
----------------------------------------
Significant level 0.05          3.250
Significant level 0.01          3.365
Significant level 0.001         3.947

QTL Map of Group#  1(t-test,trait#  1)
```

```
-----------------------------------------------------------------------
Locus            2-R     ML-R     M1       M2       t      df    P-value
-----------------------------------------------------------------------
ABC170A( 224)    0.000   0.000   5308.21  5445.01  1.247  143   0.2145
PSR154  (  32)   0.000   0.000   5351.45  5472.28  1.040  127   0.3005
Nir     ( 176)   0.123   0.137   5322.36  5435.35  0.982  130   0.3278
Amy1    (  24)   0.008   0.008   5296.98  5418.11  1.052  131   0.2945
Nar7    (   4)   0.047   0.049   5248.94  5454.21  1.826  137   0.0700
ksuA3D  (  45)   0.087   0.097   5276.11  5565.02  2.398  121   0.0180
ABG1    ( 241)   0.013   0.020   5307.42  5457.15  1.089   93   0.2791
ABC170B( 225)    0.000   0.000   5287.07  5472.71  1.694  146   0.0925
ABC175  ( 252)   0.027   0.028   5278.90  5474.73  1.794  146   0.0749
ABC163  ( 185)   0.007   0.008   5258.73  5456.13  1.796  140   0.0746
ABG388  ( 237)   0.008   0.010   5185.05  5503.51  2.729  125   0.0073
ABG379  ( 181)   0.015   0.012   5277.54  5474.83  1.796  146   0.0745
ksuD17  (  46)   0.031   0.027   5276.23  5541.46  2.344  127   0.0206
ABG474  ( 106)   0.041   0.049   5197.22  5575.13  3.459  140   0.0007
WG223B  ( 122)   0.042   0.041   5210.32  5552.49  3.185  146   0.0018
ABR335  ( 230)   0.015   0.017   5232.00  5506.82  2.538  130   0.0123
ksuA3B  (  43)   0.008   0.008   5228.79  5527.88  2.790  145   0.0060
ABG20   ( 159)   0.000   0.000   5217.85  5540.97  3.017  147   0.0030
ABA6    ( 210)   0.008   0.010   5234.36  5538.61  2.590  122   0.0108
Rrn1    (  83)   0.000   0.000   5212.28  5543.74  3.040  144   0.0028
Ubi5    ( 288)   0.021   0.019   5209.71  5547.95  3.128  144   0.0021
ABC169B( 220)    0.000   0.000   5204.42  5550.21  3.242  147   0.0015
Ubi4    ( 222)   0.023   0.028   5275.20  5557.20  2.391  124   0.0183
PSR167B(  27)    0.009   0.008   5220.59  5529.03  2.552  121   0.0119
ABG458  (  13)   0.025   0.017   5237.99  5513.76  2.482  138   0.0143
ABR331  ( 215)   0.031   0.027   5251.07  5490.91  2.004  128   0.0472
ABG387B( 236)    0.151   0.138   5230.85  5475.77  2.042  112   0.0435
PSR106  (  40)   0.045   0.044   5306.81  5434.73  1.157  145   0.2494
His3D   ( 287)   0.181   0.176   5406.90  5377.03  0.257  136   0.7974
ABC152A( 148)    0.014   0.013   5375.23  5377.05  0.017  146   0.9867
Cxp3    (  34)   0.008   0.007   5363.23  5350.09  0.110  123   0.9124
ABG378  ( 180)   0.090   0.095   5350.01  5368.24  0.153  132   0.8782
Nar1    (   1)   0.052   0.055   5415.11  5334.49  0.733  147   0.4646
ABG466  (  98)   0.007   0.008   5440.09  5335.29  0.904  137   0.3675
ABG62   ( 289)   0.046   0.052   5461.65  5341.69  1.034  125   0.3030
PSR167  (  28)   0.009   0.007   5382.32  5340.41  0.363  123   0.7172
Lth     (  25)   0.025   0.019   5426.22  5275.91  1.337  127   0.1836
```

CHAPTER 19

RESAMPLING AND SIMULATION IN GENOMICS

19.1 INTRODUCTION

Resampling techniques such as bootstrap and jackknife and Monte Carlo methods are widely applied in statistical theory and method development. This is largely due to the rapid advances in computing technology and the complexity of new statistics. Mathematical and statistical research and computing technology are coming closer together. The idea of substituting intensive computing for complex mathematical derivation is not new. Several decades ago, R.A. Fisher used permutation tests. However, the massive applications of computing intensive methods have only recently become feasible, due to the development of computing technology.

Due to the complex nature of genome data, some test statistics cannot be estimated in a parametric way or the statistics do not follow standard distributions. As discussed in earlier chapters, the log likelihood ratio test statistic for testing differences among gene orders is not distributed as standard chi-square. Also, the significance level for QTL mapping analysis cannot be precisely determined using a standard t, F or chi-square distribution, because of the multiple test problems and non-independence among the tests. Nonparametric approaches, such as resampling, reshuffling and computer simulation, have been used to solve these problems. These techniques are computationally intensive. However, as computer technology advances in speed and capacity, the computational problems become less significant.

The nonparametric approaches have been mentioned in previous chapters:

- the method for constructing confidence intervals for estimated recombination fractions is given in Chapters 4, 6 and 8
- the method for determining significant difference between two gene orders is found in Chapters 9 and 11
- the simulation approach for fitting a map function to a multiple-locus model is discussed in Chapter 10

There are many more situations in genomics where computer simulation will play a significant role.

In this chapter, the basic concepts and methodologies of resampling and computer simulation will be described. Specifically, resampling techniques such as bootstrap, jackknife and shuffling the computer simulation for generating mapping populations and some specific applications will be discussed.

19.2 RESAMPLING

A resampling technique is designed to sample from a sample. The purposes of resampling are usually to empirically estimate statistical properties (such as variance, distribution and confidence interval) of an estimator and to obtain an empirical distribution of a test statistic. There are three commonly used resampling techniques: bootstrap, jackknife and shuffling.

19.2.1 BOOTSTRAP

Bootstrap Sample

A bootstrap sample is defined as a random sample of size n drawn with replacement from a sample of n objects. Let S denote an original sample with n objects, say

$$S = (x_1, x_2, ..., x_n) \qquad (19.1)$$

The probability of drawing each of the objects, x_i, $i = 1, 2, ..., n$ is $1/n$. A bootstrap sample (S^B) from the original sample can be obtained by randomly sampling n times with replacements. The bootstrap sample is denoted by

$$S^B = (x_1^B, x_2^B, ..., x_n^B) \qquad (19.2)$$

The object (x_i) in the original sample can be a single observation or a group of observations, depending on the objectives of the resampling. For example, the objects are two-locus genotypes for each of the individuals in the mapping population when the objective of the resampling is to obtain an empirical distribution of the estimated recombination fraction between the two loci. However, if the objective is to investigate the statistical properties of the multi-locus statistics, then the objects are genotypes of an entire linkage group.

Bootstrap method can also be applied to parametric distributions. The objects for parametric distribution are the observations in the distributions with precise probabilities. Bootstrapping using parametric distribution is called a parametric bootstrap. In some ways, the parametric bootstrap is similar to the traditionally defined Monte Carlo simulation. Instead of sampling with replacement from an original sample, parametric bootstrap samples n times from a parametrically defined population.

Bootstrap Replication

Equation (19.2) denotes a single bootstrap sample. The power of the resampling techniques relies on a large number of resampling samples which can be obtained from a single original sample. If the bootstrapping process is repeated b times, then b independent bootstrap samples $S_1^B, S_2^B, ..., S_b^B$ can be obtained. Each of the b samples, S_j^B, $j = 1, 2, ..., b$, is called a bootstrap replication.

The number of bootstrap replications needed depends on the objectives of the resampling and the computing resources. In general, the more replications the better, if the computational task permits. For finding simple statistical properties of a single parameter problem, such as means and variance, a relatively small number of bootstrap replications may be sufficient. However, if the objective is to obtain an empirical distribution for a parameter, a relatively large number is needed.

Bootstrap Mean, Variance and Bias

An estimator can be obtained from each of the b bootstrap samples, $S_1^B, S_2^B, ..., S_b^B$. If $\hat{\theta}_j^B = F(S_j^B)$ denotes the estimate for the jth replication, the bootstrap mean is

$$\bar{\theta}^B = \frac{1}{b}\sum_{j=1}^{b} \hat{\theta}_j^B \qquad (19.3)$$

and the bootstrap variance is

$$\hat{V}^B = \frac{1}{b-1}\sum_{j=1}^{b} (\hat{\theta}_j^B - \bar{\theta}^B)^2 \qquad (19.4)$$

The bootstrap estimate of bias is defined as

$$Bias^B = \bar{\theta}^B - \theta \qquad (19.5)$$

where θ is the true parameter. In practice, the true parameter is usually unknown and can be replaced by the estimate from the original sample, $\hat{\theta}$.

$$\widehat{Bias}^B = \bar{\theta}^B - \hat{\theta}$$

Bootstrap Confidence Interval

If the bootstrap estimates are approximately distributed as normal variables, for a large number of bootstrap replications the bootstrap confidence interval can be

$$\bar{\theta}^B \pm z_{(1-\gamma)/2}\sqrt{\hat{V}^B} \qquad (19.6)$$

where $z_{(1-\gamma)/2}$ is a normal deviate which satisfies

$$P[z > z_{(1-\gamma)/2}] = (1-\gamma)/2$$

and γ is the confidence coefficient. If the number of the bootstrap sample size is small, then $z_{(1-\gamma)/2}$ should be replaced by the deviate from a Student t distribution with $b-1$ degrees of freedom. Equation (19.6) becomes

$$\bar{\theta}^B \pm t_{(1-\gamma)/2, b-1}\sqrt{\hat{V}^B} \qquad (19.7)$$

A bootstrap confidence interval can also be obtained using the bootstrap percentiles from the empirical distribution of the parameter. If

$$\widehat{CDF}(x) = P[\hat{\theta}_b \leq x]$$

is used to denote the cumulative distribution of the bootstrap estimates of recombination fraction, then the confidence interval with confidence coefficient γ is

$$\{CDF^{-1}[\widehat{(1-\gamma)}/2], CDF^{-1}[\widehat{(1+\gamma)}/2]\} \qquad (19.8)$$

Example: Bootstrap Approach

In Chapter 4, 1000 bootstrap samples were drawn from the data in Table 4.5. Figure 19.1 shows the empirical distributions for the bootstrap estimates

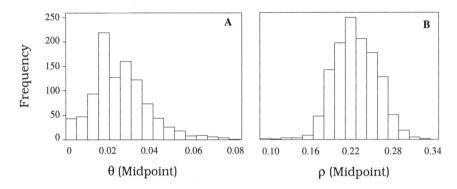

Figure 19.1 A: Empirical distribution of the ML estimator of the recombination fraction using bootstrap. B: Empirical distribution of the ML estimator of the proportion of escapes. The estimates were obtained using the Newton-Raphson iteration and the number of bootstrap replications is 1000. See Table 4.5 in Chapter 4 for information on the data.

of recombination fraction between the marker J7 and the rust resistant gene F and the proportion of disease escapes. Table 19.1 shows variances, bias and confidence intervals using parametric and bootstrap approaches. The bootstrap variances for the two parameters are close to the parametric variances. The bootstrap biases for the two parameters are sufficiently small, relative to their variances. The ratio between bias and the standard deviation (square root of variance) for recombination fraction is

$$\frac{Bias_\theta}{Sd_\theta} = \frac{0.00008}{\sqrt{0.0001666}} = 0.006$$

$$\frac{Bias_\rho}{Sd_\rho} = \frac{0.00206}{\sqrt{0.0009025}} = 0.069$$

The confidence intervals obtained parametrically or using bootstrapping are similar for this example data.

Table 19.1 Statistical properties of ML estimators for the recombination fraction and escapes for the data in Table 4.5.

	Recombination Fraction ($\hat{\theta}$)	Escapes ($\hat{\rho}$)
Parametric		
Variance	0.0001357	0.00099
95% Interval	(0, 0.0455)	(0.162, 0.286)
Bootstrap		
Variance	0.0001666	0.0009025
Bias	0.0000800	0.0020600
95% Interval (Normal)	(0, 0.048)	(0.1675, 0.2853)
95% Interval (Percentile)	(0, 0.054)	(0.1815, 0.2826)

19.2.2 JACKKNIFE

Jackknife Sample

Let S denote an original sample with n objects, say

$$S = (x_1, x_2, ..., x_n)$$

A jackknife sample is the sample, leaving out one observation at a time from the original sample, denoted by

$$S^J = (x_1, x_2, ..., x_{i-1}, x_{i+1}, ..., x_n)$$

for $i = 1, 2, ..., n$. The ith jackknife sample is a data set with the ith observation removed. For an original sample size n, jackknife can be replicated n times. Each of the n times is a jackknife replication.

Jackknife Mean, Variance and Bias

An estimator can be obtained from each of the n jackknife samples, $S_1^J, S_2^J, ..., S_b^J$. If $\hat{\theta}_i^J = F(S_i^J)$ denotes the estimate for the ith replication, the jackknife mean is

$$\bar{\theta}^J = \frac{1}{n}\sum_{i=1}^{n} \hat{\theta}_i^J \tag{19.9}$$

and the jackknife variance is

$$\hat{V}^J = \frac{n-1}{n}\sum_{i=1}^{n} (\hat{\theta}_i^J - \bar{\theta}^J)^2 \tag{19.10}$$

The jackknife estimate of bias is defined as

$$\widehat{Bias}^J = (n-1)(\bar{\theta}^J - \theta) \tag{19.11}$$

where θ is the true parameter. In practice, the true parameter is usually unknown and can be replaced by the estimate from the original sample $\hat{\theta}$.

The coefficients $(n-1)/n$ and $n-1$ in Equations (19.10) and (19.11) for the jackknife variance and bias are arbitrary. The rationale behind them is that the variation among the jackknife samples is smaller than among the bootstrap samples because the jackknife samples are n fixed data sets which are more similar to the original sample than to the bootstrap samples. The coefficients are "inflation factors" for the jackknife samples.

19.2.3 COMBINATION OF JACKKNIFE AND BOOTSTRAP

The combination of jackknife and bootstrap can be used for more complex situations. For example, we want to evaluate the impact of each observation on the statistical properties of an estimated parameter. We could design a resampling scheme to jackknife each of the observations and then bootstrap the jackknife samples. In this case, bootstrap is used to obtain the statistical properties of the parameter estimate for each of the jackknife samples. By doing this, the statistical properties can be compared among the jackknife samples which correspond to each of the observations.

In Chapter 9, a combination of jackknife and bootstrap was used to evaluate the impact of each locus on the quality of the estimated gene order. In that case, jackknife was performed on genes and bootstrap was used for each of the individuals in the mapping population. If the gene order has high confidence for the jackknife sample, then the locus corresponding to the jackknife sample may have an unfavorable impact on determining the gene order. Jackknifing also can determine the impact of the individual on gene order.

Table 19.2 shows results from the jackknifing analysis for a linkage group of *Eucalyptus* genome. The jackknife procedure, when combined with a bootstrap, is used to evaluate the impact of a locus on an overall gene order estimation. The PCO (percentage of correct order) in Table 19.2 is estimated for each locus using the bootstrap procedure when the locus is discarded from the data (jackknifing). If the PCO of the estimated gene order increases significantly when a locus is discarded from the data, then the locus has a "bad" effect on the confidence of the estimated gene order. On the other hand, if the PCO of estimated gene order decreases significantly when a locus is discarded from the data, then the locus is essential for obtaining a high confidence gene order.

Table 19.2 Jackknifing results for a linkage group of *Eucalyptus* genome.

Locus	(code)	Chi	#Missing	PCO(%)
E31G243	(1)	0.008	0	100.0
E31G246	(2)	6.126	4	100.0
E31G249	(3)	0.073	0	99.0
E31G2413	(4)	0.398	0	79.0
E31G2416	(5)	0.659	0	79.0
E31G2417	(6)	4.301	0	86.0
E31G2418	(7)	0.984	0	70.0
E31G2419	(8)	0.984	0	70.0
E31G2421	(9)	0.659	0	95.0

Chi = chi-square for single-locus segregation ratio test.
#Missing = number of missing values.
PCO(%) = average percentage of correct gene order after the locus is
    discarded.
Output from PGRI software.

19.2.4 SHUFFLING OR PERMUTATION TEST

Shuffling a Sample and Permutation

Let us consider a situation: we want to test the difference between two population means using two samples from each of the populations, denoted by

$$\begin{cases} S^X = (x_1, x_2, ..., x_n) \\ S^Y = (y_1, y_2, ..., y_n) \end{cases}$$

The estimated difference between the two sample means is $d = \bar{x} - \bar{y}$. If we know the distribution of the difference, a standard test using the distribution can be applied. However, if the distribution is unknown or complex, a standard test cannot be applied.

Let us assume that there is no difference between the two population means. The two samples are assumed to be drawn from the same population. Now, imagine putting the $2n$ observations into a bag and shaking the bag well. We can draw n observations from the bag without replacement. Two samples, one containing the n drawn observations and the other containing the remaining n observations in the bag, are obtained. These samples are called shuffling samples and the process is a permutation. The difference between the means of the two shuffling samples can be obtained. The difference does not depend on the difference between the two populations, but depends on the permutation.

Empirical Distribution of the Test Statistic

If we put all the observations back into the bag, shake the bag and do the drawing again, another permutation is obtained. If this is done p times, a distribution for the difference can be obtained. This distribution is an empirical distribution of the test statistic under the null hypothesis. Now we are ready to see where the difference between the two original samples falls in the distribution. If the difference falls in the middle of 95% of the distribution, we can conclude that the difference between the two population means are not significant at a 5% level. However, if the difference falls outside of the middle 95% of the distribution, we can conclude that the two population means differ significantly at a 5% level. The test, based on the empirical distribution obtained by the permutation, is called a permutation test. Because the permutation test was first used by R.A. Fisher to support the Student t-test, it is also called Fisher's permutation test.

The shuffling and the permutation test have been widely used in genomic analysis. For example, shuffling has been used to determine the threshold for the significance of a likelihood ratio test statistic for declaring a QTL.

Example: Permutation Test

The empirical distributions of the log likelihood ratio test statistic for the recombination fraction between the disease resistance gene and the marker and the probability of disease phenotype escapes were obtained using 10,000 permutations of the original data (Table 4.5)(Figure 19.2). The empirical distributions have longer tails than expected, when compared to the chi-square distributions. This result also shows the differences between the hypothesis tests using the parametric chi-square distribution and using the empirical distributions (Table 19.3). Chi-square values 262.02 and 162.28 are highly significant using the parametric chi-square distribution with one degree of freedom. However, the percentiles in the empirical distributions are only 99.8 and 99 for the two values. 0.2% of the 10,000 likelihood ratio test statistics are

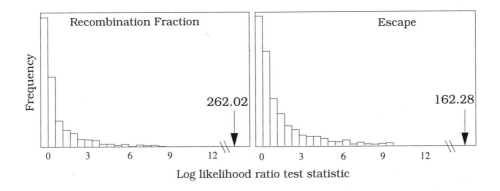

Figure 19.2 Empirical distributions of the log likelihood ratio test statistic obtained by shuffling the data of Table 4.5 10,000 times.

greater than or equal to 262.02 for the recombination fraction and 1% of the 10,000 likelihood test statistics are greater or equal to 162.28 for the escapes.

Table 19.3 A comparison of hypothesis tests using parametric chi-square distribution and the empirical distributions obtained by shuffling for the example in Table 4.5.

	Log Likelihood Ratio Test Statistic	P-value Using Chi-Square Distribution	Percentiles in the Shuffling Distributions
Recombination Fraction	262.02	< 0.000001	99.8%
Escapes	162.28	< 0.000001	99.0%

19.3 COMPUTER SIMULATIONS

19.3.1 OVERVIEW

Computer simulation is a numerical technique used to resolve complex systems. The system can be a set of procedures, an estimation methodology or a hypothesis test. Computer simulation is also called the Monte Carlo simulation. In statistics, properties of an estimator can usually be obtained parametrically if the distribution of the estimator or distribution of a function of the parameter is known and well-characterized. However, for many statistical problems, the distribution is commonly unknown and not well characterized. Computer simulation is commonly used to obtain statistical properties for relatively complex estimators. To evaluate a set of estimators or hypothesis-test procedures, a comparison between the true parameters and the estimators and an evaluation of the statistical power and the probability of false positive are needed. However, because the true parameters are usually unknown, the comparison between the true parameters and the estimators cannot be made and the true statistical power and the frequency of false positives cannot be estimated from a single or a set of experiments. This is especially true for statistics used in genomic analysis. Another disadvantage for using experimental data is that the range of the parameters is usually limited to the experimental conditions. It is usually not practical to generate experimental data for the full range of parameter space. For those problems, data can be simulated, using a computer, under different conditions of true parameters and can be analyzed to obtain the estimators and inferences on hypothesis tests. Computer simulation has the advantage of "knowing" the true parameters and of having the ability to generate data in the full range of the parameter space. The disadvantage of computer simulation is that the biological systems may be too complex to simulate.

In Section 19.2, bootstrapping was introduced as a resampling technique. Bootstrap and Monte Carlo simulations are similar in terms of their implementation. Both of them are based on repeated random sampling. The difference between them is that bootstrap is more commonly used to sample from a sample while the Monte Carlo simulation samples from a population. However, the parametric bootstrap is also designed to sample from population. So, the parametric bootstrap is much like the Monte Carlo simulation. The difference is that the parameters used for the parametric bootstrap are estimated from the sample and the parameters used for the Monte Carlo simulation are set, based on the purposes of the simulation. In terms of implementation, the parametric bootstrap and the Monte Carlo simulation are the same.

19.3.2 RANDOM SAMPLING FROM CONTINUOUS DISTRIBUTIONS

The theoretical basis of computer or Monte Carlo simulation is that any distribution can be transformed into a uniform distribution, in which the probability density function is constant in the interval (0, 1). The technique to transform the distributions is called the probability integral transformation. For a probability density

$$f(x) = \frac{dF}{dx}$$

let

$$F(x) = \int_{-\infty}^{x} f(u)\, du = y$$

Then

$$dF = \frac{f(x)}{dy/dx}dy = \frac{f(x)}{f(x)}dy = dy$$

$$\frac{dF}{dy} = f(y) = 1$$

(19.12)

So, y is distributed as a uniform random variable in the interval (0, 1). In other words, the cumulative probability of a random variable is a uniform random variable in the interval (0, 1).

Now, if a random number, y, can be generated from a uniform (0, 1), then it can be used to generate a random value for any distribution using the inverse of the probability integral transformation, that is

$$x = F^{-1}(y)$$

For example, the exponential distribution has the cumulative distribution

$$F(x) = 1 - e^{-x}$$

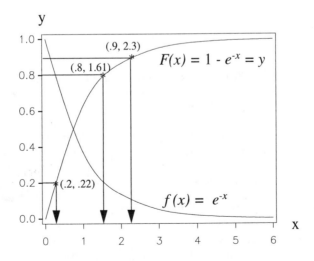

Figure 19.3 Transformation of uniform (0, 1) random values to exponential distribution random values.

If a random number, y, is drawn, then the corresponding random exponential variable is

$$x = F^{-1}(y) = -\log(1-y)$$

Figure 19.3 shows an exponential distribution. The vertical axis shows the cumulative probability, y, and the horizontal axis shows the corresponding exponential variable value, x. If three random values, 0.9, 0.8 and 0.2, are obtained from a uniform (0, 1), then the corresponding exponential values are 2.3, 1.61 and 0.22, respectively.

A random uniform variable can be generated using commonly used statistical software and computer languages. Subroutines for generating a random uniform variable are readily available for several commonly used computer languages, such as C and Fortran. Algorithms and subroutines to convert the random uniform variable to commonly used distributions are available (Stuart and Ord 1994; Press *et al.* 1986). For example, a pair of random values from uniform (0, 1), (y_1, y_2), can be transformed into a pair of standard normal random values using the Box-Müller (1958) transformation. It is

$$x_1 = \cos(2\pi y_2)\sqrt{-2\log y_1}$$

$$x_2 = \sin(2\pi y_2)\sqrt{-2\log y_1}$$

Table 19.4 Genotypes and their expected frequencies and cumulative frequencies for two independent loci.

Genotype	Frequency	Cumulative Frequency
AABB	0.0625	0.0625
AABb	0.125	0.1875
AAbb	0.0625	0.25
AaBB	0.125	0.375
AaBb	0.25	0.625
Aabb	0.125	0.75
aaBB	0.0625	0.8125
aaBb	0.125	0.9375
aabb	0.0625	1.0

19.3.3 RANDOM SAMPLING FROM DISCRETE DISTRIBUTIONS

Random samples of continuous variables with different probability distributions can be obtained by transforming the uniform (0, 1) random variable. This is also true for discrete variables. Table 19.4 shows the genotypes and their expected frequencies for two independent loci in an F2 population (produced by selfing AaBb). The third column shows the cumulative frequencies for the order of the genotypes in the table. Figure 19.4 is a graphic representation of the frequencies and the cumulative frequencies. Once the cumulative frequencies are obtained, samples can be simulated from the population. For example, three random values from the uniform (0, 1), 0.2, 0.5 and 0.9, are drawn. Their corresponding genotypes are AAbb, AaBb and aaBb, respectively. If n random values are drawn from the uniform (0, 1), then a simulated sample with n individuals is obtained from the population.

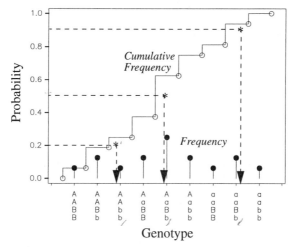

Figure 19.4 Transformation from random values from uniform (0, 1) to discrete two-locus genotypes. Locus A and locus B are independent.

Generating and analyzing a sample from a theoretical population using a computer is called a simulation replication. The number of simulation replications needed depends on the objectives of the simulation and the computing resources. In general, the more replications the better, if the computation task permits doing so.

Table 19.5 Genotypes and their expected frequencies and cumulative frequencies for two linked loci.

Locus A	Locus B	p_{ij}	$p_{ij}|_{r=0.2}$	Cumulative Frequency
AA	BB	$0.25\,(1-r)^2$	0.16	0.16
AA	Bb	$0.5r\,(1-r)$	0.08	0.24
AA	bb	$0.25\,r^2$	0.01	0.25
Aa	BB	$0.5r\,(1-r)$	0.08	0.33
Aa	Bb	$0.5\,[\,(1-r)^2 + r^2\,]$	0.34	0.67
Aa	bb	$0.5r\,(1-r)$	0.08	0.75
aa	BB	$0.25\,r^2$	0.01	0.76
aa	Bb	$0.5r\,(1-r)$	0.08	0.84
aa	bb	$0.25\,(1-r)^2$	0.16	1.0

19.4 SIMULATION OF DISCRETE MARKERS

19.4.1 A JOINT DISTRIBUTION APPROACH FOR TWO LOCI

With a given mating system, mapping population size, locations of markers on the genome and type of markers, a mapping population and its

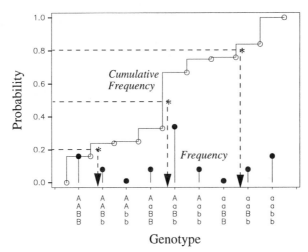

Figure 19.5 Transformation from random values from uniform (0, 1) to discrete two-locus genotypes. Locus A and locus B are linked with 20% recombination fraction.

marker genotypes can be simulated using a computer. In previous pages, a simulation of two independent markers was discussed. An extension of this approach will generate a method to simulate linked markers. We will simulate a linkage group with 5 markers. The mapping population will contain 100 individuals of classical F2 progeny (generated by selfing AaBbCcDdEe). The map distances among the loci are 20%, 15%, 5% and 10%.

There are several ways to simulate a mapping population. Consider the two loci, A and B, at one end of the linkage group. A and B are linked with 20% recombination. Table 19.5 shows the expected frequencies for the 9 genotypic classes and a graphic representation for the expected frequencies is shown by Figure 19.5. If we generate 100 random numbers from uniform (0, 1), then the corresponding 100 genotypes can be obtained.

19.4.2 A CONDITIONAL FREQUENCY APPROACH FOR MULTIPLE LOCI

The joint distribution approach is difficult to implement for a large number of loci because of the large number of possible genotypes. For k loci, there are 3^k possible genotypes for the F2 progeny. The approach to simulate one locus at a time using the conditional frequency is commonly used for simulating a large number of loci.

Again, consider the loci A and B. We can simulate locus A first. The three genotypes for locus A, AA, Aa and aa, have the expected frequencies 0.25, 0.5 and 0.25, respectively. The corresponding cumulative frequencies are 0.25, 0.75 and 1.0. If a random value from uniform (0, 1) is less than 0.25, then an AA is sampled. If the value is between 0.25 and 0.75, an Aa is sampled. If the value is greater than 0.75, an aa is obtained. Repeating the drawing 100 times, 100 random genotypes will be obtained for locus A.

For locus B, the expected genotypic frequencies, conditional on the genotypes of locus A, are shown in Table 19.6. Which genotype is sampled for locus B will be determined by the random value from the uniform (0, 1) and the genotype of locus A. For example, if a random number 0.5 is obtained and the corresponding genotype for locus A is AA, then a BB is sampled for locus B. However, if the locus A genotype is aa for the same random number, then a Bb will be sampled. Repeating the process 100 times, genotypes for locus B corresponding to the 100 individuals will be obtained.

Table 19.6 Expected frequencies of genotypes for locus B conditional on locus A and their cumulative frequencies when A and B are linked.

Locus A	Locus B	$p(B\|A)$	$p(B\|A)\|_{r=0.2}$	Cumulative Frequency
AA	BB	$(1-r)^2$	0.64	0.64
	Bb	$2r(1-r)$	0.32	0.96
	bb	r^2	0.04	1.0
Aa	BB	$r(1-r)$	0.16	0.16
	Bb	$(1-r)^2 + r^2$	0.68	0.84
	bb	$r(1-r)$	0.16	1.0
aa	BB	r^2	0.04	0.04
	Bb	$2r(1-r)$	0.32	0.36
	bb	$(1-r)^2$	0.64	1.0

For the next locus, C, the expected genotypic frequencies conditional on genotypes of locus B can be obtained using the same approach shown in Table 19.6. This procedure can also be applied to loci D and E. Here, absence of crossover interference is assumed. If crossover interference does exist, at least three loci have to be simulated simultaneously.

19.4.3 CONVERSION OF MAP DISTANCE TO RECOMBINATION FRACTION

Mapping data generated using the conditional frequencies is assumed to be absent of crossover inference. Crossover is assumed to be completely random. It is common to have map distances between the loci instead of recombination fractions. In this case, a conversion using some kind of mapping function is needed. For example, Haldane's mapping function is

$$d = \begin{cases} -0.5\log(1-2r) & if \quad 0 \le r < 0.5 \\ \infty & if \quad r \ge 0.5 \end{cases} \tag{19.13}$$

where d is Haldane's map distance in Morgans and r is the recombination fraction. The inverse of Equation (19.13) will convert the map distance into recombination fraction, which is

$$r = 0.5(1 - e^{-2d})$$

See Chapter 10 for discussion of different mapping functions.

19.5 QUANTITATIVE TRAITS: A TWO-GENE MODEL

19.5.1 TWO-GENE MODEL

To simulate a quantitative trait, a model of the trait inheritance is needed. For example, a trait may be controlled by two genes, A and B. A and B are linked r recombination units apart. On the previous pages, the ways to simulate the genes in an F2 population (produced by selfing a Bb) were described. To simulate the trait, we first simulate the genotypes and then simulate the trait based on the genotypes. We now assume that a sample with size 100 has been obtained following the methodology described before.

Ignoring epistatic interactions, a quantitative trait can be modeled using

$$y_j = \mu + a_A x_{1j} + d_A x_{2j} + a_B x_{3j} + d_B x_{4j} + \varepsilon_j \tag{19.14}$$

for an individual j in a classical F2 population, where y_j is a trait value for the jth individual in the population, x_{1j} is the dummy variable for the additive effect taking 1, 0 and -1 for genotypes AA, Aa and aa, respectively, x_{2j} is a dummy variable for the dominance effect taking 0, 1 and 0 for genotypes AA, Aa and aa, respectively, μ is the overall mean for the trait, a_A is the additive effect for gene A, d_A is the dominance effect, x_{3j}, x_{4j}, a_B and d_B are the corresponding terms for locus B and ε_j is random error for the jth individual. Table 19.7 shows the four dummy variables associated with the nine genotypic classes in the F2 population. The random error ε_j is usually composed of the residual from the gene effects. The residual may include the undetected genetic effects, the environmental effects and the simple random sampling effects.

Table 19.7 Genotypes and their expected and cumulative frequencies for two linked loci.

Locus A	Locus B	x_{1j}	x_{2j}	x_{3j}	x_{4j}	y_j^*
AA	BB	1	0	1	0	$a_1 + a_2 + e^*$
	Bb	1	0	0	1	$a_1 + d_2 + e^*$
	bb	1	0	-1	0	$a_1 - a_2 + e^*$
Aa	BB	0	1	1	0	$d_1 + a_2 + e^*$
	Bb	0	1	0	1	$d_1 + d_2 + e^*$
	bb	0	1	-1	0	$d_1 - a_2 + e^*$
aa	BB	-1	0	1	0	$-a_1 + a_2 + e^*$
	Bb	-1	0	0	1	$-a_1 + d_2 + e^*$
	bb	-1	0	-1	0	$-a_1 - a_2 + e^*$

19.5.2 MODEL SPECIFICATION

To quantify the terms of the model based on Equation (19.14), some quantitative genetics is needed. Here, a brief description of heritability, genetic variance, additive variance, dominance variance, fixed effects and random effects will be given. Please refer to quantitative genetics textbooks, such as *Introduction to Quantitative Genetics* by Falconer and Mackay (1996) and *Biometrical Genetics* by Mather and Jinks (1982) for more complete treatments.

Heritability is commonly used in quantitative genetics to quantify the importance of the genetic factors to a trait value. The broad sense heritability is defined as

$$H = \sigma_G^2 / (\sigma_G^2 + \sigma_e^2) \tag{19.15}$$

where σ_G^2 and σ_e^2 are variance components associated with the genetic effects and the residual error. The genetic variance is composed of the additive and dominance variances (σ_A^2 and σ_D^2) if the epistatic interactions are ignored, that is

$$\sigma_G^2 = \sigma_A^2 + \sigma_D^2$$

In the F2 progeny and for the model based on Equation (19.14), the additive and the dominance variances are

$$\sigma_A^2 = 0.5\,(a_A^2 + a_B^2) + (1 - 2r)\,a_A a_B$$
$$\sigma_D^2 = 0.25\,(d_A^2 + d_B^2) + 0.5\,(1 - 2r)^2 d_A d_B \tag{19.16}$$

Given a heritability of a trait H, the relationship between the genetic and residual variances is

$$\sigma_G^2 = H \sigma_e^2 / (1 - H)$$
$$\sigma_e^2 = (1 - H)\,\sigma_g^2 / H \tag{19.17}$$

For most simulations, either σ_G^2 or σ_e^2 is set to unity for simplicity of the simulations. The absolute values of the variances are not important for most of the objectives of the simulations. What is important is the relative value of σ_G^2 over σ_e^2 or the relative value of σ_e^2 over σ_G^2. For example, if σ_G^2 is set equal to one, then

$$\sigma_G^2 = 1$$
$$\sigma_e^2 = (1 - H) / H$$

The relative importances of the additive and the dominance effects can be quantified using the ratio between the two variance components, $t = \sigma_D^2 / \sigma_A^2$. The additive and dominance variances can be determined by

$$\sigma_A^2 = \sigma_G^2 / (1 + t)$$
$$\sigma_D^2 = \sigma_G^2 / (1 + 1/t) \tag{19.18}$$

Lastly, if we assume the relative sizes of the effects of genes A and B

$$k = a_A / a_B = d_A / d_B \tag{19.19}$$

the genetic effects can be obtained as

$$a_A = \sqrt{\sigma_A^2 / [0.5\,(1 + 1/k^2) + (1 - 2r) / k]}$$
$$a_B = a_A / k$$
$$d_A = \sqrt{\sigma_D^2 / [0.25\,(1 + 1/k^2) + 0.25\,(1 - 2r)^2 / k]} \tag{19.20}$$
$$d_B = d_A / k$$

19.5.3 SIMULATION OF THE TRAIT VALUES

Now we are ready to simulate the trait values for a mapping population. Assume that the genotypes for each of the individuals in the mapping population have been obtained. The genotypes can be coded as four indicative variables, using the scheme listed in Table 19.7. It is common to consider the genotypic effects for each of the genes as fixed effects. This implies that the effects are the same from individual to individual. So, the fixed part of the trait value for the model based on Equation (19.14) is

$$y^*_{j\,(fixed)} = a_A x_{1j} + d_A x_{2j} + a_B x_{3j} + d_B x_{4j} \qquad (19.21)$$

The absolute value of the trait mean is usually irrelevant for most simulation studies. What is important are the relative weights for each of the genetic effects. The trait mean is set to be zero in Equation (19.21).

The residual error is always considered to be a random effect. This means the quantity of the residual error may differ from individual to individual. A value for the residual cannot be specified, but rather a variance, σ_e^2, can. The mean of the residual is zero. A residual can be simulated for each individual as

$$e_j^* = \sigma_e Rannor\,(0, 1) \qquad (19.22)$$

where σ_e is the square root of the residual variance and $Rannor\,(0, 1)$ is a random number generated from a standard normal distribution. For example, if σ_G^2 is set equal to one, then

$$e_j^* = \sqrt{(1 - H)/H} Rannor\,(0, 1) \qquad (19.23)$$

The simulated trait value is

$$y_j^* = y^*_{j\,(fixed)} + e_j^* \qquad (19.24)$$

19.6 QUANTITATIVE TRAIT: MULTIPLE-GENE MODEL

19.6.1 MULTIPLE-GENE MODEL

A quantitative trait (when a large number of genes is assumed to control the trait) can be simulated in a way similar to that for a trait controlled by one or two genes. First, the genotypes need to be simulated. Then, the fixed part of the trait value and the random residual are simulated. For an l-locus model, the simulated trait value is

$$\begin{aligned}
y_j^* &= y^*_{j\,(fixed)} + e_j^* \\
&= \sum_{i=1}^{l} [a_i x_{(2i-1)j} + d_i x_{(2i)j}] + \sqrt{(1 - H)/H} Rannor\,(0, 1)
\end{aligned} \qquad (19.25)$$

if the epistatic interactions are ignored, where a_i and d_i are the additive and dominance effects, respectively, for the ith locus and $x_{(2i-1)j}$ and $x_{(2i)j}$ are the coefficients for the corresponding additive and dominance effects, respectively.

19.6.2 DISTRIBUTION OF GENETIC EFFECTS

The difference between simulating multiple-gene traits and single or two-gene traits is the complexity of quantifying the relative weights for the genetic effects of the genes. For the two-locus model, the relative importance of the genetic effects can be quantified as a simple ratio.

For multiple genes, the genetic effects can be considered as a geometric series

$$\sigma_G^2\,(1 - \lambda)\,[\lambda^0, \lambda^1, \lambda^2, \lambda^3, ..., \lambda^{l-1}] \qquad (19.26)$$

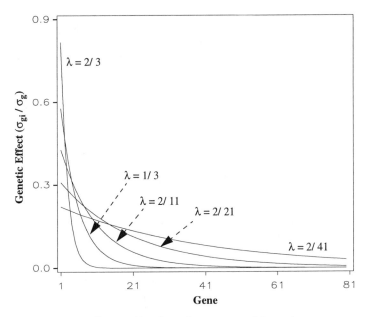

Figure 19.6 Genetic effect is distributed as a geometric series.

where σ_G^2 is the total genetic variance, λ is the parameter for the geometric series and $0 < \lambda \le 1$. The sum of the individual genetic variance is the total genetic variance

$$\sum_{i=0}^{l} \sigma_{Gi}^2 = \sum_{i=0}^{l} [\sigma_G^2 (1-\lambda) \lambda^i] = \sigma_G^2 \tag{19.27}$$

If σ_G^2 is set to 1, then the summation is unity. It is commonly defined as

$$N_e = (1+\lambda) / (1-\lambda) \tag{19.28}$$

number of effective genes. Once the genetic variance explained by the locus is obtained, the genetic effects (additive and dominance) can be determined.

Table 19.8 An example of genetic effects distributed as a geometric series when the numbers of effective loci are 2, 5, 10, 20 and 40. The total genetic variance is set equal to one.

λ	Ne	σ_{Gi}
1/3	2	0.82, 0.47, ...
2/3	5	0.58, 0.47, 0.38, 0.31, 0.26, ...
9/11	10	0.43, 0.39, 0.35, 0.32, 0.29, 0.26, 0.23, 0.21, 0.19, 0.17, ...
19/21	20	0.31, 0.29, 0.28, 0.27, 0.25, 0.24, 0.23, 0.22, 0.21, 0.20, 0.19, 0.18, 0.17, 0.16, ...
39/41	40	0.221, 0.215, 0.210, 0.205, 0.200, 0.195, 0.190, 0.185, 0.180, 0.175, ...

Table 19.8 lists some examples of the distributions of the genetic effects as the geometric series. For an effective number of 2 loci, the parameter of the geometric series is $\lambda = 1/3$ (obtained by solving $2 = (1 + \lambda) / (1 - \lambda)$). Table 19.8 contains situations of effective numbers of loci: 2, 5, 10, 20 and 40. Figure 19.6 shows the graphic distributions of the genetic effects.

It is also common to consider the genetic effects as normally distributed. This means that genes with large and small effects have low frequencies and genes with median effects have relatively high frequencies. It is also common to determine the genetic effects either arbitrarily or based on available experimental results.

19.6.3 SIMULATION OF TRAIT VALUES

Methodology for simulating mapping populations with multiple markers and quantitative traits has been discussed. Genes controlling the quantitative traits are assumed known. However, for some cases, the genes controlling the trait are considered unknown. What is known is the linkage relationship between markers and the genes. For example, to evaluate a QTL mapping approach (see Chapters 12 to 17), linkage between a QTL and a pair of markers is known. For cases like those, the simulation can be implemented as if the genes are known. The markers are not considered to simulate the trait value. When the trait values have been obtained, the identities of the genes are deleted and the markers are kept.

19.7 MULTIPLE QUANTITATIVE TRAITS

When multiple quantitative traits are considered in simulations, the quantification of the relationship among the traits and the genes controlling the traits is far more complex than for a single trait. However, as the genomic information accumulates, more and more attention will be paid to the multiple traits problem. If the multiple traits information is available and the procedure is feasible, it is logical to analyze the multiple traits simultaneously instead of analyzing them independently. When the traits are independent, each of them can be simulated independently. When the traits are not independent or are related to each other, they have been simulated simultaneously. Here, the methodology for multiple traits simulation will be introduced, using two traits as an example. For situations with more than two traits, see Johnson (1987).

19.7.1 THE MODEL

For two related traits, trait values can be modeled as

$$
\begin{aligned}
y_{1j} &= \mu_1 + \sum_{i=1}^{l_1} a_{1i} x_{1ij} + \varepsilon_{1j} \\
y_{2j} &= \mu_2 + \sum_{i=1}^{l_2} a_{2i} x_{2ij} + \varepsilon_{2j}
\end{aligned}
\tag{19.29}
$$

if dominance effects and epistatic interactions are ignored, where the subscripts '1' and '2' denote trait one and two, respectively, subscript i indicates the gene, j is the indicator for an individual in the population, y is the trait value, μ is the overall mean, a is the additive genetic effect, x is the dummy variable for the additive effect, ε is the random error and l is the number of genes controlling the trait.

19.7.2 MODEL SPECIFICATION

To quantify the genetics of the two traits and their relationship, matrices and their relationships are

$$\Sigma_p = \begin{bmatrix} \sigma_{p1}^2 & \sigma_{p12} \\ \sigma_{p12} & \sigma_{p2}^2 \end{bmatrix} = \Sigma_g + \Sigma_e = \begin{bmatrix} \sigma_{g1}^2 & \sigma_{g12} \\ \sigma_{g12} & \sigma_{g2}^2 \end{bmatrix} + \begin{bmatrix} \sigma_{e1}^2 & \sigma_{e12} \\ \sigma_{e12} & \sigma_{e2}^2 \end{bmatrix} \tag{19.30}$$

where Σ_p, Σ_g and Σ_e are variance-covariance matrices for phenotypic, genetic and environmental effects, respectively. The relationship between the two traits can be quantified by

$$\rho_p = \frac{\sigma_{p12}}{\sqrt{\sigma_{p1}^2 \sigma_{p2}^2}}, \rho_g = \frac{\sigma_{g12}}{\sqrt{\sigma_{g1}^2 \sigma_{g2}^2}} \text{ and } \rho_e = \frac{\sigma_{e12}}{\sqrt{\sigma_{e1}^2 \sigma_{e2}^2}} \tag{19.31}$$

where ρ_p, ρ_g and ρ_e are defined as the phenotypic, genetic and environmental correlations, respectively. The heritabilities for the two traits are defined as

$$H_1 = \sigma_{g1}^2 / \sigma_{p1}^2 \tag{19.32}$$
$$H_2 = \sigma_{g2}^2 / \sigma_{p2}^2$$

The genetic correlation between the two traits is caused by the genes controlling the traits being linked or the same genes having effects on both traits. Linkage is usually considered as the genetic cause; the other cause is commonly referred to as pleiotropy. Pleiotropic effects can be explained as a physiological relationship between the traits. Alternatively, the traits are influenced by the same gene products. For example, plant height and biomass may be affected by the same gene products. So, pleiotropic effects may reflect not only genetics, but also physiology. If the traits can be measured at the gene product level and pleiotropy is ruled out, genetic correlation should be caused only by linkage. Here, we consider both linkage and pleiotropy in the multiple trait simulation.

Table 19.9 Genotypes and their expected and cumulative frequencies for two linked loci.

Locus A	Locus B	p_{ij}	Trait 1	Trait 2
AA	BB	$0.25(1-r)^2$	a_1	a_2
AA	Bb	$0.5r(1-r)$	a_1	0
AA	bb	$0.25r^2$	a_1	$-a_2$
Aa	BB	$0.5r(1-r)$	0	a_2
Aa	Bb	$0.5[(1-r)^2 + r^2]$	0	0
Aa	bb	$0.5r(1-r)$	0	$-a_2$
aa	BB	$0.25r^2$	$-a_1$	a_2
aa	Bb	$0.5r(1-r)$	$-a_1$	0
aa	bb	$0.25(1-r)^2$	$-a_1$	$-a_2$

19.7.3 LINKAGE AND GENETIC CORRELATION

If the genetic correlation between two traits is the result of linkage between the genes controlling the two traits, then the key to simulating correlated traits is to find the quantitative relation between the linkage and the genetic correlation. Start with two genes, A and B, controlling traits 1 and 2, respectively. The recombination fraction between the two genes is r. Table 19.9 shows the expected genotypic frequencies for the two genes in a classical F2 population. The genetic effects for the two genes are listed in columns 4 and 5. The variances associated with the two genes are

$$\sigma_1^2 = 0.5a_1^2$$
$$\sigma_2^2 = 0.5a_2^2$$

(19.33)

The covariance between the two genes is

$$\sigma_{12} = 0.5\,(1 - 2r)\,a_1 a_2$$

(19.34)

So, the genetic correlation coefficient between the two genes is

$$\rho_{12} = \frac{\sigma_{12}}{\sqrt{\sigma_1^2 \sigma_2^2}} = \frac{0.5\,(1 - 2r)\,a_1 a_2}{\sqrt{0.5a_1^2 \times 0.5a_2^2}} = 1 - 2r$$

(19.35)

The inverse of Equation (19.35) is

$$r = 0.5\,(1 - \rho_{12})$$

(19.36)

which can be used to determine the recombination fraction between the two genes, given the genetic correlation.

Table 19.10 Composition of a genetic correlation of 0.5 with linkage.

Trait 1		Trait 2		Linkage Between
Gene	Effect	Gene	Effect	the Two Genes
1	.58	1	.43	0.1
2	.47	2	.39	0.05
3	.38	3	.35	0.3
4	.31	4	.32	0.5
5	.26	5	.29	0.5
6	0	6	.26	0.5
7	0	7	.23	0.5
8	0	8	.21	0.5
9	0	9	.19	0.5
10	0	10	.17	0.5

When multiple genes are considered, genetic correlation is due to linkage

$$\rho_g = \frac{\sigma_{g12}}{\sqrt{\sigma_{g1}^2 \sigma_{g2}^2}} = \frac{\sum_i (1 - 2r_i)\,a_{1i} a_{2i}}{\sqrt{\sum_i a_{1i}^2 \sum_i a_{2i}^2}}$$

(19.37)

where a_{1i} and a_{2i} are genetic effects of gene pair i for trait 1 and trait 2, respectively. Here, the genes are paired according to whether there is linkage between them. It should be mentioned that only two-gene linkage is considered in Equation (19.37). If more than two genes are linked with each other, the formulation will be more complex. Table 19.10 shows two related traits and their genetic relationship. Five genes are found controlling trait 1 and 10 genes are found controlling trait 2. Among them, three pairs of genes are linked. The genetic correlation due to the linkages is

$$\rho_g = \frac{\sigma_{g12}}{\sqrt{\sigma_{g1}^2 \sigma_{g2}^2}} = \frac{0.41769}{\sqrt{0.8654 \times 0.8756}} = 0.48 \qquad (19.38)$$

Equation (19.37) does not give a simple inverse for obtaining linkage from genetic correlation. For a simulation study, it is common to try many combinations of linkage and genetic effects to find a setting which fits the genetic correlation requirement and is biologically reasonable. Direct computation, such as Equation (19.38), works only for the single gene problems. The way to obtain simulation conditions for the multiple gene problems is somewhat arbitrary. However, if we assume all recombination fractions between the pairs of genes are the same, then we can obtain the recombination fraction among the gene pairs using

$$r^* = \frac{1}{2} - \rho_g \sqrt{\sum_i a_{1i}^2 \sum_i a_{2i}^2 \Big/ \left(\sum_i a_{1i} a_{2i}\right)} \qquad (19.39)$$

Furthermore, when equality for the genetic effect is assumed, Equation (19.39) reduces to

$$r^* = 0.5\,(1 - \rho_g)$$

19.7.4 MULTINORMAL DISTRIBUTION

The situation in which the genetic correlation is a result of linkage has been discussed. However, in most cases, known linkage may not be the direct cause of genetic correlation. So, information available in these situations is usually based on the quantitative quantification of the two traits, not on the

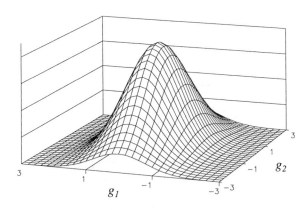

Figure 19.7 Joint distribution of two genetic effects with genetic correlation 0.6. The mean genetic effect of the second gene is two times the effect of the first gene, $E(g_2) = 2E(g_1)$.

discrete linkage relationship among genes. Simulation of the traits relies on the joint distribution of the two traits. The sampling process is not from a two-dimensional surface, but rather from a three-dimensional space. The joint distribution of the two related traits can be formulated as

$$
\begin{aligned}
f(\boldsymbol{g}) &= \frac{1}{2\pi\sqrt{|\Sigma_g^{-1}|}} \exp\left(-\boldsymbol{g}'\Sigma_g^{-1}\boldsymbol{g}\right) \\[2mm]
&= \frac{1}{2\pi\sqrt{\sigma_{g1}^2\sigma_{g2}^2(1-\rho_g^2)}} \exp\left\{-\frac{1}{2(1-\rho_g^2)}\left[\frac{g_1^2}{\sigma_{g1}^2} - \frac{2\rho_g}{\sigma_{g1}\sigma_{g2}} + \frac{g_2^2}{\sigma_{g2}^2}\right]\right\} \\[2mm]
&= \frac{1}{2\pi\sqrt{\sigma_{g1}^2\sigma_{g2}^2(1-\rho_g^2)}} \exp\left\{-\frac{1}{2(1-\rho_g^2)}\left[\left(\frac{g_1}{\sigma_{g1}} - \frac{\rho_g g_2}{\sigma_{g2}}\right)^2 + \frac{g_2^2}{\sigma_{g2}^2}(1-\rho_g^2)\right]\right\}
\end{aligned}
\tag{19.40}
$$

where

$$
\Sigma_g^{-1} = \begin{bmatrix} \dfrac{1}{\sigma_{g1}^2(1-\rho_g^2)} & \dfrac{-\rho_g}{\sigma_{g1}\sigma_{g2}(1-\rho_g^2)} \\[4mm] \dfrac{-\rho_g}{\sigma_{g1}\sigma_{g2}(1-\rho_g^2)} & \dfrac{1}{\sigma_{g2}^2(1-\rho_g^2)} \end{bmatrix}
$$

and $\left|\Sigma_g^{-1}\right| = \sigma_{g1}^2\sigma_{g2}^2(1-\rho_g^2)$

For example, Figure 19.7 shows a theoretical joint distribution of two genetic effects with genetic correlation 0.6. To simulate the traits, the following conditional genetic effects and conditional genetic variances are needed.

$$
\begin{aligned}
E(g_1|g_2) &= \frac{\rho_g g_2 \sigma_{g1}}{\sigma_{g2}} \\[2mm]
\sigma_{g_1|g_2}^2 &= \sigma_{g1}^2(1-\rho_g^2) \\[2mm]
&\text{and} \\[2mm]
E(g_2|g_1) &= \frac{\rho_g g_1 \sigma_{g2}}{\sigma_{g1}} \\[2mm]
\sigma_{g_2|g_1}^2 &= \sigma_{g2}^2(1-\rho_g^2)
\end{aligned}
\tag{19.41}
$$

19.8 SIMULATION OF DATA WITH MULTIPLE GENERATIONS

19.8.1 INTRODUCTION

The ability to simulate data with multiple generations enables many powerful applications in genomic research. Examples of applications are simulation studies of how efficient use of genomic information in plant and animal improvement requires the simulation of multiple generation breeding schemes and studies on molecular marker based disease diagnosis in humans may need to simulate the co-inheritance of the disease genes and molecular markers in multiple generation pedigrees.

The methodology for simulating data with a single generation was previously described. For a single or a two-locus model, the data can be generated using the expected cumulative frequencies for the possible genotypes. When multiple loci are considered, the expected cumulative frequencies for the locus conditional on the previous locus can be used. For this case, no interference is assumed. To generate data with multiple generations, the methodology for the single generation problems can be readily used, once the expected genotypic frequencies are obtained. So, the key to simulating data with multiple generations is to determine the expected frequencies from generation to generation.

Several approaches are available for generating the expected genotypic frequencies for multiple generations. Here, I introduce the concept of generation transition matrix. The theory involving a generation matrix was developed several decades ago (Fisher 1949). The multiple generations of breeding can be characterized as a Markov chain. What is needed are the genotypic frequencies at a starting generation and a transition matrix. The most important characteristic of the Markov chain (for the multiple generations breeding problem) is that the expected genotypic frequencies of a generation $t+1$ depend on only the observed genotypic frequencies for generation t and not on the genotypic frequencies for earlier generations.

19.8.2 SINGLE GENE

Consider a single locus first. For a single locus A, allelic frequencies for A and a are 0.5. If a heterozygous individual is selfed, the expected genotypic frequencies are 0.25, 0.5 and 0.25 for AA, A and aa, respectively, in the next generation. If the progeny are selfed again, the expected frequencies in the next generation can be represented using matrix notation; they are

$$P_{t+1} = TP_t = \begin{bmatrix} 1 & 0.25 & 0 \\ 0 & 0.5 & 0 \\ 0 & 0.25 & 1 \end{bmatrix} \begin{bmatrix} \hat{p}_{AA} \\ \hat{p}_{Aa} \\ \hat{p}_{aa} \end{bmatrix} = \begin{bmatrix} \hat{p}_{AA} + 0.25\hat{p}_{Aa} \\ 0.5\hat{p}_{Aa} \\ 0.25\hat{p}_{Aa} + \hat{p}_{aa} \end{bmatrix} \tag{19.42}$$

where

$$P_t = \begin{bmatrix} \hat{p}_{AA} \\ \hat{p}_{Aa} \\ \hat{p}_{aa} \end{bmatrix},$$

$$T = \begin{bmatrix} 1 & 0.25 & 0 \\ 0 & 0.5 & 0 \\ 0 & 0.25 & 1 \end{bmatrix}$$

$$\text{and } P_{t+1} = \begin{bmatrix} \hat{p}_{AA} + 0.25\hat{p}_{Aa} \\ 0.5\hat{p}_{Aa} \\ 0.25\hat{p}_{Aa} + \hat{p}_{aa} \end{bmatrix}$$

are the vector containing the genotypic frequencies for generation t, the generation transition matrix and the vector containing the expected genotypic frequencies for the generation $t+1$, respectively. The first column of the transitional matrix shows that the probability of producing the AA genotype is 1.0 when an AA individual is selfed. The second column shows that the probabilities of producing AA, Aa and aa genotypes are 0.25, 0.5 and 0.25, respectively, when an individual with genotype AA is selfed. By the same token, the third column shows that the probability of producing an aa genotype is 1.0 when an individual aa is selfed. The cumulative probability for the three genotypes in the generation $t + 1$ is

$$
P_C = \begin{bmatrix} 1 & 0 & 0 \\ 1 & 1 & 0 \\ 1 & 1 & 1 \end{bmatrix} P_{t+1} = \begin{bmatrix} 1 & 0 & 0 \\ 1 & 1 & 0 \\ 1 & 1 & 1 \end{bmatrix} \begin{bmatrix} \hat{p}_{AA} + 0.25\hat{p}_{Aa} \\ 0.5\hat{p}_{Aa} \\ 0.25\hat{p}_{Aa} + \hat{p}_{aa} \end{bmatrix}
$$

$$
= \begin{bmatrix} \hat{p}_{AA} + 0.25\hat{p}_{Aa} \\ \hat{p}_{AA} + 0.75\hat{p}_{Aa} \\ 1 \end{bmatrix}
$$

(19.43)

The genotypes can then be simulated using the cumulative probabilities for the next generation.

19.8.3 TWO LOCI

When two loci are considered in a simulation, the transition matrix is more complex than the matrix for the single locus situation. Let us consider the two loci, A and B, linked r recombination units apart. A simulated population of selfing genotype AaBb is obtained. This population is defined as generation t. The observed frequencies for the nine possible genotypes in the simulated data are

$$
P_t = \begin{bmatrix} P[AABB] \\ P[AABb] \\ P[AAbb] \\ P[AaBB] \\ P[AaBb] \\ P[Aabb] \\ P[aaBB] \\ P[aaBb] \\ P[aabb] \end{bmatrix} = \begin{bmatrix} \hat{p}_1 \\ \hat{p}_2 \\ \hat{p}_3 \\ \hat{p}_4 \\ \hat{p}_5 \\ \hat{p}_6 \\ \hat{p}_7 \\ \hat{p}_8 \\ \hat{p}_9 \end{bmatrix}
$$

(19.44)

The transition matrix for the two-locus model is

$$
T = \begin{bmatrix} 1 & 0.25 & 0 & 0.25 & 0.25(1-r)^2 & 0 & 0 & 0 & 0 \\ 0 & 0.5 & 0 & 0 & 0.5r(1-r) & 0 & 0 & 0 & 0 \\ 0 & 0.25 & 1 & 0 & 0.25r^2 & 0.25 & 0 & 0 & 0 \\ 0 & 0 & 0 & 0.5 & 0.5r(1-r) & 0 & 0 & 0 & 0 \\ 0 & 0 & 0 & 0 & 0.5[(1-r)^2 + r^2] & 0 & 0 & 0 & 0 \\ 0 & 0 & 0 & 0 & 0.5r(1-r) & 0.5 & 0 & 0 & 0 \\ 0 & 0 & 0 & 0 & 0.25r^2 & 0 & 1 & 0.25 & 0 \\ 0 & 0 & 0 & 0.25 & 0.5r(1-r) & 0 & 0 & 0.5 & 0 \\ 0 & 0 & 0 & 0 & 0.25(1-r)^2 & 0.25 & 0 & 0.25 & 1 \end{bmatrix}
$$

(19.45)

The expected genotypic frequencies for next generation are

$$
P_{t+1} = TP_t = \begin{bmatrix}
1 & 0.25 & 0 & 0.25 & 0.25\,(1-r)^2 & 0 & 0 & 0 & 0 \\
0 & 0.5 & 0 & 0 & 0.5r\,(1-r) & 0 & 0 & 0 & 0 \\
0 & 0.25 & 1 & 0 & 0.25r^2 & 0.25 & 0 & 0 & 0 \\
0 & 0 & 0 & 0.5 & 0.5r\,(1-r) & 0 & 0 & 0 & 0 \\
0 & 0 & 0 & 0 & 0.5\,[\,(1-r)^2+r^2\,] & 0 & 0 & 0 & 0 \\
0 & 0 & 0 & 0 & 0.5r\,(1-r) & 0.5 & 0 & 0 & 0 \\
0 & 0 & 0 & 0 & 0.25r^2 & 0 & 1 & 0.25 & 0 \\
0 & 0 & 0 & 0.25 & 0.5r\,(1-r) & 0 & 0 & 0.5 & 0 \\
0 & 0 & 0 & 0 & 0.25\,(1-r)^2 & 0.25 & 0 & 0.25 & 1
\end{bmatrix}
\begin{bmatrix}
\hat{p}_1 \\ \hat{p}_2 \\ \hat{p}_3 \\ \hat{p}_4 \\ \hat{p}_5 \\ \hat{p}_6 \\ \hat{p}_7 \\ \hat{p}_8 \\ \hat{p}_9
\end{bmatrix}
\qquad (19.46)
$$

The cumulative frequencies can be obtained using the matrix notation

$$
P_C = \begin{bmatrix}
1 & 0 & 0 & 0 & 0 & 0 & 0 & 0 & 0 \\
1 & 1 & 0 & 0 & 0 & 0 & 0 & 0 & 0 \\
1 & 1 & 1 & 0 & 0 & 0 & 0 & 0 & 0 \\
1 & 1 & 1 & 1 & 0 & 0 & 0 & 0 & 0 \\
1 & 1 & 1 & 1 & 1 & 0 & 0 & 0 & 0 \\
1 & 1 & 1 & 1 & 1 & 1 & 0 & 0 & 0 \\
1 & 1 & 1 & 1 & 1 & 1 & 1 & 0 & 0 \\
1 & 1 & 1 & 1 & 1 & 1 & 1 & 1 & 0 \\
1 & 1 & 1 & 1 & 1 & 1 & 1 & 1 & 1
\end{bmatrix} P_{t+1}
\qquad (19.47)
$$

Now, data involving multiple generations and two linked loci can be simulated.

Table 19.11 Expected frequencies of genotypes for locus B conditional on locus A and their cumulative frequencies when A and B are linked.

Generation t	Generation $t+1$		
	AA	Aa	aa
AA	$\begin{bmatrix} 1 & 0.25 & 0 \\ 0 & 0.5 & 0 \\ 0 & 0.25 & 1 \end{bmatrix}$		
Aa	$\begin{bmatrix} 1 & (1-r)^2 & 0 \\ 0 & r(1-r) & 0 \\ 0 & r^2 & 1 \end{bmatrix}$	$\begin{bmatrix} 1 & r(1-r) & 0 \\ 0 & (1-r)^2+r^2 & 0 \\ 0 & r(1-r) & 1 \end{bmatrix}$	$\begin{bmatrix} 1 & r^2 & 0 \\ 0 & r(1-r) & 0 \\ 0 & (1-r)^2 & 1 \end{bmatrix}$
aa			$\begin{bmatrix} 1 & 0.25 & 0 \\ 0 & 0.5 & 0 \\ 0 & 0.25 & 1 \end{bmatrix}$

19.8.4 MULTIPLE LOCI

When more than two loci are considered in simulations, the transition matrix, such as in Equation (19.45), will be large and not practical. For these cases, conditional transition matrices can be used to generate multiple generation data. For the two adjacent loci, A and B, linked with recombination fraction r, the transition matrix for locus B conditional on the genotypes of locus A for both generations t and $t+1$ is shown in Table 19.11. The multiple loci can be simulated one at a time and the next locus is simulated conditional on the previous one.

EXERCISES

Exercise 19.1

Computer simulation is a powerful tool for resolving many complex statistical issues in theoretical biology. However, computer simulation has been criticized by both experimental biologists and theoretical statisticians. Some experimental biologists think that computer simulation is a "numbers game" without biological basis. Some theoretical statisticians think that the result of a computer simulation is not theoretically rigorous. What are your own thoughts on computer simulation? Pay attention to the following factors.

(1) The relationship between the models upon which the computer simulation is based and the biological reality.

(2) Whether, when a theoretical model is too complex to solve, a computer simulation approach solves the model by simplifying the model or by keeping the model as it is.

(3) Computational intensity.

Exercise 19.2

Computer simulation has been used to validate experimental results. For example, data can be generated under a wide variety of experimental conditions and assumptions. If the experimental results agree with the simulation results, then the experimental results are often thought to be validated. Can simulation results be validated by experiments?

Exercise 19.3

Computer simulation and statistics are tools for resolving biological inferences when the inferences cannot be resolved by biological data alone. Can you think of any case where the relationship is reversed (where biology is a tool for solving complex computation and statistical problems)?

Exercise 19.4

Permutation has been used to construct empirical distributions of test statistics under a null hypothesis. The success of permutation is largely dependent on whether the permutation can be implemented under the true null hypothesis and implemented without altering the data structure. In construction of the empirical distribution of a test statistic for QTL mapping, linkage relationships between potential QTLs and markers should be broken and the linkage relationship among the markers should be maintained. Outline the necessary steps of permutation to obtain empirical distribution of a t-test statistic for QTL mapping, using the single-marker analysis in a backcross progeny.

additive effect see genetic effect

amplified fragment length polymorphism (AFLP) polymorphism generated by an approach combining restriction enzyme digestion and PCR amplification of DNA. The polymorphism can be used as genetic marker.

algorithm a set of programmable procedures or strategies for solving computational problems.

allele one form of a gene or a genetic marker.

alternative hypothesis see hypothesis test.

average effect of gene substitution average effect on the trait of one allele being replaced by another.

bacterial artificial chromosome (BAC) a type of DNA vector which can have inserts of approximately 100kb.

backcross a cross between an offspring and one of its parents. In genomics, it refers to a cross between a heterozygote and a homozygote, *e.g.*, a cross between an individual with genotype Aa and an individual with genotype AA or aa.

bias difference between the expectation of an estimator ($\hat{\theta}$) and the true parameter (θ), which is $E(\hat{\theta}) - \theta$. Bias can be positive (overestimated), negative (underestimated) or zero (unbiased).

blotting method for transferring DNA, RNA or protein fragments from gels to durable nitrocellulose or nylon membranes. The individual fragments become immobilized on the membrane. DNA is transferred in a Southern blot, while RNA is transferred in a Northern blot.

bootstrap a resampling technique. A bootstrap sample is defined as a random sample drawn with replacement from an original sample. Bootstrap can also be applied to parametric distributions (called parametric bootstrap).

bootstrap confidence interval confidence interval built using the distribution of a large number of bootstrap estimates. It can be estimated using normal deviate assuming normal distribution or using percentiles of the empirical distribution.

bridging the process of obtaining linkage maps for some segments of the genome by jointly analyzing data from multiple populations and analyzing the other parts of the genome independently for each mapping population

broad sense heritability see heritability

bulk segregant analysis (BSA) an effective strategy for finding genes controlling simple inherited traits, such as disease resistant genes. The BSA technique involves pooling DNA samples from individuals similar in some way into bulks. A polymorphic marker which is clearly different between the two bulks may be linked to the differing chatacteristic between the two sets of individuals.

cDNA a single-stranded DNA molecule produced from an RNA template.

centiMorgan (cM) an unit for measuring genetic distance. A Morgan is 100cMs. A cM is approximately equivalent to a 1% recombination value if double and high levels of crossovers are ignored.

centromere a region of chromosome at which sister chromatids are attached during cell division.

chiasma a cross-shaped structure between nonsister chromatids visible in cell division. It has been believed to be cytological evidence of homologous crossover between nonsister chromatids.

chromosome rearrangement a chromosomal structure change, such as inversion or translocation. It is usually associated with abnormal biological development.

chromosome walking a multiple step strategy to sequence a large section of DNA. Each step involves the isolation and sequencing of overlapping DNA clones.

codominance when all alleles of a gene are expressed fully in heterozygotes. In genomics, alleles of a gene or of a marker can be identified in heterozygotes.

codons three-nucleotide sequence that codes one amino acid in protein synthesis.

coefficient of coincidence (C) number of observed double-crossover events divided by the expected number. The quantity of $1 - C$ is defined as crossover interference.

C	Interference
$C > 1$	negative
$C < 1$	positive
$C = 0$	complete
$C = 1$	absent

comparative mapping a genetic mapping strategy for transferring genomic information across species, based on genome homology among the species, *e.g.*, between mouse and human.

composite interval mapping a QTL mapping procedure using partial regression. The independent variables in the model include variables for the interval and variables for the other parts of the genome to control genetic background effects.

computer simulation a process to generate data using defined models. In genomics, it has been used to solve complex statistical problems.

conditional probability probability of an event given a condition. For example, probability that individuals have AA genotype given that the individuals have BB genotype is the probability of AA conditional on BB.

confidence coefficient see confidence interval.

confidence interval If 95% of the time in a series of repeated experiments, the true parameter is contained within an interval between L_1 and L_2, the interval is defined as a confidence interval. L_1 and L_2 are defined as confidence limits and 95% is the confidence coefficient. Confidence intervals can be built parametrically, using an estimator and its variance and nonparametrically, using the empirical distribution of the estimator.

confidence intervals for locus order Possible genome positions for a locus is the confidence interval for the genome position of the locus. For example, if it is 95% certain that locus A is located between genome positions 4 and 7, then between genome positions 4 and 7 is the confidence interval for the genome position of locus A. If a confidence interval is built for every locus in a linkage group, then a confidence interval is obtained for the locus order.

confidence limits see confidence interval.

cosmid a type of DNA vector which can have inserts of between 30 - 40kb.

coupling see linkage phase

crossover chromosome segment exchange between nonsister chromatids of homologs during meiosis. The result of the crossover is the existence of non-parental chromosomes in meiotic products. It is the basis of recombination-based genomic analysis.

crossover hot spot genome position where crossovers occur with a higher frequency than expected.

crossover interference non-independence of crossover events between adjacent genome segments (see also coefficient of coincidence).

DNA (Deoxyribonucleic Acid). the genetic material and main focus of molecular genetics and genomics.

DNA marker a small piece of DNA that can be identified and used as genetic marker.

DNA vector a cloning vehicle for replicating DNA fragments. Commonly used vectors are bacterial artificial chromosome (BAC), cosmid, phage and yeast artificial chromosome (YAC).

dominance (1) property that the phenotype of a heterozygote is indistinguishable from a homozygote, because phenotype of one allele is masked by a dominant allele. (2) departure from additive genetic model for a quantitative trait.

dominant marker genetic marker showing dominant inheritance. In genomics, most PCR-based markers are dominant markers, *e.g.*, presence or absence of a PCR product.

double crossover when two crossover events happen in adjacent genome segments.

EM algorithm an iterative procedure for obtaining the maximum likelihood estimator for problems of missing data. The procedure includes two steps, expectation and maximization. It has been used extensively in genomic analyses such as recombination fraction estimation and multilocus model construction.

epistatic interaction non-independence between the effect of one gene and others.

exon a DNA segment that remains in the final, processed mRNA. Introns are removed during processing

expressed sequence tag (EST) a sequence tagged site (STS) with a direct relationship to an expressed gene.

expectation a statistical term referring to the average value of a variable over the whole distribution.

F2 progeny generated by selfing an F1 individual or crossing two genetically identical individuals. In genomic analyses, it is commonly referred to as the progeny of a cross between two heterozygotes, *e.g.*, progeny of a cross between individuals with genotype Aa for gene A with two alleles, A and a.

false negative event resulting from accepting a wrong null hypothesis.

false positive event resulting from rejecting a correct null hypothesis.

fluorescence *in situ* hybridization (FISH) a technique combining traditional cytology and DNA hybridization to visualize by light microscopy, the location where a DNA segment with specific sequences hybridizes. The predecessor of FISH, called ISH, used radioactive isotope-labeled probes.

framework locus locus with interval support of at least lod 3.

framework map a genetic map composed of all framework loci. The gene order for the framework map is 1,000 times more likely than any other order.

gene a unit of heredity. In a population of individuals of a sexually reproducing species, a single gene is passed from generation to generation following simple Mendelian inheritance. At the molecular level, a gene is a segment of DNA which encodes or minapulates protein synthesis.

genetic by environment interactio the concept that genes can have different expressions or effects in different environments.

genetic correlation relativeness between two traits due to genetics. It is estimated using genetic variances and covariances of the traits.

genetic effect difference of an individual value from the population mean as a result of genetic difference for a quantitative trait. Genetic effect due solely to average effects of gene substitution is defined as additive genetic effect. Departure from the additive model is result of dominance or epistatic interactions (epistasis).

genetic linkage nonrandom segregation of genes or markers because they are located closely.

genetic map a linear arrangement of genes or genetic markers obtained based on recombination.

genetic marker an inheritable character easily discerned and used for identification of an individual and for genetic mapping. Commonly used genetic markers include morphological, protein and DNA markers. The marker can be a known function gene, an unknown function DNA fragment or a inheritable phenotype.

genome the complete set of DNA carried by a gamete.

genome database database containing genomic information.

genome informatics the study of management, analysis and exchange of genomic information.

genomic library a collection of all the DNA in a species genome, broken into random DNA fragments, inserted into DNA vectors.

genomics the study of genome composition, structure and functions, which can be classified into classical genomics (crossover-based), physical genomics (DNA sequence-based) and genome informatics.

genotype genetic constitution of an individual.

Hardy-Weinberg Equilibrium a state in which gene and genotypic frequencies remain constant from generation to generation in a large random mating population without mutation, selection or migration.

heritability a statistical quantity measuring the proportion of phenotype variation being explained by genetic variation for a quantitative trait. The total genetic variance divided bythe phenotypic variance is defined as broad sense heritability. The additive genetic variance divided by the phenotypic variance is defined as narrow sense heritability.

heterogeneity test test for differences among populations.

heterozygosity state of an individual having 2 different alleles of a gene.

homology genome similarity.

hypothesis test a statistical test for rejecting or accepting a null hypothesis. If the null hypothesis is rejected, the alternative hypothesis might be accepted.

information content minus the expectation of the second derivative of log likelihood function or support function with respect to one parameter. It is a matrix for multiple parameter situations.

interval mapping a set of procedures using two adjacent markers to estimate genetic effects and genome locations of genes controlling quantitative traits.

interval support the base 10 logarithm of a likelihood ratio between the likelihood of a locus being in a particular interval of the linkage map and the highest likelihood for placing the locus in any other intervals.

intron a noncoding DNA segment.

ISH (*in situ* hybridization) see fluorescence *in situ* hybridization.

jackknife A jackknife sample is a series of samples, leaving out one observation at a time from the original sample.

least squares estimation estimation based on minimization of the sum of squares of the difference between an observed and a predicted value.

likelihood function the joint density of observed data as a function of parameters. For discrete variables, it is a function of joint probability, given values of parameters. The logarithm of likelihood function is commonly used.

likelihood ratio test statistic two times the natural logarithm of the ratio ($2\log[L_1/L_0]$) between likelihood (L_1) of θ taking a value $\hat{\theta}$ which maximizes the likelihood, e.g. a maximum likelihood estimator

and likelihood (L_0) under the null hypothesis, e.g. $\theta = 0.5$. It is distributed as chi-square with degrees of freedom being the difference between the numbers of estimated parameters of the two hypotheses when sample size is large. It is used extensively in genomic data analyses.

linkage grouping a set of procedures for finding which linkage group a locus belongs to. A group of loci inherited together, according to certain criteria of linkage statistics, is defined as a linkage group, and can correspond to a whole or a part of a chromosome.

linkage map bridging the process of obtaining linkage maps for some segments of the genome by jointly analyzing data from multiple populations and analyzing the other parts of the genome independently for each mapping population.

linkage phase a term used to denote chromatid locations of alleles of two linked loci. Two alleles located on same sister chromatid are linked in coupling phase. Two alleles located on different chromatids are linked in repulsion phase.

linkage map pooling the process of obtaining a whole linkage map by jointly analyzing data from multiple mapping populations.

locus ordering a set of procedures for obtaining locus order. Locus order is defined as the linear arrangement of genes or genetic markers in a linkage group.

lod score base 10 logarithm of the ratio between likelihoods under the null and alternative hypotheses.

map density a term which can be quantified using the maximum gap between adjacent markers or the number of markers in a unit of the genome.

map distance the expected number of crossovers between two loci. It is additive for multiple loci.

map function a mathematical function converting recombination fraction to map distance.

mapping population a genetically defined population used to obtain data for gene mapping.

marker coverage a term which can be quantified using the proportion of genome covered by markers or the maximum genome segment length between two adjacent markers.

maximum likelihood estimator the estimator of the parameter which maximizes the likelihood function with respect to the parameter.

megagametophyte female gametophyte develops from a megaspore. It is haploid and has maternal genotype.

meiosis a type of cell division that produces four daughter cells, each having half of the number of chromosomes of the progenitor cell. Meiosis occurs during sexual reproduction.

Mendelian laws include the law of segregation and the law of independent assortment.

microsatellite repetitive DNA with repeats ranging in size from 1 to 6 bp. It is also referred to as simple sequence repeat (SSR).

minisatellite tandem repeats of sequences ranging in size from 9 to 100 bp. The number of repeats varies (usually less than 1000). Minisatellite markers are commonly referred to as VNTRs.

mitosis cell division producing two genetically identical cells from a single cell.

mixture linkage phases occur when parents producing progeny for linkage analysis of a gene have the loci in different linkage phases.

multiple responses regression regression models with multiple dependent variables.

narrow sense heritability see heritability.

negative interference see crossover interference.

Newton-Raphson Iteration an iterative approach for solving equations. It is commonly used to obtain maximum likelihood estimator in genomic data analyses.

northern blotting see blotting

nucleotide monomeric unit of DNA and RNA polymers.

nucleotide sequence linear order of nucleotides in RNA or DNA molecules. For DNA, the sequence consists of the nitrogenous bases adenine (A), cytosine (C), guanine (G) and thymine (T) in various orders.

null hypothesis see hypothesis test.

outcrossing crossing of an individual with others. Outcrossing rate is the portion of offspring that are progeny of outcrossing.

parameter a quantity specifying a population. In practical statistics, parameters are estimated from samples. For example, recombination fraction is a parameter that can be estimated from samples.

pedigree ancestral history shown using diagram.

penetrance trait trait in which only a portion of individuals with the genotype show the expected phenotype. The proportion is called penetrance. 1 - penetrance is sometimes called "escape".

percentage of correct order (PCO) mean probability that a gene or marker is located at its correct genome position. In practical genomic data analyses, PCO can be estimated using resampling techniques, such as bootstrap.

permutation test hypothesis tests using empirical distribution of test statistic. The empirical distribution is obtained by permuting the original sample. The permuted sample is considered to be a sample from the population under the null hypothesis.

phase-unknown linkage analysis a set of procedures for linkage detection and recombination fraction estimation when linkage phase is unknown. Phase-unknown differs from mixture linkage phases.

phenotype observed characteristics of an individual. It can be the result of the interactions of the genotype of the individual and environment. If environment has no effect on the characteristic then phenotype is the same for all individuals with the same genotype.

physical map assembly determination of a linear order for cloned DNA fragments. It can be achieved using nucleotide sequences, restriction site fingerprints, *etc.*, of the fragments.

physical map linear order of genes or DNA fragments. A physical map is commonly a map which contains ordered overlapping cloned DNA fragments. The cloned DNA fragments are usually obtained using restriction enzyme digestion.

polymerase chain reaction (PCR) a technique for amplifying a target portion of a DNA molecule. Some genetic markers used in genomic analyses, such as RAPD and AFLP, are PCR-based markers.

polymorphism different, detectable alleles for a gene or marker in a population.

polymorphism information content (PIC) a statistic used to quantify the amount of polymorphism and is a function of allelic frequencies of the gene or marker.

pooling the process of obtaining a whole linkage map by jointly analyzing data from multiple mapping populations

population genetics study of gene frequency variation at a population level.

power of the test see statistical power.

probe a short fragment of DNA or RNA labeled radioactively or in other ways, such as fluorescently. Probes are used to identify genes by hybridization.

probability of false negative probability of accepting a wrong null hypothesis, *e.g.*, accepting independence between two genes when they are linked.

probability of false positive probability of rejecting a correct null hypothesis, *e.g.*, rejecting independence between two genes when they are located on two chromosomes.

protein marker marker using differences in the charge and size of proteins, such as isozyme marker.

pseudo-backcross or pseudo-testcross a strategy for screening and using genes or markers with backcross configuration at single locus level in heterozygous F1 (cross between two heterozygotes parents at multiple-locus level) for genomic analyses.

QTL mapping a set of procedures for detecting genes controlling quantitative traits (QTL) and estimating their genetic effects and genome locations.

QTL resolution precision of QTL locations obtained using QTL mapping.

quantitative genetics study of quantitative trait inheritance.

quantitative trait trait having value with continuous distribution.

quantitative trait loci (QTL) genes controlling quantitative traits.

random amplified polymorphic DNA (RAPD) process in which a single arbitrarily chosen short oligonucleotide can be used as a primer to amplify genome segments flanked by two appropriately oriented primer-binding sites. If polymorphism exists in the binding sites among different genotypes or the fragment length differs at same site from genotype to genotypes, then a RAPD marker is obtained.

random drift gene frequency change from generation to generation due to random sampling.

recessive allele an allele whose phenotype is masked by another allele.

recombinant a gamete or an individual which is not a parental type as the result of genetic crossover in meiosis.

recombination process of crossover that creates a recombinant.

recombination fraction ratio between recombinant gametes produced by meiotic events and total meiotic events. It is commonly estimated by maximizing likelihood functions which are built using the observed genotypic frequencies in mapping populations and the expected genotypic frequencies as functions of recombination fractions.

repulsion see linkage phase.

resampling sampling from a sample.

restriction enzyme enzyme that makes sequence specific cuts in DNA.

Sanger method method developed by Sanger for obtaining DNA sequence.

score the first derivative of a natural logarithm of a likelihood function with respect to parameters.

sib-pair method an approach for QTL mapping using regression of the square of sib-difference of the trait on identity by decent for a marker for the sibs.

simple sequence repeat (SSR) see microsatellite.

sequence tagged sites (STS) a short unique fragment of DNA (~300 bp).

simulated annealing an optimization algorithm developed using analogies to the thermodynamics of liquid crystallization.

sister chromatids the two identical DNA strands of a chromosome.

Southern blotting see blotting

statistical genomics study of the statistical issues of genomics.

statistical power probability of rejecting the null hypothesis when the alternative hypothesis is true. It is also referred as power of the test.

support function a natural log likelihood function.

synteny genome similarity.

tandem repeats identical DNA segments located in clusters on the genome.

telomere DNA at ends of chromosome.

trans dominant linked marker (TDLM) two dominant markers closely linked in repulsion phase.

transcription the process of converting DNA to RNA.

translation the process of converting RNA to polypeptide.

transmission disequilibrium test (TDT) an approach to find markers linked to disease genes based on the idea that marker alleles associated with a disease have a high probability of being transmitted to affected offspring.

traveling salesman problem(TSP) the problem of finding the shortest cyclical itinerary for a traveling salesman who must visit each city, optimally once, given a symmetric matrix of distances among a set of cities.

variable number of tandem repeats (VNTR) see minisatellite.

yeast artificial chromosome (YAC) a type of DNA vector with inserts of approximately 250kb.

BIBLIOGRAPHY

Aarts, E. and J. Korst. 1989. Simulated Annealing and Boltzman Machines. John Wiley & Son, New York.

Abramowitz, M. and I.A. Stegun (editors). 1970. Handbook of Mathematical Functions. Dover Publications, New York.

Agresti, A. 1990. Categorical Data Analysis. John Wiley & Sons, New York.

Ahn, S.N., J.A.Anderson, M.E. Sorrells and S.D. Tanksley. 1993. Homologous relationships of rice, wheat and maize chromosomes. Mol Gen Genet 241:483-490.

Ahn, S.N. and S.D. Tanksley. 1993. Comparative linkage maps of the rice and maize genomes. Proc Natl Acad Sci USA. 90:7980-7984.

Allard, R.W. 1956. Formulas and tables to facilitate the calculation of recombination values in heredity. Hilgardia 21:10.

Allard, R.W. 1963. Evidence for genetic restriction of recombination on the lima bean. Genetics 48:1399-1395.

Allard, R.W. 1988. Genetic changes associated with the evolution of adaptedness in cultivated plants and their progenitors. J Hered 79:225-238.

Allard, R.W., A.L. Kahler and B.S. Weir. 1972. The effect of selection on estebase allozymes in a barley population. Genetics 72:489-503.

Amerson, H. personal communications.

Arondel, V., B. Lemieux, I. Hwang, S. Gibson, H.M. Goodman and C.R. Somerville. 1992. Map-based cloning of a gene controlling omega-3 fatty acid desaturation in arabidopsis. Science 258:20 Nov.

Baase, S. 1988. Computer Algorithms: Introduction to Design and Analysis. Second Edition. Addison-Wesley Publishing Company, Reading, Ma.

Bailey, N.T.J. 1961. Introduction to the mathematical theory of genetic linkage. Oxford: Clardon Press.

Barendse, W. et al. 1994. A genetic linkage map of the bovine genome. Nature Genetics 6:277-235.

Basten, C.J., B.S. Weir and Z.B. Zeng. 1994. ZMAP-A QTL cartographer. Proceedings of the 5th World Congress on Genetics Applied to Livestock Production 22: 65-66.

Beavis, W.D. and D. Grant. 1991 A linkage map based on information from four F2 populations of maize(Zea mays L.). Theor Appl Genet 82:636-644.

Bigwood, D.W., Broome, C.R. and Mitchell, P.A. 1992. Overview of PGD: the USDA Plant Genome Database Sysyem. Plant Genome I, The International Conference on the plant Genome, San Diego,CA. pp:7.

Boehnke, M. 1986. Estimating the power of a proposed linkage study: A practical computer simulation approach. Am J Hum Genet 39:513-27.

Boehnke, M. 1994. Limits of resolution of genetic linkage studies: Implications for the positional cloning of human disease genes. Am J Hum Genet 55:379-390.

Boehnke, M., N. Arnheim, H. Li, and F.S. Collins. 1989. Fine structure genetic mapping of human chromosomes using the polymerase chain reaction on single sperm: Experimental design considerations. Am J Hum Genet 45: 21 - 32.

Bohning, D. and P. Schlattmann. 1992. Computer-assisted analysis of mixtures (CA Man): statistical algorithms. Biometrics 48:283-303.

Bonierbale, M.W., R.L. Plaisted and S.D. Tanksley. 1988. RFLP maps based on a common set of clones reveal modes of chromosomal evolution in potato and tomato. Genetics 120:1095-1103.

Bonierbale, M.W., R.L. Plaisted and S.D. Tanksley. 1993. A test of the maximum heterozygosity hypothesis using molecular markers in tetraploid potatoes. Theor Appl Genet 86:481-491.

Botstein, D., R.L. White, M. Skolnick and R.W. Davis. 1980. Construction of a genetic linkage map in man using restriction fragment length polymorphisms. Am J Hum Genet 32:314-331.

Bradshaw, H.D, Jr. and R.F. Stettler. 1995. Molecular genetics of growth and development in *Populus*: IV. Mapping QTLs with large effects on growth, form and phenology traits in a forest tree. Genetics 139:963-973.

Bridges, C.B. 1915. A linkage variation in *Drosophila*. J Exp Zool 19:1-12.

Bridges, C.B. and K.S. Brehme. 1944. The mutants of *Drosophila melanogaster*. Carnegie Insttute of Washington Publisher, 552, Washington DC.

Briscoe, D., J.C. Stephens and S.J. O'Brien. 1994. Linkage disequilibrium in admixed populations: applications in gene mapping. J of Heredity 85:1.

Brooks, L.D. and R.W. Marks. 1986. The organization of genetic variation for recombination in *Drosophila melanogaster*. Genetics 114:525-447.

Buetow, K.H. 1987. Multipoint gene mapping using seriation. II. Analysis of simulated and empirical data. Am J Hum Genet 41 189-201.

Buetow, K.H. and A. Chakravarti. 1987. Multipoint gene mapping using seriationI General methods. Am J Hum Genet 41:180-188.

Buetow, K.H., T. Shiang, P. Yang, Y. Nakamura, G.M. Lathrop, R. White, J.J. Wasmuth, S. Wood, L.D. Berdahl, N.J. Leysens, T.M. Ritty, M.E. Wise and J.C. Murray. 1991. A detailed multipoint map of human chromosome 4 provides evidence for linkage heterogeneity and position-specific recombination rates. Am J Hum Genet 48:911-925.

Burr, B. and F.A. Burr. 1991. Recombinant inbreds for molecular mapping in maize: theoretical and practical considerations. Trends Genet 7:55-60.

Caldecott, R.S. and L. Smith. 1961. A study of X-ray induced chromosomal aberrations in barley. Cytologia 17:224-242.

Campbell, M. personal communications.

Cann, J.R. 1990. Analysis of the Gel Electrophoresis of looped protein-DNA complexes by computer simulation. J Mol Biol 216:1067-1075.

Cardon, L.R. and D.W. Fulker. 1994. The power of interval of quantitative trait loci, using selected sib pairs. Am J Hum Genet 55:825-833.

Carey, G. 1986. A general multivariate approach to linear modeling in human genetics. Am J Hum Genet 39:775-786.

Carey, G. and J. Williamson. 1991. Linkage analysis of quantitative traits: Increased power by using selected samples. Am J Hum Genet 49:786-796.

Carlson, W.R . 1977. The cytogenetics of corn, in Corn and Corn Improvement, edited by G.F. Sprague. American Society of Agronomy, Madison, Wisc.

Carter, T.C. and D.S. Falconer. 1951. Stocks for detecting linkage in the mouse and the theory of their design. J Genet 50:307-23.

Cavalli-Sforza, L.L. and M.C. King. 1986. Detecting linkage for genetically heterogeneous diseases and detecting heterogeneity with linkage data. Am J Hum. Genet. 38:599-616.

Cederberg, H. 1985. Recombination in other chromosomal regions than the interval subjected to selection, in lines of *Neurospora crassa* selected for high and for low recombination fraction. Hereditas 103:89-97.

Ceppellini, R., M. Siniscalco and C.A.B. Smith. 1955. The estimation of gene frequencies in a random-mating population. Am J Hum Genet 20:97- 115.

Chakraborty, R. 1987. Further considerations of difficulties of estimating familial risks from pedigree data. Hum Hered 37:222-28.

Chakraborty, R., M.D. Andrade, S.P. Daiger and B. Budowle. 1992. Apparent heterozygote deficiencies observed in DNA typing data and their implications in forensic applications. Ann Hum Genet 56:45-57.

Chakraborty, R. and P.E. Smouse. 1988. Recombination of haplotypes leads to biased estimates of admixture proportions in human populations. Proc Natl Acad Sci USA. 85:3070-3074.

Chakraborty, R. and K.M. Weiss, 1988. Admixture as a tool for finding linked genes and detecting that differenc from allelic association between loci. Proc Natl Acad Sci USA. 85:9119-9123.

Chakravarti, A., J.A.Badner and C.C. Li. 1987. Tests of linkage and heterogeneity in Mendelian diseases using identity by descent scores. Genet Epidemiol 4:255-266.

Chakravarti, A., C.C. Li and K.H. Buetow. 1984. Estimation of the marker gene frequency and linkage disequilibrium from conditional marker data. AmJ Hum. Genet. 36: 177-86.

Chakravarti, A. and S.A. Slaugenhaupt. 1987. Methods for studying recombination on chromosomes that undergo nondisjunction. Genomics 1 : 35 -42.

Charlesworth, B. 1975. Recombination modification in a fluctuating environment. Genetics 83:181-195.

Charlesworth, B. 1990. Mutation-selection balance and the evolutionary advantage of sex and recombination. Genet Res Camb 55:199-221.

Charlesworth, B. and D. Charlesworth. 1985a. Genetic variation in recombination in *Drosophila*. I. Response to selection and preliminary genetic analysis. Heredity 54:71-83.

Charlesworth, B. and D. Charlesworth. 1985b. Genetic variation in recombination in *Drosophila*. I. Genetic analysis of a high recombination stock. Heredity 54:85-98.

Chinnici, J.P. 1971a. Modification of recombination fraction in *Drosophila*. I. Selection for increased and decreased crossing-over. Genetics 69:71-83.

Chinnici, J.P. 1971b. Modification of recombination fraction in *Drosophila*. II. The polygenic control of crossing-over. Genetics 69:85-96.

Churchill, G.A. and R.W. Doerge. 1994. Empirical threshold values for quantitative trait mapping. Genetics 138: 963-971.

Clarke, L. and J. Carbon. 1976. A colony bank containing synthetic Col E1 hybrid plasmids representative of the entire *E. coli* genome. Cell 9:91-99.

Clerget-Darpoux, F., C. Bonaiti-Pellie and J. Hochez. 1986. Effects of misspecifying genetic parameters in lod score analysis. Biometrics 42:393-99.

Cockerham, C.C. 1954. An extension of the concept of partitioning hereditary variance for analysis of covariance among relatives when epistasis is present. Genetics 39:859

Cockerham, C.C and B.S. Weir. 1983. Linkage between a marker locus and a quantitative trait of sibs. Am J Hum Genet 35:263-73.

Coe, E.H. 1992. Comprehensive Mapping of the Maize Genome and the Mutual Role of the Maize Database. Plant Genome I, The International Conference on the Plant Genome, San Diago, Ca. pp:7.

Coe, E.H., D.A. Hoisington and M.G. Neuffer. 1984. Linkage map of corn (maize) In Genetic Maps, ed. S.J. O'Brien, Cold Spring Harbor Laboratory, New York, pp: 491-507.

Collins, F.S. 1995. Positional cloning moves from perditional to traditional. Nature Genetics 9 Apr.

Collins, F.S. and D. Galas. 1993. A new five-year plan for the US human genome project. Science 262:1 Oct.

Comell, J.A. 1990. Experiments with mixtures. John Wiley and Sons, New York

Conley, E.C. 1992. Mechanism and genetic control of recombination in bacteria. Mutation Research 284:75-96.

Cook-Deegan, R. 1995. The Gene War: Science, Politics, and the Human Genome. W. W. Norton & Company, New York.

Crawford, S.L. 1994. An application of the laplace method to finite mixture distributions. J of the American Statistical Association 89:425.

Darnell, J., H. Lodish and D. Baltimore. 1986. Molecular Cell Biology. Scientific American Books, New York.

Darvasi, A. and M. Soller. 1994. Optimum spacing of genetic makers for determining linkage between marker loci and quantitative trait loci. Theor Appl Genet 89:351-357.

Darvasi, A. and M. Soller. 1995. Advanced intercross lines, an experimental population for fine genetic mapping. Genetics 141: 1199-1207.

Darvasi, A., A. Weinerb, V. Minke, J. Weller and M. Soller. 1993. Detecting marker-QTL linkage and estimating QTL gene effect and map location using a saturated genetic map. Genetics 134:943-951.

Darvasi, A. and J. Weller. 1992. On the use of the moments method of estimation to obtain approximate maximum likelihood estimates of linkage between a genetic marker and a quantitative locus. Heredity 68:43-46.

Das, K. 1955. Cytogenetic studies of partial sterility in X-ray irradiated barley. Indian Jour Genetic and Plant Breeding 15:99-111.

Davisson, M.T. and E.C. Akeson. 1993. Recombination suppression by heterozygous tobertsonian chromosome in the mouse. Genetics 133:649-667.

Dawson, D.V., E.B. Kaplan and R.C. Elston. 1990. Extensions to sib-pair linkage tests applicable to disorders characterized by delayed onset. Genetic Epidemiology 7:45-466.

Dempster, A.P., N.M. Laird and D.B. Rubin. 1977. Maximum likelihood from incomplete data via the EM algorithm. JR Statist. Soc. 39B: 1 - 38.

Dietrich, W., H. Katz, S.E. Lincoln, H.S. Shin, J. Friedman, N.C. Dracopoli and E.S. Lander. 1992. A genetic map of the mouse suitable for typing intraspecific crosses. Genetics 131:423-447.

Doerge, R.W. 1995. The relationship between the LOD score and the analysis of variance- F-statistic when detecting QTLs using single markers. Theor Appl Genet 90: 980-981.

Donis-Keller, H., P. Green, C. Helms, S. Cartinhour, B. Weiffenbach, K. Stephens, T.P. Keith, D.W. Bowden, D.R. Smith, E.S. Lander, D. Botstein, G. Akots, K.S. Rediker, T. Gravius, V.A. Brown, M.B. Rising, C. Parker, J.A. Powers, D.E. Watt, E.R. Kauffman, A. Bricker, P. Phipps, H. Muller-Kahle, T.R. Fulton, S. Ng, J.W. Schumm, J.C. Braman, R.G. Knowlton, D.F. Barker, S.M. Crooks, S.E. Lincoln, M.J. Daly and J. Abrahamson. 1987. A genetic linkage map of the human genome. Cell 51:319-37.

Dudley, J.W. 1993 Molecular markers in plant improvement: manipulation of genes affecting quantitative traits. Crop Sci. 33: 660-668.

Edwards, A.W.F. 1984. Likelihood. Cambridge: Cambridge University Press.

Edwards, J.H. 1987. The locus ordering problem. Ann. Hum. Genet. 51:251-258.

Edwards, J.H. 1990. The linkage detection problem. Ann. Hum. Genet. 54:253-275.

Edwards, J.H. 1994. Comparative genome mapping in mammals. Current Opinion in Genetics and Development 4:861-867.

Edwards, M.D, T. Helentjaris, S. Wright and C.W. Stuber. 1992. Molecular-marker-facilitated investigations of quantitative trait loci in maize. 4. Analysis based on genome saturation with isozyme and restriction fragment length polymorphism markers. Theor Appl Genet 83:765-774.

Efron, B. and R.J. Tibshirani. 1993. An Introduction to the Bootstrap. Chapman & Hall, New York.

Elston, R.C. 1979. Major locus analysis for quantitative trails. Am J Hum Genet 31:655-661.

Elston, R.C. 1984. The genetic analysis of quantitative trait differences between two homozygous lines. Genetics 108:733-744.

Elston, R.C. 1986 Probability and paternity testing. Am J Hum Genet 39:112-122.

Elston, R.C., V.T. George and F. Severtson. 1992. The Elston-Stewart algorithm for continuous genotypes and environmental factors. Hum Hered 42:16-27.

Elston, R.C. and K. Lange. 1975. The prior probability of autosomal linkage. Ann Hum Genet 38:341-50.

Elston, R.C. and J. Stewart. 1971. A general model for the analysis of pedigree data. Hum Hered 21:523-42.

Elston, R.C. and J. Stewart. 1973. The analysis of quantitative traits for simple genetic models from parents, Fl and backcross data. Genetics 73:695-711.

Ewens, W.J. and R.S. Spielman. 1995. The transmission/disequilibrium test: history, subdivision, and admixture. Am J Hum Genet 52:455-464.

Fain, P.R., E. Wright, H.F. Willard, K. Stephens and D.F. Barker. 1989. The order of loci in the pericentric region of chromosome 17, based on evidence from physical and genetic breakpoints. Am J Hum Genet 44:68-72.

Falconer, D.S. and T.F.C. Mackay. 1996. Introduction to Quantitative Genetics. Fourth Edition. Longman Scientific & Technical, New York.

Falk, C.T. 1989. A simple scheme for preliminary ordering of multiple loci: application to 45 CF families. In: Elston, Spence, Hodge and MacCluer (ed.), Multipoint mapping and linkage based upon affected pedigree members. Genetic Workshop 6. Liss, New York, 17-22.

Falk, C.T and P. Rubinstein. 1987. Haplotype relative risks: an easy reliable way to construct a proper control sample for risk calculations, Am J Hum Genet 51:227-233.

Fehr, W.R. 1987. Principles of Cultivar Development. Macmillan Publishing Company, New York.

Felsenstein, J. 1979. A mathematically tractable family of genetic mapping functions with different amounts of interference. Genetics 91:769-75.

Fisher, R.A. 1949. Statistical Methods for Research Workers, 11th ed. Oliver and Boyd, Edburgh, London.

Frankel, W.N. 1995. Taking stock of complex trait genetics in mice. Trends Genet. 11:471-477.

Friebe, B. and B.S. Gill. 1995. Chromosome banding and genome analysis in diploid and cultivated polyploid wheats. In P.P. Jauhar (ed.) Methods of Genome Analysis in Plants. pp 39-60. CRC Press, Boca Raton, FL.

Fulker D.W and Cardon LR. 1994. A sib-pair approach to interval mapping of quantitative trait loci. Am J Hum Genet 54:1092-1103.

Fulker, D.W., S.S. Cherny and L.R. Cardon. 1995. Multipoint interval mapping of quantitative trait loci, using sib pairs. Am J Hum Genet 56:1224-1233.

Gallant, A.R. 1987. Nonlinear Statistical Models. John Wiley & Sons, New York.

Goffinet, B., P. Loisel and B. Laurent. 1992. Testing in normal mixture models when the proportions are known. Biometrica 79: 842-846.

Goldberg, D.E. 1989. Genetic Algorithms in Search, Optimization, and Machine Learning. Addison-Wesley, New York.

Goldgar, D.E. 1990. Multipoint analysis of human quantitative genetic variation. Am J Hum Genet 47:957-967.

Goldgar, D.E. and P.R. Fain. 1988. Models of multilocus recombination: Nonrandomness in chiasma number and crossover positions. Am J Hum Genet 43:38-45.

Goldgar, D.E., P.R. Fain and W.J. Kimberling. 1989. Chiasma-based modcls of multilocus recombination: Increased power for exclusion mapping and gene ordering. Genomics 5:283-290.

Goldgar, D.E., P. Green, D.M. Parry and J.J. Mulvihill. 1989. Multipoint linkage analysis in neurofibromatosis type I: An international collaboration. Am J Hum Genet 44:6-12.

Goldgar, D.E. and R.S. Oniki. 1992. Comparison of a multipoint identity-by-descent method with parametric multipoint linkage analysis for mapping quantitative traits. Am J Hum Genet 50:598-606.

Gonick, L. and M. Wheellis. 1991. The Cartoon Guide to Genetics. Harper Collins Publishers, New York.

Grattapaglia, D. and R. Sederoff. 1994. Genetic linkage maps of *Eucalyptus grandis* and *Eucalyptus urophylla* using a pseudo-testcross: mapping strategy and RAPD markers. Genetics 137:1121-1137.

Grattapaglia, D., F.L. Bertolucci and R. Sederoff. 1995. Genetic mapping of QTLs controlling vagetative propagation in *Eucalyptus grandis* and *Eucalyptus urophylla* using a pseudo-testcross: mapping strategy and RAPD markers. Theor Appl Genet 90:933-947.

Griffiths, A.J.F., J.H. Miller, D.T. Suzuki, R.C. Lewontin and W.M. Gelbart. 1993. An Introduction to Genetic Analysis, Fifth Edition, Freeman and Company, New York.

Groover, A., M.Devey, M. Fiddler, T. Lee, J. Megraw, T. Mitchell-Olds, T. Sherman, B. Vujcic, C. Williams and D. Neale. 1994. Identification of quantitative trait loci influencing wood specific gravity in an outbred pedigree of loblolly pine. Genetics 138:1293-1300.

Guo, S.W. and E.A. Thompson. 1991. Monte Carlo estimation of variance component models for large complex pedigrees. IMA Journal of Mathematics Applied in Medicine & Biology. 8:171-189.

Gyapay, G., J. Morissette, A. Vignal, C. Dib, C. Fizames, P. Millasseau, S. Marc, G. Bernardi, M. Lathrop and J. Weissenbach. 1994. The 1993-94 Genethon human linkage map. Nature Genetics 7:246-249.

Haldane, J.B.S. 1919. The combination of linkage values and the calculation of distances between the loci of linked factors. J Genet 8:299-309.

Haldane, J.B.S. and C.A.B. Smith. 1947. A new estimate of the linkage between the genes for colour-blindness and haemophilia in man. Ann Eugen 14: 10-31.

Haley, C.S. 1991. Use of DNA fingerprints for the detection of major genes for quantitative traits in domestic species. Animal Genetics 22:259-277.

Haley, C.S. 1995. Livestock QTLs - bringing home the bacon? Trends in Genetics 12:488-492.

Haley, C.S. and S.A. Knott. 1992. A simple regression method for mapping quantitative trait loci in line crosses using flanking markers. Heredity 69: 315-324.

Haley, C.S. and S.A. Knott. 1992. A simple regression method for mapping quantitative trait loci in line crosses using flanking markers. Heredity 69:315-324.

Haley, C.S., S.A. Knott and J.M. Elsen. 1994. Mapping quantitative trait loci in crosses between outbred lines using least squares. Genetics 136: 1197-1207.

Hallauer, A.R. and J.B. Miranda, Fo. 1988. Quantitative Genetics in Maize Breeding. Second Edition. Iowa State University Press, Ames, Iowa.

Hanson, W.D. 1963. Heritability. In: W.D. Hanson and H.F. Robinson (eds.), Statistical Genetics and Plant Breeding. NAS-NRC Publ. 982.

Hartl, D.L. 1988. A Primer of Population Genetics. Sinauer, Sunderland, MA.

Haseman, J.K. and R.C. Elston. 1972. The investigation of linkage between a quantitative trait and a marker locus. Behav Genet 2:3-19.

Hastbacka, J., A.D.L. Chapelle, M.M. Mahtani, G. Clines, M.P. Reeve-daly, M. Daly, B.A. Hamilton, K. Kusmi, B. Trivedi, A. Weaver, A. Coloma, M. Lovett, A. Buckler, I. Kaitila and E.S. Lander. 1994. The diastrophic dysplasia gene encodes a novel sulgate transporter:positional cloning by fine-structure linkage disequilibrium mapping. Cell 78:1073-1087.

Hastings, P.J. 1992. Mechanism and control of recombination in fungi. Mutation Research 284:97-110.

Hauge, B.M., S.M. Hanley, S. Cartinhour, J.M. Cherry, H.M. Goodman, M. Koornneef, P. Stam, C. Chang, S. Kempin, L. Medrano and E.M. Meyerowitz. 1993. An integrated genetic/RFLP map of the *Arabidopsis thaliana* genome. The Plant Journal 3(5): 745-754.

Hayes, P.M., B.H. Liu, S.J. Knapp, F. Chen, B. Jones, T. Blake, J. Franckowiak, D. Rasmusson, M. Sorrells, S.E. Ullrich, D. Wesenberg and A .Kleinhofs. 1993. Quantitative trait locus effects and environmental interaction in a sample of North American barley germplasm. Theor Appl Genet 87: 392-401.

Hedrick, P.W. 1987. Gametic disequilibrium measures: proceed caution. Genetic 117:331-341.

Helentjaris, T., M.A.T. Cushman and R. Winkler. 1992. Developing a genetic understanding of agronomy traits with complex inheritance. p.397-406. In: Y. Dettee, C. Dumas and A. Gallais (eds.), Reproductive Biology and Plant Breeding. Springer-Verleg Berlin Heidelberg.

Helentjaris, T., G. King, M. Slocum, C. Siedenstrang and S. Wegman. 1985. Restriction fragment polymorphisms as probes for plant diversity and their development as tools for applied plant breeding. Plant Molecular Biology 5:109-118.

Helentjaris, T., M. Slocum, S.Wright, A. Schaefer and J. Nienhuis. 1986. Construction of genetic linkage maps in maize and tomato using restriction fragment length polymorphisms. Theor Appl Genet 72: 761-769.

Helentjaris, T., D.F. Weber and S. Wright. 1986. Use of monosomics to map cloned DNA fragments in maize. Genetics 83:6035-6039.

Heslop-Harrison, J.S. and T. Schwarzacher. 1995. Genomic Southern and in situ hybridization for plant genome analysis. In P.P. Jauhar (ed.) Methods of Genome Analysis in Plants. pp 163-180. CRC Press, Boca Raton, FL.

Hill, W.G. and B.S. Weir. 1994. Maximum-likelihood estimation of gene location by linkage disequilibrium. Am J Hum Genet 54:705-714.

Hoisington, D.A. and E.H. Coe. 1989. Methods for correlating RFLP maps with conventional genetic and physical maps in maize. In: T. Helentjaris and B. Burr (eds.) Current Communications in Molecular Biology. Cold Spring Harbor Laboratory Press, Cold Spring Harbor, NY. pp 19-24.

Holland, J.H. 1975. Adaptation in Natural and Artificial Systems. University of Michigan Press, Ann Arbor.

Holliday, R. 1964. A mechanism for gene conversion in fungi. Genet Res 5:282-303.

Hulbert, S.H., T.E. Richter, J.D. Axtell and J.L. Bennetzen. 1990. Genetic mapping and characterization of sorghum and related crops by means of maize DNA probes. Proc Natl Acad Sci USA 87: 4251-4255.

Jansen, R.C. 1992. A general mixture model for mapping quantitative trait loci by using molecular markers. Theor Appl Genet 85:252-260.

Jansen, R.C. 1993. Maximum likelihood in a generalized linear finite mixture model by using the EM algorithm. Biometrics 49:227-231.

Jansen, R.C. 1993. Interval mapping of multiple quantitative trait loci. Genetics 135:205-211.

Jansen, R.C. 1996. A general Monte Carlo method for mapping multiple quantitative trait loci. Genetics 142: 305-311.

Jansen, R.C. and P. Stam. 1994. High resolution of quantitative traits into multiple loci via interval mapping. Genetics 136: 1447-1455.

Jeffreys, A.J, R. Neumann and V. Wilson. 1990. Repeat unit sequence variation in minisatellites: A novel source of DNA polymorphism for studying variation and mutation by single molecule analysis. Cell 60:473-85.

Jeffreys, A.J., V. Wilson and S.L. Thein. 1985. Individual-specific "fingerprints" of human DNA. Nature 316:76-79.

Jensen, J. and J.H. Jorgensen. 1975. The barley chromosome 5 linkage map. Hereditas 80: 5-16.

Jensen, J. and J.H. Jorgensen. 1975. The barley chromosome 5 linkage map II. extension of the map with four loci. Hereditas 80:17-26.

Jiang, C. and Z.B. Zeng. 1995. Multiple trait analysis of genetic mapping for quantitative trait loci. Genetics 140:1111-1127.

Jiang, C. and Z.B. Zeng. 1997. Mapping quantitative trait loci with dominant and missing markers in various crosses from two inbred lines. Genetika. (in press).

Jinks, J.L. and J.M. Perkins. 1972. Predicting the range of inbred lines . Heredity 28:399-403.

Johnson, M.E. 1987. Multivariate Statistical Simulation. John Wiley & Sons, New York.

Julier, C., R.N. Hyer, J. Davies, F. Merlin, P. Soularue, L. Briant, G. Cathelineau, I. Deschamps, J.I. Rotter, P. Froguel, C. Boitard, J.I. Bell and G.M. Lathrop. 1991. Insulin-IGF2 region on chromosome 11p encodes a gene implicated in HLA-Dr4-dependent diabetes susceptibility. Nature 354:14.

Kammerer, C.M. and J.W. MacCluer. 1988. Empirical power of three preliminary methods for ordering loci. Am J Hum Genet 43:964-970.

Kaplan, N.L., W.G. Hill and B.S. Weir. 1995. Likelihood methods for locating disease genes in nonequilibrium populations. Am J Hum Genet 56:18-32.

Kaplan, N.L. and B.S. Weir. 1992. Expected behavior of conditional linkage disequilibrium. Am J Hum Genet 51:333-343.

Kaplan, N.L. and B.S. Weir. 1995. Are moment bound on the recombination fraction between a marker and a disease locus too good to be true? Allelic association mapping mapping revisited for simple genetic diseases in the Finnish population. Am J Hum Genet 57:1486-1498.

Karlin, S. and U. Liberman. 1978. Classifications and comparisons of multilocus recombination distributions. Proc Natl Acad Sci USA 75:6332-36.

Karlin, S. 1984. Theoretical aspects of genetic map functions in recombination processes. In Human population genetics: The Pittsburgh symposium, edited by A Chakravarti, 209-28. Van Nostrand Reinhold, New York.

Kasha, K.J. 1961. Inversion in chromosome 5. Barley Newsletter 4:16.

Keats, B., J. Ott and M. Conneally. 1989. Report of the committee on linkage and gene order. Cytogenet Cell Genet 51:459-502.

Keats, B.J., S.L. Sherman, N.E. Morton, E.B. Robson, K.H. Buetow, H.M. Cann, P.E. Cartwright, A. Chakravarti, U. Franvke, P.P. Green and J. Ott. 1991. Guideline for human linkage maps: an international system for human linkage maps (ISLM, 1990). Genomics 9:557-560.

Kempthorne, O. 1957. An Introduction to Genetic Statistics. John Wiley. New York.

Kidwell, M.G. 1972a. Genetic change of recombination value in Drosophila melanogaster. I. Artificial selection for high and low recombination and some properties of recombination modifying genes. Genetics 70:419-432.

Kidwell, M.G. 1972b. Genetic change of recombination value in Drosophila melanogaster. II. Simulated natural selection. Genetics 70:433-443.

Kirkpatrick, S., C.D. Gelatt and M.P. Vecchi. 1983. Optimization by simulated annealing. Science. 220:671-680.

Kleinhofs, A., A. Kilian, M.A. Saghai Maroof, R.M. Biyashev, P.M. Hayes, F.Q. Chen, N. Lapitan, A. Fenwick, T.K. Blake, V. Kanazin, E. Ananiev, L. Dahleen, D. Kudrna, J. Bollinger, S.J. Knapp, B.H. Liu, M. Sorrells, M. Heun, J.D. Fanckowiak, D. Hoffman, R. Skadsen and B.J. Steffenson. 1993. A molecular, isozyme, and morphological map of the barley (Hordeum vulgare) genome. Theor Appl Genet 86: 705-712.

Knapp, M., G. Wassmer and M.P. Baur. 1995. The relative efficiency of the Hardy-Weinberg equilibrium-likelihood and the conditional on parental genotype-likelihood methods for candidate-gene association studies. Am J Hum Genet 57:1476-1485.

Knapp, S.J. 1991. Using molecular markers to map multiple quantitative trait loci: Models for backcross, recombinant inbred, and doubled haploid progeny. Theor Appl Genet 81: 333-338.

Knapp, S.J. 1994. Mapping quantitative trait loci. p.58-96. In: R.L. Phillipes and I.K. Vasil (eds.), DNA-based Markers in Plants. Kluwer Academic Publishers.

Knapp, S.J. and W.C. Bridges. 1990. Using molecular markers to estimate quantitative trait locus parameters: Power and genetic variances for unreplicated and replicated progeny. Genetics 126: 769-777.

Knapp, S.J., W.C. Bridges and D. Birkes. 1990. Mapping quantitative trait loci using molecular marker linkage maps. Theor Appl Genet 79: 583-592.

Knapp, S.J., W.C. Bridges and B.H. Liu. 1992. Mapping quantitative trait loci using nonsimultaneous and simultaneous estimators and hypothesis tests. In: J.S. Beckmann and T.S. Osborn (eds.), Plant Genomes: Methods for Genetic and Physical Mapping, pp. 209-237. Kluwer Academic Publishers, Dordrecht, The Netherlands.

Knapp, S.J., J.L. Holloway, W.C. Bridges and B.H. Liu. 1995. Mapping dominant markers using F2 matings. Theor Appl Genet 91:74-81.

Knapp, S.J. and B.H. Liu. 1992. Mapping genetic markers using matings between non-inbred individuals. Plant Genome I, The International Conference on the Plant Genome, San Diego, Ca. pp:59.

Knott, S.A. and C.S. Haley. 1992. Maximum likelihood mapping of quantitative trait loci using full-sib families. Genetics 132:1211-1222.

Knott, S.A. 1994. Prediction of the power of detection of marker-quantitative trait locus linkages using analysis of variance. Theor Appl Genet 89:318-322.

Knott, S.A. and C.S. Haley. 1992. Aspects of maximum likelihood methods for the mapping of quantitative trait loci in line crosses. Genet Res 60:139-151.

Koo, F.K.S. 1958. Pseudo-isochromosomes produced in *Avena strigosa* Schreb. by ionizing radiations. Cytologia 23:109-111.

Kosambi, D.D. 1944. The estimation of map distances from recombination values. Ann Eugen 12: 172-75.

Kruglyak, L. and E.S. Lander. 1995. Complete multipoint sib-pair analysis of qualitative and quantitative traits. Am J Hum Genet 57:439-454.

Kruglyak, L. and E.S. Lander. 1995. A nonparametric approach for mapping quantitative trait loci. Genetics 139: 1421-1428.

Kruglyak, L. and E.S. Lander. 1995. High-resolution genetic mapping of complex traits. Am J Hum Genet 1212-1223.

Lalouel, J.M. 1977. Linkage mapping from pair-wise recombination data. Heredity 38:61 -77.

Lande, R. and R. Thompson. 1990. Efficiency of marker-assisted selection in the improvement of quantitative traits. Genetics 124: 743-756.

Lander, E.S. and D. Botstein. 1986a. Strategies for studying heterogeneous genetic traits in humans by using a linkage map of restriction fragment length polymorphisms. Proc Natl Acad Sci USA 83:7353-57.

Lander, E.S. and D. Botstein. 1986b. Mapping complex genetic traits in humans: New methods using a complete RFLP linkage map. Cold Spring Harbor Symp Quant Biol 5 1: 49-62.

Lander, E.S. and D. Botstein. 1989. Mapping Mendelian factors underlying quantitative traits using RFLP linkage maps. Genetics 121:185-199.

Lander, E.S. and P. Green. 1987. Construction of multilocus genetic linkage maps in human. Proc Natl Acad Sci USA. 84:2363-2367.

Lander, E.S. and P. Green. 1991. Counting algorithms for linkage: correction to morton and collins. Ann Hum Genet 55:33-38.

Lander, E.S., P. Green, J. Abrahamson, A. Barlow, M.J. Daly, S.E. Lincoln and L. Newburg. 1987. MAPMAKER: An interactive computer package for constructing primary genetic linkage maps of experimental and natural populations. Genomics I : 174-81.

Lander, E.S. and N.J. Schork. 1994. Genetic dissection of complex traits. Science 265:2037-2048.

Lange, K. 1986. The affected sib-pair method using identity by state relations. Am J Hum Genet 39:148-50.

Lange, K. 1986. A test statistic for the affected-sib-set method. Am J Hum Genet 40:283-290.

Lange, K. and R.C. Elston. 1975. Extensions to pedigree analysisI Likelihood calculations for simple and complex pedigrees. Hum Hered 25:95- 105.

Lange, K. and M. Boehnke. 1982. How many polymorphic genes will it take to span the human genome? Am J Hum Genet 34:842-45.

Lange, K. and M. Boehnke. 1983. Some combinatorial problems of DNA restriction fragment length polymorphisms. Am J Hum Genet 35: 177-92.

Lange, K. and M. Boehnke. 1992 Bayesian methods and optimalexperimental design for gene mapping by radiation hybirds. Ann Hum Genet 56:119-144.

Lange, K., M. Boehnke, D.R. Cox and K.L. Lunetta. 1995. Statistical methods for polyploid radiation hybrid mapping. Genome Research 5:136-150.

Lange, K. and S. Matthysse. 1989. Simulation of pedigree genotypes by random walks. Am J Hum Genet 45:959-70.

Lange, K., M.A. Spence and M.B. Frank. 1976. Application of the lod method to the detection of linkage between a quantitative trait and a qualitative marker: A simulation experiment. Am J Hum Genet 28:167-73.

Lange, K., D. Weeks and M. Boehnke. 1988. Programs for Pedigree Analysis: MENDEL, FISHER, and DGENE. Genet Epidemiol 5:471-72.

Lange, K. and D.E. Weeks. 1989. Efficient computation of LOD scores: Genotype elimination, genotype redefinition, and hybrid maximum likelihood algorithms. Ann Hum Genet 53:67-83.

Lanzov, V., I. Stepanova and G. Vinogradskaja. 1991. Genetic control of recombination exchange frequency in Escherichia coli K-12 Biochimie 73:305-312.

Lathrop, G.M., J. Chotai, J. Ott and J.M. Lalouel. 1987. Tests of gene order from three-locus linkage data. Ann Hum Genet 51:235-49.

Lathrop, G.M., P. O'Connell, M. Leppert, Y. Nakamura, M. Farrall, L.C. Tsui, J.M. Lalouel and R. White. 1989. Twenty-five loci form a continuous linkage map of markers for human chromosome 7. Genomics 5:866-73.

Lathrop, G.M., A.B. Hooper, J.W. Huntsman and R.H. Ward. 1983. Evaluating pedigree data: I. The estimation of pedigree error in the presence of marker mistyping. Am J Hum Genet 35:241-62.

Lathrop, G.M., J.M. Lalouel, C. Julier, and J. Ott. 1984. Strategies for multilocus linkage analysis in humans. Proc Natl Acad Sci USA 81:3443 46.

Lathrop, G.M., J.M. Lalouel, C. Julier and J. Ott. 1985. Multilocus linkage analysis in humans: detection of linkage and estimation of recombination. Am J Hum Genet 37:482-498.

Lathrop, G.M., J.M. Lalouel and R.L. White. 1986. Construction of human linkage maps: likelihood calculations for multilocus linkage analysis. Genetic Epidemiology 3:39-52.

Lehmann, E.L. 1986. Testing Statistical Hypothesis. Second Edition. Wadsworth & Brooks, Pacific Grove, CA.

Lehmann, E.L. 1991. Theory of Point Estimation. Second Edition. Wadsworth & Brooks, Pacific Grove, CA.

Levine, R.P. 1955. Chromosome structure and the mechanism of crossing over. Proc Nat Acad Sci USA 41:727-730.

Lewin, B. 1994. Genes V. Oxford University Press, Oxford.

Lewontin, R.C. 1988. On measures of gametic disequilibrium. Genetics 120:849-852

Li, C.C. 1955. Population Genetics. The University of Chicago Press, Chicago.

Li, C.C, D.E. Weeks and A. Chakravarti. 1993. Similarity of DNA fingerprints due to chance and relatedness. Hum Hered 43:45-52.

Liberman, U. 1986. A general reduction principle for genetic modifiers of recombination. Theoretical Population Biology 30:341-371.

Liberman, U. and S. Karlin. 1984. Theoretical models of genetic map functions. Theoretical Population Biology 25:331-46.

Lichter, P., C.J.C. Tang, K. Call, G. Hermanson, G.A. Evans, D. Housman and D.C. Ward. 1990. High resolution mapping of human chromosome 11 by *in situ* hybridization with cosmid clones. Science 247:64-69.

Lin, S. 1965. Computer solutions of the traveling salesman problem. Bell System Technical Journ. 44:2245-2269.

Lincoln, S.E. and E.S. Lander. 1992. Systematic detection of errors in genetic linkage data. Genomics 14:604-610.

Liu, B.H. and S.J. Knapp. 1990. A FORTRAN program for the Mendelian analysis of segregation and linkage based on log-likelihood ratios. J. Hered. 81(5):407.

Liu, B.H. and S.J. Knapp. 1992. GMENDEL2.0, a software for gene mapping. Oregon State University.

Liu, B.H. and S.J. Knapp. 1992. QTLSTAT1.0, a software for mapping complex trait using nonlinear models. Oregon State University.

Lu, Y.Y. and B.H. Liu. 1995. PGRI, a software for plant genome research. Plant Genome III. San Diego, CA.

Luo, Z.W. and M.J. Kearsey. 1991. Maximum likelihood estimation of linkage between a marker gene and a quantitative trait locus. II. application to backcross and doubled haploid populations. Heredity 66:117-124.

Manly, K.F. and E.H. Cudmore, Jr. 1996. New versions of MAP Manager genetic mapping software. Plant Genome IV (Abs.) p:105.

Mark, K.T. and A.J. Morton. 1993. A modified lin-Kernighan traveling-salesman heuristic. Operation Research Letters 13:127-132.

Martinez, O. and R.N. Curnow. 1992. Estimation the locations and the sizes of the effects of quantitative trait loci using flanking markers. Theor Appl Genet 85:480-488.

Mather, K. 1938. The Measurement of Linkage in Heredity. Methuen & Co. Ltd, London.

Mather, K. and J.L. Jinks. 1971. Biometrical Genetics (2nd edn), Chapman & Hall, London.

Mayo, O. 1987. The Theory of Plant Breeding, second edition. Clarendon Press, Oxford.

Mazur, B.J. and S.V. Tingey. 1995. Genetic mapping and introgression of genes of agronomic importance. Current Opinion in Biotechnology 6:175-183.

McCouch, S.R. and R.W. Doerge. 1995. QTL mapping in rice. Trends Genet. 11:482-487.

McCullagh, P. and J.A. Nelder. 1989. Generalized Linear Models, Second Edition. Chapman & Hall, London.

McGill, C.B., B.K. Shafer, K. Derr and J.N. Strathern. 1993. Recombination initiated by double-strand breaks. Curr Genet 23:305-314.

Meselson, M.S. and C.M. Radding. 1975. A general model for genetic recombination.. Proc Natl Acad Sci USA 72:358-361.

Metropolis, N., A.W. Rosenbluth, M.N. Rosenbluth and A.H. Teller. 1953. Equation of state calculations by fast computing machines. J Chem Phys 21:1087-1092.

Michelmore, R.W., L. Paran and R.V. Kesselli. 1991. Identification of markers linked to disease resistance genes by bulked segregant analysis: a rapid method to detect markers in specific genome regions using segregating populations. Proc Natl Acad Sci USA 88:9828-9832.

Michelmore, R.W. and D.V. Shaw. 1988. Character dissection. Nature 335:672-673.

Miller, D.L. and J.F. Pekny. 1991. Exact solution of large asymmetric traveling salesman problems. Science 251:754-761.

Mood, A.M., F.A. Graybill and D.C. Boes. 1974. Introduction to the Theory of Statistics. McGraw-Hill, Inc., New York.

Moreno-Gonzalez, J. 1992. Genetic models to estimate additive and non-additive effects of marker-associated QTL using multiple regression techniques. Theor Appl Gcnct 85: 435-444.

Morgan, T.H.. 1928. The theory of genes. Yale University Press, New Haven, Conn.

Morris, R. 1955. Induced reciprocal translocations involving homologous chromosomes in maize. Amer. Jour. Bot. 42:546-551.

Morton, N.E. and A. Collins, Counting algorithms for linkage. Ann Hum Genet 54:103-106.

Muller, J. 1916. The mechanism of crossing over. Am. Nat. 50:193-207.

Mullis, K.B. 1990. The unusual origin of the polymerase chain reaction. Sci Amer 262:56-65.

Murray, J.C. and K.H. Buetow. 1992. The chromosome 4 workshop report. Genomics 12:857-858.

Murray, M., J. Cramer, Y. Ma, D. West, J. Romero-Severson, J. Pitas, S. Demars, L. Vilbrandt, J. Kirshman, R. McLeester, J. Schilz and J. Lotzer. 1988. AgriGenetics maize RFLP linkage map. Maize Genet. Coop. Newsl. 62:89-91.

Neale, D.B. and C.G. Williams. 1991. Restriction fragment length polymorphism mapping in conifer and applications to forest genetics and tree improvement. Can J For Res 21:545-554.

Nei, M. 1987. Molecular Evolutionary Genetics. Columbia University Press, New York.

Nelder, J.A. and R. Mead. 1965. A simplex method for function minimization. Computer J 7:308-313.

Newton, K.J. and D. Schwartz. 1980. Genetic basis of the major malate dehydrogenase isozymes in maize. Genetics 96:424-442.

Nienhuis, J., T. Helentjaris, M. Slocum, B. Ruggero and A. Schaefer. I 987. Restriction fragment length polymorphism analysis of loci associated with insect resistance in tomato. Crop Sci 27: 797-803.

Nordheim, E.V., D.M. O'Malley and R.P. Guries. 1983. Estimation of recombination frequency in genetic linkage studies. Theor Appl Genet 66:313-321.

Nowak, R. 1995. Entering the postgenome era. Science 270:368-371.

O'Malley, D. personal communications.

Old, R.W. and S.B. Primrose. 1994. Principles of Gene Manipulation: An Introduction to Genetic Engineering. Blackwell Science, Oxford.

Olson, J.M. and M. Boehnke. 1990. Monte Carlo comparison of preliminary methods for ordering multiple genetic loci. Am J Hum Genet 47:470-482.

Olson, J.M., L. Hood, C. Cantor and D. Botstein. 1989. A common language for physical mapping of the human genome. Science 245:1434-35.

Ott, J. 1974. Estimation of the recombination fraction in human pedigrees: Efficient computation of the likelihood for human linkage studies. Am J Hum Genet 26:588-97.

Ott, J. 1977. Counting methods (EM algorithm) in human pedigree analysis: Linkage and segregation analysis. Ann Hum Genet 40:443-54.

Ott, J. 1979. Maximum likelihood estimation by counting methods under polygenic and mixed models in human pedigrees. Am J Hum Genet 31:161-75.

Ott, J. 1989. Statistical properties of the haplotype relative risk. Genetic Epidemiology 6:127-130.

Ott, J. 1991. Analysis of human genetic linkage, revised edition. The Johns Hopkins University Press, Baltimore.

Ott, J. 1992. Strategies for characterizing highly polymorphic markers in human gene mapping. Am J Hum Genet 51:283-290.

Ott, J. and G.M. Lathrop. 1987. Goodness-of-fit tests for locus order in three-point mapping. Genet Epidemiol 4:51-57.

Padberg, M. and G. Rinaldi. 1987. Optimization of a 532-city symmetric traveling salesman problem by branching and cut. Operation Research Letters, 6:1-7.

Paterson, A.H., S. Damon, J.D. Hewitt, D. Zamir, H.D. Rabinowitch, S.E. Lincoln, E.S. Lander and S.D. Tanksley. 1991. Mendelian factors underlying quantitative traits in tomato: comparison across species, generations, and environments. Genetics 127:181-197.

Paterson, A.H., J.W. Deverna, B. Lanini and S.D. Tanksley. 1990. Fine mapping of quantitative trait loci using selected overlapping recombinant chromosomes, in a interspecific cross of tomato. Genetics 124:735-742.

Paterson, A.H., E.S. Lander, J.D. Hewitt, S. Peterson, S.E. Lincoln and S.D. Tanksley. 1988. Resolution of quantitative traits into Mendelian factors by using a complete RFLP linkage map. Nature 335: 721-726.

Penrose, L.S. 1935. The detection of autosomal linkage in data which consist of pairs of brothers and sisters of unspecified parentage. Ann. Eugen. 6: 133-38.

Plomion, C, BH Liu, and DM O'Malley. 1996. Genetic analysis using trans dominant linked markers in an F2 family. Theoretical and Applied Genetics. 93:1083-1089.

Press, W.H., S.A. Teukolsky, W.T. Vetterling and B.P. Flannery. 1992. Numerical Recipes in C: the A.rt of Scientific Computing. Cambridge University Press, New York.

Punnen, A.P. 1992. Traveling salesman problem under categorization. Operations Research Letters 12:89-95.

Rafalski, J.A. and S.V. Tingey. 1993. Genetic diagnostics in plant breeding: RAPDs, microsatellites and machines. Trends Genetics 9:275-280.

Rao, D.C., B.J. Keats, J.M. Lalouel, N.E. Morton and S. Yee. 1979. A maximum likelihood map of chromosome 1. Am J Hum Genet 31:680-696.

Rao, D.C., B.J. Keats, N.E. Morton, S. Yee and R. Lew. 1978. Variability of human linkage data. Am J Hum Genet 30:516-529.

Rebai, A., B. Goffinet and B. Nangin. 1994. Approximate thresholds of interval mapping tests for QTL detection. Genetics. 138:235-240.

Rebai, A., B. Goffinet and B. Mangin. 1995. Comparing power of different methods for QTL detection. Biometrics 51:87-99.

Reiter, R.S., J.G. Coors, M.R. Sussman and W.H. Gabelman. 1991. Genetic analysis of tolerance to low-phosphorus stress in maize using restriction fragment length polymorphisms. Theor Appl Genet 82:561-568.

Reiter, R.S., J.G.K. Williams, K.A. Feldmann, J.A. Rafalski, S.V. Tingey and P.A. Scolnik. 1992. Global and local genome mapping in arabisopsis thaliana by using recombinant inbred lines and random amplofied polymorphic DNAs. Genetics 89:1477-1481.

Risch, N. 1990. Linkage strategies for genetically complex traits I: Multilocus models. Am J Hum Genet 46:222-228.

Risch, N. 1990. Linkage strategies for genetically complex traits II: the power of affected relative pairs. Am J Hum Genet 46:229-241.

Risch N. 1990. Linkage strategies for genetically complex traits III: the effect of marker polymophism on analysis of affected relative pairs. Am J Hum Genet 46:242-253

Ritland, K. and S. Jain. 1981. A model for the estimation of outcrossing rate and gene frequencies using n independent loci. Heredity 47:35-52.

Ritter, E., C. Gebhardt and F. Salamini. 1990. Estimation of recombination frequencies and construction of RFLP linkage maps in plants from crosses between heterozygous parents. Genetics 125:645-654.

Robertson, D.S. 1985. A possible technique for isolating genomic DNA for quantitative traits in plants. J Theor Biol 17: 1- 1 0.

Robertson, D.S. 1989. Understanding the relationship between qualitative and quantitative genetics. p.81-87. In: T. Helentjaris and B. Burr (eds.), Development and Application of Molecular Markers to Problems in Plant Genetics. Cold Spring Harbor Laboratory.

Rodolphe, F. and M. Lefort. 1993. A multi-marker model for detecting chromosomal segments displaying QTL activity. Genetics 134:1277-1276.

Russel, W.A. and C.R. Burnham. 1950. Cytogenetic studies of an inversion in maize. Scient Agr 30:93-111.

Russell, K.W. 1992. Database Development and Information Support for the USDA Plant Genome Research Program. Plant Genome I, The International Conference on the Plant Genome, San Diego, Ca. pp:13.

Sall, T. 1991. Genetic control of recombination in barley II: recombination between the hordein loci in three different genotypes. Hereditas 115:13-16.

Sall, T. and B.O. Bengtsson. 1989. Apparent negative interference due to variation in recombination frequencies. Genetics 122:935-942.

Sanger, F., S. Nicklen and A.R. Coulson. 1977. DNA sequencing with chain-terminating inhibitors. Pro Natl Acad Sci USA 74:5463-5467.

Sax, K. 1923. The association of size differences with seed-coat pattem and pigmentation in *Phaseolus vulgaris*. Genetics 8:552-560.

Schork, N.J. 1993. Extended multipoint identity-by-descent analysis of human quantitative traits: efficiency, power, and modeling considerations. Am J Hum Genet 53:1306-1319.

Searle, S.R. 1971. Linear Models. John Wiley and Sons, New York

Sederoff, R.R. personal communications.

Shao, J. and D. Tu. The Jackknife and Bootstrap. Springer, New York. 1995.

Shapiro, R. The Human Blueprint: The Race to Unlock the Secrets of Our Genetic Code. A Bantam Book/St. Martin's Press, New York. 1992.

Simchen, G. and J. Stamberg. 1969. Fine and coarse controls of genetic recombination. Nature 222 April.

Smith, C.A.B. 1957. Counting methods in genetical statistics. Ann Hum Genet 21:254-76.

Smith, C.A.B. 1963. Testing for heterogeneity of recombination fraction values in human genetics. Ann Hum Genet 27: 175-82.

Smith, C.A.B. 1975. A non-parametric test for linkage with a quantitative character. Ann Hum Genet 38:451.

Smith, C.A.B. 1989. Some simple methods for linkage analysis. Ann Hum Genet 53:277- 83.

Smith, C.A.B. 1990. Probabilities of orders in linkage calculations. Amm Hum Genet 54:339-363.

Sokal, R.R. and F.J. Rohlf. Biometry. Second Edition. Freeman and Company, New York. 1981.

Soller, M. and J.S. Beckmann. 1987. Cloning quantitative trait loci by insertional mutagenesis. Theor Appl Genet 74:369-378.

Soller, M. and J.S. Beckmann. 1990. Marker-based mapping of quantitative trait loci using replicated progenies. Theor Appl Genet 80: 205-208.

Soller, M., T. Brody and A. Genizi. 1976. On the power of experimental design for the detection of linkage between marker loci and quantitative loci in crosses between inbred lines. Theor Appl Genet 47:35-39.

Southern, E.M. 1975. Detection of specific sequences among DNA fragments separated by gel electrophoresis. J Mol Biol 98:503- 17.

Spielman, R.S., R.E. McGinnis and W.J. Ewenst. 1993. Transmission test for linkage disequilibrium: the insulin gene region and insulin-dependent diabetes mellitus(IDDM). Am J Hum Genet 52:506-516.

Stam, P. 1993. Construction of integrated genetic linkage maps by means of a new computer package: JoinMap. The Plant Journal 3(5): 739-744.

Stewart, J. 1992. Genetics and biology: a comment on the significance of the Elston-Stewart algorithm. Hum hered 42:9-15.

Stuart, A.and J.K. Ord. 1994. Kendall's Advanced Theory of Statistics, Volume I: Distribution Theory. Edward Arnold, London.

Stuber, C.W. 1995. Mapping and manipulating quantitative traits in maize. Trends Genet 11:477-481.

Stuber, C.W., S.E. Lincoln, D.W. Wolff, T. Helentjaris and E.S. Lander. 1992. Identification of genetic factors contributing to heterosis in a hybrid from two elite maize inbred lines using molecular markers. Genetics 132: 823-839.

Sturt, E. 1976. A mapping function for human chromosomes. Ann Hum Genet Lond 40:147.

Sturtevant, A.H. 1913. The linear arrangement of six sex-linked factors in *Drosophila*, as shown by their mode of association. J Exp Zool 14:43-59.

Sturtevant, A.H. 1931. Known and probable inverted sections of the autosomes of *Drosophila melanogaster*. Carnegie Institute of Washington Publisher, 421:1-27, Washington DC.

Swanson, C.P, T. Merz, and W.J.Young. Cytogenetics: The Chromosome in Division, Inheritance, and Evolution. Second Edition. Prentice-Hall, Englewood Cliff, NJ. 1981.

Szostak, J.W., T.L. Orr-Weaver, R.J. Rothstein and F.W. Stahl. 1983. The double-strand-break repair model for recombination. Cell 33:25-35.

Takahashi, N., K. Yamamoto, Y. Kitamura, S.-Q. Luo, H. Yoshikura and I. Kobayashi. 1992. Nonconservative recombination in *Escherichia coli*. Genetics 89:5912-5916

Tanksley, S.D. 1983. Molecular markers in plant breeding. Plant Mol. Biol. Reporter 1: 3-8.

Tanksley, S.D. 1993. Mapping polygenes. Ann Rev Genet 27:205-233.

Tanksley, S.D., M.W. Ganal and G.B. Martin. 1995. Chromosome landing: a paradigm for map-based gene cloning in plants with large genomes. Trends Genet 11:63-68.

Tanksley, S.D., M.W. Ganal, J.P.rince, M.C. de Vicente, M.C. Bonierbale, M.W. Broun, T.M. Fulton, J.J. Giovannoni, S. Grandillo, G.B. Martin, R. Messeguer, J.C. Miller, L. Miller, A.H. Paterson, O. Pineda, M.S. Roder, R.A. Wing, W. Wu and N.D. Young. 1992. High density molecular linkage maps of the tomato and potato genomes. Genetics 132:1141-1160.

Tanksley, S.D. and J. Hewitt. 1988. Use of molecular markers in breeding for soluble solids content in tomato - a re-examination. Theor Appl Genet 75:811-823.

Tanksley, S.D. and J.C. Nelson. 1996. Advanced backcross QTL analysis: a method for the simultaneous discovery and transfer of valuable QTLs from unadapted germplasm into elite breeding. Theor Appl Genet 92:191-203.

Terwilliger, J.D. 1992 A haplotype-based "haplotype relative risk" approach to detecting allelic associations. Hum Hered 42:337-337.

Terwilliger, J.D. and J. Ott. 1994. Handbook of Human Genetic Linkage. The Johns Hopkins University Press, Baltimore.

Thompson, E.A. 1986. Pedigree analysis in human genetics. Johns Hopkins University Press, Baltimore.

Thompson, E.A. 1987. Crossover counts and likelihood in multipoint linkage analysis. IMA Journal of Mathematics Applied in Medicine & Biology. 4:93-108.

Thompson, E.A. 1988. Two-locus and three-locus gene identity by descent in pedigrees. Ima J of mathematics aplied in medicine &biology 5, 261-279.

Thomson, G. 1995. Mapping disease genes: family-based association studies. Am J Hum Genet 57:487-498.

Tingey, S.V., J.A. Rafalski and J.G.W. Williams. 1992. Genetic Analysis With RAPD Markers. Proceedings of the symposium on Applications of RAPD Technology to plant Breeding. Minneapolis, Minnesota. pp:3-8.

van Ooijen, J.W. 1992. Accuracy of mapping quantitative trait loci in autogamous species. Theor Appl Genet 84:803-811.

van Ooijen, J.W. and C. Maliepaard. 1996. MAPQTL version 3.0: Software for the calculation of QTL position on genetic map. Plant Genome IV (Abs.) p105.

Waterman, M.S. 1995. Introduction to Computational Biology. Chapman and Hall, London.

Watson, J.D. 1990. The human genome project: past, present, and future. Science 248:44

Watson, J.D., M. Gilman, J. Witkowski and M. Zoller. Recombinant DNA. Second Edition. Scientific American Book, New York. 1992.

Weber, D. and T. Helentjaris. 1989. Mapping RFLP loci in maize using B-A translocations. Genetics 121:583-590.

Weeks, D.E. and K. Lange. 1987. Preliminary ranking procedures for multilocus ordering. Genomics 1:236-242.

Weeks, D.E. and K. Lange. 1988. The affected-pedigree-member of linkage analysis. Am J Hum Genet 42:315-325.

Weeks, D.E. and K. Lange. 1992. A multilocus extension of the affected-pedigree-member method of linkage analysis. Am J Hum Genet 50:859-868.

Weeks, D.E., G.M. Lathrop and J. Ott. 1993. Multipoint mapping under genetic interference. Hum Hered 43:86-97.

Weeks, D.E., J. Ott and M. Lathrop. 1994. Detection of genetic interference: simulation studies and mouse data. Genetics 136:1217-1226.

Weir, B.S. 1979. Inferences about linkage disequilibrium. Biometrics 35:235-254.

Weir, B.S. 1996. Genetic Data Analysis II. Sinauer, Sunderland, MA.

Weir, B.S. and C.C. Cockerham. 1979. Estimation of linkage disequilibrium in randomly mating populations. Heredity, 42:105-111.

Weissenbach, J., G. Gyapap, C. Dib, A. Vignal, J. Moressette, P. Millasseau, G. Vaysseix and. M Lathrop. 1992. A second generation linkage map of the human genome based on highly informative microsatellite loci. Nature 359:794-802.

Weller, J.I. 1986. Maximum likelihood techniques for the mapping and analysis of quantitative trait loci with the aid of genetic markers. Biometrics 42:627-640.

Weller, J.I., M. Soller and T. Brody. 1988. Linkage analysis of quantitative traits in an intrespecific cross of tomato (*Lycopersicon esculentum* x *Lycopersicon pimpinellifolium*) by means of genetic markers. Genetics 118:329-339.

White, R.L., J.M. Lalouel, Y. Nakamura, H. Donis-Keller, P. Green, D.W. Bowden, P. Mathew, D.F. Easton, E.B. Robson, N.E. Morton, J.F. Gusella, J.L. Haines, A.E. Retief, K.K. Kidd, J.C. Murray, G.M. Lathrop and H.M. Cann. 1990. The CEPH consortium primary linkage map of human chromosome 10. Genomics 6:393-412.

Whitkus, R., J. Doebley and M. Lee. 1992. Comparative genome mapping of sorghum and maize. Genetics 132:1119-1130.

Wilcox, P.L., H.V. Amersom, E.G. Kuhlman, B.H. Liu, D.M. O'Malley and R.R. Sederoff. 1996. Detection of a major gene for resistance to fusiform rust disease in loblolly pine by genomic mapping. Pro Nalt Acad Sci USA 93:3859-3864.

Williams, C.G. and D.B. Neale. 1992. Conifer Wood Quality and Marker-aided selection: a case study. Can J For Res 22: 1009-1017.

Williams, J.G.K., A.R. Kubelik, K.J. Livak, J.A. Rafalski and S.V. Tingey. 1990. DNA polymorphisms amplified by arbitrary primers are useful as genetic markers. Nucleic Acids Research 18:6531-6535.

Wilson, S.R. 1988. A major simplification in the preliminary ordering of linked loci. Genet. Epidemiol. 5:75-80.

Wricke, G. and W.E. Weber. 1986. Quantitative Genetics and Selection in Plant Breeding. Walter de Gruyter, Berlin.

Young, N.D. and S.D. Tanksley. 1989. Restriction fragment length polymorphism maps and the concept of graphical genotypes. Theor. Appl. Genet. 77:95-101.

Zeng, Z.B. 1993. Theoretical basis of separation of multiple linked gene effects on mapping quantitative trait loci. Proc Natl Acad Sci USA 90:10972-10976.

Zeng, Z.B. 1994. Precision mapping of quantitative trait loci. Genetics 136:1457-1466.

Zhao, L.P., E. Thompson and R. Prentice. 1990. Joint estimation of recombination fractions and interference coefficients in multilocus linkage analysis. AmJ Hum. Genet. 47:255-65.

AUTHOR INDEX

Subject Index

A

additive effect 32
adenine 37
AFLP 77
algorithms 284
    branch-and-bound 287
    EM algorithm 113, 149
    Metropolis algorithm (MA) 286
    seriation 285
    simulated annealing 286
    traveling salesman problem 284
allelic frequency 23
    covariance 148
    estimation 147
analysis of variance 34
    QTL mapping 394, 404
asymptotically normal 130
automation
    automated scoring system 81
    robotic-assisted assay 81
average effect of gene substitution 30

B

BAC 52
backcross 11, 59
    linkage analysis 171
    QTL mapping 390, 418, 448
bias 121
    nonparametric 122
    parametric 122
bias-corrected estimator 130
binomial distribution 90
bootstrap 124, 546
    combining with jackknife 549
    confidence interval 547
    mean, variance and bias 547
    replication 546
    sample 546
Box-Müller transformation 554
breeding value 31
broad sense heritability 35, 558
bulk segregant analysis 502
    false negative 504
    false positive 503

C

C-band technique 64
cDNAs 41
cell division 12
centi-Morgan 305, 318
chiasma 14
chi-square distribution 90
chi-square partition 164
chromosomal rearrangement
    inversion 20, 47
    translocation 20, 47
chromosome banding technique 64